Robotic Exploration of the Solar System

Part 2: Hiatus and Renewal 1983–1996

Paolo Ulivi with David M. Harland

Robotic Exploration of the Solar System

Part 2: Hiatus and Renewal 1983–1996

Published in association with
Praxis Publishing
Chichester, UK

Dr Paolo Ulivi
Cernusco Sul Naviglio
Italy

Dr David M. Harland
Space Historian
Kelvinbridge
Glasgow
UK

SPRINGER–PRAXIS BOOKS IN SPACE EXPLORATION
SUBJECT *ADVISORY EDITOR*: John Mason B.Sc., M.Sc., Ph.D.

ISBN 978-0-387-78904-0 Springer Berlin Heidelberg New York

Springer is a part of Springer Science + Business Media (*springer.com*)

Library of Congress Control Number: 2007927751

Front cover image: Copyright David A. Hardy/www.astroart.org/STFC

Cover design: Jim Wilkie
Copy editing: David M. Harland
Typesetting: BookEns Ltd, Royston, Herts., UK

Printed in Germany on acid-free paper

Contents

In Part 1

Now in Part 2

Illustrations

Front cover: The Ulysses spacecraft passing through the tail of Comet Hyakutake

Rear cover: The Galileo spacecraft on its IUS stage in Earth parking orbit

Chapter 4

Tables

Foreword

The series *Robotic Exploration of the Solar System* by P. Ulivi and D. M. Harland is, first of all, a monumental chronicle of the amazing adventure that in the last 50 years allowed mankind to visit and understand the immense and eerie domain of the solar system, with its hidden nooks and unexpected peculiarities, providing data, images and in some cases samples. The story is told with an extraordinary amount of factual and technical details, mostly arranged to trace each project from its conception to engineering design, to construction of the spacecraft, execution of the actual mission, data analysis and, finally, publication of the results. Most of these details are not known even to the communities of experts: temporary reports, especially if technical, are seldom published and are easily forgotten or lost. The style of this series is one of first class journalism: the story unfolds in a fascinating and easy-going way, without difficult digressions at the physical and engineering level. But the content is in no way superficial or vague: the accuracy of the information is confirmed not only by its exhaustive quantitative level, but also by the supporting primary documents quoted in the bibliography. Any future historical study of space exploration will have to be based on this chronicle. Much of its content refers to details of the instrumentation on each spacecraft, and to the manner in which the mission was accomplished. The design, making and testing of instruments for use in space is not an easy task. Conditions in space are often prohibitive, as, for instance, near the Sun, owing to its radiation and solar wind. Systems must reliably function for years without any check and repair. Extraordinary sensitivities for various physical quantities, like very weak magnetic fields and high-energy particles, are required. The possibility of storing on board very large amounts of data, processing it and sending it back to Earth is an essential condition for success. To reproduce space conditions on the ground to test systems is difficult, if not impossible.

I have been a Principal Investigator of the Ulysses mission, which is described in this volume. Launched in 1990, it conducted for the first time a deep exploration of the solar system environment outside the ecliptic plane in which most of the planets orbit the Sun – with outstanding results, as announced in the journal *Nature* on 3 July 2008. In the near future, after 18 years, its operation will terminate, not because of instrument problems, but because its radioisotope fuel is nearly exhausted.

The word 'robotic' in the title of this series points to an important controversy in space exploration: is direct human involvement necessary, or even advisable? For example, is the International Space Station commendable from the scientific point of view? I am clear on this point: the extraordinary developments in remote-sensing, software and control make a human presence on an orbiting machine for exploration useless for most of the time, costly and dangerous. Even when the round-trip time of a radio signal from Earth takes hours – such as in the descent of the Huygens probe to the surface of Titan, Saturn's large satellite (a mission that will be discussed in the next volume of this series) – an unmanned probe can work very well, even though the control from Earth is delayed and an immediate reaction to unforeseen conditions impossible. The system on Huygens, on the basis of pre-planned choices, was able to decide autonomously which actions to take on the basis of the physical conditions it encountered in the descent.

The word 'exploration', usually romantically understood as the strenuous efforts of daring and often irresponsible people to survey unknown lands and civilizations, has acquired another meaning: instruments provide us with eyes and sensors far more powerful and penetrating than our own senses, supported by a vast memory capacity. The accounts in this series impressively confirm this view. This leads me to my final topic: the use of robotic space probes in the solar system to understand the structure of space and time. As the *Oxford English Dictionary* explains, the primary meaning of the verb 'to explore' is to investigate; to survey an unknown land is secondary. Most emphatically, the main purpose of the exploration of the solar system is not the sheer collection and cataloguing of images and data in very great quantities; it is the rational understanding of the structure, the history and the functioning of the physical objects that they refer to. In 1958, at the beginning of space exploration of the solar system, the conceptual framework was already set up and well accepted: first, planets and other large bodies move according to the laws of gravitation devised by Isaac Newton and applied to an exceedingly refined degree by mathematicians in France and England in the nineteenth and twentieth centuries; secondly, the origin of the planetary system in the collapse of a rotating interstellar cloud of gas and dust, at the centre of which the Sun began to shine 4.56 billion years ago, was a well established scenario. Space exploration did not change this general framework, but it opened up unexpected windows and led to extraordinary discoveries, two of which I shall quote. Planets and their satellites are not point-like, as assumed in the Newtonian model; their finite size gives rise to new forces and tidal effects that significantly influence the evolution of the system, and these have been extensively investigated with space probes. In 1979 Voyager 1 discovered a few active volcanoes on Io, one of Jupiter's moons. In fact, their existence had been predicted by S.J. Peale and his collaborators at the University of California at Santa Barbara, on the basis of tidal forces exerted on Io by the nearby moons Europa and Ganymede. Space probes have also allowed immense progress in the investigation of planetary atmospheres, in particular on their composition, their evolution, and how they are maintained or replenished in spite of their continuous loss to space. Again, the traditional laws of chemistry and physics are not under question here; but no theory can predict or even explain the wealth of interlocking phenomena and complex behaviours, which often can be revealed and understood only with in-situ observations. A striking example is the recent discovery of extensive water activity on the surface of Mars in the geological past; of course, this has a bearing on the possible presence of life. But acceptance of physical laws can never be uncritical; indeed, the statement that a natural law is correct is idle and logically inconsistent, as there is no way to test it; one can only say, in the negative, that a given physical law is self-contradictory or conceptually inadequate, or that it disagrees with observations. It is well known, for example, that the Newtonian law of gravity works very well in most cases, but on both counts it is unacceptable. Minor anomalies in the motions of planets and the propagation of light in the solar system that are inexplicable by it are a quantitative consequence of the theory of general relativity announced by Albert Einstein in 1915; this theory is the currently accepted framework. The large computer programs used to predict and control the motions of interplanetary probes are in fact based on a fully relativistic mathematical scheme, and they include as an essential part the appropriate corrections to Newtonian theory to take account of relativity. A major question faced by theoretical physicists is: how, and at what quantitative level is general relativity violated? Space probes play a very important role in addressing this fundamental issue. They orbit the Sun at very large distances in an environment which is practically empty, and free from Earth's gravity and mechanical disturbances like microseisms. The sophistication of measurements using space probes of time intervals, distances and relative velocities is improving all the time, and such measurements have allowed the predictions of general relativity to be tested to a very high degree of accuracy. Remarkably, more than 90 years after its discovery, Einstein's theory is still unchallenged; but the assault is mounting, with a number of new missions in preparation to explore the deep nature of gravitation. An important experiment was carried out in 2002 by the Cassini spacecraft, which was cruising through interplanetary space to Saturn. Its radio system and a specially built antenna at NASA's Deep Space Network complex at Goldstone, California, enabled the relative velocity between them to be measured to an unprecedented accuracy, and made possible a new test of a relativistic effect of the Sun's gravitational field on the propagation of radio waves. No discrepancy from the prediction of general relativity was detected. It is quite remarkable that space probes are able not only to explore the mechanisms by which the objects in the solar system work, but also to investigate the very nature of space and time.

Bruno Bertotti
Dipartimento di Fisica Nucleare e Teorica
Università di Pavia (Italy)

Author's preface

The first part of *Robotic Exploration of the Solar System* ended with launches in 1981, but related missions in flight at that time through to their completion. This second part covers missions launched between 1983 and 1996, employing the same "spotter's guide to planetary spacecraft" approach. While the period covered is short, and was marked by a frustrating hiatus with rare missions, it saw the debut of new players, the decline of another, and a number of triumphs and failures. It was also marked by the 'Christmas tree' approach to planetary exploration which on the one hand caused a dearth of planetary missions and on the other hand a number of missions that produced an overwhelming return of results, not all of which were able to be included in this book. The period was also shaped by some peculiar external conditions: the American emphasis on human spaceflight and Shuttle flights, which deprived planetary missions of badly needed funds; the Challenger accident which derailed those few projects that had managed to survive; and finally the Strategic Defense Initiative, which provided technology for the low-cost revolution in deep-space missions of the 1990s. The low-cost approach, too, would soon dramatically show its shortcomings, but these will be left to future volumes in the series.

Paolo Ulivi
Milan, Italy
July 2008

Acknowledgments

As usual, there are many people that I must thank. First, I must thank my family for their support and help. I found invaluable support from the library of the aerospace engineering department of Milan Politecnico, and the Historical Archives of the European Union, as well as members of the Internet forums in which I participate. Special thanks go to all of those who provided documentation, information, and images for this volume, including Giovanni Adamoli, Nigel Angold, Luciano Anselmo, Bruno Besser, Michel Boer, Bruno Bertotti, Robert W. Carlson, Dwayne Day, David Dunham, Kyoko Fukuda, James Garry, Giancarlo Genta, Olivier Hainaut, Brian Harvey, Ivan A. Ivanov, Viktor Karfidov, Jean-François Leduc, John M. Logsdon, Richard Marsden, Sergei Matrossov, Don P. Mitchell, Jason Perry, Patrick Roger-Ravily, Jean-Jacques Serra, Ed Smith, Monica Talevi and David Williams; I apologize if I have inadvertently left out anyone. I also thank all of my friends. In addition to all of those already mentioned in the first volume, I must add my work colleagues Attilio, Claudio, Erika, Ilaria, Massimiliano, Paolo, Rosa and Teresa. I particularly thank Giorgio B., whose enthusiasm makes me feel like there are people out there still interested in these subjects.

I must thank David M. Harland for his support in reviewing and expanding the subject, and Clive Horwood and John Mason at Praxis for their help and support. I must thank Bruno Bertotti for sharing with me some of his recollections of working as scientist on these missions and for writing the Foreword. And I am grateful to David A. Hardy of www.astroart.org for the cover art, which was originally made for the Particle Physics and Astronomy Research Council of the UK government. Although I have managed to identify the copyright holders of most of the drawings and photographs, in those cases where this has not been possible and I deemed an image to be important in illustrating the story, I have used it and attributed as full a credit as possible; I apologise for any inconvenience this may create.

The most special thank-you of course goes to Paola, the wonderful brown-eyed planet of which I am the sputnik.

4

The decade of Halley

THE CRISIS

By the end of the 1970s the American program of solar system exploration was in disarray. After the success of Viking, and with the Voyager and Pioneer Venus missions underway, it appeared to some that planetary exploration had achieved its goals and, consequently, there was little left to do. In addition, many other factors conspired against launching further missions. Chief among them was the fact that the National Aeronautics and Space Administration (NASA) spent so much of its budget on human spaceflight, and in particular the Space Shuttle, which the agency had 'sold' to Congress by promising that the high development cost (projected at about $5 billion) would be offset in service by partial re-usability and the high rate of flights (as many as 60 per year). In fact, the Shuttle cost almost twice as much to develop and proved to be capable of at most a dozen flights per year, and the actual degree of re-usability and turn-around time left a great deal to be desired. To cover the Shuttle overruns, NASA cut into the budgets of its scientific programs, creating such havoc that these took almost a decade to recover. Another reason for the crisis in the planetary exploration program was America's détente with the Soviet Union, which fostered cooperation rather than competition in space. But planetary science gained little if any advantage from it, and the rapprochement declined in the early 1980s. In the meantime, NASA shifted its scientific focus away from planetary exploration towards terrestrial studies and astronomy, in particular approving the development of the Large Space Telescope, which would later become the Hubble Space Telescope, as the first in a series of space-based 'Great Observatories' that would, between them, cover the electromagnetic spectrum from the far infrared to gamma-ray wavelengths. Finally, in the face of budgetary austerity, Congress was unsympathetic to proposals for planetary missions costing $500 million – although at that time this was less than the procurement cost of almost any program by the Department of Defense. Thus, as the 1980s began, NASA and the Jet Propulsion Laboratory (JPL) of the California Institute of Technology, which as a result of a NASA reorganization remained the only facility building planetary probes, had just three missions approved for development, none of which was on a particularly solid financial footing. They were the Venus Orbiting Imaging Radar, the Galileo Jupiter orbiter and probe, and the out-of-ecliptic International Solar Polar Mission. In the meantime, the principal source of fresh data would be the 'Grand Tour' which Voyager 2 would conduct, with terrestrial observatories filling in gaps, for example by serendipitously discovering that Uranus possesses a ring system. Still, the hiatus would mean that, for the first time in 18 years, no fresh data on the solar system would be collected in 1982; and unless things changed nor would there be any in 1983, 1984 or 1985.[1]

The situation worsened when Ronald Reagan became US president in 1981 and promptly sought to cut federal spending in many areas, including civilian space. As a result, one of the planetary missions in development was scaled back, another was canceled, and consideration was given to closing JPL's Deep Space Network, the worldwide network of antennas that provided communication with all probes in deep space – which would in turn mean ending the Voyager 2 mission at Saturn. James Beggs, the incumbent NASA administrator, pointed out that "elimination of the planetary exploration program [would] make the JPL in California surplus to our needs". At the same time, Reagan's science adviser, George Keyworth, floated the suggestion of completely eliminating planetary missions for 10 years so as to enable NASA to focus on getting the Shuttle into service and then using this to conduct a variety of more worthwhile missions. The proponents of such a myopic viewpoint were unconcerned by the difficulty JPL would face in maintaining its institutional knowledge of how to design, build and operate a planetary spacecraft, in order to enable it to pick up the program after a decade of inactivity.[2,3]

As NASA and JPL struggled to keep alive those planetary missions which were underway, and to fend off threats to the budgets for the development of new ones, the Soviet Union continued its own program. The exploration of Venus, which had proved to be within the capabilities of the relatively unreliable but rugged Soviet technology would continue, at least in the short term, while an effort was underway to resume missions to Mars – which had been abandoned after a secret 'War of the Worlds' debate in the 1970s. Of course, by this time, the Superpowers had come to realize that planetary missions no longer had the propaganda value which they had delivered in the early 1960s.[4] Nevertheless, such activities remained popular with the public.

Finally, new entrants in the space arena were set to steal the show from both the financially strapped United States and the technically limited Soviet Union. After 20 years of considering possible deep-space missions, Europe was gearing up to fly one. This program capitalized on the cooperative programs between the individual member nations (France, Germany, the United Kingdom, Italy, Austria, etc) with both of the Superpowers. And ever since launching its first satellite in 1970 Japan had also been studying possible deep-space missions, and now had the capability to join in.

THE FACE OF VENUS

Having successfully imaged Venus at ground level using Veneras 9, 10, 13 and 14, the logical next step for the Soviet Union was to place a spacecraft into orbit to use an imaging radar to observe the surface through the enshrouding clouds and create a topographical map.[5]

Imaging radars, or synthetic-aperture radars (SAR) as they are more correctly called, record the Doppler shift and time delay of returned echoes of short pulses of microwave energy from a surface, and combine them to produce a high-resolution 'image', with each picture element (pixel) assigned a brightness proportional to the energy returned by the particular combination of Doppler shift and time delay for that point. The returned energy is influenced by surface slope, degree of roughness on the scale of the wavelength of the illumination pulse, and dielectric properties of the surface material. By extensive computer processing, the points collected as the spacecraft travels along its trajectory can be used to synthesize (hence the name) or simulate the observations of a much larger antenna. The illuminated 'footprint' is offset to one side of the ground track, because otherwise it would not be possible to discriminate between echoes coming from the left side and those from the right side. Such was the computing power needed to process SAR data, however, that when NASA's first radar satellite, named Seasat, was launched in 1978, it was predicted that it would take *75 years* to process all the data from the planned 3-year mission. Compared to other applications, the analysis of the data from a spaceborne radar had to take into account a number of additional factors, including the fact that orbital motion and ionospheric effects introduced Doppler shifts and phase scintillations.[6,7]

In the early 1970s two teams, one at Ames Research Center, the other at JPL, started to study a dedicated Venus mapping-radar mission. Ames proposed to adapt the Pioneer Venus spacecraft that it was developing, while the JPL proposal, which was named VOIR (Venus Orbiting Imaging Radar, but also "to see" in French), envisaged a new spacecraft using a radar system equipped with a large parabolic antenna such as on the Pioneers and Voyagers which were to explore the outer solar system, or alternatively a linear phased-array antenna. To minimize the orbit-insertion burn Ames intended to put its spacecraft into an elliptical orbit, but JPL wanted a circular orbit so that all the data would be collected at the same altitude and thus simplify the data reduction and analysis, even although this would greatly complicate the orbit-insertion process and would require the craft to have larger propellant tanks. Although some scientists argued that terrestrial radio-telescopes would soon be able to obtain data similar to that expected from an orbiting radar, at a much lower cost, in 1977 NASA adopted the VOIR proposal. In fact, the Arecibo telescope in Puerto Rico had recently achieved a resolution as fine as 100 meters in a few selected areas of the planet.

Meanwhile, a series of experiments were conducted to refine the SAR concept. Between 1977 and 1980, JPL tested its planetary synthetic-aperture radar by flying NASA's Convair 990 'Galileo II' aircraft over the forests of Guatemala and Belize, and demonstrated that the radar could penetrate the foliage to reveal ancient roads,

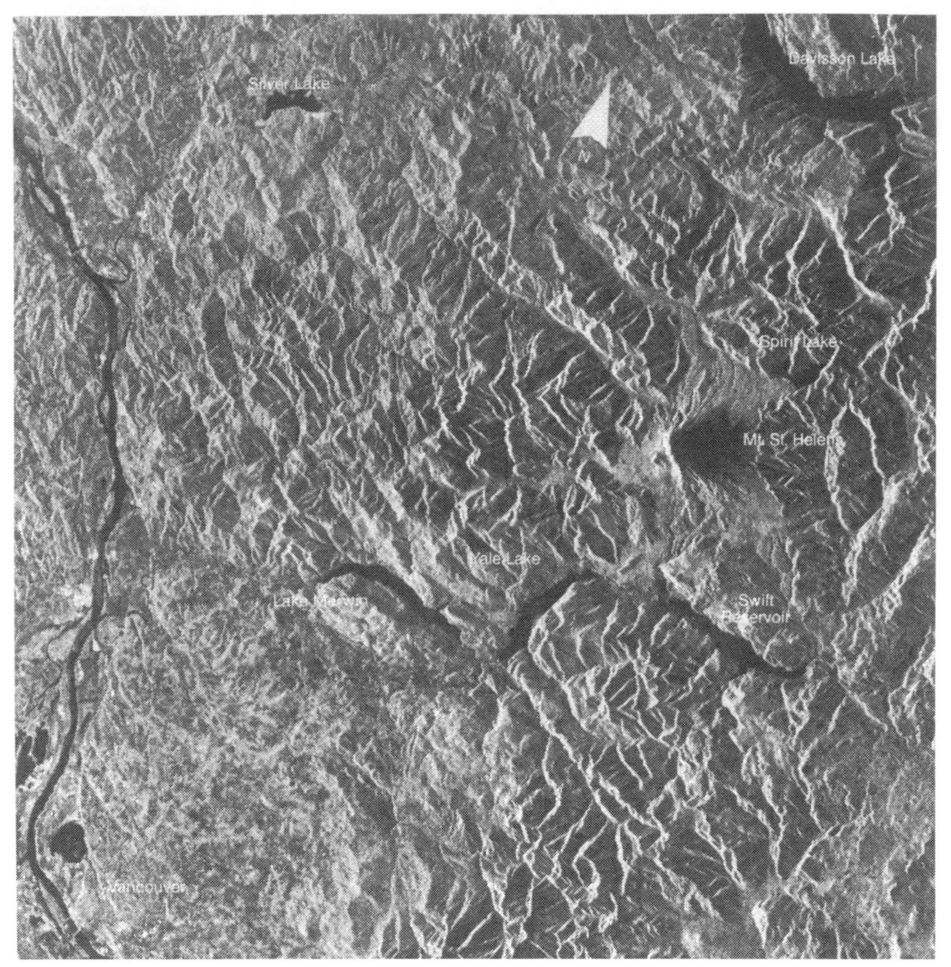

A Seasat synthetic-aperture radar image of a section of the Cascade Range in the western United States featuring Mount St. Helens. The Venus Orbiting Imaging Radar (VOIR) would have returned images of Venus at a comparable resolution. (JPL/NASA/Caltech)

stone walls, terraces and agricultural canals, in the process providing insights into the Mayan civilization and its economic structure (and, by coincidence, furthering the centuries-old association between the Mayans and the planet Venus).[8,9] JPL's Seasat, which was America's first civilian radar imaging satellite, was launched in June 1978 but it was crippled after 105 days by the failure of an electrical slip-ring connector. Nevertheless, its data greatly impressed oceanographers. It also showed why SAR was popular with the military: Seasat was reputedly capable of detecting the bow shocks of submerged submarines and also of the prototypes of 'stealth' airplanes crossing water.[10] Meanwhile, the Pioneer Venus Orbiter was compiling a preliminary radar map of Venus with a resolution of 150 km.[11]

The antenna of JPL's prototype planetary radar protruding from the rear fuselage of the 'Galileo II' airplane prior to a NASA flight over the Guatemalan forest.

In early 1980 Martin Marietta Aerospace, Hughes Aircraft and Goodyear Aerospace submitted proposals for the development of the VOIR spacecraft and its synthetic-aperture radar, and the project was included in the agency's 1981 budget – although with its launch postponed from May–June 1983 to May 1986.[12] Martin Marietta was eventually selected to build the spacecraft, while Hughes, which had worked in an analogous role on the Pioneer Venus Orbiter, would supply the radar.

The plan was for VOIR to be launched by the Space Shuttle and released in low Earth orbit, then boosted by a Centaur stage on a trajectory that would reach Venus in November 1986, whereupon the spacecraft would enter orbit and undertake a 5-month survey mission that would map the entire surface at 600 meters resolution and certain areas at somewhat higher resolution, and provide a global topographic and gravimetric map. The result would hopefully be a leap in knowledge of Venus to match that of Mars after Mariner 9. This would provide context for the pictures taken at ground level by the Venera landers, and the geological analyses derived from them, and would identify processes that were not evident in the low-resolution radar map provided by the Pioneer Venus Orbiter. In fact, transferred to Earth, the resolution of the Pioneer Venus Orbiter's radar would have missed the largest river basins, including the Mississippi and the Amazon; would have washed out some of the most geologically important mountain ranges, including the American Rockies, the Alps and Mount Everest in the Himalayas; and, even worse, would not have shown the continental margins, knowledge of which is the key to understanding the processes which have shaped the terrestrial crust. To minimize its cost, VOIR was to reuse as many components from previous missions as possible: the solar panels were spares left over from Mariner 10, the electronics were from Voyager, the radar altimeter was from the Pioneer Venus Orbiter and the imaging radar from Seasat. Compared to synthetic-aperture radars carried by aircraft, that of VOIR operated at the longer wavelength of about 25 cm, which was better able to penetrate the dense Venusian atmosphere without substantial attenuation of the signal. The spacecraft was to carry several other instruments. One was to be a microwave radiometer to measure the amount of energy radiating from various depths in the atmosphere and to determine the temperature and how much sulfur dioxide, sulfuric acid and water vapor were present. An airglow spectrometer and photometer would observe the upper atmosphere and ionosphere to study the circulation of the atmosphere in this region. A Langmuir probe would measure the temperature and distribution of ions and electrons in the ionosphere, as a quadrupole mass spectrometer monitored the composition, temperature and concentration of neutral gases. The final instrument would measure the temperature and density of ions in the ionosphere. On reaching the planet, the spacecraft would first enter an elliptical polar orbit, then circularize this by using either a conventional engine or the novel technique of aerobraking in which it would fire its engine to lower the periapsis of its orbit into the fringe of the atmosphere and then exploit atmospheric drag on successive passes to lower its apogee to the desired altitude. Although this technique had been pioneered in Earth orbit by the Atmospheric Explorer C satellite in 1973, it was nevertheless a risky maneuver. Its attraction was that it would enable the mass of the VOIR spacecraft to be limited to 850 kg. After circularization, by January or February 1987, VOIR would jettison the aerobraking shield, raise its periapsis from the atmosphere and start its primary mission, using its high-gain antenna to relay the data in real-time at 1 Mbps; fully 500 times the data rate of the Pioneer Venus Orbiter. The resulting map would provide almost global coverage, including one of the poles. In addition, for about 30 seconds on each orbit the radar would image a swath 10 km wide and 200 km long at higher resolution. In total, such 'spot data' would cover about 2 per cent of the surface. The primary mission was to last 120 days, or half of a Venusian day. An extension of up to a year was possible, so as to fill in gaps in the coverage and map the other pole, and to provide a detailed gravimetric survey which would enable geophysicists to estimate the thickness of the crust and place constraints on the size of the planet's core (if any) and on the rigidity of the mantle.[13,14,15] Overall, it promised to be a tremendous mission.

But in 1981 the incoming Reagan administration decided to scale down federal spending, and NASA was told to cancel one major program. The cost of VOIR was then estimated at $680 million, and the launch had been slipped again, this time to March 1988, so NASA reluctantly canceled it.[16]

In the Soviet Union the Lavochkin bureau, which had specialized in planetary and lunar missions since 1965, had in 1976 started work on a Venus orbiter that would carry a synthetic-aperture radar to map the radio reflectivity and topography of the surface. In 1977 further studies were supported by the Academy of Sciences, the Ministry of General Machine Building (a vast organization whose innocuous name 'hid' the space industry) and the Ministry of Radio Production, and contracts were awarded to make a suitable radar system. Although the 'Kometa' bureau led by

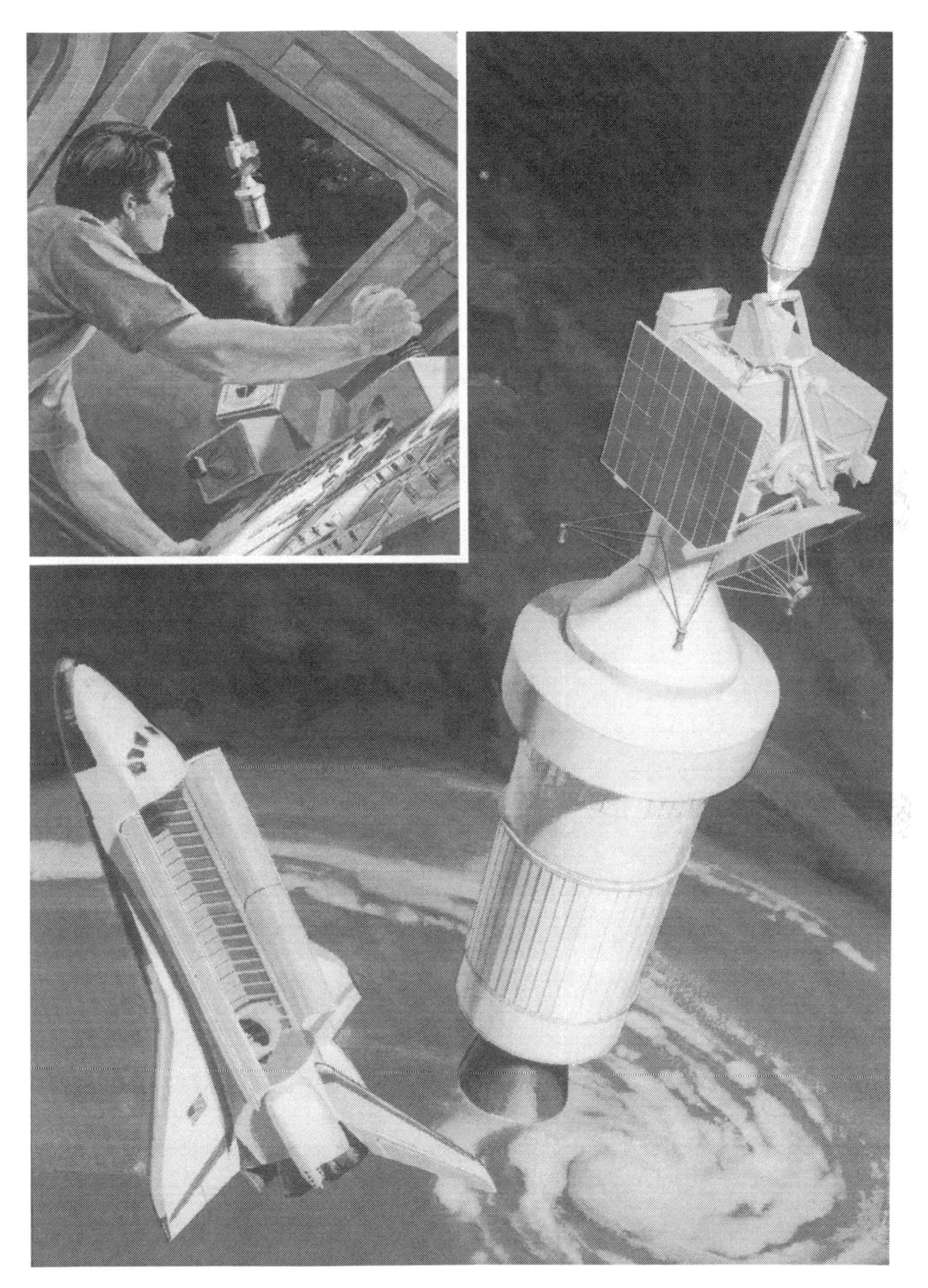

Impressions of the Venus Orbiting Imaging Radar (VOIR) spacecraft showing its deployment by Shuttle, and (inset) ignition to leave Earth orbit.

The Venus Orbiting Imaging Radar (VOIR) spacecraft showing (inset) aerobraking into a circular Venus orbit, and in its mapping configuration.

Anatoli I. Savin was more capable, it was overburdened with work on military projects such as the US-A and US-P satellites, which are known in the West as the RORSAT (Radar Ocean Reconnaissance Satellite) and EORSAT (Electronic Ocean Reconnaissance Satellite) respectively. Instead, the task of developing the planetary radar went to the MEI (Moskovskiy Energeticheskiy Institut; Moscow's Power Institute) led by Alexei F. Bogomolov. The development of the radar proved more difficult and protracted than expected, with problems concerning in particular the data storage system. Unlike VOIR, the Soviet spacecraft would not return data in real-time, but would instead record it during the periapsis passage of its eccentric orbit and transmit it to Earth at apoapsis. Modifications to the standard Venera bus were extensive but straightforward: the cylindrical core was lengthened by 1 meter to accommodate an additional 1,000 kg of fuel for the orbit-insertion maneuver, the nitrogen attitude control system was provided with 114.2 kg of gas instead of 36 kg in order to perform the many attitude changes required by the operational plan, two extra solar panel sections increased the total collection area to 10 m^2 to power the radar, and the diameter of the parabolic antenna was increased to 2.6 meters to boost the data rate from 6 to 100 kbps. The plan was to beat VOIR by launching two of these spacecraft (assigned the model designation 4V-2) in 1981, using the same window as Veneras 13 and 14. The spacecraft were ready in the spring of 1981, but the radar was not. MEI suggested that one 4V-2 be launched in 1981 and the other in 1983, because this would enable the remaining time to be devoted to testing and integrating a single unit. In the end, however, the radar was not ready in time, and both 4V-2 spacecraft were delayed to 1983.[17,18,19,20]

In 1979 rumours began to circulate in the West that the Soviets would soon send a radar imaging orbiter to Venus, but American space officials remained skeptical that the Soviets would be able to produce in a short time a flightworthy planetary radar that required less power than that of the RORSAT, which had a 3-kW nuclear reactor.[21] Neither did they believe the Soviets to possess the technology to operate a synthetic-aperture radar on a spacecraft, in particular the computing power. The CIA (Central Intelligence Agency), which had for years tried to 'listen in' to Soviet spacecraft, initially from a purpose-built intercept site in Ethiopia and later from an undisclosed friendly Western nation, planned to detect the scientific telemetry from any radar-equipped spacecraft. The stated objectives of this effort were three-fold: (1) to learn something of the Soviet military radar imaging capabilities, (2) to assist with planning the Venus Radar Mapper, which was the lower-cost successor to the VOIR mission, and (3) to provide ideas for future SETI (Search for Extraterrestrial Intelligence) experiments – as with a SETI signal, the exact frequencies and times of Soviet transmissions were not known. Of course, the CIA was at odds to explain the relationship between intercepting Soviet planetary telemetry and US national security, and the SETI connection must have made this even more complicated.[22]

The Polyus-V (Pole-Venus) synthetic-aperture radar comprised the antenna and the electronics, which were in a toroidal hermetic compartment. The entire system weighed 300 kg. The antenna was a 6 × 1.4-meter parabolic cylinder. It was fitted at the top of the spacecraft, with its axis displaced 10 degrees to the main axis – this corresponding to the vertical direction with respect to the planet. The antenna was built in three foldable sections to enable it to fit inside the shroud of the Proton launcher. It operated at a wavelength of 8 cm. Alongside the radar was a smaller 1-meter parabolic antenna for the 6-km-footprint radar altimeter that was capable of measuring the vehicle's altitude to within 50 meters. Since the launch window of 1983 was more favorable than that of 1981 it was possible to increase the payload, and for the first time an instrument supplied by one of the Soviet Union's close fraternal neighbours was included. This infrared Fourier spectrometer was developed in East Germany and managed by the local Academy of Sciences under the aegis of the Interkosmos organization, whose program also covered sounding rockets, scientific and application satellites and human space flights. The instrument was based on an analogous spectrometer flown on terrestrial meteorological satellites in the Meteor series, and slightly different versions were supplied for the two 4V-2 spacecraft. Its spectra would enable the temperature and composition of the Venusian atmosphere to be measured.[23] The payload suite was completed by an infrared radiometer, six cosmic-ray sensors and a detector for solar plasma. It has also been reported that an Austrian magnetometer was carried, but this is probably a confusion with such an instrument on Veneras 13 and 14.[24]

The two 4V-2 spacecraft would be placed into similar, near-polar orbits around Venus with periods of 24 hours and their 1,000-km periapses at about 60°N. When a spacecraft's altitude dropped below 2,000 km on approaching periapsis, it was to switch on its imaging radar and record data for a swath 150 km wide and 6,000 to 7,000 km long oriented in the direction of travel. As it climbed towards apoapsis, the spacecraft would turn to point its antenna at Earth and download this data. One orbit later, Venus would have turned 1.48 degrees on its axis and the radar swath would cover an area displaced with respect to the previous pass, enabling it to map the entire surface poleward of 30°N during a 243-day axial rotation. Although the ground resolution of 1–2 km (diminishing perpendicular to the orbital track) would be similar to that attained by terrestrial radars, the spacecraft would be able to map the northern polar regions which were not accessible from Earth.

Venera 15, the first of the 4V-2 spacecraft, weighed 5,250 kg at launch and was dispatched on 2 June 1983 into a heliocentric orbit ranging between 0.71 and 1.01 AU. Venera 16, slightly heavier at 5,300 kg, set off on 7 June and entered a similar orbit with an aphelion of 1.02 AU. Partially confirming that the long-awaited radar orbiters had been launched, TASS announced that they did not carry landers, and were to go into orbit around the planet. Venera 15 performed course corrections on 10 June and 1 October, and Venera 16 on 15 June and 5 October.[25,26]

At 03:05 UTC on 10 October Venera 15 began its braking burn, and entered an initial 1,021 × 64,689-km, 23h 27m orbit at 87.5 degrees to the Venusian equator. It was only the third Soviet spacecraft to enter orbit around Venus, and the fourth overall – the other one being NASA's Pioneer Venus Orbiter. Venera 16 began its braking burn 4 days later, at 06:22 UTC. The parameters of its orbit have not been published in detail, but 1,600 × 65,200 km and 80 degrees of inclination have been cited. Two days after entering orbit Venera 15's East German Fourier spectrometer was activated to take 20 preliminary spectra. In the following months it examined both the night- and day-side of the planet, and in addition to the carbon dioxide that

The Venera radar-mapping orbiters had a stretched Venera bus, larger solar panels, a larger high-gain antenna dish and (at the top) the Polyus-V synthetic-aperture radar, here shown in its deployed configuration.

This is one of the first Venera radar images to be released, showing what appears to be either a volcanic structure or an impact crater in the north polar region of Venus.

comprises most of the atmosphere it found water vapor, sulfur dioxide and sulfuric acid. Moreover, the temperatures enabled profiles to be drawn for altitudes ranging between 60 and 90 km.[27] The infrared radiometer reportedly observed several 'hot' spots which it was speculated might mark the locations of active volcanoes.[28]

Venera 15 activated its synthetic-aperture radar on 16 October, and the first data was received that same day by the 64-meter deep-space communication antenna at Medvezkye Ozyora (Bear Lake), which had been built to augment the 70-meter antenna at Yevpatoria in Crimea. The swath covered a 1 million km^2 area near the north pole. To the astonishment of the engineers and scientists, the spacecraft was returning low-resolution 'preview' images in addition to the raw data, because the MEI, unbeknownst to the other players, had added a 'quick look' image processor to the onboard system! After tests and orbital adjustments, routine mapping started on 11 November 1983 and concluded on 10 July 1984, during which a total area of 115 million km^2 was covered.[29,30] Because the orbital ground track did not actually pass over the pole, one spacecraft was turned 20 degrees to the side once every few weeks in order to inspect the most northerly area.[31] Meanwhile, after 21 years of trying, the CIA succeeded in detecting scientific telemetry from Soviet probes. At the same time, the Kremlin, led by Yuri Andropov, the former chairman of the KGB, became concerned about publishing the data, lest this reveal too much about the Soviet military radar capabilities. US officials, skeptical of Soviet capabilities in this field, eagerly awaited the first published data from the mission.[32,33]

Although no landforms resembling terrestrial lithospheric plates were found, the imagery indicated that the Venusian surface is subjected to tectonism. In particular, fractures as wide as 2,000 meters and tens of kilometers in length were suggestive of extensional stresses. In addition, 'tessera' or 'parquet' terrain characterized by cross-cutting ridges and grooves, each 10 to 20 km long and a few kilometers wide, was observed. This was a morphology unique to the planet. Several large and small tesserae were surrounded by smooth lava plains. Small volcanic domes between 2 and 15 km in diameter were very common, often occurring in groups up to 80 km across which hinted at the existence of volcanic 'hot spots', although there was no connection with tectonism. In addition, much larger domes or 'ovoids' suggested the crust had been pushed up by 'plumes' in the mantle. In the north polar region there was an elevated terrain standing as high as 5,000 meters above the mean radius of the planet.

Between 12 and 25 January 1984 Venera 16's track took it over the Maxwell Montes – the only feature on the planet to be named after a male: the physicist James Clerk Maxwell. This range of mountains had been discovered by terrestrial radar, but owing to its high latitude it had been viewed obliquely. The orbital view provided a vertical perspective. Tantalizing details were obtained of the peripheral areas, which are covered by parallel ridges and grooves of compressive origin, and also of Cleopatra Patera, a large depression 100 km in diameter, 200 km from and 2.5–4.5 km below the summit at an elevation of 11.5 km. Altimetry of Cleopatra Patera obtained on a track which chanced to run across it revealed the presence of a nested 60-km crater. Although the structure was suggestive of a collapsed volcanic caldera, an impact origin could not be ruled out. The plains adjacent to the Maxwell Montes in almost every direction possessed a morphology similar to the lunar maria (which are basaltic lava flows) and a smoothness which suggested that they were formed comparatively recently.

The radar coverage included the northern part of Beta Regio (the second feature on the planet to be discovered by terrestrial radar surveys) which was characterized

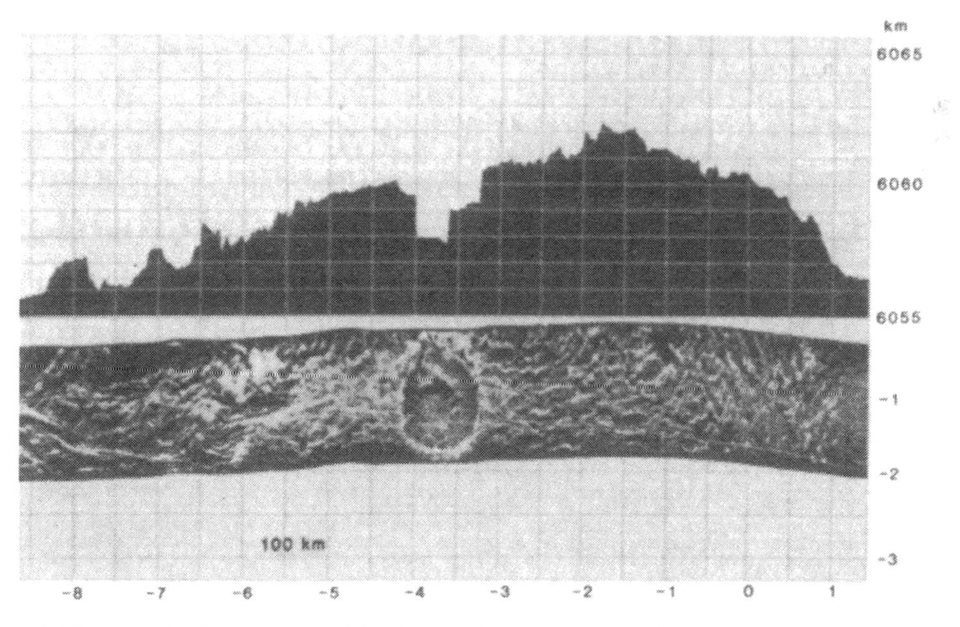

A Venera radar imaging and altimetry swath running across Cleopatra Patera, on the flank of the Maxwell Montes.

as a radar-bright 'continent'. The new imagery showed that it had both hummocky and smooth terrain. The Venera 9 landing site proved to be in hummocky terrain; a fact which was confirmed by the 'ground truth'. In the area covered by the images, a total of some 150 structures resembling impact craters were recognizable, and an analysis indicated a paucity of craters with diameters less than 20 km. It is likely that all impactors capable of making craters smaller than this are destroyed in the very dense atmosphere. The orbital imagery overlapped some of the areas surveyed by Arecibo, which allowed comparison of the same features seen at different radar look-angles. In particular, flat volcanic plains returning radio echoes at low look-angles were almost invisible to Arecibo, but easily recognizable in the spacecraft imagery. Combined processing of altimetric and synthetic-aperture radar data also yielded the radio-wave reflectivity and mean slope angles of the various Venusian surface features.[34,35,36,37,38,39,40,41,42]

On 15 June 1984 Venus was occulted by the Sun at superior conjunction, and signals from Venera 15 were tracked by the 70-meter antenna at Yevpatoria and by a 25-meter antenna near Moscow for several days in order to investigate the solar plasma.[43] The date of the final transmission by Venera 15 has not been published, but it is reported to have run out of attitude control propellant and been shut down in March 1985. Venera 16 returned cosmic-ray data until 28 May 1985.[44,45] It is

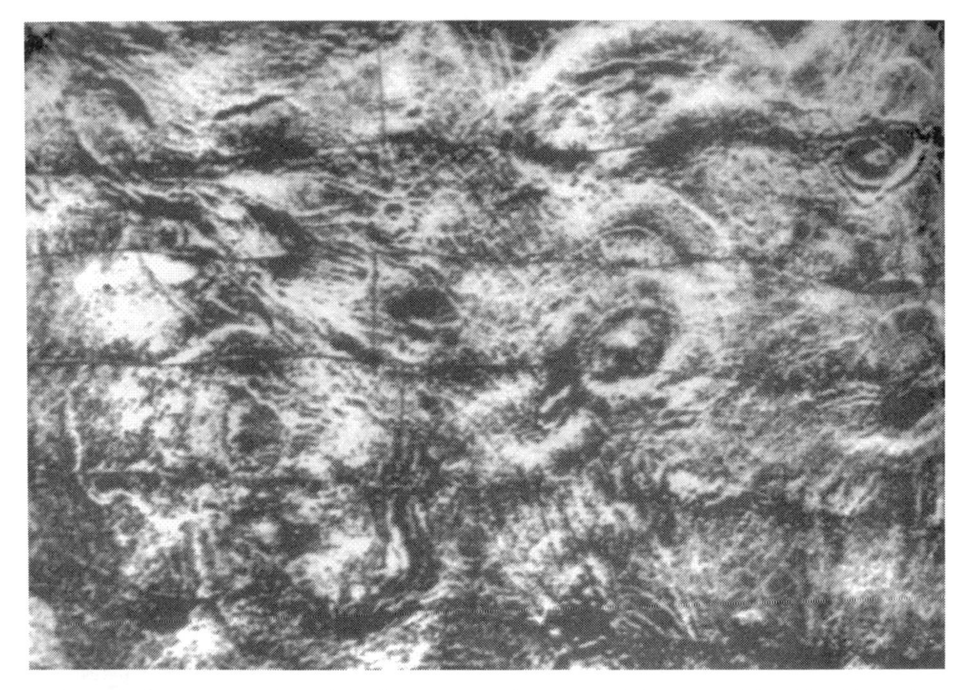

A 1,000-km-wide section of the Venusian surface located between Sedna Planitia and Bell Regio. The circular volcanic structures were called 'arachnoids' because of their resemblance to spider webs.

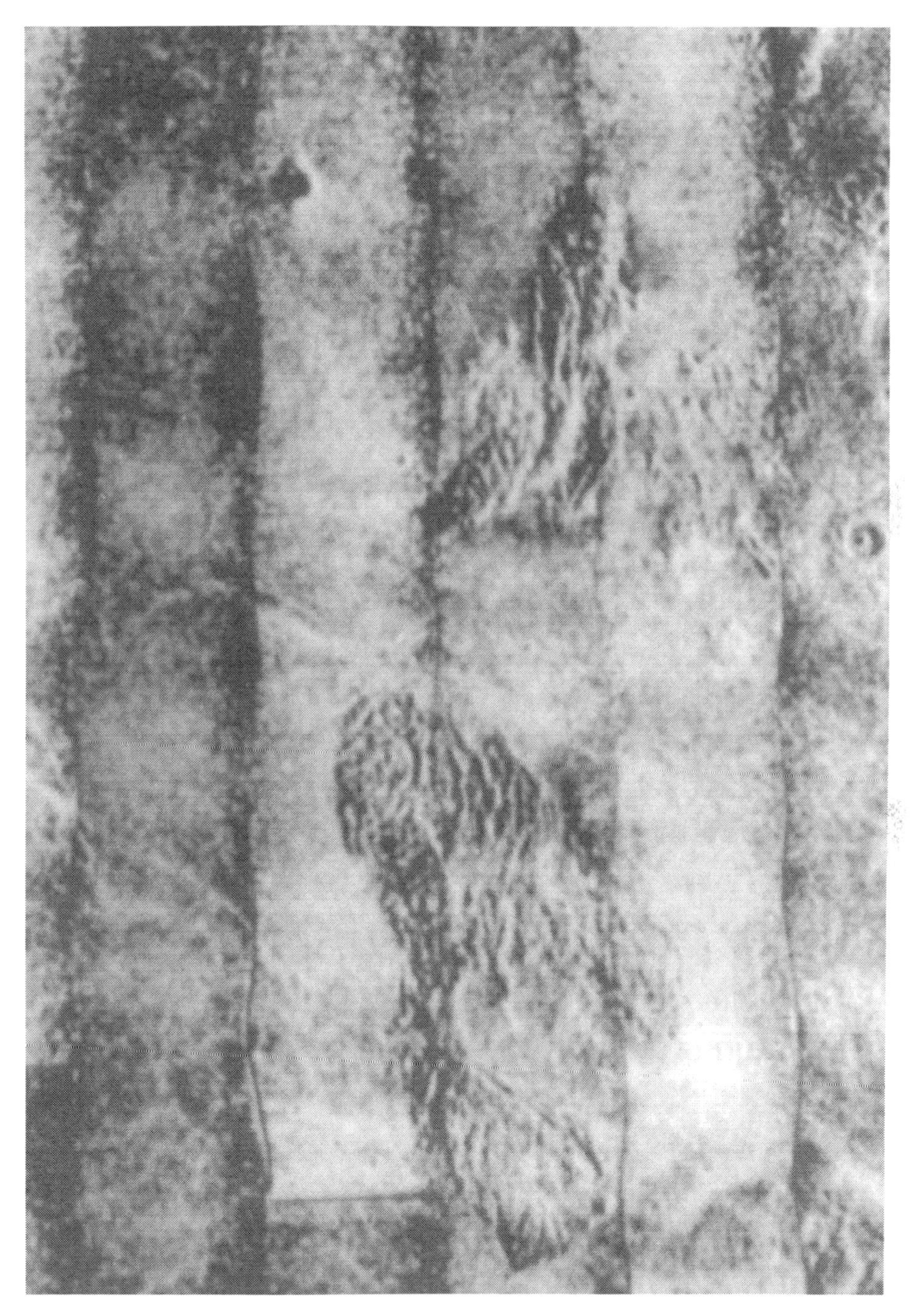

Sedna Planitia, with underlying structures showing through a flat lava flow.

possible that the orbit of at least one of the spacecraft was to be lowered in order to take higher resolution radar data, but if this was planned it was never done.[46]

THE MISSION OF A LIFETIME

Despite the interest of scientists and engineers in sending a mission to a comet, by the end of the 1970s no space agency had approved such a project. However, with comet Halley, the most famous and historically important such object, due to reach the perihelion of its 76-year orbit in February 1986 several agencies began to give serious consideration to sending spacecraft to inspect it.

Although calculations have accurately traced the orbit of Halley's comet back as far as 1404 B.C., when Egyptian civilization was at its zenith, the first reliable observations were made in China in 240 B.C., at the time of Qin Shi Huang, "the Unifier", who in 221 B.C. united the realms of ancient China and started the Qin dynasty. During the next two millennia the passages of the comet were recorded by a number of civilizations, who often associated it with traumatic events such as the defeat of Attila the Hun in 451 and the landing of the Normans in England in 1066, on which occasion its appearance was depicted in the Bayeux Tapestry. Its periodic nature was established by Edmund Halley in 1695, demonstrating for the first time the astronomical nature of comets, and, true to his prediction, it was recovered in 1758. It was then extensively observed in 1835 and in 1910, and expectations were high for its return in 1986, even although by the comet's standards it would not be a favourable apparition due in part to the fact that when at perihelion it would be at a high southern declination, and also because it would be present in daylight.[47,48] In terms of celestial mechanics, Halley's orbit is typically cometary: it is extremely eccentric, ranging from beyond the orbit of Neptune at 30 AU to within 0.587 AU of the Sun, with the result that its velocity relative to the Sun at perihelion is very high. Furthermore, its motion is retrograde, which means that it travels in the opposite direction to the planets. This meant that its velocity relative to Earth would be extremely high. It would therefore be a difficult target for a spacecraft to reach. But interest was high; in part owing to its historical significance, but also because in terms of its brightness and rates of production of gas and dust it was more like a long-period comet than its short-period sisters, and, most importantly, because its well-defined ephemeris would enable the orbit of the spacecraft to be precisely set up.

Halley's comet first raised the interest of the astronautics community in 1967, when the Lockheed Missile and Space Company in the United States made the first study of interception and rendezvous missions. A spacecraft could be placed into a similar orbit around the Sun to the comet, to make a slow-speed rendezvous with it. However, since the spacecraft would inherit the direction of Earth's travel around the Sun, it would have to maneuver into a retrograde orbit. There were a number of options for achieving this: (1) by firing a conventional chemical engine to deliver a brief impulse near aphelion, although the propulsion requirements to perform such a maneuver would be extremely high; (2) by firing a low-thrust engine for a long time to deliver a small but constant thrust to shape and then reverse the orbit; (3) a near-

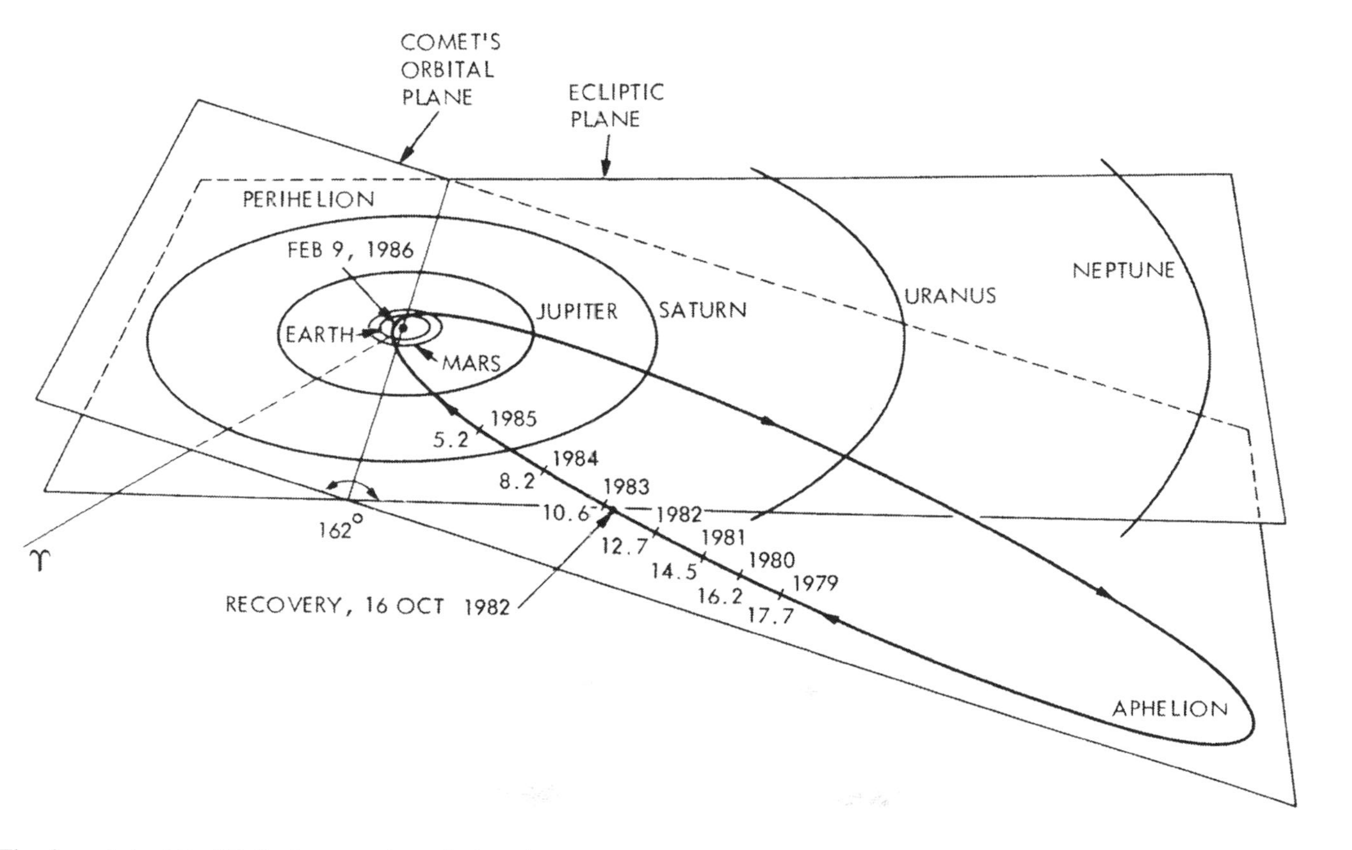

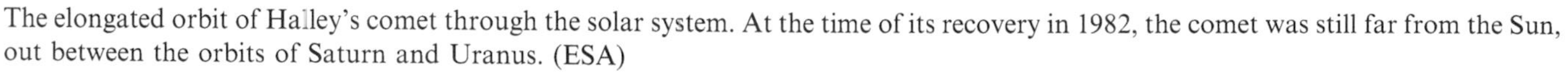
The elongated orbit of Halley's comet through the solar system. At the time of its recovery in 1982, the comet was still far from the Sun, out between the orbits of Saturn and Uranus. (ESA)

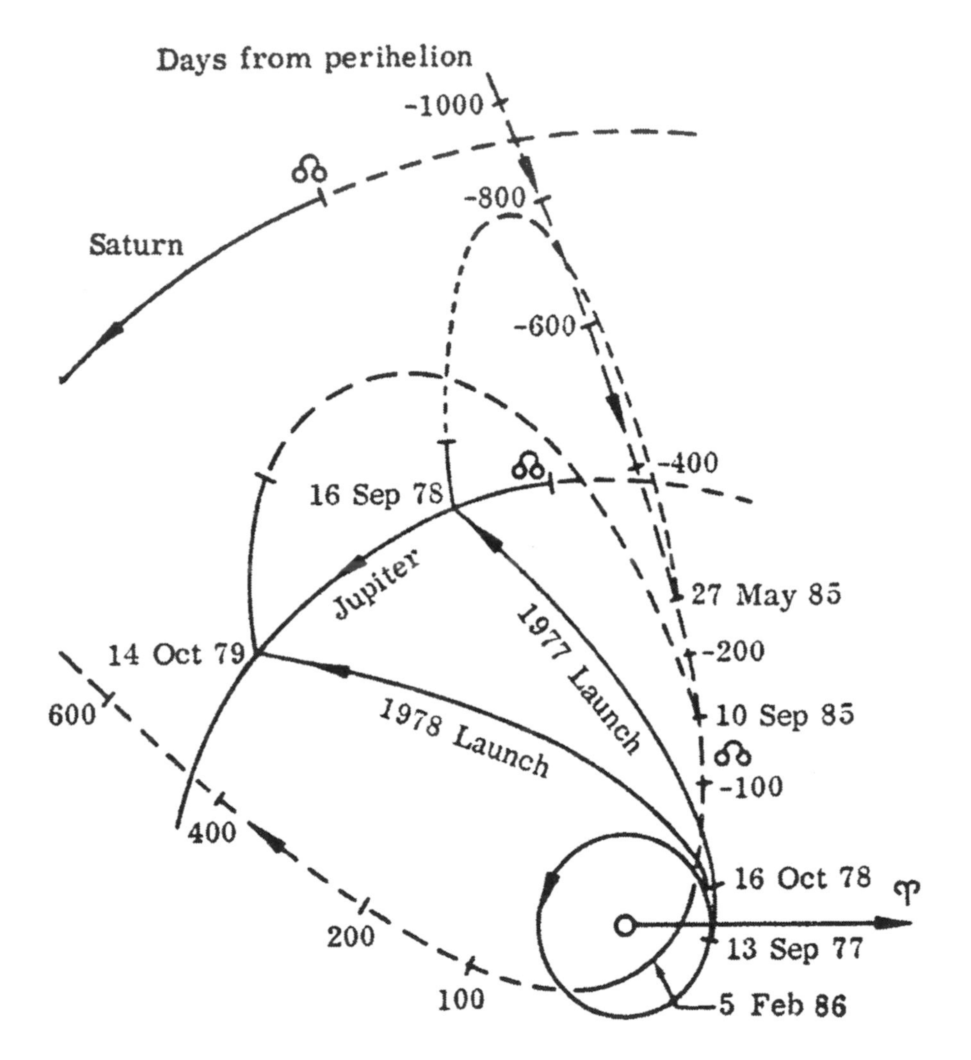

Two possible trajectories that would allow a spacecraft to rendezvous with Halley's comet following a Jupiter gravity-assist that would not only incline the spacecraft's orbit relative to the ecliptic but also make its motion retrograde, to match that of the comet. (Reprinted from: Michielsen, H.F., "A Rendezvous with Halley's Comet in 1985–1986", Journal of Spacecraft, 5, 1968, 328–334)

polar flyby of Jupiter or Saturn. Rejecting the first option owing to its deep-space propulsion requirements, NASA then elaborated on the alternatives. A small spacecraft could be launched in 1977 or 1978 by a Saturn V with a Centaur upper stage and a Jovian 'slingshot' used to deflect the probe into a retrograde orbit that would intersect the comet's orbit 5–8 months prior to its perihelion. A small burn would then put the craft into a Halley-centric orbit. Beside requiring the expensive Saturn V launch vehicle, for which production was limited, this plan was rendered

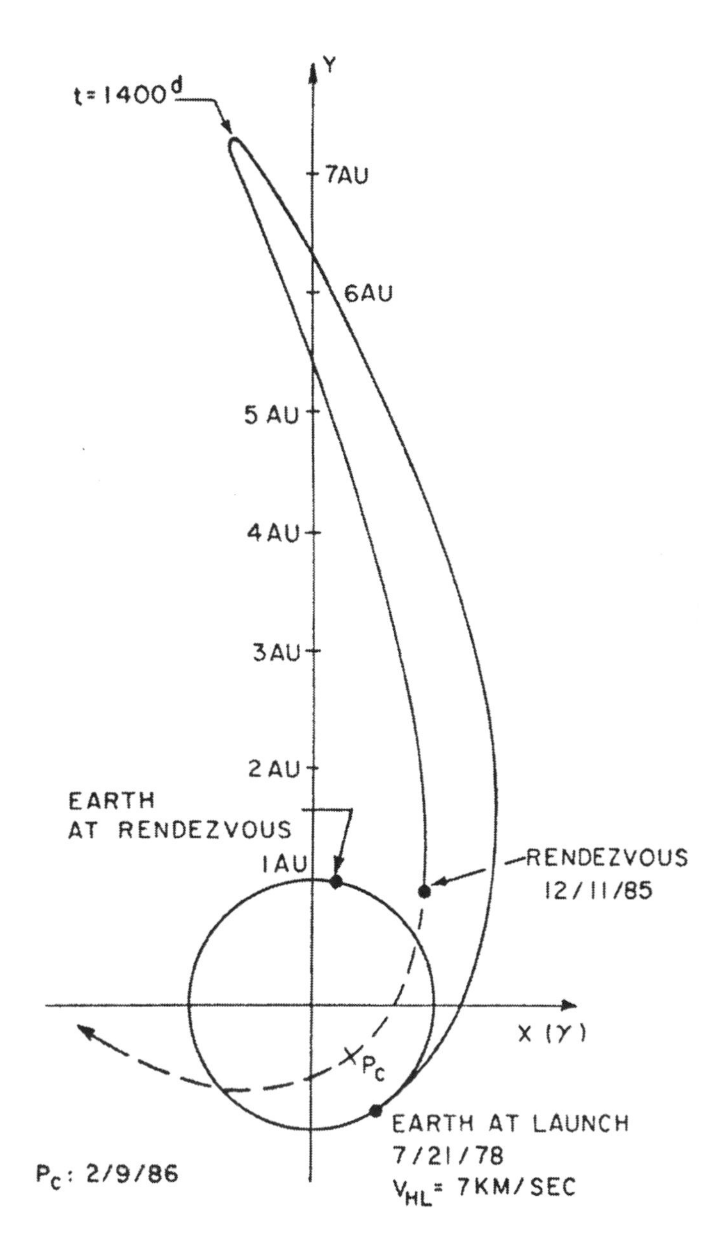

The complex trajectory of an electric-propulsion Halley rendezvous mission. In this case, the motion reversal would be performed by the engine, and would occur quite far out from the Sun. (Reprinted from: Friedlander, A.L., Niehoff, J.C., Waters, J.I., "Trajectory Requirements for Comet Rendezvous", Journal of Spacecraft, 8, 1971, 858–866)

unrealistic by the long time of at least 7 years in flight prior to the encounter. The alternative was a probe equipped with ion thrusters powered by either solar panels or a nuclear generator. After being placed into an eccentric solar orbit, this would gradually reduce its speed, and finally, when far from the Sun, make a maneuver to adopt retrograde motion. On heading back towards the inner solar system, it would encounter the comet 2–3 months prior to its perihelion. But such a flight would still take about 7 years, and because solar panels would generate very little power at the heliocentric distance of the reversal maneuver they would need to be inordinately large. A nuclear spacecraft would be able to run its ion engine essentially all of the time, which would cut the flight time to less than 3 years, but it would involve the development of a nuclear generator for use in space.

A simpler option was to intercept rather than rendezvous with the comet. This would involve placing the spacecraft into a much less energetically expensive orbit in the plane of the ecliptic which would pass through one of the points, known as nodes, where the comet's orbit intersected the ecliptic. As it approached perihelion, Halley would reach the ascending node on 8 November 1985, crossing the ecliptic from south to north at a heliocentric distance of 1.8 AU, which was in the asteroid belt. After perihelion on 9 February 1986, Halley would reach the descending node on 10 March at a heliocentric distance of 0.85 AU. Although both nodes were good prospects for an interception, the spacecraft's speed relative to the comet would be very high – in excess of 60 km/s. The ascending node presented a slightly smaller relative speed at encounter but a larger distance from the Sun, which would require a more powerful escape stage. The descending node would involve a lesser escape speed but would result in a higher relative speed at encounter. This was the case because the spacecraft needed only to be injected into an orbit similar to that of Earth, with a period of 10 rather than 12 months. Launch windows for ascending node encounters existed in either February or July 1985, but descending node encounters could be dispatched only in July and August 1985.[49,50] Of course, higher energy ballistic flights existed. For example, a launch in January 1986 could result in an encounter in April at the somewhat reduced relative speed of 46 km/s, when Halley was in opposition and closest to Earth; and a Jovian slingshot could also provide an encounter at only 15 km/s, although this would not occur until well into the 1990s and would be at a large heliocentric distance (over 15 AU) by which time the comet would have resumed its dormant state.[51]

NASA was not alone in studying Halley missions. In 1973 the European agency ESRO (European Space Research Organization), which during the previous decade had studied a mission to a comet, stressed that its future scientific program should include a mission to comet Halley, possibly using solar-electric propulsion.[52] Unfortunately, despite the importance of cometary studies to Soviet astronomers, as evidenced by the number of comets that bear Russian and Ukrainian names, it is not known which (if any) plans were under discussion in the USSR at that time.

In the mid-1970s the number of studies picked up. JPL became aware of one by Jerome Wright, an engineer at the Battelle Memorial Institute who was working on a NASA contract, which showed that a low-speed rendezvous with Halley could be achieved using solar radiation pressure as the means of propulsion. This involved

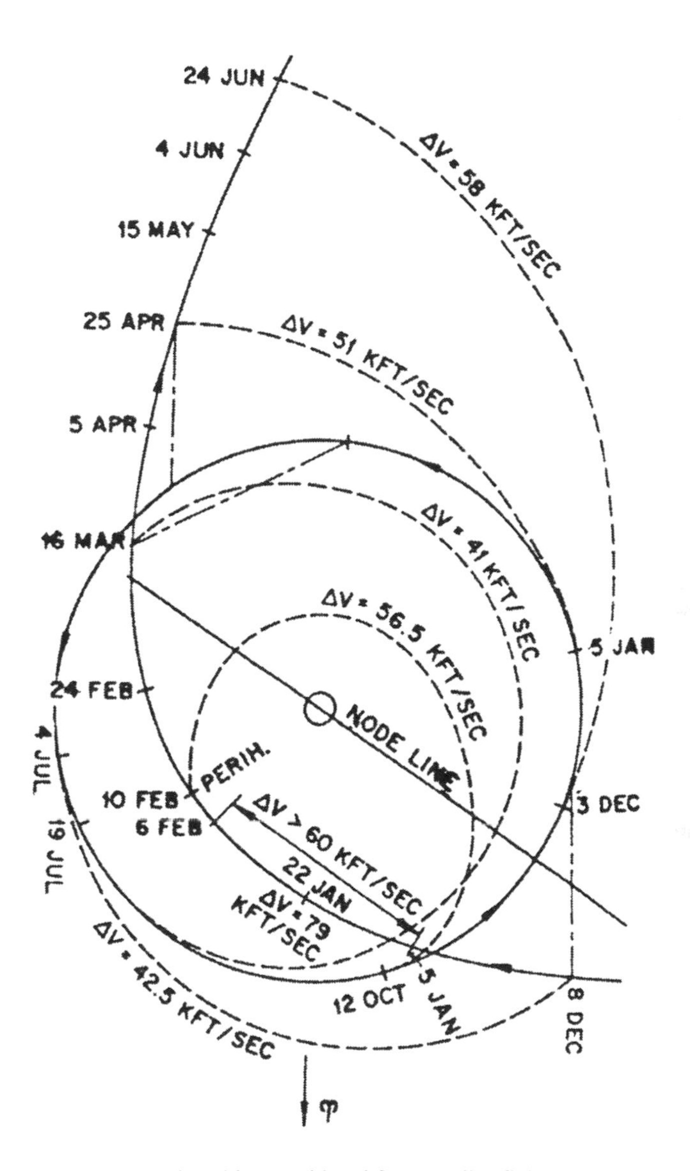

A diagram of various ballistic orbits considered for a Halley flyby, taken from the first paper to deal with such missions. Note that the orbit which meets the comet nearest to the time of its nodal passage on 16 March would involve the smallest velocity change (marked ΔV) at injection. This trajectory would be flown by both ESA's Giotto and Japan's Suisei missions. (Reprinted from: Michielsen, H.F., "A Rendezvous with Halley's Comet in 1985–1986", Journal of Spacecraft, 5, 1968, 328–334)

collecting the pressure of solar radiation using a large but very thin sheet of plastic and metal. Wright showed that a 'light sail' launched by the Space Shuttle in late 1981 or early 1982 could undertake a mission profile similar to that of a spacecraft using an ion engine powered by a nuclear reactor. Such sails exploit a consequence of Maxwell's laws of electromagnetism. Maxwell envisaged light, and electromagnetic radiation in general, in terms of packets of particles which, on striking an obstacle, transfer momentum to it. We don't experience this 'radiation pressure' in everyday life because it is extremely small, but experiments since the early 1900s using large yet lightweight objects verified its existence. Solar sails were apparently a Russian invention, their being mentioned by Konstantin Tsiolkovskii, the father of Russian astronautics, and also by Fridrikh Tsander, the pioneer who predicted that vehicles would cross interplanetary space using "tremendous mirrors of very thin sheets".[53] They were then rediscovered by US engineers during the 1950s, being promoted as a more efficient way than a chemical rocket to travel to the planets.[54,55] However, radiation pressure was still poorly understood and often neglected at the start of the 'space age' – so much so that when Explorer 12 was launched in 1961 its spin had been expected to decrease, but it increased due to the pressure of solar radiation on its four paddle-like solar panel.[56] Nor were solar radiation pressure perturbations appreciated when an incorrect value of the Astronomical Unit was calculated from tracking Pioneer 5 in 1960. Small sails were installed at the tips of the solar panels of Mariner 2 and Mariner 4 for stabilization, but they proved to be ineffective. By the mid-1970s, when Wright conducted his study, the use of solar sails as a means of propulsion was an untested technique, because very little theoretical and almost no practical work had been carried out. The most practical experiment had been on a much smaller scale, and had involved using solar radiation pressure to control the attitude of Mariner 10.

Although the theoretical feasibility of 'solar sailing' had been proved, the actual feasibility of building a large sail and deploying it in space had not. To remedy this deficiency, JPL started an in-depth analysis, and a solar sail Halley spacecraft was included in the 1976 'Purple Pigeons' study of planetary missions that were likely to capture the interest of the public – it was rated as the 'purplest' of all.[57] The initial design funded by a $5.5 million NASA grant was for a square sail 800 meters on a side, supported by four crossbeams (possibly deployed by astronauts in Earth orbit) rigged for strength and carrying an 800-kg probe at the center. Four smaller vanes at the corners would provide directional and attitude control. Such a sail would be visible to the naked eye in daylight from Earth for months following deployment! During the first 250 days of flight it would spiral in towards the Sun, then it would crank up its orbital inclination by 20 degrees every 60-day revolution. Nine months later it would be traveling in a retrograde orbit. It would intercept Halley in early 1986. The sail would then be jettisoned, and the spacecraft would study the comet at close distance and low relative speed through perihelion passage.[58,59,60] Once the concept had been proved, it was expected that solar sailing would facilitate a great variety of missions to the planets of the inner solar system, to near-Earth asteroids, and to return samples of comets. The potential of this technique is illustrated by the fact that a Halley-like solar sail could deliver a payload of approximately *10 tonnes*

Dr Tsung-Chi Tsu of the Westinghouse Research Laboratory with a model of his proposed late-1950s solar sail spacecraft. (BIS)

to Mercury! Feasibility studies were awarded to private companies, in particular to investigate materials. However, in view of the many unknowns associated with the deployment of a boom and sail on such a scale, the square design was replaced in early 1977 by the 'heliogyro', whose sail consisted of many large rectangular strips arranged like the blades of a helicopter. This would be simpler to deploy, since the strips could be unreeled from storage drums; a task which would be assisted by the centrifugal force of the spacecraft's spin. After deployment, the 12-km-long blades would be able to be oriented so as to maintain the spin, thereby contributing to the spacecraft's stabilization.[61,62,63]

Competing against the solar sail probe was a more conservative concept using a solar-electric propulsion module. This technology had been under study by NASA and US industry for almost 15 years and so, even though still experimental, it was deemed to be more mature than a solar sail. As a result, in September 1977 NASA announced that it had chosen electric propulsion for future interplanetary missions, including the Halley rendezvous. Unfortunately, the Halley project was short-lived: owing to the overruns in developing the Space Shuttle and a projected cost for the Halley mission in excess of $500 million, it was not included in the 1979 budget, which meant that it could not meet its 1982 launch window. However, even if the spacecraft had been developed, it is doubtful that the Space Shuttle, which did not

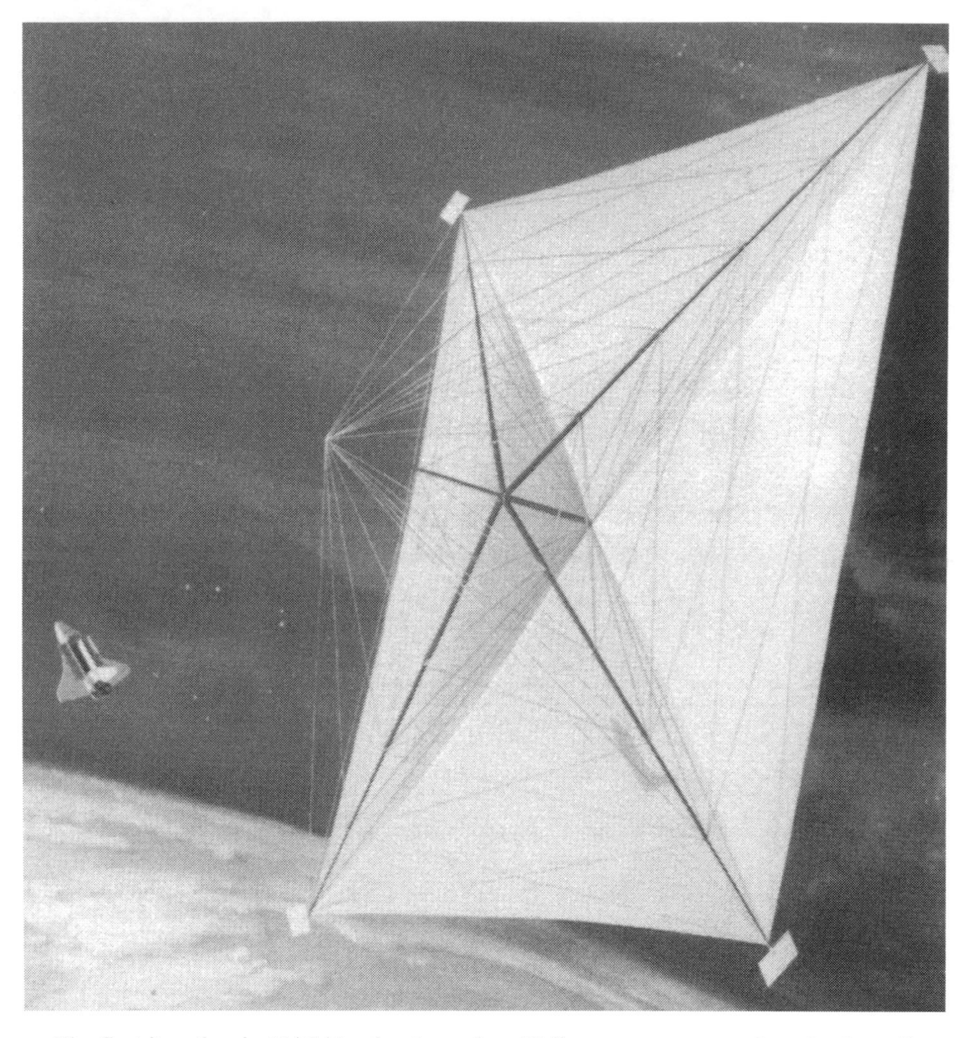

The first iteration in NASA's plan to explore Halley was an unprecedented solar sail rendezvous mission. (JPL/NASA/Caltech)

make its first test flight until April 1981, would have been able to meet the narrow planetary launch window.[64]

Although NASA's Halley rendezvous was canceled, by the first half of 1978 the agency's Comet Science Working Group had devised a new proposal which would address the same scientific objectives: to observe a known short-period comet for long enough to determine the physical and chemical nature of its nucleus, its coma and its tail; to observe the physical changes at various heliocentric distances; and to characterize its interaction with the interplanetary medium. On this plan the flyby of Halley would be followed by a rendezvous with comet Tempel 2, a well-known relatively bright object with a 5-year period that had been observed at most returns

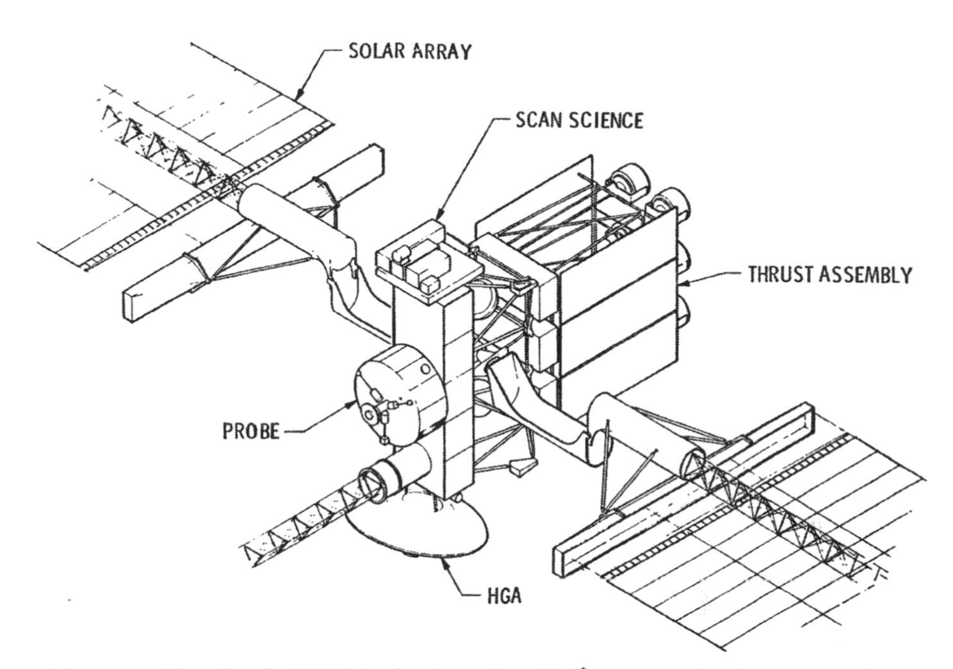

The second iteration in NASA's plan to explore Halley was a solar-electric propulsion flyby that would release a European piggy-back probe. The main spacecraft would go on to rendezvous with Tempel 2. (ESA)

since its discovery in 1873; this having been selected because it offered one of the best opportunities for a low-speed rendezvous during the 1980s.[65] Although a small probe was to be released during the Halley flyby to make a closer inspection of that comet, the primary scientific objectives would be addressed at Tempel 2. Because the US science community considered this 'fast' Halley flyby to be second-rate science, the European Space Agency (ESA) was invited to supply the daughter probe. As a result, the Halley Flyby/Tempel 2 Rendezvous was also known as the International Comet Mission (ICM). The 2,700-kg US spacecraft would comprise a propulsion module with six ion thrusters fueled by 870 kg of mercury, a large mission module housing all the systems, and the science platform. The scientific payload would be extensive, with instruments to measure the surface composition and temperature of the nucleus, the composition of the coma and its interactions with the solar wind. Particles of dust would be collected by sticky surfaces and analyzed on board. The imaging system was to be derived from that developed for the Galileo spacecraft, modified to make it 'dust proof', with cameras to take distant wide-angle views of the coma and close narrow-angle pictures of the nucleus. Each camera was to have a wheel with as many as 20 filters: some for color imaging, some for polarization studies, and others at wavelengths characteristic of the ions expected to be present in the coma. The ion thrusters – which would be used almost continuously – would be powered by a pair of very large solar arrays which would give the spacecraft a total span in excess of 64 meters and provide up to 25 kW at 1 AU. The European probe was to be a based on the ISEE 2 (International Sun–Earth Explorer) satellite, would weigh 150–250 kg and have seven instruments, including a magnetometer, a dust analyzer and an optical photometer. Since the probe would be spin stabilized, it was not to undertake imaging. As it was not expected to survive the encounter, it would be powered either by batteries or by a combined battery-solar transducer system.

The mission was to be launched during a short 10-day window in late July 1985 by a Space Shuttle, and sent on its way by an 'all-solid' Interim Upper Stage (IUS, renamed the Inertial Upper Stage when the 'Space Tug' for which it was to provide a stop-gap was canceled). Once released, the spacecraft's ion engine would start to thrust in order to arrange a Halley encounter at a heliocentric distance of 1.53 AU in late November 1985. Some 15 days before encounter, the piggy-back European probe would be spun up and released on a trajectory that would pass no more than 1,500 km from the comet's nucleus. The still-thrusting mothership would fly by the comet at a range of 130,000 km, taking low-resolution pictures of its inner coma and nucleus. Instruments on the probe would be activated by a timer several hours prior to the predicted moment of closest approach, returning data in real-time to the main spacecraft for relay to Earth. The instruments would operate until either the probe was destroyed by dust or its power ran out. After the Halley encounter, the main spacecraft would thrust to intercept the orbit of Tempel 2 in July 1988, some 2 months ahead of the comet's perihelion. It would approach first to a 'dust safe' distance of several thousand kilometers, and then, if the prospects of survival were good, it would close to 100 km and finally to just 50 km. The Tempel 2 rendezvous would last at least a year, and the spacecraft would study the physical and chemical state of the nucleus and the coma. There would be several phases of investigation, conducted at various distances from the nucleus. In particular, from 10 days before the comet's perihelion through to 30 days after it, during which time the dust coma would be at its densest, the spacecraft would temporarily recede from the nucleus. If possible, it would then attempt to land on the nucleus in order to characterize the strength of the surface. If it became evident that the spacecraft would not be able to be launched in time to make the Halley flyby, the intention was to send it directly to Tempel 2.[66,67,68,69]

To support the cometary missions, JPL proposed that an International Halley Watch be organized to coordinate observations made using a variety of techniques, both on the ground and in space. Its tasks would include planning, proposing and scheduling individual observations, then archiving, publishing and disseminating the data. It would also issue a handbook designed to ensure that observation made by professionals and amateurs adhered to a common standard.[70]

NASA presented the Halley Flyby/Tempel 2 Rendezvous for inclusion in its budget in 1979, but because other missions were being proposed at the same time – including VOIR – the strategy was to push for the development of the propulsion module to be in the 1980 budget, and the mission proper in 1981. But the agency's finances were still being soaked up by the development of the Space Shuttle, and in November 1979 work on the propulsion module was halted. The European partners learned of this in a trade magazine! Although it was possible that a simple ballistic Halley flyby might be approved at a later date, there was little enthusiasm for this because the data likely to result from such a fast flyby was deemed "unacceptable" and "scientifically inadequate".[71,72,73,74]

The first agency to approve a Halley mission was a latecomer to the field of space exploration. Having had its aircraft industry obliterated by the Second World War, Japan built a world-class aerospace industry as soon as it was permitted to do so. It tried to launch a satellite in 1966, but did not succeed until 1970. Meanwhile, an unusual situation developed, with two agencies being formed: ISAS (Institute of Space and Astronautical Sciences) was an offshoot of the University of Tokyo that specialized in small launchers and scientific spacecraft; NASDA (National Space Development Agency) built larger launchers and application satellites.[75] In the late 1970s both agencies began to study deep-space missions, with NASDA focusing on a relatively complex lunar orbiter. In 1973 an ISAS delegation visited NASA's headquarters and learned that the American agency, despite years of planning, was unlikely to stage a cometary mission in the near future.[76] Probably spurred by this, ISAS started to study whether the Mu-3S launcher, which could deliver 300 kg to low-Earth orbit, could send a small probe to intercept comet Halley in 1985–1986. Several profiles were studied for encounters at the ascending and descending nodes of the comet's orbit, and also the option of a slingshot at Venus. A post-perihelion encounter at the descending node was chosen, in part for its relatively small energy requirements and simple deep-space navigation, but also, and remarkably, due to a constraint imposed by the powerful Japanese fishing lobby which limited launches from Japan to two windows per year so as not to "scare" fish – only the descending node encounter was consistent with this constraint.[77] The Japanese Halley mission was officially approved in 1979, with 6 years to go before launch. It was decided at some point to launch two almost identical spacecraft: the Planet-A probe proper, and a technology demonstrator, provisionally named MS-T5 (Mu Satellite-Test 5), that would be launched some months earlier in order to test the spacecraft bus, its operating techniques and the 'escape' capability of the Mu-3SII launcher. This was an uprated version of the Mu-3S with two strap-on boosters, able to deliver some 770 kg to low-Earth orbit. If a kick stage was included, as would be the case on an interplanetary mission, the escape payload was less than 150 kg. It would mark the first time that an all-solid rocket was used to launch a deep-space mission – although the NASA/MIT Sunblazer project in the 1960s had envisaged using an all-solid Scout launcher.

Apart from their payloads, the two spacecraft built by the Nippon Electronics Corporation (NEC) were identical. A central thrust tube supported an electronics platform, and a cylindrical solar array providing between 67 and 104 W of power made them disk-shaped, 70 cm tall and 140 cm in diameter. During the cruise they would be spin stabilized at 6.3 rpm, but could be slowed down to 0.2 rpm using a momentum wheel 'buffer' to facilitate the use of imaging instruments for up to 15 hours per day. Two spherical titanium tanks held a total of 10 kg of hydrazine to slow the spin after separation from the launcher and to perform attitude and course correction maneuvers. The jets with a thrust of 3 N provided up to 50 m/s of total velocity change. Sun and Canopus sensors provided attitude determination, and the control system ensured that the spin axis remained perpendicular to the orbit plane.

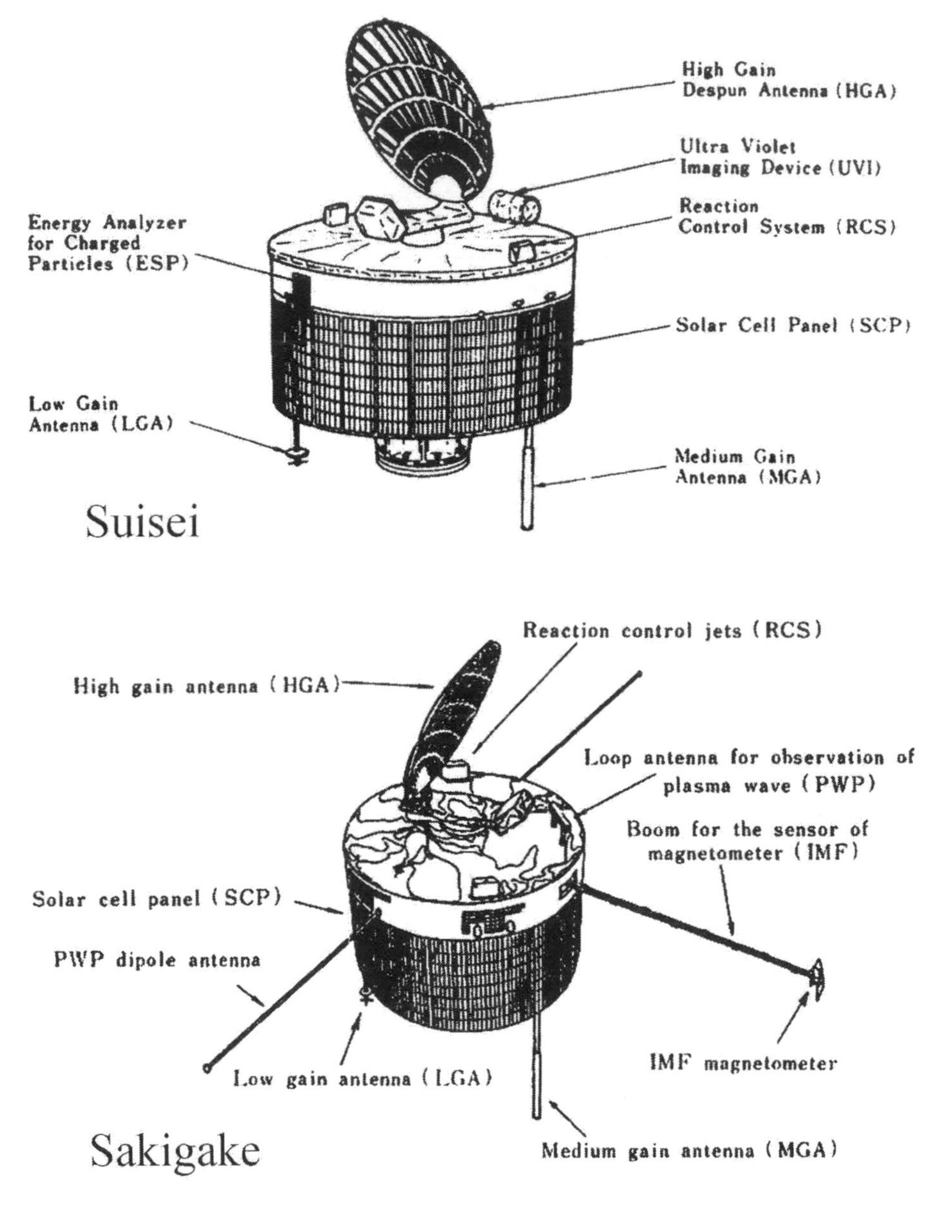

The structurally identical Planet-A (Suisei) and MS-T5 (Sakigake) spacecraft which Japan launched on Halley flyby missions. (JAXA)

Communications were by low- and medium-gain antennas, or by a despun 80-cm mesh antenna capable of data rates of up to 64 bps. The total height of each probe, including its antennas, was about 2.5 meters. To reduce the spacecraft's mass to fit the launch vehicle's performance, carbon composites and lightweight aluminum honeycomb were used extensively for the structural components – so much so that MS-T5 had a launch mass of just 138.1 kg and Planet-A was 139.5 kg. To further save mass, it had been decided not to fit Planet-A with shielding against dust, and to approach no closer than 200,000 km to the nucleus of the comet, since a study had indicated that there would be very little dust at this distance. To communicate with the spacecraft, ISAS built its own Deep Space Tracking Center with a 64-meter antenna in Usuda, a radio-quiet valley 170 km from Tokyo. It was completed in October 1984.[78] Each spacecraft had a scientific suite with a total mass of about 12 kg. The MS-T5 precursor, which was to pass 5 million km sunward of Halley in order to calibrate the observations by its twin, had instruments to characterize the interplanetary medium. An ion detector measured the direction, speed, density and temperature of the solar wind, a plasma wave probe with a 10-meter-span dipole antenna measured electric fields, and an onboard search-coil magnetometer and a triaxial magnetometer at the tip of a 2-meter pantograph boom measured magnetic fields. Planet-A had only two instruments. An electrostatic analyzer provided data on charged particles, and an ultraviolet imager based on the camera of the Kyokko (Exos-A) astronomical satellite provided up to six images per day and photometry of the hydrogen cloud which developed to surround a comet in the months around its perihelion passage, in order to characterize the variation of the production rate of water molecules with heliocentric distance. The imager comprised a mirror optical system with a focal length of 100 mm, an ultraviolet image intensifier, and a 122 × 153 pixel CCD (Charge Coupled Device). Owing to data processing constraints, it was not possible to run the two instruments simultaneously. There was a 1-megabit memory to store data while out of contact with the ground station.[79,80,81]

In parallel with studies of the Halley Flyby/Tempel 2 Rendezvous with the US, European scientists and engineers worked on another Halley flyby devised in 1979 by Professor Giuseppe Colombo of Padua University. This HAPPEN (Halley Post Perihelion Encounter) mission was actually a pair of related proposals: HAPPEN 1 would start with a 2-year study of Earth's magnetotail, then boost out of Earth orbit to encounter Halley at its descending node in March 1986. HAPPEN 2 would send a spacecraft carrying three daughter probes to fly by Halley. While the joint studies with NASA were underway, HAPPEN was retained as a backup option.[82] It was reassessed in January 1980 after the joint venture collapsed, but rejected owing to the modest scientific interest in flying a spin-stabilized spacecraft down the tail of the comet millions of kilometers from its nucleus. Over the ensuing months the proposal was split into a magnetospheric satellite (later rejected) and a spacecraft capable of penetrating the coma of the comet to image its nucleus. The project also acquired a new name. Colombo drew attention to an article published the previous year in which it was speculated that while painting the Star of Bethlehem in a 1303 fresco of the Scrovegni Chapel in Padua the Florentine artist Giotto di Bondone had been inspired by Halley's 1301 apparition. It was decided to name the mission Giotto.[83]

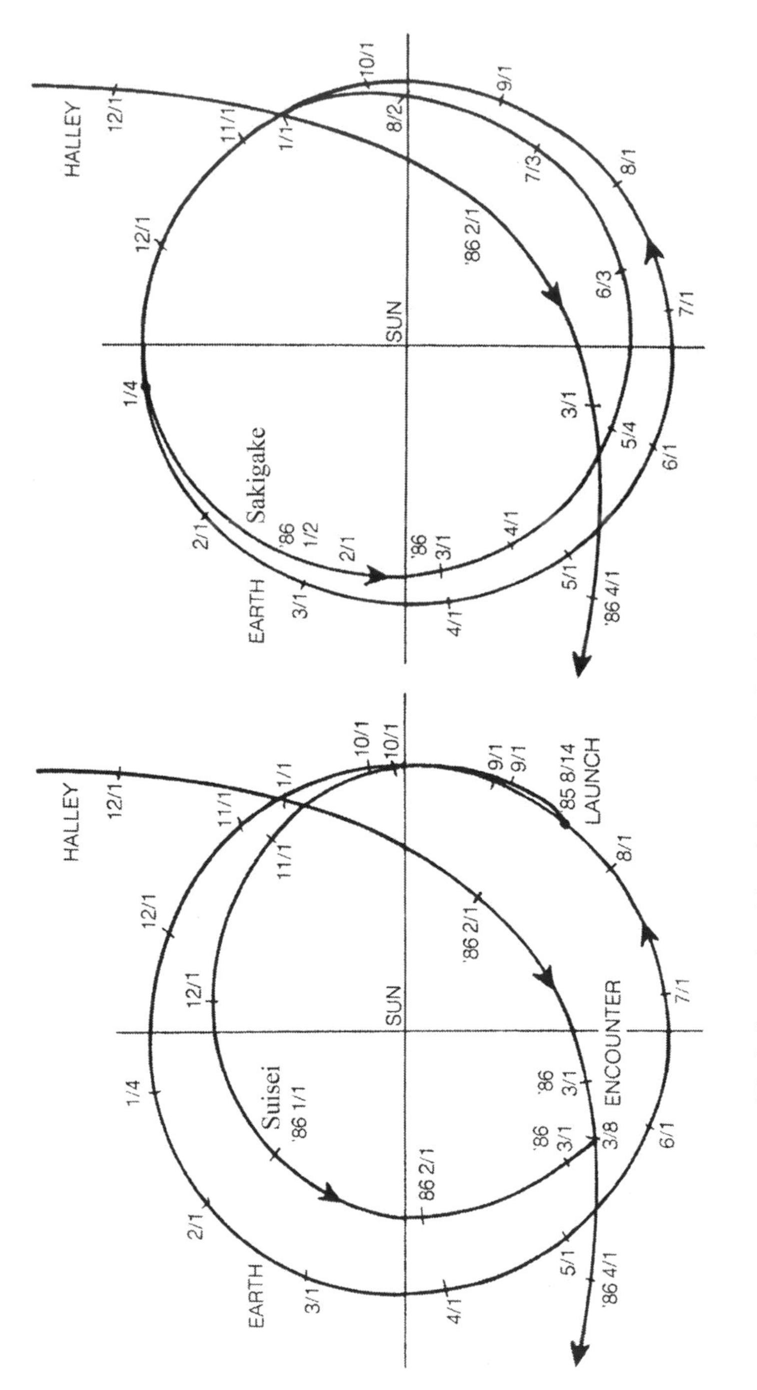

The heliocentric orbits of the Planet-A (Suisei) and MS-T5 (Sakigake) missions.

On 8 July 1980 ESA's science committee approved the mission, despite criticism from France, which was one of the most influential member states. It was to be launched in July 1985. NASA offered to launch the spacecraft. The prospect of using a medium-weight Delta was undermined by America's determination to phase out 'expendable' launchers in favour of the Space Shuttle, and the Europeans were reluctant to use the Shuttle. In fact, Europe was determined to prove that it could mount a deep-space mission using its own resources – all the more so after having been strongly criticized by those US scientists who considered a 'fast' flyby to be inadequate. Consequently, the American contribution to Giotto would be limited to cooperation on a few instruments and (as it eventually turned out) communications via the Deep Space Network.[84,85,86,87]

For reasons of cost and schedule, it was decided to base Giotto on the structure of the GEOS magnetospheric satellite built by British Aerospace, with the addition of a high-gain antenna and a dust shield. It would be spin stabilized at 90 rpm at orbit insertion but slowed to 15 rpm once on its way, and would have no booms or other appendages that would be exposed to the impact of cometary particles. In its final design, Giotto was built around a central aluminum thrust tube on which were mounted three platforms: the uppermost one for the high-gain antenna, the middle one with avionics and four tanks of hydrazine for attitude and orbit control, and the lowest for scientific payload and the star mapper which would provide attitude determination. To enable the spacecraft not only to survive but also to operate deep in the coma, it would require to be protected against specks of dust striking at high speed. Remarkably, the solution to this problem had been devised in 1946 by the American astronomer Fred Whipple of the Harvard College Observatory while studying meteor streams. The development by the Nazis of the V2 missile raised the prospect that at a date not too far in the future humans would set out to explore space. Whipple proposed that "space vessels" could be protected from meteoroids simply by placing thin sheets of metal several centimeters in front of their primary structure – a small meteoroid would be vaporized and rendered harmless before it could reach the spacecraft; the risk of damage would be limited to the larger, rarer, particles. The design could be further improved by using several shields in cascade. "Whipple shields" – he patented the idea – protected large manned spacecraft like the Skylab space station.[88,89,90] Of course, Whipple was also the originator of the 'dirty snowball' theory of the makeup of comets.[91] On Giotto, the Whipple shield was on the bottom platform, and comprised a 1-mm-thick aluminum bumper sheet (also called the "sacrificial shield") that was held by a dozen struts 23 cm in front of a Kevlar-reinforced plastic plate 13.5 mm thick. The front shield would intercept dust particles with masses up to 0.1 gram, and shatter larger particles whose debris would be absorbed by the second shield. The spacecraft was expected to be able to survive the impact of 1-gram particles striking at a relative velocity of 70 km/s.

It was decided early on to piggy-back Giotto on an Ariane 3 launch vehicle with a commercial communications satellite, release both in an elliptical geostationary transfer orbit and use a kick motor to boost the spacecraft by 1.4 km/s to enter the desired heliocentric orbit. However, Giotto's launch window constraints meant that no payload could be found to share the launch, and eventually it was decided to use a

The Giotto spacecraft about to undergo solar thermal tests. The white cylinder in the foreground is the camera baffle, the upward inclined black object at the left is the star mapper. Both would be heavily damaged by dust during the flyby. (ESA)

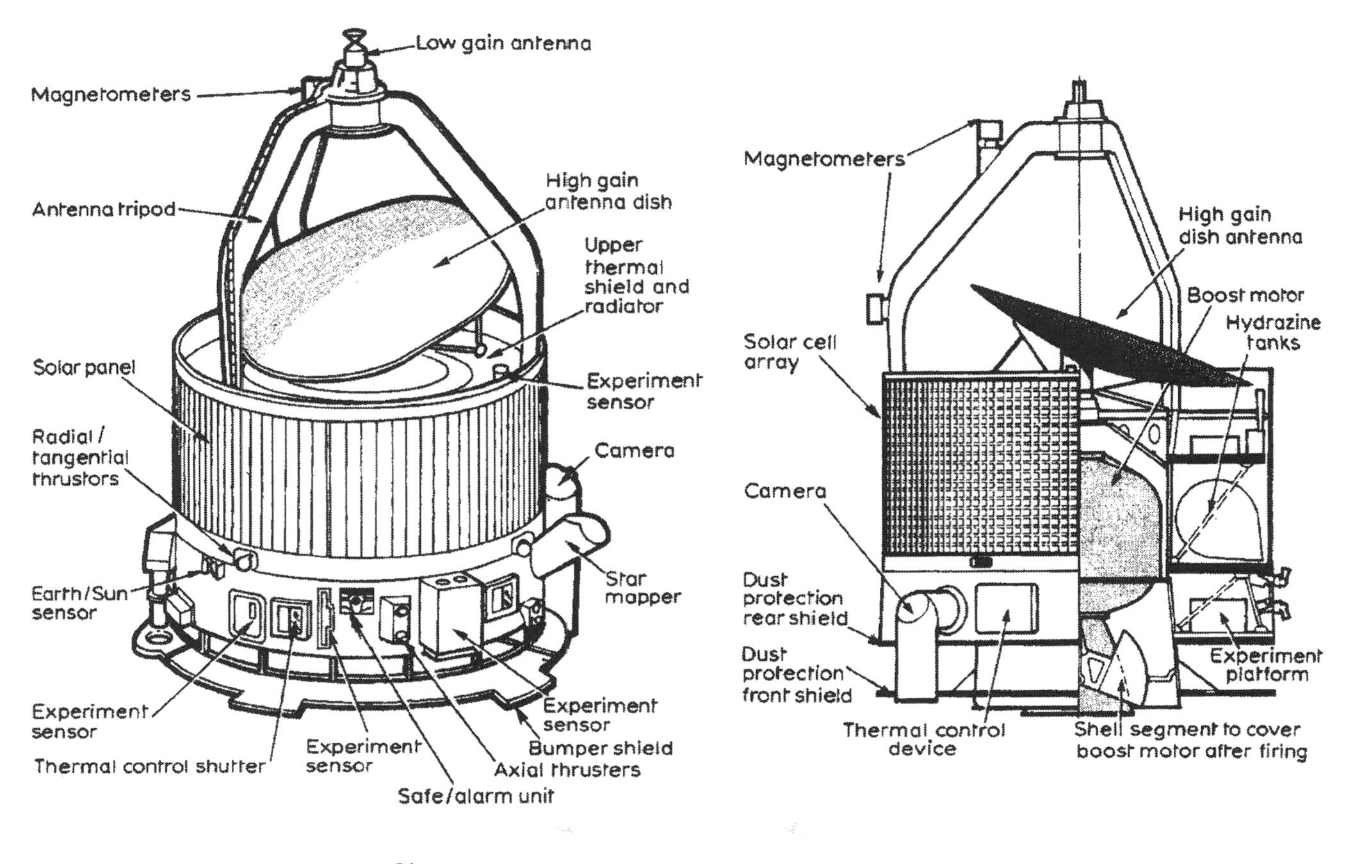

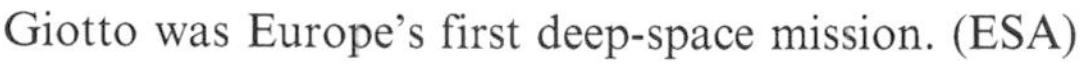

Giotto was Europe's first deep-space mission. (ESA)

dedicated Ariane 1 which, even although it was the least powerful of the series, was too powerful for the lone spacecraft. The design of Giotto had advanced to the point that it was no longer possible to alter the launch profile using the kick motor, so this was retained and techniques devised to shed some of the launcher's surplus energy. The Ariane 1, which first flew on Christmas Eve in 1979, thereby became the first European rocket to demonstrate a planetary capability.[92,93] The kick motor was the solid-propellant MAGE 1S (Moteur d'Apogée Geostationaire Européen; European Geostationary Apogee Motor) built by a European consortium – and in view of the origin of Giotto's name, it was apt. It was housed inside the thrust tube, and because it was not designed to be discarded after its burn a clever 'eyelid' was installed to seal the nozzle to preclude dust particles from finding their way into the spacecraft's interior through the empty casing.

The high-gain antenna and the tripod holding its feed were placed on top of the drum-shaped spacecraft, opposite the dust shield. The feed was fixed on the body, while the 1.47-meter-diameter parabolic reflector was mated to a motor that would precisely counteract the spacecraft's rotation in order to maintain the beam aimed at Earth. When during its development this motor showed a tendency to stick, it took a whole year to rectify the issue. The S- and X-Band systems provided a maximum data rate from the payload during the Halley encounter of about 29 kbit/s. In fact, Giotto had been tailored specifically for the geometry of the encounter with Halley at its descending node, when the spacecraft's spin axis, and hence its dust shield, would be oriented against the relative velocity of the spacecraft with respect to the comet. To point at Earth, the antenna's axis would have to be set 44.3 degrees from the spin axis. The alignment requirements were quite strict, since a discrepancy of just 1 degree would break the X-Band link. Such a misalignment could be caused, for example, by dust particles massing several tenths of a gram impacting near the rim of the bumper shield. Two low-gain antennas were also provided. Because there was a significant chance that Giotto would not survive its passage through the coma, it was to send its data in real-time and there was no data recorder – although some of the particles and field instruments had in-built data storage capabilities. An array of solar cells capable of providing up to 285 W of power formed the cylindrical outer shell. Giotto had a total height of 296.4 cm and a maximum diameter of 186 cm at the front bumper shield, which was slightly oversized in order to tolerate impacts if a strike were to misalign the spin axis, but only 181.4 cm across the main body. Its 985-kg launch mass included 374 kg of solid propellant and 69 kg of hydrazine and helium pressurant, making it the heaviest ESA scientific spacecraft yet flown. In view of the short development time and the fact that there would be no second chance to reach Halley if the July 1985 launch window were missed, the spacecraft was to employ as many proven technologies as possible, with new items being kept to a minimum – the latter including the star mapper for attitude determination, the eyelid to seal the kick motor, and the system to despin the antenna.[94,95,96]

Giotto's scientific payload was selected in January 1981. It comprised no fewer than 10 instruments having a total mass of 59 kg. The most important (and at 13.5 kg the heaviest) was the Halley Multicolor Camera (HMC) for high-resolution imaging of the comet's nucleus. Because conventional vidicon sensors would not

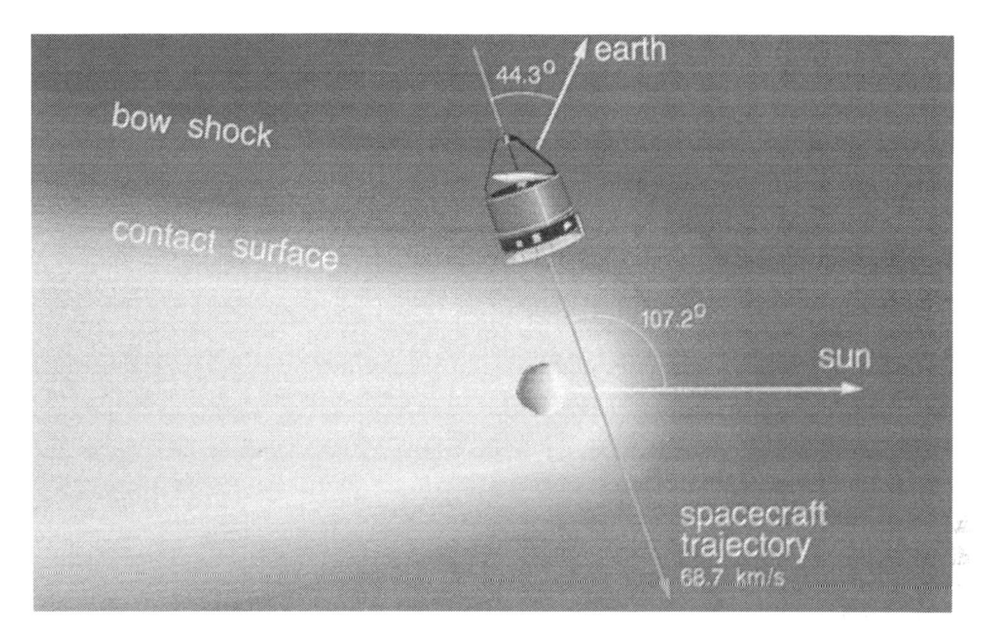

The geometry of Giotto's flyby of Halley's comet. The angle between the velocity of the spacecraft and the direction to Earth dictated the fixed angle of the high-gain antenna. (ESA)

provide clear images from a spinning spacecraft owing to the time taken to integrate the signal to produce a pixel, it was decided to use a pair of much faster CCDs, each of which was divided into two sections, and to read out only a single line of each portion. The camera would produce a one-dimensional scan of the image, and the spacecraft's spin would build up the second dimension. Three of the four slits had fixed red, blue and clear filters, and the fourth had a wheel with 11 narrow-band, wide-band and polarizing filters. The exposure time could range from 6 milliseconds to just 57 microseconds. The sensors were illuminated by a Ritchey-Chretien telescope with an aperture of 16 cm and a focal length of 998 mm. To avoid exposing the delicate optics to damage from cometary dust, the telescope was oriented sideways towards a 45-degree lightweight aluminum mirror protected by a Kevlar baffle which could rotate through 180 degrees in order to track the nucleus. A complex control system having as much computing power as the rest of the spacecraft was to recognize the nucleus as the brightest spot in the vicinity, predict the offset and spin phase of the next picture and move the mirror at every rotation by the amount required to keep the nucleus in the field of view – even allowing for an unpredictable wobble of the spacecraft. Thanks to this system, it was estimated that the best images would yield a 13-meter "surface" resolution at closest approach. The camera team boasted that the camera was capable of taking a picture of the pilot of a jet aircraft passing 160 meters away at the speed of sound![97,98,99,100] It was built by a European consortium of German, French, Belgian and Italian universities and research centers, together with the Ball Aerospace Corporation in America (having

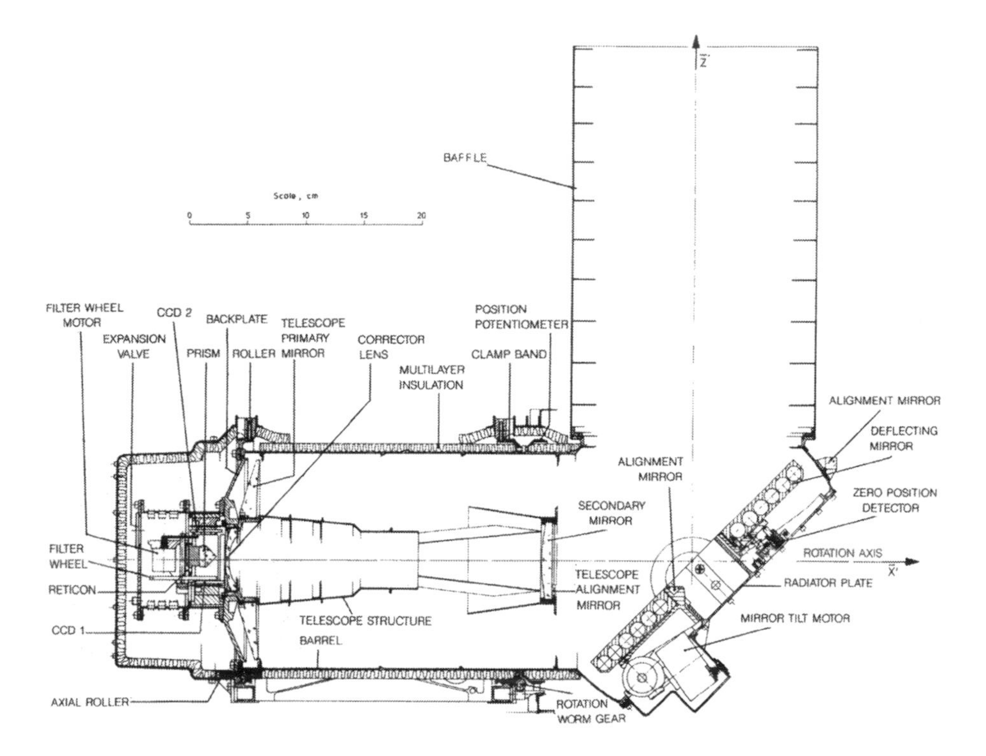

The Halley Multicolor Camera of the Giotto spacecraft. Note the mirror set at an angle of 45 degrees and the baffle that was eventually shattered by dust impacts. (ESA)

beaten a similar French-JPL proposal). By 1983, however, the development of the camera was impaired by a severe financial crisis, and it became evident that unless the budget was increased Giotto would have to fly without its main instrument! In fact, the development was further impaired by technical problems and the camera was not able to be delivered until several weeks before the launch was due, and even then it had not been fully tested and debugged.[101]

The second 'optical' instrument was a photopolarimeter that would look through the cometary dust, with the spacecraft's spin causing the polarizer to scan through 360 degrees to monitor the emissions of various gaseous species typical of comets. A magnetometer measured the magnetic field using two sensors. As there were no booms, the sensors were sited as far as possible from the main metallic structure of the spacecraft – on the carbon-fiber antenna feed tripod. In fact, to ensure that the instrument would make meaningful observations the magnetometer team was made responsible for the magnetic cleanliness of the spacecraft.[102] A German dust mass spectrometer analyzed the composition of plasma generated by the vaporization of dust particles on a metallic target. This instrument was designed on the assumption that cometary dust would resemble "Brownlee particles", which were microscopic fluffy dust specks that had been collected by aircraft flying at high altitude, had a chemistry similar to some types of meteorite and were suspected to be of cometary origin. However, because no facility existed to accelerate such particles to speeds approaching the 70 km/s of the Halley encounter it proved particularly difficult to calibrate the instrument.[103] A network of five independent acoustic and capacitance sensors mounted on both shields were to record the flux of dust motes with masses up to milligrams. A neutral mass spectrometer would measure the elemental and isotopic composition of the comet's coma so as to determine the 'parent' molecules issued by the nucleus. In this regard, although in the 'dirty snowball' model it was widely assumed that water was the primary constituent of a comet, water itself is spectroscopically unobservable from the Earth's surface, and it was 1974 before it was confirmed in comet Bradfield by observations at radio wavelengths. Cometary ions would be characterized by an ion mass spectrometer, while the interactions between the cometary environment and the solar wind would be studied by three plasma and energetic particle instruments, one built in Ireland, one in the UK and one in France, having a total of seven sensors to observe the ion energy distribution in the solar wind, the cometary shock and the near-nucleus environment, and to characterize the cometary ions carried by the solar wind. The Irish instrument was named EPONA (Energetic Particle Onset Admonitor) after a Celtic goddess. The other two instruments were known respectively as the Johnstone Plasma Analyzer (JPA) and Rème Plasma Analyzer (RPA) after their principal investigators. Finally, the cometary ionosphere along the line of sight to Earth and the change in Giotto's velocity due to dust impacts could be measured by monitoring the radio carrier. In the end, however, the radio sounding experiment was not done because the project managers decided not to operate the dual-frequency transmission system at closest approach.[104,105]

Routine data from Giotto would be received by an ESA antenna at Carnarvon in Australia, but owing to the lack of dedicated European deep-space antennas, ESA asked the large Parkes radio-telescope in Australia to assist during the encounter, which was agreed in exchange for ESA-paid upgrades and a payment of $200,000; lest this seem expensive, NASA's Deep Space Network wanted $10 million for tracking. Later, the NASA network also agreed to contribute (initially at a more sensible price, and eventually for free) to ensure continuous coverage of the days around closest approach.[106]

The third and last country to approve a Halley mission was the Soviet Union. In 1977 the USSR and France had signed a memorandum covering the exploration of Venus, focusing in particular on the Venera 84 mission that replaced the joint Eos atmospheric balloon mission and was to deploy a smaller French-designed balloon. This would introduce the UMVL (Universalnyi Mars, Venera, Luna; Universal for Mars, Venus and the Moon) spacecraft. The 10-meter balloon would carry 28 kg of instruments for meteorological and atmospheric studies, including a VLBI (Very Long Baseline Interferometry) transponder to enable terrestrial antennas to track the balloon very accurately as it was carried around the planet by the super-rotating atmosphere. After it had released its capsule, the spacecraft was to enter orbit and use some 80 kg of scientific instruments mounted on a scan platform to observe the circulation of the atmosphere, its composition and interactions with the solar wind. Jacques Blamont, the French scientist in charge of the program, was asked by JPL to

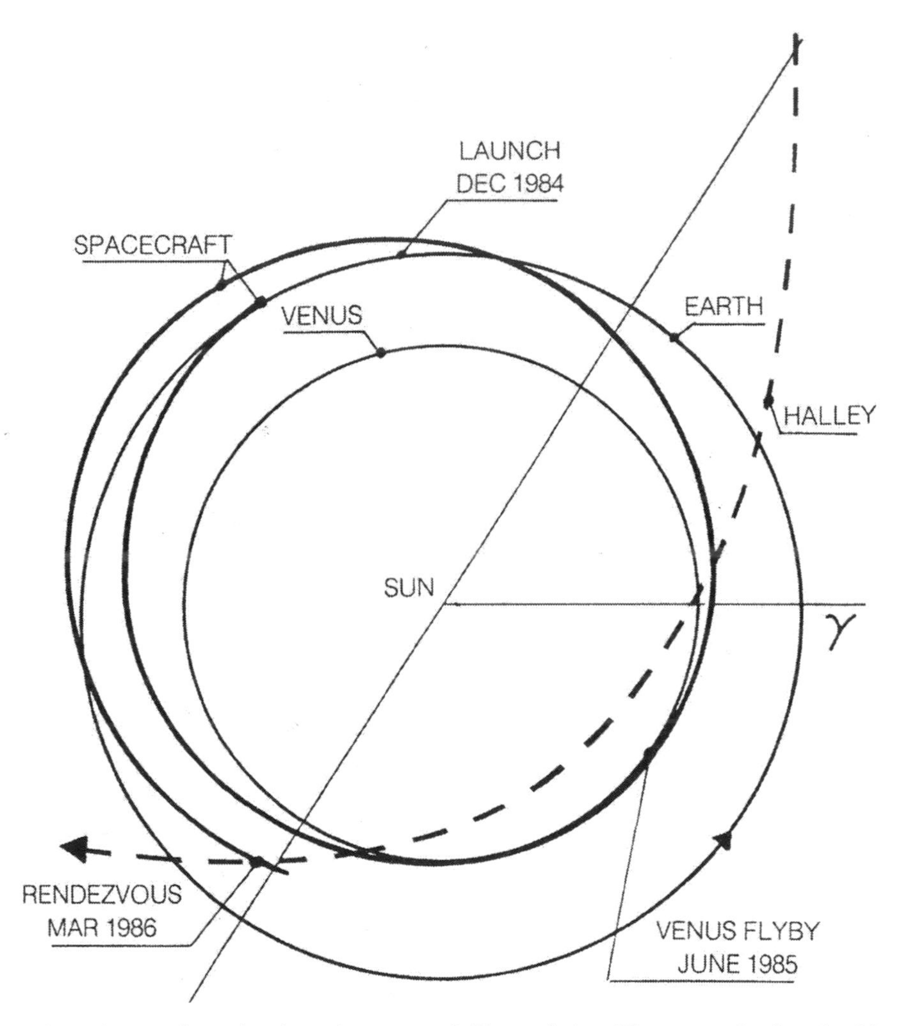

The trajectory flown by the twin-spacecraft Vega mission. They were the first (and in fact the only) Soviet spacecraft to employ the gravity-assist technique to reach a second target. (ESA)

invite the Soviets to join the International Halley Watch. This invitation spurred Soviet studies of possible space missions involving the comet. In contrast to Earth, which offered a poor view of the comet during its perihelion passage, Venus was a much better vantage point because the comet would approach within 40 million km of the planet and shine at magnitude –0.7. Initially, therefore, the Soviets simply envisaged having their spacecraft orbiting Venus turn to make ultraviolet studies of the comet, just as NASA planned to do using its Pioneer Venus Orbiter.[107,108]

Then, however, celestial navigators found that a spacecraft sent to Venus during the December 1984 window could exploit a slingshot with the planet to intercept the comet at its descending node in March 1986. It would be the first multi-target Soviet mission, as well as the first to use the gravity-assist technique (although this had been intended for the canceled 5M mission to return a sample from the surface of Mars). Moreover, it would mark the longest duration Soviet planetary mission. Driven by the wishes of the scientists – quite likely for the first time in the Soviet planetary program – the Venera 84 mission was completely overhauled to use no fewer than four spacecraft. Two orbiters (designated 5VS by the Soviets) would be launched first, then two Halley flyby spacecraft (5VP) each of which would release a Venera entry capsule carrying a French balloon. The spacecraft would now revert to the Venera bus that had been introduced in the mid-1970s. The Venus–Halley mission, designated Vega by the Soviets as a contraction of Venera–Galley (there being no 'H' in the Russian alphabet), was proposed in the spring of 1980, and on 22 August was endorsed by both the Soviet Academy of Sciences and the Ministry of General Machine Building – albeit in a reduced form: the orbiters were deleted, as were the French balloons, which were replaced by standard Venera landers and by smaller balloons which the Soviets invited the French to supply, but the CNES (Centre National d'Etudes Spatiales) withdrew from the project.[109,110,111,112]

Only 4 years remained to refine the mission plan, modify the spacecraft, design and test instruments and integrate the entry probes. Building on the experience of flying foreign instruments on Soviet satellites and spacecraft, Roald Sagdeev, the director of IKI (Institut Kosmicheskikh Isledovanii; Institute for Cosmic Research), took the bold step of suggesting that international cooperation be sought on a broad basis, including the United States. A total of 125 kg had initially been allocated to the scientific payload, but it proved possible to double this to 240 kg and to carry 14 separate instruments.[113] The highest scientific priority of the mission was given to determining the shape, size, volume and state of rotation of the comet's nucleus. For this reason, a flyby at a range of 8,000–9,000 km was selected, this being deep enough into the dusty inner environment for the requisite imaging, yet survivable. In addition, the cameras would monitor the evolution and dynamics of the jets that emit material from the surface, and the behaviour of this material in the coma in the immediate vicinity of the nucleus.

The television experiment and its data processing system were to be built jointly by the Soviet Union, Hungary and France – which actually built the optics, based on a Soviet design. It comprised a 150-mm f/3-focal-length wide-angle camera that was restricted to a wavelength range corresponding to the red part of the spectrum, and a 1,200-mm f/6.5 narrow-angle camera equipped with six filters ranging from the visible to the far infrared. Since no one knew what the nucleus would look like, the exposure times were to be determined automatically. The maximum theoretical resolution at closest approach would be about 150 meters. The performance of the computers used to prepare the data from this and the other instruments prior to its transmission to Earth was greatly improved by the use of Western electronics. One component that the Soviets were unable to obtain was a CCD imaging chip for the camera, because this was a restricted technology. However, a Leningrad enterprise managed to develop an entirely Soviet version with a 512 × 512 pixel matrix and a performance similar to its Western counterparts. In addition, the lack of experience

A mockup of the Vega spacecraft, including the atmospheric capsule containing the balloon and lander for Venus. Apart from the addition of the scan platform (which is not visible in this picture), the main difference from the Veneras which delivered landers was the adoption of the larger solar panels developed for the radar mappers. (Courtesy of Jean-François Leduc)

of the Hungarian team assigned the responsibility for developing the electronics for the camera meant they had difficulty designing and building a system that would survive the launch vibrations and would function correctly in the vacuum of space. The Soviets assisted the Hungarians, even providing a vacuum chamber for testing. Nevertheless, ESA later 'poached' some members of the Hungarian team to assist with the troubled Giotto camera. Although Vega was not the first Soviet planetary mission to use electronic rather than film imagers (because a 'push broom' camera had been introduced by the mission to Mars in 1973) it was probably the first to be capable of returning data in real-time at 64 kbit/s, although a tape recorder was also carried for later replay of the same data.[114]

The Vega spacecraft had an infrared spectrometer supplied by the French that was to provide radiometry of the nucleus in order to measure its temperature, and spectrometry of the inner coma to determine its composition. It was also equipped with a 3-channel spectrometer built by a Bulgarian–Soviet–French team which was sensitive in the infrared, visible and ultraviolet. A 2-sensor gas detector supplied by West Germany was to measure the density of neutral (non-ionized) cometary gas in the vicinity of the nucleus. Three instruments would study the dust particles. Two of these were to be built by the Soviets and would employ conventional acoustic or plasma sensors. The third would detect the surge in voltage as a block of polarized polymer was 'cored' by impacting dust, the surge being proportional to the mass of the particle. This instrument was provided by John A. Simpson of the University of Chicago. It had been turned down for Giotto because the proposal was received too late, and IKI had invited Simpson to fly it on Vega, with the German Max Planck Institute acting as an intermediary and NASA providing about $300,000 in funding and support to gain the export licence needed to send it to the Soviet Union. To ensure that the Soviets would acquire no insight into restricted US technology, the associated hardware was reportedly built mostly using "off the shelf" electronics. A fourth instrument counted the impacts of the largest dust particles on the Whipple shield.[115,116,117] A cometary dust mass spectrometer was supplied as a joint venture between West Germany, the USSR and France, and was a slightly redesigned form of a Giotto instrument. In fact, to avoid controversy and facilitate a more complete analysis, the three almost identical instruments on Giotto and the two Vegas would use different materials for the target that would vaporize the dust particles.[118] Other experiments included a plasma instrument with an array of six detectors, some of which were to be pointed towards the comet and the others towards the Sun, and an energetic particle detector developed jointly by Hungary, the USSR, West German and ESA.[119,120] Austria provided a magnetometer based on the experiment flown on Veneras 13 and 14. This had three flux-gate sensors mounted on a boom on one of the solar panels, and a fourth sensor a meter closer in to enable the signal to be corrected for the presence of the spacecraft.[121] Finally, two plasma wave analyzers were carried, one built jointly by ESA, France and the USSR for high-frequency waves, and the other developed by the USSR, Poland and Czechoslovakia for low frequency waves. The high-frequency instrument used Langmuir probes on booms at the ends of the two solar panels. The low-frequency instrument was mounted on a Y-shaped boom.[122,123] Although Poland had been an enthusiastic member of the

Interkosmos organization, it had only a marginal involvement in the Vega mission as a secondary partner in the plasma wave experiment; quite possibly as a result of the social turmoil in that country during the early 1980s that was opening the first crack in the 'iron curtain'.

Unlike the previous Soviet spacecraft that had body-mounted experiments with fixed viewing angles which required the entire craft to maneuver to make targeted observations, the Vega (5VK) had a stabilized scan platform that massed 82 kg and had 2 degrees of freedom. This held the instruments that required to be accurately pointed at the comet, such as the camera and the optical spectrometers. In fact, at the time of closest approach, the comet's relative motion would be almost 1 degree per second! The platform and its targeting and stabilization system were developed jointly by the USSR and Czechoslovakia, in preference to a heavier Soviet model – the fact that the Soviet alternative twice failed its vibration tests undoubtedly was a factor in the decision! The platform could slew through 220 degrees in azimuth and nod up and down through 60 degrees.[124] It proved particularly difficult to design a sensor that would maintain the platform aimed at the comet, simply because no one knew how the cometary nucleus would appear in terms of its brightness relative to the inner coma. An 8-element photodiode array was installed, but while this would be reliable and fast its 'dumb' software would simply point the platform at its field of view's center of brightness. A better option was to use the wide-angle camera as an optical sensor to zero-in on the nucleus. To achieve this, IKI scientists derived a model of the nucleus from early observations of Halley since its recovery heading inbound.

On the few previous occasions when foreign scientists had provided instruments for Soviet planetary missions, these had simply been handed over for installation on the spacecraft and the contributors had played no further role until (if they were fortunate) they were given their data. But the external involvement in Vega was so great that the Soviets made available an old test model of the 1973 Mars spacecraft to enable the international teams to participate in the integration work.[125,126,127] The 5VK was a modification of the 4V-1 spacecraft. The new solar panels of the radar-equipped Venera orbiters were adopted, and a two- and three-layer Whipple shield 1 mm in thickness was fitted around the body and on its base to block particles of dust massing up to 0.1 grams. An impact on such a spacecraft could be particularly serious because if the pressurized body were to be pierced this could rapidly lead to a malfunctioning of the electronics. To facilitate the extended mission duration, the operational lifetime of the spacecraft was increased from 1 year to 450 days.

Each 5VK was to carry a standard spherical entry capsule in which there would be a modified lander and a small aerostatic station (AS, Aerostatnaya Stantsiya). The geometry of the slingshot at Venus to send the spacecraft on to Halley required the landers and balloons to enter the atmosphere on the night-side of the planet. On the one hand this meant that there would be no point in equipping the landers with either cameras or photometers (apparently, no consideration was given to refitting the floodlights which were carried by the original Venera 9 and 10 landers), but on the other hand it would extend the missions of the balloons since the helium which inflated them would not start to leak until sunrise. The aerodynamic stability of the

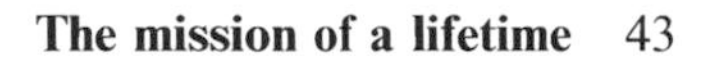

Another view of the Vega spacecraft, showing the scan platform and some of its instruments. An inflated balloon probe is visible in the background.

landing module was further improved by the adoption of a conical surface beneath the circular stabilizer (the 'sawtooth stabilizer' of Veneras 13 and 14 was deleted). The landers carried a number of new and redesigned instruments.[128] Two platinum wire thermometers and three barometers were the basic suite to profile the pressure and temperature during the descent, while a French ultraviolet spectrophotometer obtained spectra of a sample of gas that was to be admitted into a 85-cm-long tube on one side which would be illuminated by a strobe lamp, before light was fed to a spectrometer inside the hermetic body. An optical spectrometer would measure the size, number and refraction index of aerosol particles, both near the probe during the descent and in passing through a sampling chamber. The instrument could also function as a photometer to measure ambient light. Another photoelectric aerosol-particle counter would take measurements to determine the structure of the cloud deck. A Soviet–French mass spectrometer would purge an atmospheric sample of carbon dioxide to isolate the aerosol particles and determine their composition. An X-ray fluorescence analyzer and an improved Sigma-3 gas chromatograph would also analyze the composition of the aerosols. Because of its growth in volume and capabilities, the gas chromatograph could not be fitted inside the lander's body, so it was mounted in a thermally controlled pod on the landing ring. A 'double-bell' shaped humidity sensor was also mounted on the landing ring. This completed the atmospheric suite. The surface package comprised a gamma-ray spectrometer, an improved form of the Venera drill system and X-ray fluorescence soil analyzer to collect a small sample of the surface material and transfer it to an internal chamber for chemical analysis, and the type of penetrometer used by Veneras 11 to 14 but probably fitted with encoders capable of surviving the high temperature to enable it to return its data directly instead of being monitored by the camera.[129,130,131] Even although the landing would be in darkness, there was a solar panel to measure the illumination at the surface – probably carried simply because it formed part of the standardized payload of the previous landers.

After many French-led iterations starting with the 1970s Eos balloon proposal, in 1980 the French declined to make the Vega aerostat (which was mostly Soviet-built) but they did supply part of its scientific payload. The aerostatic system was to be developed by the Dolgoprudenskii bureau, which specialized in balloons, but owing to the technical issues presented by the peculiar Venusian environment this task reverted to the Lavochkin bureau. The only modifications required to install the balloon into the existing entry sphere were mounting brackets and fiberglass guides to ensure that it separated cleanly from the heat shield. The gondola and the deflated balloon were housed inside a toroidal compartment installed around the lander's cylindrical antenna. This compartment was to split into two halves. The upper part had spherical helium tanks at its periphery and plumbing for the inflation system, in addition to the 35-m^2 parachute. The lower part was tethered to the gondola and was to serve as ballast during the initial descent and during the inflation, and then be jettisoned once the aerostat had reached the desired cruising altitude. Of course, the inflation would have to be sufficiently rapid to ensure that the balloon did not sink into the hotter lower atmosphere before it became buoyant. The 1-meter-long gondola comprised three modules in a vertical stack, linked by flexible data cables.

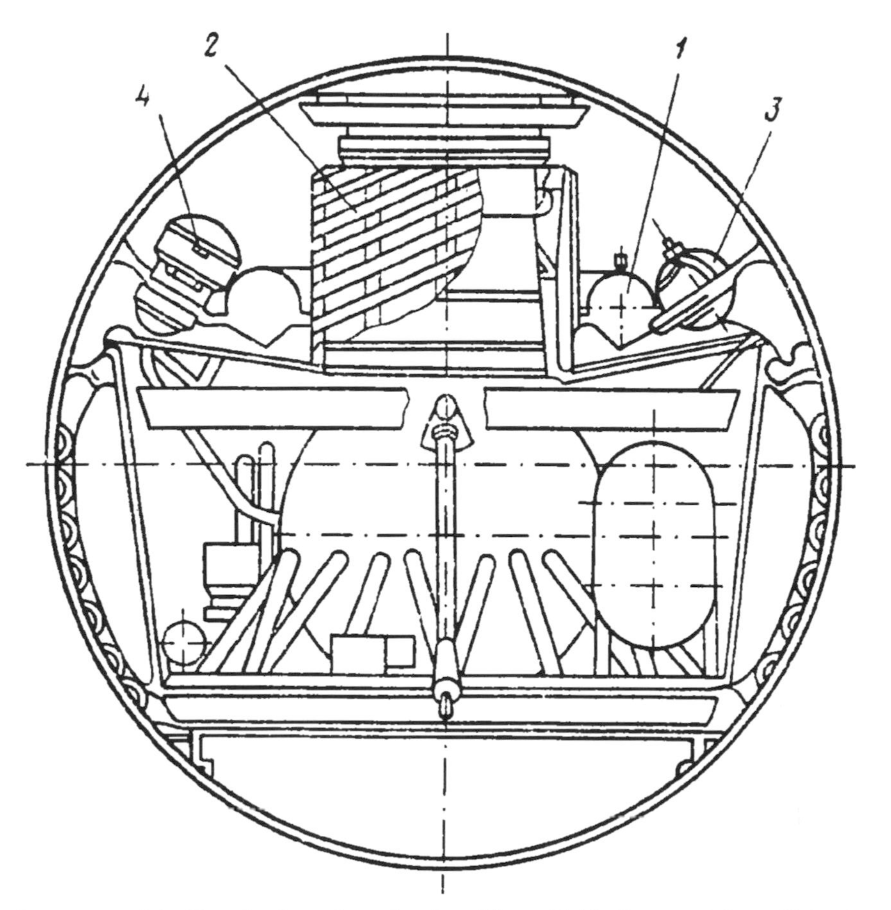

A cutaway of the Vega lander sphere showing: (1) toroidal balloon container, (2) lander antenna, (3) helium inflation tank and (4) parachute container. Note also the additional aerodynamic stabilizer of the lander. (From Sagdeev, R., et al., "The Vega Balloon Experiment", Soviet Astronomy Letters, 12, No.1, 1986, 3–5; reprinted with permission of the American Institute of Physics)

On top there was a helical antenna 37 cm long. Next was the 40.8 × 14.5 × 13-cm scientific instrument and transmission compartment. A 24-cm carbon-fiber boom held atmospheric instruments, including a thermometer and a vertical anemometer equipped with a small polypropylene propeller to detect ascending and descending winds of the order of 2 m/s. There was also a barometer and an ambient light meter and lightning detector. The bottom module, 9 × 14.5 × 15 cm in size, comprised a 1-kg battery pack rated at 250 W-hr and a French nephelometer (with American input) to measure the size of the cloud particles. The entire gondola was covered with a special paint to protect it from the corrosive atmosphere. A 13-meter tether linked the gondola to the spherical 3.54-meter-diameter radio-transparent balloon. In view of the fact that the balloon was to be inflated by very-low-viscosity helium, it proved difficult to develop a balloon fabric that would be both pressure-tight and corrosion resistant. A Teflon-coated fabric was used, with the envelope being built in slices, like an orange, with the joints between the slices covered with layers of a special paint.

The mass of the balloon probe was 21 kg (or more precisely 20.82 kg for Vega 1 and 21.11 kg for Vega 2). This including a 6.7-kg gondola, 2 kg of helium and 12 kg for the balloon, tether, etc. The total mass of the aerostat, including its ballast and the inflation system, etc, did not exceed 110 kg. Two factors would limit the operational life: one was the seepage of helium from the balloon, but it was hoped that the pressure would be able to be maintained for about 5 days; the other, more critical, factor was the battery pack, which could support a duration of just 46 to 52 hours. However, during this time the probes would be carried about 15,000 km by the fierce super-rotating wind.[132,133,134,135]

Although the transmission system had an output of just 4.5 W, its wavelength of 18 cm was situated in a standard astrophysical range to enable the telemetry to be received by any of the world's large radio-telescopes. In fact, the French had been made responsible for the interferometric tracking experiment in which many radio-telescopes thousands of kilometers from each other would locate the balloon by the measured phase difference of its radio carrier, thereby enabling its position to be calculated to within 10 km and its velocity to within 3 km/h – despite it being more than 100 million km from Earth! The balloon probe would poll its instruments one by one every 15 seconds and store the data in a 1,024-bit memory which would be read out to Earth at 4 bps over a 5-minute communication session. This was to be done every 30 minutes for the first 10 hours of the flight, then again from the 22nd to the 34th hour, but once per hour during other periods. Mixed with these sessions would be 5-minute sessions during which it would transmit only two tones which bracketed the central 18-cm wavelength, in order to undertake the interferometric tracking experiment.

Thanks to the fact that a western space agency (CNES) was leading the balloon tracking, it was possible to obtain support for the interferometric experiment from countries which had no direct cooperation agreements with the USSR, in particular the United States. In fact, the American–Soviet agreement for the Mars missions of the 1970s had been left to expire, and relations between the two Superpowers were at their worst for decades: it was the time of the Soviet invasion of Afghanistan, of Soviet-imposed marshal law in Poland, of the shooting down of a Korean airliner, of the American 'Star Wars' effort, and of the deployment by both sides of missiles in Europe. The French connection not only enabled the communications antennas at the three stations of NASA's Deep Space Network to participate in tracking the Vega balloons, but also the giant Arecibo radio-telescope and antennas across the continental United States. Support was also provided by Canada, Brazil, Germany (using the 100-meter Effelsberg radio-telescope), the UK (Jodrell Bank), Sweden, and South Africa, which, as an apartheid state was the 'pariah' of the international community. Remarkably, however, none of the 14 antennas outside of the Soviet Union which tracked the balloons was in France itself! These stations added to six Soviet antennas at sites including Yevpatoria and Simeis in Crimea, Medvezkye

Ozyora, Pushchino, Ulan Ude and Ussurisk in the Soviet Far East, which was built for Vega.[136] Whereas NASA had positioned the stations of its Deep Space Network in California, Australia and Spain (and for a certain period also in South Africa) in order to be able to maintain continuous contact with its interplanetary missions, the Soviet's had no foreign deep-space communication facilities. In fact, despite having built an antenna for the Molniya telecommunication satellites in Cuba, and often replenishing their tracking ships at the port of La Havana, the Soviets appear not to have considered building a deep-space antenna on that Caribbean island. If they had done so, a Cuban antenna would have been able to work with Yevpatoria and Ussurisk to yield a coverage very similar to that provided by the Deep Space Network.[137]

NASA and JPL did not give up so easily on an American Halley mission. Dual ballistic missions to Halley using slingshots from either Venus or Earth had been studied as a low-cost alternative. After the Earth flyby, additional encounters with other comets (namely Tempel 2, Borrelly, Kopff, Tuttle–Giacobini–Kresak, Faye and others) were found possible. A mission that would reuse a spare spacecraft of Mariner 10 vintage was studied, as was one using a second Galileo spacecraft.[138] A final American attempt was JPL's Halley Intercept Mission (HIM). In comparison to the proposals for solar sails and ion propulsion, it was very conventional, calling for a spacecraft using existing technology that would be launched in the summer of 1985 by either a Titan rocket or the Space Shuttle/IUS combination and would fly a ballistic trajectory. After observing the comet for 5 months in order to accurately measure its position, the spacecraft would encounter it in late March 1986 close to its descending node and fly through the coma about 1,000 km ahead of the nucleus, imaging this using a 'periscope' system resembling that of Giotto. The spacecraft would be protected by a double-layer Whipple shield, and would retract its solar panels during the encounter.[139,140] Celestial navigators pointed out that a spacecraft which encountered Halley in March 1986 could be maneuvered onto a path which, after 6 orbits, exactly equivalent to 5 years, would return to the Earth's vicinity. If it were to employ its shield to collect dust and gas samples, these could then either be delivered directly to the Earth's surface using a re-entry capsule or maneuvered into low-Earth orbit to be retrieved by a Space Shuttle. Although this Halley Earth Return (HER) mission gained backing from the US National Research Council, its scientific value was diminished by the fact that its material would comprise at best "atomized samples" embedded in the metallic collector, as opposed to the pristine source material.[141,142] The cost of HIM was estimated at $355 million, while HER would cost $400 million. A major obstacle to the effort to sell these missions was that US comet scientists had earlier dismissed the 'fast' flyby option as unworthy. Alternative funding options were considered, including private subscription, but in the end work on HIM was cancelled when it was not recommended for inclusion in NASA's budget. A key criticism was that (contrary to what the scientists involved claimed) it was unlikely to yield results better than those from Giotto.[143] Thus, the United States became the only major space power not to send a mission to Halley. However, several of its spacecraft in Earth orbit and elsewhere in the inner solar system would be able to make observations. Pioneer Venus Orbiter would have the

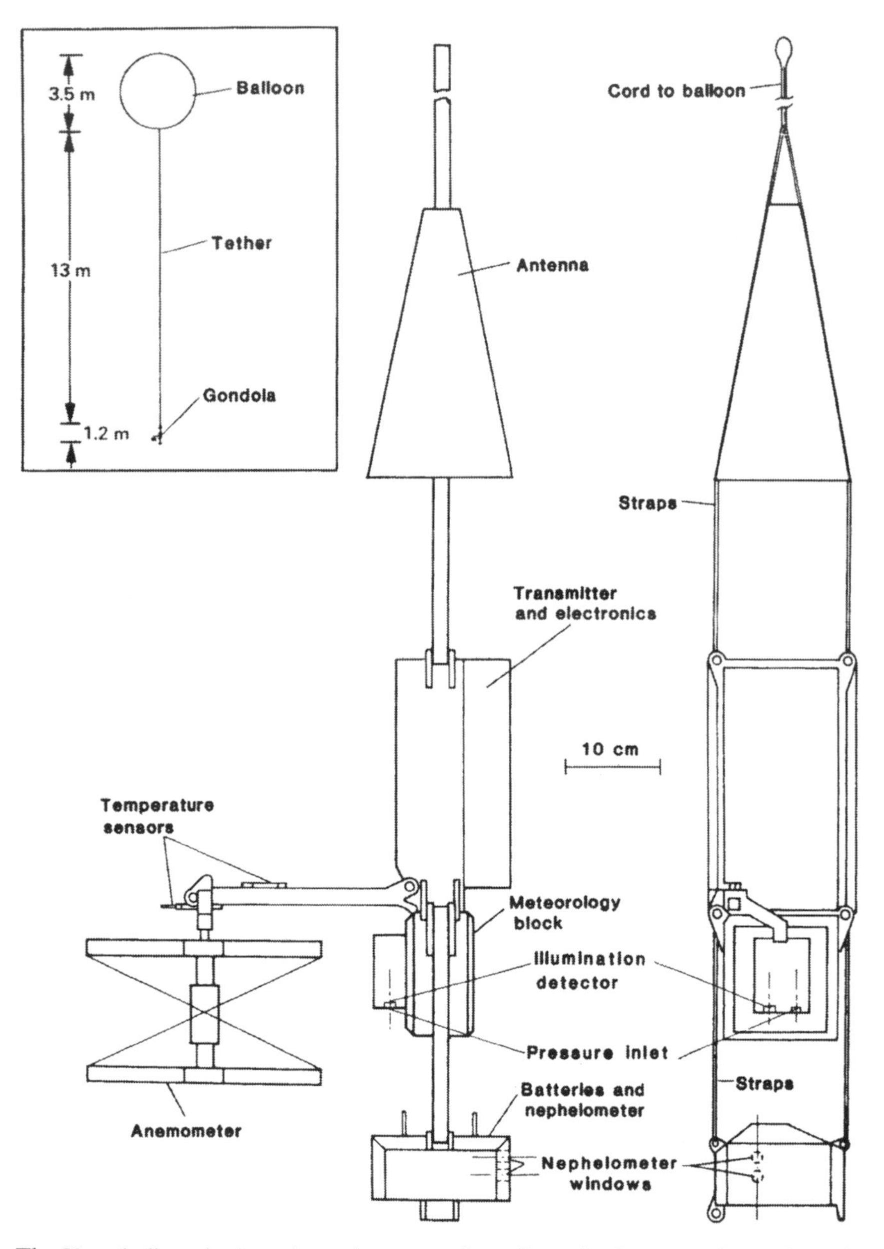

The Vega balloon is the only probe ever to have flown in the atmosphere of another planet. (From Kremnev, R.S., et al., "VEGA Balloon System and Instrumentation", Science, 231, 1986, 1408–1411; reprinted with permission of the AAAS)

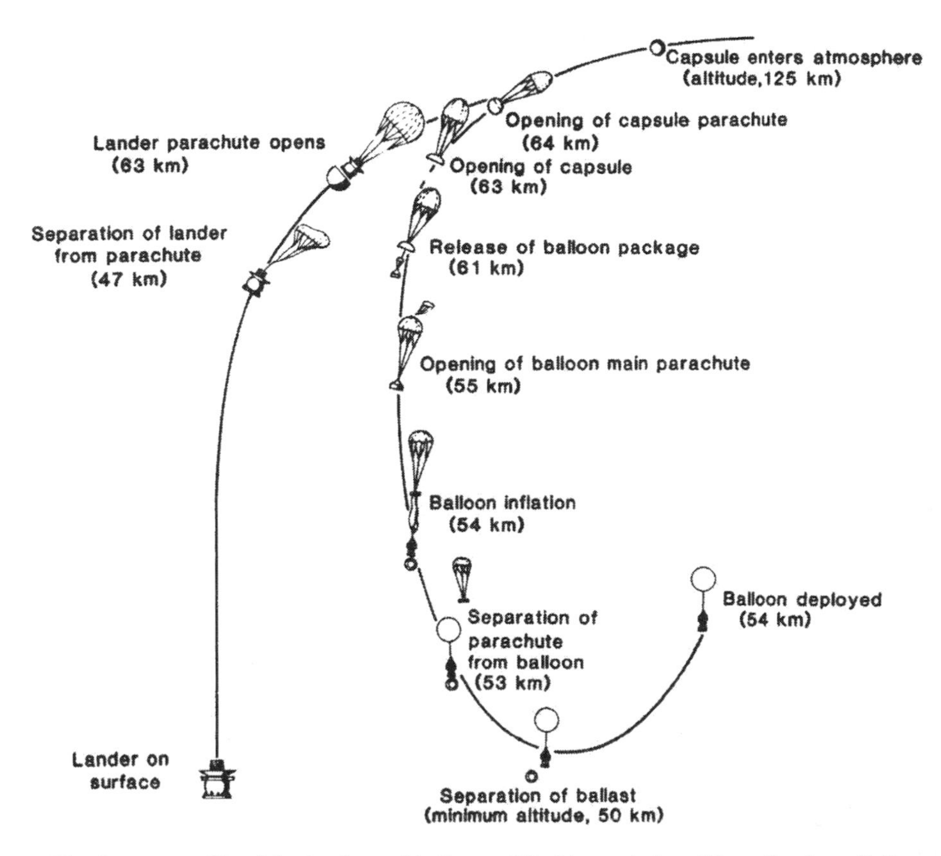

The descent profile of the lander and balloon of the Vega mission. (From Sagdeev, R.Z., et al., "The VEGA Balloon Experiment", Science, 231, 1986, 1407–1408; reprinted with permission of the AAAS)

best view. By sheer luck, the 1960s-vintage Pioneer 7 solar wind monitor would make a rather close approach to the comet. Furthermore, if they were still working (in the event, they were not) one of the Helios probes would be 50 million km from Halley at its perihelion, and the Vikings in orbit around Mars and on that planet's surface would be 78 million km from it in April 1986. It was also discovered that one of the many trajectories studied for the delayed Galileo Jupiter mission (an option that involved a 2-year heliocentric orbit designed to return the spacecraft to Earth for a slingshot) would put Galileo within 30 million km of the comet in late 1985, but it would be able to undertake only long-distance imaging from a viewing angle that was not particularly favorable. The option of retargeting Galileo for a closer flyby was studied, but when the cost of such a diversion was calculated ($300 million) it proved to be not much less than that for launching a dedicated mission![144,145]

After years of investigating missions to Halley, all that remained for America was the International Halley Watch, the possibility of observing the comet from the yet to be launched Space Telescope, and from instruments on well-timed Shuttle flights.

As initially envisaged, the Space Telescope would probably have dedicated a large fraction of its observing time to cometary studies during the whole of Halley's apparition. While on the one hand the telescope's narrow field of view would have restricted it to studying localized phenomena within a few thousand kilometers of the nucleus, on the other hand the clarity of the view in space would have revealed detail difficult to discern from the ground. In the event, its launch was delayed until after the Halley apparition. This left two Shuttle flights to observe the comet. The first, mission STS-51L in early 1986, was to deploy a free-flying satellite to carry out ultraviolet and photographic observations on the days around perihelion in order to measure the rates of production of several chemical species. Then STS-61E would use three ultraviolet telescopes mounted in its payload bay to observe the comet at about the time that the international armada of probes encountered it.[146,147,148] This time the plan was derailed by the tragic loss of the Shuttle Challenger and her crew as STS-51L lifted off on 28 January 1986, prompting the grounding of the fleet for the next two and a half years.

The various Halley missions had put the space agencies in a rather paradoxical situation: ESA and the Soviet Union would dedicate three spacecraft with similar payloads to similar and overlapping missions, and JPL, having had all its ambitious plans rejected, was now desperately trying to get a similar mission. Some degree of coordination was clearly needed. First the International Halley Watch, and later the Inter-Agency Consultative Group (IACG) for space exploration sponsored by the United Nations, worked to coordinate ESA, NASA, ISAS and the USSR to ensure that they performed complementary rather than duplicative observations. The most important initiative was the 'Pathfinder' project, whereby Giotto, as the last of the armada to reach Halley, would be able to exploit the measurements taken only days before by the twin Vegas to determine the location of the nucleus within the coma and so establish the range of its close encounter. In fact, owing to the 'rocket-like' effects of the jets that emitted material from the surface, the small angular distance of Halley from the Sun at perihelion as viewed from Earth and the uncertainty of the terrestrial measurements at that distance, the position of the nucleus would not be able to be specified to better than 3,000 km. Calculations showed that the Vega data should reduce this uncertainty in the ephemeris to a mere 125 km. It would be a truly international effort, with the Deep Space Network locating the positions of the Vegas by using their interferometric transponders, the Soviets providing data on where the scan platforms were pointing and where the comet was in the field of view of the cameras, NASA and ESA computing the revised orbit of the comet and finally ESA ordering course corrections to Giotto and providing its camera with pointing instructions.[149]

As comet Halley sped towards perihelion, detailed studies of its orbit and of the forthcoming passage were published, and its history was reconstructed, identifying records and observations for ancient apparitions that had hitherto escaped scrutiny. The comet had last been seen in June 1911, and the first attempt to recover it was made in November 1977 by Mount Palomar, at which time it was calculated to be over 19 AU from the Sun; but it was not found. Opinions on when Halley would be recovered ranged from an optimistic 1982 to a pessimistic 1984; the latter less than 2

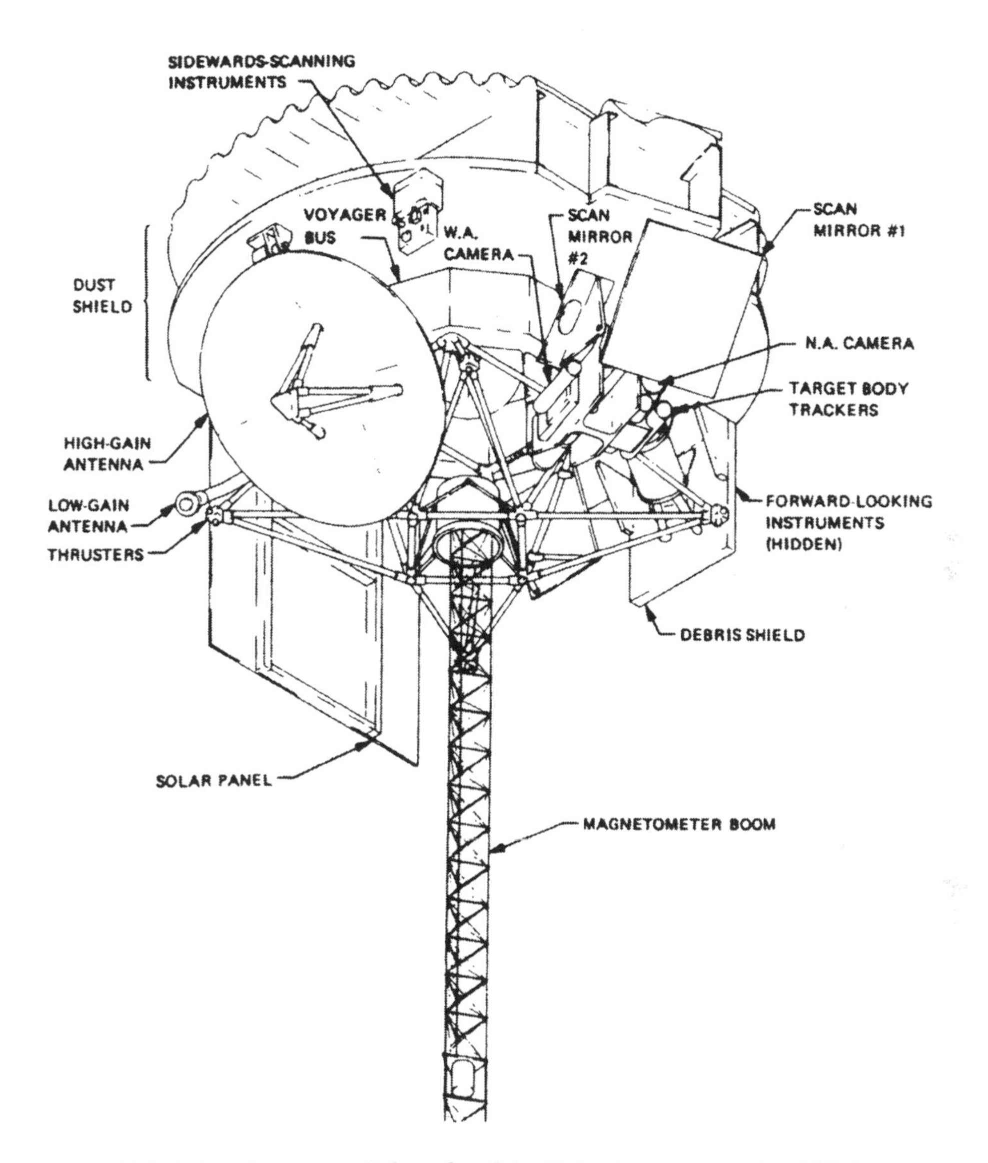

NASA designed a spacecraft for a fast-flyby Halley Intercept Mission (HIM).

years before perihelion. In the early 1980s the largest astronomical observatories replaced their traditional photographic cameras with sensitive CCD detectors which enabled fainter objects to be detected with shorter exposures. It was thanks to one such detector that on 16 October 1982 Mount Palomar spotted the bare nucleus of comet Halley, some 11 AU from the Sun and only 60,000 km from the predicted position. Never before in its four predicted returns had Halley been detected so far from the Sun.[150,151]

In the end NASA settled for observing Halley's comet using instruments carried on a Shuttle in Earth orbit.

BALLOONS TO VENUS

In late 1984, after 4 years of development, the two Vega spacecraft were ready, but it was a close result because the Soviets apparently underestimated the complexity of the mission and its management; so much so that at one point it was thought that they might miss the launch window for Venus, in which case they would have to be stripped of their atmospheric capsules and be launched directly towards Halley in the same window as Giotto and Planet-A in mid-1985.[152] A Proton, the standard Soviet planetary launcher, lifted off on 15 December with Vega 1, which weighed 4,920 kg. After spending 70 minutes in parking orbit, the spacecraft was injected into a heliocentric orbit ranging between 0.70 and 0.98 AU, heading for Venus. On 21 December Vega 2 followed in an orbit ranging between 0.70 and 1.00 AU. As a result of the international participation in the mission, for the first time Westerners were allowed to visit the Baikonur cosmodrome to watch a Proton launch. In fact, this was also the first time that the main launch vehicle of the Soviet planetary and lunar programs, introduced into service in the late 1960s, was actually shown on Soviet television. Nevertheless, even although the Proton no longer had a military role, the coverage was not allowed to disclose the actual ascent trajectory or the times of staging.[153]

One problem identified early on was that as the particles and fields instruments to study interplanetary space were switched on one by one, the right-hand boom of the plasma wave experiment failed to deploy on both spacecraft. However, the engineers were patient and the booms deployed after the course corrections (on

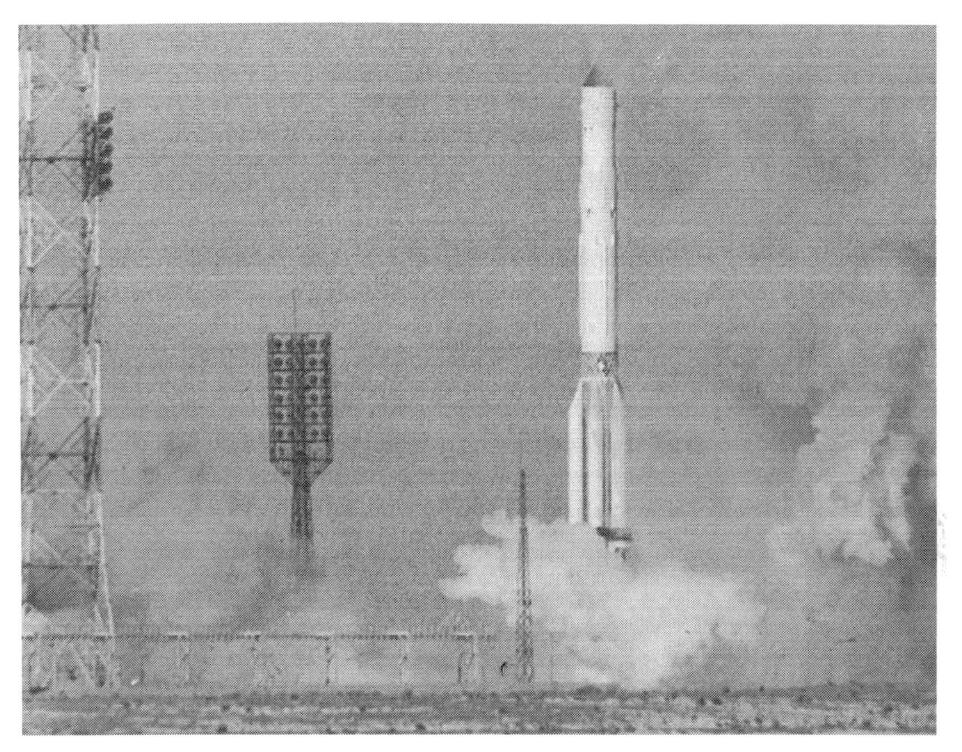

The launch of Vega 2. The Proton launcher had dispatched planetary missions since 1969, but this was the first time that it was shown publicly.

20 December for Vega 1) that put the vehicles on trajectories to impact the night-side of Venus, as required to deliver the atmospheric capsules.[154]

Despite the tense relationship between the United States and the Soviet Union, which was compounded by a small scandal when the American press learned that the spacecraft had a US-built instrument, in early 1985 the interferometric network began to take shape. On 22 January both the antenna at Medvezkye Ozyora in the Soviet Union and the Deep Space Network station at Goldstone in California were able to lock onto Vega 1. This was the first time that an American tracking station had *officially* tracked a Soviet spacecraft in deep space. On 18 February five of the foreign radio-telescopes locked on and demonstrated the localization system using distant quasars as celestial references.[155] On 9 June 1985, when some 650,000 km from Venus, Vega 1 released the spherical entry capsule and then fired its engine to transform its collision course into a 39,000-km flyby, for a slingshot that would result in an encounter with Halley 9 months later. Remarkably, Venera 16, in orbit, had been turned off less than 2 weeks earlier!

The Vega 1 capsule entered the Venusian atmosphere on 11 June at a speed of 11 km/s and an angle of 19 degrees below the local horizon. As viewed from Earth, the entry point was close to the dark limb of the planet's 'half phase' illumination. After the initial aerodynamic braking, during which the deceleration load reached 400 g, the heat shield split open at an altitude of 65 km to expose the lander and the aerostat. Soon thereafter, the compartment with the aerostat opened, the parachute deployed and the gondola and balloon were drawn out. The inflation of the balloon started at an altitude of 54 km, and the parachute was released. Upon reaching an external pressure of 900 hPa (at about 50 km), the empty helium tanks and other ballast were jettisoned and, 15–25 minutes after entry, the aerostat began to rise to its buoyant cruising altitude. Meanwhile, at 02:21:41 UTC, the antennas at both Yevpatoria and the Deep Space Network station at Canberra picked up its signal.[156] The early temperature and pressure data showed that after releasing its ballast the balloon rose rapidly for about 30 minutes until it reached an altitude of 53.6 km, where pressure was 535 hPa and the temperature was about 30°C. This was right in the middle of the cloud layers, and would enable the balloon to be carried along by the super-rotating wind.

Meanwhile, the lander, having released its parachute at a height of 47 km, was freely falling. It had begun to report data immediately upon being extracted from the entry shield. About 15 minutes prior to landing, it suffered a major problem. The strong turbulence and wind buffeting led the 8-sensor accelerometer to believe that the landing had occurred. This initiated the surface activity, starting with the drill, which collected air! Remarkably, this occurred near the altitude at which all of the Pioneer Venus atmospheric probes had experienced instrument failures, suggesting that they may have suffered intense vibration. At 03:03 UTC, the lander set down on the surface at 7.2°N, 177.8°E (which would also be the approximate coordinates of the starting point of the balloon's flight). Other references place the center of the target ellipse at 8.10°N, 175.85°E, which is on Rusalka Planitia somewhat to the north of the elevated Aphrodite Terra and 1,000 km west of Sapas Mons. Near the center of the ellipse is a gently sloping volcanic dome some 10 km in diameter.[157] Whereas the earlier landers had come down on upland plains or smooth lowlands, the Vegas were targeted at a piedmont in Aphrodite Terra where an upland plain transitioned to a mountainous area, and where rocks from the plains were expected to be intermixed with those from the massifs. The surface temperature was 460°C, and the pressure was 93,400 hPa. The probe survived on the surface for 56 minutes (other sources say only 21 minutes). Owing to the absence of cameras and to the premature activation of the surface experiments, it carried out a reduced program that included a gamma-ray analysis of the composition of nearby rocks.

After the lander ceased to transmit, all attention switched to the balloon, which had achieved its cruising altitude. At the start of the cruise, the Doppler changes to the frequency of the received signal indicated that the balloon was over a region of turbulence. In fact, sometimes during the brief transmission sessions the signal was lost as the nacelle swung back and forth and caused the antenna beam to miss the Earth. Just short of 32 hours into the flight, the photometer recorded an increase in the light level, indicating that sunrise was imminent. But as measured by tracking it crossed onto the day-side only after 34 hours, by which time it was 8,000 km from the entry point. The balloon encountered turbulence again towards the end of the flight. The communication session scheduled at 01:7:21 UTC on 12 June did not occur because by then the batteries were exhausted.[158] In the 46 hours during which the balloon functioned, it traversed 109 degrees of longitude and moved 31 degrees onto the day-side; a distance of 11,600 km. The tracking network was not equipped to measure the latitude of the balloon, but this was assumed to be essentially fixed at 7.3°N. The data suggested that at the balloon's cruising altitude the wind had an average speed of 69 m/s.[159,160]

Two days after the Vega 1 balloon fell silent, Vega 2 released its capsule and set up the flyby slingshot to head for Halley. The lander and balloon separated, and at 03:01 UTC on 15 June the lander settled at 6.45°S, 181.08°E, some 1,500 km south of its predecessor. Other references state 7.14°S, 177.67°E, in the transition zone between Rusalka Planitia and the eastern edge of Aphrodite Terra. In radar images, the target ellipse includes a radar-bright fractured plain and, towards the northeast, a radar-dark flat plain.[161] The surface temperature was 452°C, and the pressure was 86,000 hPa. Although this was cooler than at the Vega 1 site, the pressure was also lower, and since there is a steep inverse thermal gradient with altitude this implied that the Vega 2 landing site was more elevated.[162] This time the drill was activated correctly, and successfully performed its analysis in the first 3 minutes following landing. At 57 minutes, the Vega 2 lander's surface mission lasted only 1 minute longer than its predecessor.

The scientific results from the Vega landers were, as usual, quite extensive. The thermometers and barometers were the first instruments to report, activating at an altitude of 63.6 km. Although the Vega 1 thermometers failed during the descent, it was possible to get the surface temperature by extrapolating other data. The Vega 2 data was remarkable in that it showed a thermal inversion at high altitude, reaching a minimum of about –20°C at an altitude of 62 km, with the temperature sharply increasing above and below this level.[163] The optical spectrometers operated from 63 km down to 30 km (Vega 1) or 32 km (Vega 2) and generally confirmed the atmospheric structure reported by the previous landers, including the presence of a triple main cloud deck. However, there were substantial differences in the number and size of the droplets.[164] The aerosol particle-size counters operated only down to an altitude of 47 km, but because of a loss of sensitivity were restricted to detecting the largest of particles. The density profiles in both cases were remarkable in that, despite being obtained 4 days apart and at sites separated by 1,500 km, they were extremely similar, indicating that the cloud-layer structure was both pervasive and stable. The only major difference concerned the uppermost layer, which was much less dense where Vega 2 entered.[165] Meanwhile, the gas chromatographs reported that there was almost 1 milligram of sulfuric acid aerosol for every cubic meter of the atmosphere. However, the instrument also indicated that the chemistry of the aerosol was more complicated than just water and sulfuric acid. The concentration of sulfuric acid was corroborated by the Vega 1 mass spectrometer (this instrument seems to have failed on Vega 2) which confirmed that sulfur and sulfur-bearing molecules form the main constituent of the high altitude clouds, and also detected sulfur dioxide and chlorine. The X-ray fluorescence aerosol analyzer reported the presence of phosphorus in addition to the other compounds.[166,167,168] The humidity analyzers started at an altitude of 62 km and reported the percentage of water in the atmosphere down to 25 km.[169] Each French ultraviolet spectrometer returned about

30 full-resolution spectra of the atmosphere down to the surface. While designing the instrument, the scientists had expected that the most abundant gas that it would find would be sulfur dioxide, but it seemed that another gas was actually present. This was not definitively identified, but it could well have been some form of molecular sulfur.[170] The only direct rock analysis performed (by Vega 2) indicated that the site resembled a terrestrial gabbro, which is a basalt that solidifies underground rather than after being extruded onto the surface. This was confirmed by the gamma-ray spectrometers, which returned fairly similar data for the abundances of potassium, uranium and thorium which were consistent with either gabbro or a tholeitic basalt similar to that which is produced at mid-ocean ridges on Earth. Vega 2's sampling of the flank of a mountain massif drew to a conclusion the preliminary survey of Venusian geology: Veneras 8 and 13 had analyzed upland rolling plains, Venera 14 had sampled a flat lowland plain and Veneras 9 and 10 had landed on young shield volcanoes.[171,172]

Meanwhile, Vega 2's balloon was blown by a 66 m/s wind over 105 degrees of longitude (35 degrees onto the day-side) at an assumed constant latitude of 6.6°S. Starting 33 hours after entry, it encountered strong downward gusts. These peaked at 36 hours, some 2 hours after crossing onto the day-side, when the data indicated the balloon to be in a strong descending current that drew it down to a level where the pressure was 900 hPa, which was equivalent to a drop of almost 3 km. It is not known what caused these gusts, but at that time the balloon was at 98°E longitude and crossing some of the tallest mountains of Ovda Regio in Aphrodite Terra and it is possible that this topography was responsible.[173] Vega 1 experienced turbulence only at the start and end of its flight. The photometer on Vega 2 suffered some kind of malfunction, but its corrupted data still showed the light of dawn some 3 hours before the balloon crossed the terminator. At no time did either balloon detect any lightning. The instrument on Vega 2 reported one burst of light, but this occurred at dawn and is therefore considered to be spurious. No data was received from the Vega 2 nephelometer, and the data from this instrument on Vega 1 was difficult to interpret owing to an inability to calibrate it. Nevertheless, the data showed that the balloon never emerged from the cloud layers into a totally clear patch of sky.

All 20 terrestrial antennas were able to perform their monitoring of the balloons. Although they flew at similar latitudes on either side of the equator, Vega 2 found the temperatures at similar pressure levels to be somewhat lower. The reason for this was not apparent, but it showed the dynamics of the atmosphere to be much more complex than believed. The barometric data showed that the balloons floated mostly at pressures in the range 535 to 620 hPa, slowly descending as they lost helium. It has been estimated that they lost no more than 5 per cent of their gas during their missions, and did not suffer any major leaks.[174] Their transmissions ceased when they drained their batteries, and it is not known where they crashed after finally either losing buoyancy or bursting.[175,176]

The balloon mission happened at a turning point in Soviet history. In June 1985, as international teams were tracking the balloons in the Venusian atmosphere, the new Communist Party Secretary Mikahil S. Gorbachev issued a call for a reform of Soviet science and technology, saying that it (and indeed Soviet society in general)

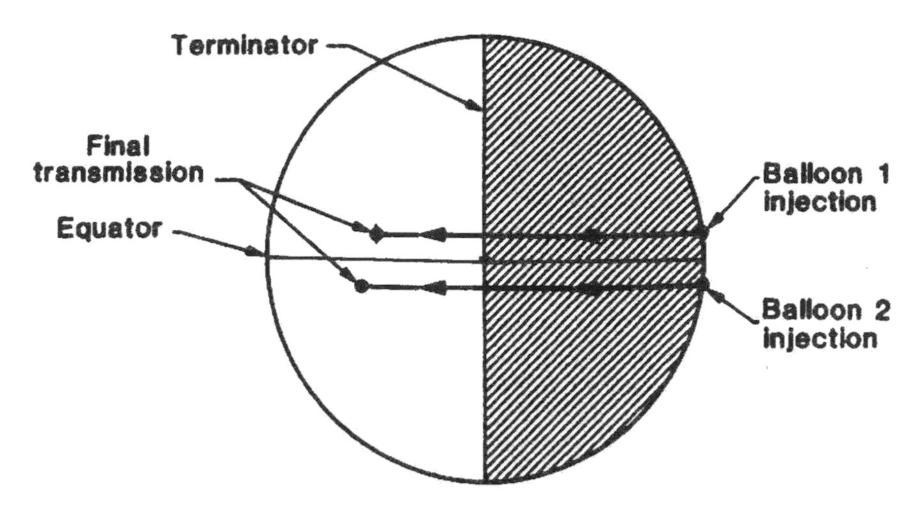

As the super-rotating wind carried the Vega balloons from their entry points near the center of the night-side across the terminator into daylight, they traversed the half-illuminated disk of Venus as seen from Earth. The latitude of the balloons is taken as constant. (From Sagdeev, R.Z., et al., "The VEGA Balloon Experiment", Science, 231, 1986, 1407–1408; reprinted with permission of the AAAS)

should be opened to the wider world. Sagdeev, the IKI director, became one of the champions of this approach. This marked the start of the brief era of "perestroika", and the beginning of the end of the Soviet Union.[177]

Other than acting as relays for their landers, the Vega spacecraft made very few observations of Venus, simply calibrating some of their spectroscopic instruments. As no results appear to have been published, it is possible that there were no radio-occultation soundings of the atmosphere during the limb crossings.[178] Indeed, the scan platform was in its 'transport' position, held against the side of the spacecraft; it was not scheduled to be released until a month before the Halley encounter. And in any case there would have been little to see on the night-side of the planet!

On 25 and 29 June 1985 respectively, Vegas 1 and 2 corrected their slingshot trajectories in order to refine their encounters with Halley, placing Vega 1 into an orbit ranging between 0.72 and 1.07 AU; Vega 2's orbit must have been similar.

Remarkably, no probes have landed on Venus since the landers delivered by the Vegas, and, despite their success, the aerostats remain the only probes ever to have been 'flown' in the atmosphere of another planet. Of course, other missions were announced to be under study in the USSR, using either the old 5V bus or the new UMVL to deliver a variety of innovative payloads, but none advanced beyond the preliminary study phase. These include an atmospheric probe to collect data during the descent, a simplified Venera lander capable of collecting images of the surface as soon as it had broken through the cloud deck in order to gain a bird's-eye view of the terrain, an improved form of the Vega balloon capable of operating for up to a year, and new concepts such as a pair of balloons joined by a 20-km-long tether, and tethered kites which would fly at different altitudes in order to use the different wind speeds to remain aloft for several weeks or months.[179] These missions were to be followed in 1998 by an orbiter which would release 10 penetrators to establish a network of long-lived seismometers. However, following the collapse of the Soviet Union the launch of this mission was indefinitely deferred. It was rumored at that time that a number of flightworthy Venera landers were in storage, and that the newly formed Russian Space Agency had offered to sell them to foreign agencies for a few million dollars each.[180]

TWO LIVES, ONE SPACECRAFT

While the international armada was in transit to Halley, another probe made the first targeted encounter with a comet. ISEE 3 (International Sun–Earth Explorer) was derived from the Interplanetary Monitoring Platform design. It was a 16-sided drum that was 1.77 meters in diameter and 1.58 meters in height, and was spun at 20 rpm for stability. Two radio science antennas and a medium-gain antenna for a pair of redundant 5-W transmitters protruded from the ends, giving the spacecraft a total length of 14 meters. Four radial antennas for plasma wave and radio science sensors gave it a span of 92 meters. A plan to include a high-gain antenna was not pursued. Four 3-meter-long radial booms held a magnetometer and plasma wave sensors. Two bands of solar cells supplied up to 182 W of power at the start of the mission. A dozen 18-N thrusters and an initial load of 89 kg of hydrazine provided attitude and trajectory control. The 479-kg spacecraft had no fewer than 104 kg of scientific instruments positioned along its 'equator', including a solar wind plasma experiment, magnetometers, low-, medium- and high-energy cosmic-ray detectors, a plasma wave instrument, a plasma composition instrument, proton, cosmic-ray, X-ray and electron detectors, a radio-astronomy experiment and gamma-ray burst detectors. In addition, solar studies were to be conducted by observatories on Earth in support of the mission.

The mission profile was new, as ISEE 3 would be the first spacecraft to enter a 'halo' orbit centered on the L1 point of the Sun–Earth system. In 1772 the French–Italian mathematician Joseph Louis de Lagrange proved that a gravitational system comprised of two large bodies (for example the Sun and Earth) and a third body of comparatively negligible mass would include five equilibrium or 'libration' points, later named Lagrangian points. In the case of the Sun–Earth system, one of these points (the L1 point) is located 1.5 million km on the sunward side of Earth, and another (L2) on the anti-sunward side. A spacecraft placed precisely at the L1 point could sample the solar wind before it reached Earth, but it would appear directly in front of the Sun as viewed from Earth and communicating with it would be highly unreliable. However, it is also possible to orbit *around* the L1 point as if there were a gravitating mass present. Because a spacecraft in such a halo orbit would not be in solar conjunction, communications would be easier. A number of attempts were made between the mid-1960s and 1970 to gain approval to operate a solar probe in this manner. In 1972 it was decided to make an L1 probe the third mission of the joint NASA–ESRO International Sun–Earth Explorer program designed to conduct

The International Sun–Earth Explorer 3 spacecraft undergoing ground preparations. It was later renamed the International Cometary Explorer (ICE).

a series of coordinated observations of how the Earth's magnetosphere interacted with the interplanetary medium. ISEE 3 would be stationed in halo orbit to monitor the interplanetary medium and solar wind upstream of Earth to provide NASA's ISEE 1 and ESRO's ISEE 2, operating in high elliptical orbits around Earth, with up to an hour's notice of changes.[181,182] ISEE 3 was launched by a Delta rocket on 12 August 1978, reached the desired halo orbit on 20 November, and was operated by NASA's Goddard Space Flight Center. In the ensuing years it provided data on the solar wind, cosmic rays, solar storms, gamma-ray bursts, and a variety of other cosmic phenomena. After its 3-year primary mission three options were considered for an extended mission:

- To operate it where it was for as long as possible (its consumables were sufficient to maintain the halo orbit for 10 years).
- To exit the halo orbit in order to explore the little-known tail of the Earth's magnetosphere for up to a year, and thereafter return to the L1 point.
- To explore the tail of the Earth's magnetosphere and then escape to explore one of either comet Halley or comet Giacobini–Zinner.

Calculations by JPL showed the Halley option to be impractical, since without a high-gain antenna on the spacecraft the data rate at the distance from Earth at the time of the encounter with the comet would be too slow to transmit the data in real time, and there was no tape recorder to store the data for later replay. A proposal to relay through one of the two Japanese spacecraft was considered, but dismissed. In comparison, the Giacobini–Zinner encounter would be more favourable, in that, at 71 million km, not only would it be much closer to Earth, it would also take place at

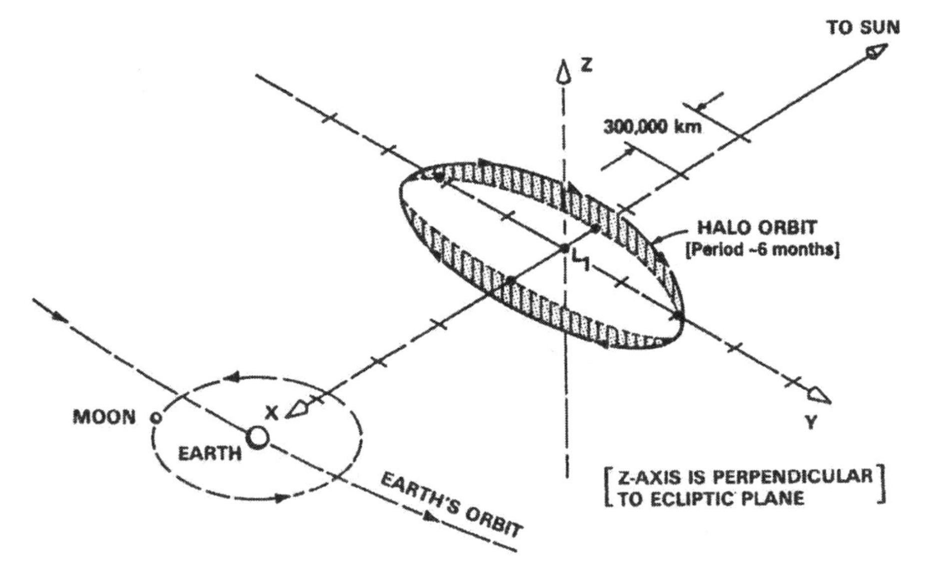

ISEE 3 was initially placed into a 'halo' orbit around the L1 Lagrangian point of the Sun–Earth system.

least 6 months before the other missions reached Halley. The data would provide a useful point of comparison with that from Halley. And making the historic first targeted encounter with a comet would compensate to some degree for the fact that there was no American mission to Halley. Moreover, if the encounter was timed to occur when Giacobini–Zinner was near the zenith of the non-steerable 300-meter-diameter Arecibo antenna, which was the world's largest radio-telescope, then the effective data rate from the spacecraft could be maximized.[183] This comet was first spotted by Michel Giacobini on 20 December 1900 and became relatively bright. It had an orbital period of about 6.5 years, and was recovered on 23 October 1913 by Ernst Zinner. They share its name. The 1985 apparition would be its 11th observed perihelion.[184] It was known to be the parent body of the meteor stream which had spectacular displays in 1933 and 1948, when Earth crossed the comet's orbit only a few weeks after its passage.[185] An article published in 1985 reported observations which suggested that the nucleus is a flattened ellipsoid with an equatorial diameter of about 2.5 km and a polar diameter only one-eighth of that, spinning every 1.66 hours. Unfortunately, as ISEE 3 did not have an imaging system it would be unable to verify this.[186] Although the spacecraft had not been built for a cometary mission, eight of its instruments could make useful observations. A significant advantage of Giacobini–Zinner over Halley was that its shorter period and non-retrograde orbit would result in a much slower relative speed (21 km/s) at encounter. Even although this comet was known to be relatively free of dust, the slow encounter would go some way towards compensating for the fact that the spacecraft was not shielded against dust impacts.

Robert Farquhar, the NASA astrodynamics expert who had designed the halo orbit concept and had developed early studies for cometary missions, proposed a Giacobini–Zinner flyby at a science meeting in February 1982, thereby igniting a debate about the merits of flying such a mission using a spacecraft and instruments which had not been designed for the purpose.[187] On 10 June 1982 ISEE 3 left its position up-Sun of Earth and set off towards the tail of the Earth's magnetosphere, which is down-Sun, to undertake the first part of its extended mission. NASA had not yet approved the Giacobini–Zinner plan (and Farquhar had yet to work out how to get the spacecraft into heliocentric orbit!) but there was a general understanding that if this were to be deemed impractical, after investigating conditions in the tail of the magnetosphere the spacecraft would return to the L1 halo orbit. As support for the cometary mission increased it was endorsed by both NASA's Solar System Exploration Committee and the National Academy of Sciences, and on 30 August 1982 the agency gave its approval. Most of the $3 million budget would be spent on upgrading the antennas of the Deep Space Network and equipping Arecibo to receive the spacecraft's faint signal. In October ISEE 3 passed just inside the orbit of the Moon (which was the closest that it would come to Earth) heading for its first pass through the geotail in February 1983. This region had been sparsely explored: a few satellites had ventured into it as far out as 80 Earth radii (Re) and Pioneers 7 and 8 had made passes at 1,000 and 500 Re respectively, but no data existed for the intermediate range. In particular, the Pioneer data had not clarified whether the tail was intact at large distances from Earth or shredded into separate filaments. During its various passes, which reached as deep as 240 Re, ISEE 3 discovered a structure which was surprisingly similar to that near Earth, and with an internal structure the same as that near Earth. Moreover, it found that auroral storms sent large 'bubbles' of magnetically confined plasma (called 'plasmoids') sweeping down the tail "like drops of water falling from a dripping faucet".[188,189] A trajectory had been devised in which a series of lunar encounters alternating between the leading and trailing edges of the Moon would extend a geotail pass into an escape maneuver which would dispatch the spacecraft to meet Giacobini–Zinner. The first such flyby was on 30 March. It made a second flyby on 23 April and then headed for a second geotail pass in June.[190] There were flybys on 27 September and 21 October. The fifth and final flyby on 22 December 1983 skimmed over the lunar surface at an altitude of just 120 km, and the spacecraft spent 28 minutes in the Moon's shadow. Because the battery had failed in 1981 there was concern that the hydrazine would freeze while the spacecraft was in darkness, but the temperature decrease was smaller than expected and the spacecraft emerged unscathed, now in a heliocentric orbit ranging between 0.93 and 1.03 AU. Soon thereafter, NASA renamed the spacecraft the International Cometary Explorer (ICE).[191]

Since imaging was not intended, there was no requirement for ICE to intercept Giacobini–Zinner on the illuminated side of its nucleus. It could be sent to explore any region of the comet that the scientists wished, and it was decided to study this comet's distinctive long narrow plasma tail at a point some 10,000 km downstream of the nucleus where the tail had already formed but had not yet emerged from the coma, to investigate the interactions between the magnetic fields in the tail and the solar wind. Because of the geometric requirements of the armada of spacecraft that would encounter Halley, this region would not be able to be sampled in the case of that comet. A clever experiment had been designed to record the bursts of plasma created as dust motes were vaporized on hitting the body of the spacecraft. Impacts could also be observed by using the pressurized cells in a charged-particle detector as a 'puncture' detector. A software upgrade was also devised that would increase the sensitivity of the plasma composition instrument beyond hydrogen, helium and other lightweight ions in the solar wind, in order to analyze the composition of ions in the comet's tail.[192]

The targeting for Giacobini–Zinner was based on the ephemeris calculated after its 1979 apparition, but uncertainties remained due to the unpredictable rocket-like forces a comet suffers as jets emit the material that forms the coma and tail. The predictions for its 1985 return could therefore easily be off by several hours. If the error were to be as much as a day, the remaining propellant would be insufficient to retarget for the revised encounter position. Giacobini–Zinner was recovered in April 1984 by the 4-meter telescope at the Kitt Peak Observatory in America, nearly a year and a half before its perihelion passage on 5 September 1985. Fortunately for ICE, these first observations indicated that the ephemeris was off by a mere 0.01 days, or about 15 minutes.[193] In April 1985, as the comet drew near to Earth, a series of observations were made to estimate the unpredictable perturbations, and it was calculated that if nothing was done ICE would pass by at a range of about 172,000 km. A trajectory correction in May was canceled and replaced by a pair of

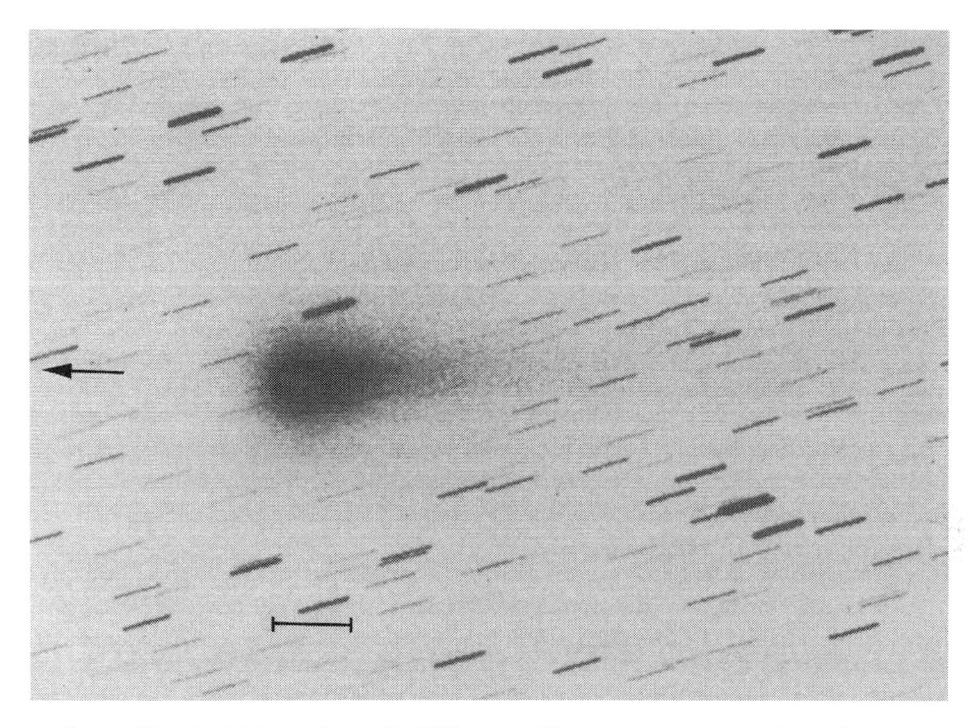

Comet Giacobini–Zinner during its 1972 return. The arrow points towards the Sun, and there is a faint ion tail visible in the anti-Sun direction. At the comet's distance from Earth the scale-marker spanned 100,000 km, more than ten times ICE's flyby range. (CNRS Observatoire de Haute-Provence)

maneuvers on 5 June and 9 July designed not only to reduce the range to the comet at the encounter but also to produce the best possible coverage by the Deep Space Network. Further precise observations showed that the spacecraft would pass by 8,000 km from the nucleus, which was now actively degassing, and some 600 km from the axis of the tail. On 8 September, 3 days before the encounter, a 2.3-m/s correction was made to regain the axis of the tail. A last-minute maneuver to open the range to 10,000 km was deemed excessive (and too risky) and was not attempted. Meanwhile amateur and professional astronomers, satellites, the Pioneer Venus Orbiter and the Japanese spacecraft heading for Halley were monitoring the comet and its associated meteor stream.[194,195]

The first hint of the presence of the comet was noted early on 10 September, at a range of no less than 2.3 million km, when the plasma wave experiment detected electric field turbulence due to cometary ions 'picked up' by the solar wind. Then, 20 hours prior to the encounter, high-energy ions were detected, apparently mostly water molecules that had been issued by the nucleus and ionized by solar radiation. The influence of the comet became greater from 800,000 km inward. The presence of a bow shock in the solar wind had been predicted. Starting at about 09:10 UTC on 11 September, 188,000 km from the axis of the tail, the plasma wave instrument reported a strong bow shock phenomena. The magnetometer did not sense this, but it noted a lot of noise and a gradual rise in the magnetic field at about 09:30. A transition region some 50,000 km thick was then traversed that was characterized by a very turbulent plasma which drove the instruments almost off the top of their scales. After crossing this turbulence the spacecraft found itself in a calmer sheath, within which was the tail. It entered the plasma tail at 10:50, and over the next 20 minutes traversed its 25,000-km width. The magnetometer saw clear evidence that magnetic field lines captured by the solar wind were draped behind the coma and, by confining the plasma tail, maintained its narrow shape. The magnetic field lines were expected to form two lobes of opposite polarity, separated by a 100-km-thick 'neutral sheet' that had no magnetic field at all. Serendipitously, the spacecraft's path took it from one lobe across the neutral sheet into the other lobe. Meanwhile, other instruments monitored the number, velocity and temperature of the charged particles in the plasma tail. It was a very 'cold' plasma (in astrophysical terms) and the electron density increased dramatically. Although the acquisition technique of the plasma composition instrument meant that it spent the entire time inside the tail stepping through its sampling range, it was able to provide the first in-situ analysis of the composition of a comet's tail, detecting many water and a smaller number of carbon monoxide ions – consistent with the 'dirty snowball' model of a cometary nucleus. There was also an unidentified constituent, maybe sodium or magnesium atoms. At 11:10 the spacecraft exited the tail unscathed. As expected, in this comet the dust environment was quite benign, with the plasma wave experiment detecting a dust grain hit once every few seconds. No attitude disturbance from dust strikes was noted, nor was there any loss of output from the solar cells. Phenomena were observed in reverse order outbound from the comet. The encounter ended when the spacecraft crossed the bow shock again at about 12:20. The closest approach to the nucleus was calculated to have occurred at 11:02, and at a range of 7,682 km. The encounter had been tracked by the Deep Space Network, by Arecibo, and also by the Japanese deep-space antenna at Usuda; and to increase the received power from a distance 50 times that intended when the spacecraft was designed, the data was sent simultaneously using both its redundant transmitters.[196,197,198,199,200,201,202,203,204]

After its encounter with Giacobini–Zinner, the spacecraft's distance from Earth progressively increased. On 28 March 1986 it passed 0.21 AU up-Sun of Halley, and by reporting on the state of the unperturbed solar wind upstream of that comet it yielded data to assist in interpreting the data from the instruments of the inbound international armada. It also detected cometary oxygen ions picked up by the solar wind at distances more than three times greater than those detected by Giotto and the Soviet Vegas. One-fifth of ICE's remaining propellant was used to execute two course corrections in 1986 that targeted it for a lunar flyby on 10 August 2014. On 5 May 1997 NASA terminated ICE operations and support, but the transmitter was purposefully left on in order to allow further tracking, as was done in 1999 when it traveled behind the Sun and the radio signal was used to probe the corona.[205] Then the spacecraft (which was still fully functional) was put into a state of hibernation that will last until at least 2010.

Again, three options have been identified for further extending the ICE mission

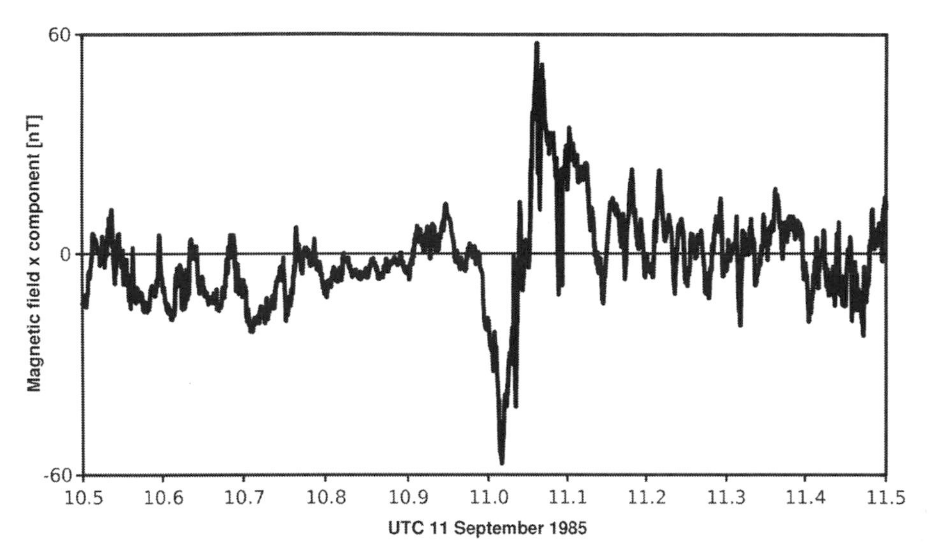

The x-component of the magnetic field as reported by ICE over an interval of one hour centered on the time of closest approach to Giacobini–Zinner at 11:02 UTC. The crossing of the neutral sheet is marked by the change of sign of the magnetic field.

after its August 2014 lunar flyby, providing it is still working. The simplest option would be to return the spacecraft to its station in the L1 halo orbit, 32 years after it left it. Another option would be to place it into a highly elliptical Earth orbit whose apogee could be lowered by aerobraking passes through the upper atmosphere until it could be retrieved so that its coating of cometary material could be analyzed and the spacecraft finally donated to the Smithsonian National Air and Space Museum. Finally, ICE could be targeted to make a second flyby of Giacobini–Zinner on 19 September 2018. The original rationale for this option was that it would encounter Giacobini–Zinner a fortnight before NASA's CONTOUR (Comet Nucleus Tour) could do so if that mission were to be extended, but unfortunately CONTOUR was lost soon after its launch in 2002.[206]

"BUT NOW GIOTTO HAS THE SHOUT"

While the Soviet Vega probes were approaching their first target, Venus, the other three spacecraft of the international Halley armada were being prepared for launch. The first off would be Japan's MS-T5 'precursor' probe. The role of the Japanese spacecraft in the armada was to provide preliminary estimates of the comet's water production rates, to study jets and outbursts activity, and to monitor the state of the solar wind during the other encounters. The twin spacecraft were integrated during the summer of 1984, then taken to the Kagoshima launch site at the southern tip of the island of Honshu. Bad weather and problems with its rocket delayed the launch of MS-T5 by three days, but it set off on 7 January 1985 and was directly injected

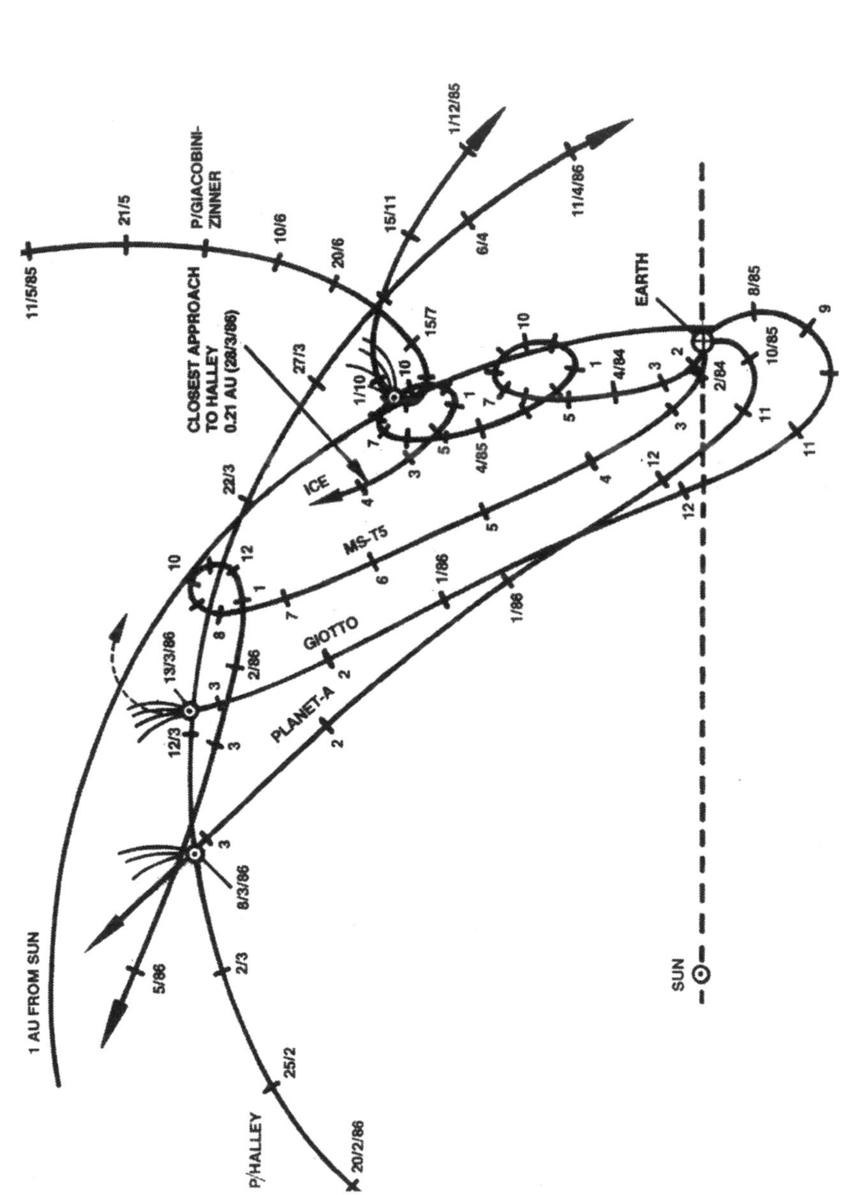

The orbits of the ICE, Giotto, Suisei (Planet-A) and Sakigake (MS-T5) spacecraft, and the comets Giacobini–Zinner and Halley relative to a frame of reference that rotates with Earth. (ESA)

The Japanese Halley spacecraft were launched individually by Mu-3SII rockets, which comprised the three solid propellant stages of the Mu-3S augmented by two strap-on boosters. (ISAS/JAXA)

into a heliocentric orbit ranging between 0.817 and 1.014 AU, with a period of 320 days. It was then renamed 'Sakigake' (Forerunner or Pioneer), although it was also sometimes referred to as SS-10 (Scientific Satellite No. 10). It was the first probe to be sent into deep space by a country other than the two Superpowers. The delay in launching had the effect of increasing the Halley flyby range by about 3 million km, to 7.6 million km, but this was able to be reduced to 6.99 million km by two course corrections made on 10 January and 14 February. The first 6 weeks were spent on engineering tests of the systems for orbit determination, communication, attitude and trajectory control. Sakigake deployed its booms and other devices between 19 and 20 February, and then activated and calibrated its instruments and settled down to monitoring its environment.[207,208]

Giotto arrived in French Guiana in South America in the late spring of 1985. On its travels within Europe during its development, the spacecraft had survived a fire, a strike, a snow storm and a road accident![209] It was placed on its rocket on 25 June and, following a brief hold due to the weather, launched without a hitch at 08:23:16 UTC on 2 July, 10 minutes into the 1-hour-long slot on the first day of the launch window. Remarkably, this would be the only deep-space launch using a 'classical' Ariane before this family of vehicles was superseded by the differently constructed Ariane 5 in the late 1990s. After a day and a half in parking orbit, and while out of contact with the ground, Giotto fired its kick stage to enter a heliocentric orbit that ranged between 0.731 and 1.078 AU. There was provision for a substantial course correction early in the flight to compensate for the erratic performance intrinsic to a solid-fuel motor, but the motor behaved perfectly. As a result, Giotto had such a margin of propellant that people began to consider the possibility, never mentioned before, of planning an extended mission – providing, of course, that it survived the fierce 'sandblasting' that it was sure to suffer during its Halley encounter.[210] On 27 August it made a 7.4-m/s correction to its trajectory to move the encounter point within 4,000 km of the best estimate of the comet's position. At about that time, the Deep Space Network began to make interferometric observations of the two Soviet Vegas, which were now beyond Venus, to precisely determine their positions for the Pathfinder experiment.[211]

The final spacecraft of the Halley armada to be launched was Planet-A, but this was not done until after the Sakigake precursor had demonstrated the performance of the launcher in the deep-space role and the viability of the spacecraft's systems. On 18 August 1985 Planet-A was launched and placed into a heliocentric orbit that ranged between 0.683 and 1.013 AU and had a projected encounter distance of just 210,000 km, which was almost perfect. It was renamed 'Suisei' (Comet) and listed as SS-11. In fact, the escape maneuver was so accurate that the only correction was executed relatively late in the flight, on 14 November, and the 12-m/s burn reduced the flyby range to 151,000 km for an encounter at a relative speed of 73 km/s. The ultraviolet imager was activated one month into the mission. After calibrating it by observing the Earth and stars that shine strongly in this portion of the spectrum, an attempt was made to image Giacobini–Zinner at the time of the ICE encounter, but the comet was not detected until several days later.[212] The charged-particle package was activated on 27 September 1985 to begin routine monitoring of the solar wind.

Giotto marked the first (and indeed last) time that the original form of the Ariane launcher dispatched an interplanetary spacecraft. (ESA)

Despite Giotto being the first ESA deep-space mission, it had only a few minor glitches. In addition to taking routine particles and fields data, it made two or three impromptu science observations per week during the cruise. The first instrument to be activated was the camera, on 10 August; and in September calibration pictures were taken of Jupiter and the star Vega (alpha Lyrae) to test the performance of the optics. On 18 and 23 October it took pictures of Earth from a range of 20 million km. Although the planet spanned only 27 pixels, it was possible to make out clouds over Australia, Asia and Antarctica. This involved rotating the camera's baffle to an unusual (and potentially dangerous) position to view backwards "over Giotto's shoulder", in the process casting a shadow on the body-mounted solar cells. It was noted early in the flight that despite the effort to make the spacecraft magnetically clean, the magnetometer would still be severely influenced by the strong magnetic

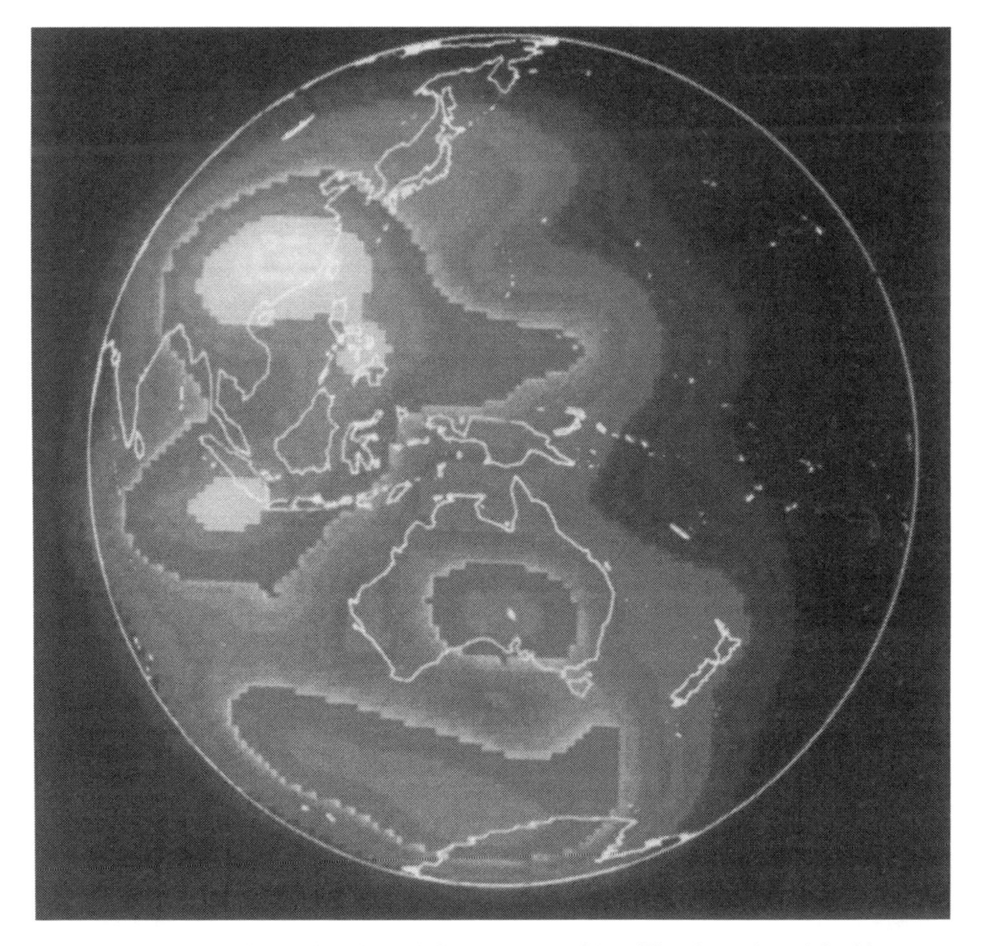

In October 1985 Giotto viewed Earth from a range of 20 million km. The original image spanned just 27 pixels. (ESA)

fields created by the motors onboard to despin the high-gain antenna and to point the camera's mirror.[213,214,215]

Meanwhile Sakigake's monitoring of the solar wind made it possible to observe a 'tail disconnection' event in which Halley's ion tail was ripped off by a burst of high-velocity solar wind plasma. Such events usually happen when a comet crosses the neutral sheet which separates the two polarities of the solar magnetic field, with the magnetized tail finding itself in a region of opposite polarity to the head of the comet; Sakigake's observations showed that this is not the only situation in which a comet can be stripped of its ion tail.[216]

A small crisis arose in late January 1986 when the Parkes radio-telescope had to discontinue tracking Giotto for several days in order to support Voyager 2 as this spacecraft made the Uranus flyby of its Grand Tour. A backup antenna in Perth in Western Australia was assigned to cover the gap, but it lost track for several hours. Although JPL's Goldstone antenna came to the rescue, a rumor circulated that ESA had lost "the bird".[217,218] One day before Halley reached perihelion on 9 February, Giotto and the Vegas reported an energetic solar flare.[219] The Vegas initiated their Halley encounter activities on 10 February. At that time Vega 1 performed its final course correction. The equivalent maneuver by Vega 2 a week later was canceled. On 12 and 15 February respectively, these spacecraft unlatched and deployed their scan platforms and began to calibrate the instruments mounted on them; including imaging Jupiter and Saturn as a means of assessing the pointing in preparation for the Pathfinder experiment. Also on 12 February, Giotto executed its second course correction, this time of just 0.566 m/s, to aim for a point on the sunward side of the nucleus of Halley for optimal imaging. Meanwhile, no fewer than 35 observatories in the Soviet Union were producing a grand total of around 2,700 measurements of the comet's position to enable its orbit to be computed with sufficient accuracy for the times of the spacecraft encounters to be predicted to within 10 or 20 seconds.[220] Of course, Halley was also being monitored daily by many other telescopes around the world, most notably by the European Southern Observatory at La Silla in Chile, which collected positional data as well as images to document its behavior during most of February and March.[221] A study of the morphology of the jets, envelopes, spirals, etc, on plates taken during the 1910 apparition had resulted in a map of the most active sites of gas and dust emission at that time. Realizing that the brightest features in the vicinity of the nucleus would probably be the jet plumes, the Soviet engineers hurriedly reprogrammed the Vega cameras to recognize and reject jets so that these would not seduce them off the nucleus itself.[222,223,224] Regular imaging by Vega 1 began with a 90-minute session at 06:10 UTC on 4 March when 14 million km from the comet. Giotto, lagging behind, took its first pictures on that same day from a range of 59 million km. As Vega 1 closed within about 10 million km later in the day, it sensed the comet's presence for the first time, in the form of energetic particles. This outer region consisted of neutral molecules (mostly water) from the nucleus which had been ionized by solar radiation and picked up and accelerated by the solar wind. Vega 1 had a second imaging session on 5 March, by which time its range to the comet had halved.[225]

Encounter day for Vega 1 was 6 March. At 03:46 UTC, as the spacecraft closed

A telescopic image of Halley's comet taken on 8 March 1986, around the time of the spacecraft encounters. (ESO)

within about 1.1 million km of the comet, the magnetometer noted a sudden rise in the strength of the magnetic field and the plasma instruments a sharp enhancement of extremely low-frequency waves. This was the unmistakable signature of a bow shock. It was some 10,000 km thick. Once through it, Vega 1 entered a region that was called a 'cometosheath' (by analogy with the structure of a magnetosphere). It was not immediately clear what was causing this but, whatever it was, the process had created an environment very different to the mild bow transition that ICE had found near Giacobini–Zinner. Unlike planetary bow shocks such as those of Earth, Jupiter or Saturn, which mark where the plasma of the solar wind 'piles up' in front of an obstacle, the bow shock of Halley's comet may well have been caused by the solar wind becoming 'loaded up' with cometary ions – recall that the solar wind is composed primarily of protons and electrons, whereas a water ion is much heavier. The main encounter sequence started at a distance of 760,000 km from the nucleus, and was to run for 4 hours 50 minutes. This included the 20 minutes during which the spacecraft would be deeply immersed in the coma, and the moment of closest point of approach to the nucleus. One surprise came at 637,000 km, inbound, when the American dust counter detected the first particles of cometary origin; they were extremely small. In fact, the coma would prove to contain many more small dust particles than expected. Meanwhile, the inner coma and nucleus continued to grow in the camera's field of view, and 20 minutes before the encounter the nucleus (or material in its immediate vicinity) was finally resolved into a handful of pixels. As Vega 1 closed within 50,000 km, the viewing and illumination phase angles started to change dramatically. Strangely, the little that could be seen of the nucleus gave the impression of two bright objects adjacent to each other. As the spacecraft sped by, the nucleus seemed to have a bulge, but it was not possible to tell whether this was due to topography or a particularly dense jet of dust. The images seemed to be a bit fuzzy. At the time, it was reported that the camera was slightly out of focus but it was later realized that the fuzziness arose because the nucleus was embedded in a particularly dense cocoon of dust. At a range of 28,600 km the spacecraft lost the first of two of its instruments to the thickening dust, when the European plasma wave analyzer suddenly fell silent.[226] Another surprising discovery was made at a range of about 15,000 km, when the plasma instrument indicated the spacecraft had entered a region in which there were no solar ions, only cometary ions. This was duly named the 'cometopause'. The boundary was also noted by other instruments in the form of spikes of energetic particles. It can be thought of as a thick magnetic barrier which divides the solar wind (which is being decelerated by cometary ions behind the bow shock) from the non-magnetized cometary plasma which resides in the inner coma. In fact, the strongest recorded magnetic field was registered while crossing this boundary.[227,228]

At 07:20:06 UTC Vega 1 reached the point of closest approach to the nucleus of the comet, calculated at 8,890 km with an uncertainty of 45 km, which it sped past at the amazingly fast relative speed of 79.2 km/s. The University of Chicago's dust detector indicated that the spacecraft passed through a dense dust jet at this time: whereas the instrument had been reporting fluxes of the order of 100 particles per second, in a matter of seconds the rate surged to a peak of 4,000 per second.[229] In a

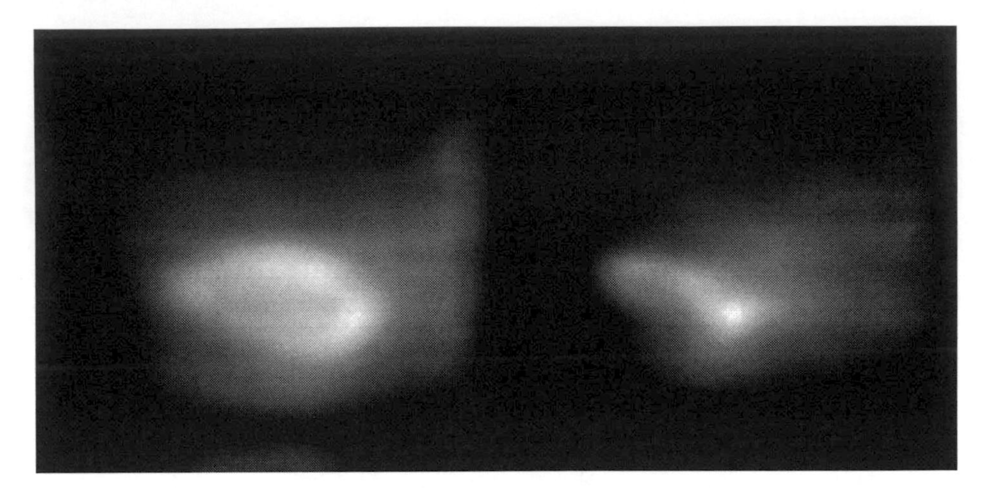

Two of the best pictures of the nucleus of Halley's comet taken by Vega 1 during its encounter.

related event at a range of about 45,000 km heading out, the cometary gas detector observed surges in the density of gas which lasted up to 10 seconds, indicating a jet at least 500 km across.[230] To conclude the encounter, Vega 1 again crossed the bow shock at a range of 1.1 million km from the nucleus, less than 10 hours after it had crossed it on the inbound leg.[231] As a result of the encounter dust strikes robbed the solar panels of 55 per cent of their available power and disabled two instruments, both of which were mounted on the unprotected solar panels. At the conclusion of the communication session for the encounter sequence, the spacecraft lost its 3-axis stabilization. But control was regained the next day, and further observations were made of the comet as the range increased until a scan platform orientation error on 8 March prevented more images from being taken.[232,233,234,235]

The second spacecraft to reach Halley was Suisei. Its solar wind instrument had been activated on 27 September 1985 and the ultraviolet camera had begun to take pictures in mid-November, when the comet was still 250 million km from the Sun. The observations in November and December showed that the hydrogen cloud that surrounded the comet changed in brightness in a cyclic manner from one day to the next, and further analysis revealed that it was 'breathing' with a period of about 2.2 days (some 53 hours). This rhythm was inferred to correspond to the spin period of the nucleus. This inference was supported by the possibility of a similar period in the 1910 observations.[236] Intriguingly, to Suisei the hydrogen cloud appeared to be made of several concentric shells. From the spacecraft's perspective the position of the comet in the sky was so close to the Sun by 10 January 1986 that ultraviolet observations had to be suspended; but they were able to be resumed on 9 February, the day on which the comet reached perihelion. Profiles of the density of hydrogen in the cloud suggested traces of fine structure. When the spacecraft penetrated the hydrogen cloud several days prior to the closest point of approach the imager was switched to its photometric mode, as there was no point continuing to take pictures.

In photometric mode, the distribution of hydrogen inside the cloud was mapped as a function of distance from the nucleus. Two days before the encounter, the camera was switched off to give downlink preference to the charged-particle instrument. Unfortunately, real-time observations on the day of the encounter, 8 March, could be received only when the deep-space antenna at Usuda had a line of sight to the spacecraft. The observations started at a distance of 200,000 km from the nucleus, just prior to closest approach at 13:06 UTC. As a result, the bow shock and other inbound phenomena were missed. Despite the range of the encounter, the cometary environment was completely different to the solar wind as hitherto observed, and Suisei suffered attitude disturbances at 159,800 and 174,900 km which changed the orientation of its spin axis and the period of its spin, indicating that it was struck by a milligram- and a microgram-sized particle. There had been a major outburst on the comet the previous day and evidently much of the dust was still in the vicinity of the nucleus.[237,238] The solar wind instrument reported for the final 34 minutes of the inbound leg and for 4 hours outbound. Around the time of closest approach, it observed how the solar wind was deflected and decelerated around the comet, and noted the presence of ions of cometary origin: water, carbon monoxide and carbon dioxide. After the range had increased to about 420,000 km several hours after the encounter there was an abrupt change in the plasma flow, probably corresponding to crossing the bow wave. In the days following the encounter the comet was again monitored by the ultraviolet camera, in particular collecting 58 hours of continuous photometric data starting on 21 March which revealed at least two major and four minor outbursts. Imaging was discontinued in mid-April, when the hydrogen cloud was too faint to record. Continuous photometry revealed a number of outbursts that seemed to be correlated with reports by other spacecraft in the armada of variations in the dustiness of the coma.[239,240,241]

As Vega 1 and Suisei receded from Halley, Vega 2 was making its approach. It began to image the comet on 7 March at a range similar to that at which its partner

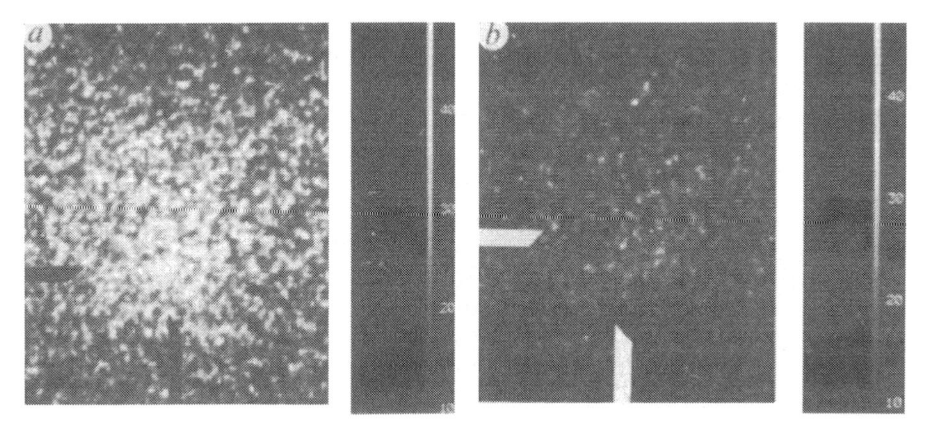

Two ultraviolet views of Halley's comet taken by Suisei: at minimum brightness on 25 February 1986 (right) and at maximum brightness on 28 February 1986 (left). (ISAS/JAXA)

did so, taking 100 pictures in order to further refine the position of the nucleus. The encounter sequence was executed during a 5-hour 20-minute communications window on 9 March. A smooth increase in plasma waves and plasma temperature about 1.5 million km from the nucleus signaled the inbound crossing of a mild bow shock. This was very different to the corresponding feature reported by Vega 1. In fact, the coma appeared to be much more 'sedate' than it had been; so much so that dust was not detected until the range had reduced to 280,000 km, less than half as far as for Vega 1.[242] Indeed, very little dust was encountered until the particle densities surged at 150,000 km. There was a similar surge some 50,000 km from the nucleus on the outbound leg. These events were probably due to crossing the paraboloid of dust which solar radiation pressure shaped into an apex some 45,000 km from the nucleus.[243] The fact that the coma was less dusty throughout the Vega 2 encounter was interpreted to mean that the less-active side of the nucleus was facing the Sun, which supported a rotational period for the nucleus of about 52 hours because this would place the Vega 2 encounter about one and a half cycles after that of Vega 1. When the main processor for pointing the scan platform failed 32 minutes prior to the flyby, the spacecraft switched over to the simpler backup sensor and took some 700 pictures. However, considering that the coma in close to the nucleus was relatively clear, the images were not as good as they could have been. The spacecraft crossed the cometopause at a range similar to that of its predecessor, into a region of cold ("almost stagnant") cometary ions. Serendipitously, in these conditions the plasma instrument was able to operate as a mass spectrometer and analyze the composition of this isolated environment. Most abundant were water ions, followed by carbon dioxide-generated ions. Remarkably, there was a spike in the spectrum that could reasonably be attributed to iron. Iron had never been identified before in optical spectra of Halley's comet, and in fact it had been detected previously only in comets whose perihelia were very close to the Sun. Shortly after this, however, the plasma experiment fell silent.[244]

The two Vega encounters were timed so that they would occur in sight of Soviet deep-space antennas, and for this reason they occurred at exactly the same time of the day. Vega 2 flew by the nucleus at 07:20:00 UTC, at a range of 8,030 km and a relative speed of 76.8 km/s. The sensors of the magnetometer were mounted on an unprotected boom, and the outermost three were lost right at closest approach; only the calibration sensor located closer to the body of the spacecraft survived.[245] As in the case of Vega 1, the high-frequency plasma wave experiment was also partially damaged at closest approach.[246] Several minutes later the acoustic sensor of one of the dust detectors was lost. Worst of all, it lost 80 per cent of the output of its solar panels. Ironically, despite the coma being relatively dust-free, Vega 2 suffered the greatest damage! In fact, although the Soviet scientists and engineers who worked on the mission had been assisted by the nation's nuclear weapons designers, whose knowledge of plasmas must have been quite extensive, it seemed that in designing the spacecraft inadequate attention had been given to dust-generated plasma, since many of the instrument failures appeared to have been caused by the accumulation of electric charge and the ensuing arcing, rather than to direct impacts. Despite its rough ride, Vega 2 was able to perform two imaging sessions on 10 and 11 March, at distances of 7 and 14 million km respectively.[247,248]

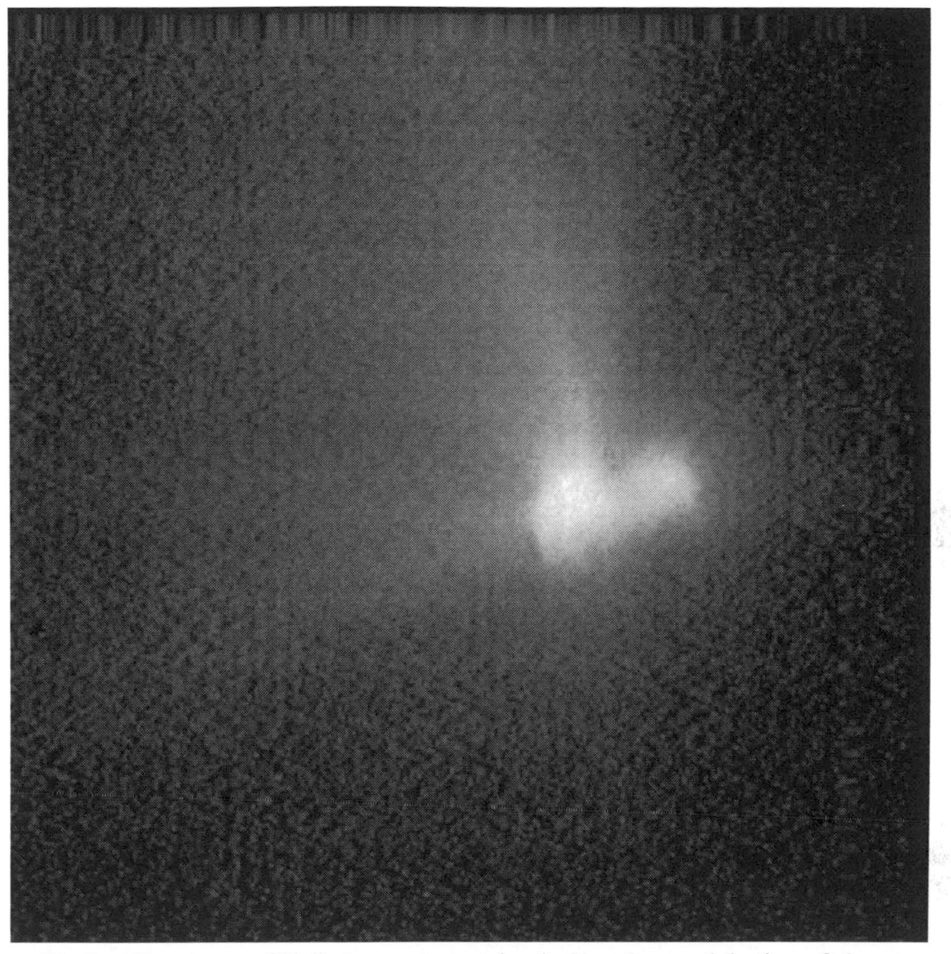

The best Vega image of Halley's comet was taken by Vega 2 around the time of closest approach. It clearly shows the peanut shaped nucleus with jets emerging from it.

A total of some 1,500 images of Halley were returned by the Soviet mission, the best of which were taken by Vega 2 whilst within 9,000 km of the nucleus. Vega 1 only returned pictures taken using three filters (red, near-infrared and visible), and they were cropped to a 128 × 128 pixel window which was centered on where the scan platform's pointing system thought the nucleus was located. Vega 2 switched to the backup pointing system, and since this was less accurate it made the camera return the full frames of images taken using four filters. The early pictures showed the characteristic parabolic shape of the coma extending in front of the nucleus. As the range reduced ever more detail became visible: initially marked asymmetries, and then the actual jets emitted by the still-invisible nucleus. For Vega 1, the coma was so dusty that the profile of the nucleus was not evident until closest approach. When processed on Earth, the images showed this mysterious object, whose longest axis appeared to be more or less pointing at the spacecraft at the moment of closest approach, as distinctly roundish ("potato shaped") in form. But it was evident from the fuzzier pictures taken prior to and after closest approach that the nucleus was elongated, and at that time the larger end was facing the spacecraft. Because when Vega 2 flew by 72 hours later there was less dust between the spacecraft and the nucleus, it was able to obtain sharper images. Unfortunately, most of the images were overexposed due to the failure of the primary pointing software. Only a handful were deemed to be fully usable. The best ones showed a "peanut shaped" object 14 km in length and 7.5 km in width. Only a few details could be discerned, but it appeared to have an irregular shape. The fact that it was not a cluster of objects flying in formation supported the proposal advanced by Whipple more than three decades earlier that the nuclei of comets are 'dirty snowballs'. The pictures taken within 20 minutes of closest approach indicated that some parts of the nucleus were much more active than others. In contrast to the Whipple model, the surface of this nucleus was not uniformly sublimating, but was releasing material from just a few active spots. One of the best pictures was taken by Vega 2 at a range of 8,030 km 15 seconds before closest approach, and it showed the irregular nucleus as having two bright centers and five (perhaps six) narrow jets of material. A remarkable result from measuring the size and brightness of the nucleus was that the albedo of its surface was just 4 per cent: as black as coal. In other words, this made Halley the darkest solar system object known, rivaled only by the mysterious dark leading hemisphere of Saturn's moon Iapetus and by the small satellites of Uranus that had been discovered only a few weeks earlier by Voyager 2. This came as a considerable surprise, because the nucleus had been expected to be a bright icy body about 6 km across. The fact that it was significantly larger than this was explicable by its being considerably darker. Although the few correctly exposed Vega 2 pictures revealed the shape of Halley's nucleus and the fact that its surface was selectively active, it was not possible to infer much about the individual surface features.[249,250,251]

The cryogenic system designed to chill the infrared spectrometer on Vega 2 had a leak, so this instrument did not provide any data from Halley. The corresponding instrument on Vega 1 was sent an erroneous command which put it into calibration mode and caused a 30-minute blackout during the closest approach phase! Even so, it was able to report an emission center spanning several kilometers. Working in its spectrometric mode, the instrument detected an emission feature in the coma with a signature interpreted as the C–H bond of hydrocarbon molecules. One of the most important and unique Vega results (because no other spacecraft in the international armada had an infrared instrument) was the temperature of the nucleus, which was measured at 300–400K; considerably higher than predicted by models of pure water ice but lower than that calculated for a totally black surface. However, if the icy nucleus was masked by a thin crust of black insulating material the conflicting evidence could be reconciled. In this 'friable sponge' model, only a fraction of the heat from the Sun would reach the comet's interior, because the dark crust would radiate a significant fraction of it back to space in the infrared. As the water vapor that was formed leaked away, it took with it dust and other material. In some cases, however, the crust cracked to expose the ice to large-scale sublimation, producing a jet. According to Soviet scientists, the thickness of this dark crust might be only a few centimeters, and in places possibly as little as several millimeters.[252] If organic molecules were present in the ice, then this crust could be characterized as a 'lag' that built up on the surface as the nucleus shrank with successive perihelion passes. Like the infrared spectrometer, the 3-channel spectrometer did not work perfectly. That of Vega 1 suffered an electrical failure, and that of Vega 2 lost the ultraviolet channel. Nevertheless, the instrument was able to detect the signature of water and its dissociated OH radical, various compounds of carbon, including carbon dioxide (which seemed to be the second most abundant 'parent molecule'), the dissociation products of methane and ammonia and the CN radical – it was the discovery of the latter during Halley's 1910 apparition that gave rise to fears that when Earth passed through the comet's tail this might poison our atmosphere. Precise production rates were determined for each species. Vega 2 measured a water production rate deep in the coma of about 40 tonnes per second, but there appeared to be more OH than could be explained solely by water production. Sulfur (which had been discovered to be present in comets during IRAS–Araki–Alcock's remarkably close approach to Earth in 1983) may also have been detected.[253 254]

Strong narrow jets of dust, noted as abrupt discontinuities in the dust flux, were injected into a broad cone centered on the Halley–Sun line, and whose apex closely matched the source of the jets seen in the pictures. One of the most striking results of the dust detectors was that Vegas 1 and 2 both recorded dust of differing mass as the spacecraft traveled a few thousand kilometers. This meant that solar radiation pressure must be highly efficient in segregating particles by their masses, and must rapidly broaden an initially narrow jet. Indeed, the smaller particles are eventually removed from the coma and swept down the dust tail. Contrary to expectation, dust particles were detected at sizes right down to the smallest mass that the instrument was capable of registering. Nevertheless, the cometary environment was dominated by particles having masses at the heavier end of the range.[255,256,257] The dust fluxes suggested production rates of the order of tens of tonnes per second at the time of the Vega 1 encounter but only half as much when Vega 2 arrived 3 days later. In fact, the high dust-to-gas ratio in the coma prompted the suggestion that rather than being 'dirty snowballs', comets are actually 'snowy dirtballs'.[258]

The mass spectrometer on Vega 1 obtained more than 1,000 spectra of the dust. However, owing to serious voltage problems, possibly due to overworking the scan platform, Vega 2 returned only a few hundred spectra. The results showed that the particles could be divided into three families depending on their composition: one reminiscent of carbonaceous chondrite meteorites, rich in potassium, magnesium, calcium and iron; another considerably enriched in carbon and nitrogen; and a third of water, water ice and carbon dioxide ice.[259] During the 1-hour period centered on the time of closest approach, the cometary plasma was 'sounded' using a double-frequency radio (in the same way as was done in the case of the solar corona when a spacecraft was on the opposite side of the Sun from Earth) to measure the density of electrons along the line of sight. In principle this experiment could also measure how a spacecraft was slowed upon being struck by a dust particle, but no example of this was detected.[260]

Even although it suffered from technical problems which marred its scientific results, the Vega mission was a great success that added to the growing impression that the Soviet Union had regained the lead in space from the United States, whose civilian space program was in chaos after the loss of the Space Shuttle Challenger. However, the apparent technological superiority of the Soviet Union was shattered a few weeks later when a reckless experiment at the Chernobyl nuclear power plant produced history's worst nuclear accident.[261]

The fourth probe to arrive was Sakigake, reaching the point of closest approach at 04:18 UTC on 11 March. During a continuous 13-hour session around this time the magnetometer reported several changes in the polarity of the magnetic field, probably due to the spacecraft crossing the heliospheric neutral sheet. Surprisingly, while making its crossing the comet did not experience a tail disconnection event, possibly owing to the relative geometry of the solar magnetic field and the comet's ion tail. In addition to reporting the undisturbed solar wind about 4 hours upstream of the comet for the other members of the Halley armada, Sakigake detected low-frequency plasma waves and magnetic disturbances that probably resulted from the comet's presence.[262,263,264,265]

Upon learning from the Vegas and Suisei that the inner coma was a very dusty environment, the Giotto team grew gloomy. Because Giotto was to fly much closer to the nucleus than its predecessors, if it passed through one of the jets reported by Vega 1 then it would very likely be destroyed. It was impossible to chart a course through the dust jets, because they were so poorly defined and so variable. But if the rates of production of gas and dust were proportional, the good news was that monitoring by the International Ultraviolet Explorer satellite suggested that the rate of gas production should be near its minimum as Giotto made its flyby.[266] Various rehearsals of the 4-hour encounter had been conducted during the 8-month cruise. By the final rehearsal on 10 March the spacecraft, the payload and the ground team were all confirmed to be ready. It was on this date that Giotto's flyby distance was decided, based on the results of the Pathfinder project. The current trajectory would produce a flyby at about 700 km from the nucleus. The camera scientists wanted a flyby between 500 and 1,000 km, and no closer lest the control system lose track of it. In contrast, the particles and fields scientists would have preferred a closer pass, even at the cost of losing the spacecraft. As a compromise, it was decided to reduce the range to 540 km with an uncertainty of 40 km. Accordingly, early on 12 March Giotto thrusted for 32 minutes to trim its speed by 2.5 m/s.[267]

The solar wind was fairly quiet in the days before the encounter, which made it easier to predict when Giotto should reach the cometary bow shock; if indeed this existed. On the afternoon of 12 March, at a distance of 7.8 million km, a plasma analyzer became the first instrument to note the presence of the comet by detecting hydrogen ions which had been picked up by the solar wind. Shortly thereafter, the energetic particles experiment detected pick-up ions at 7.5 million km. The second plasma analyzer then detected interactions between the comet and electrons in the solar wind. Early on 13 March the spacecraft turned to face its dust shield precisely forward, and fully activated its particles and fields suite. Magnetic field variations signaling the presence of the comet were first detected at a range of 2 million km. At

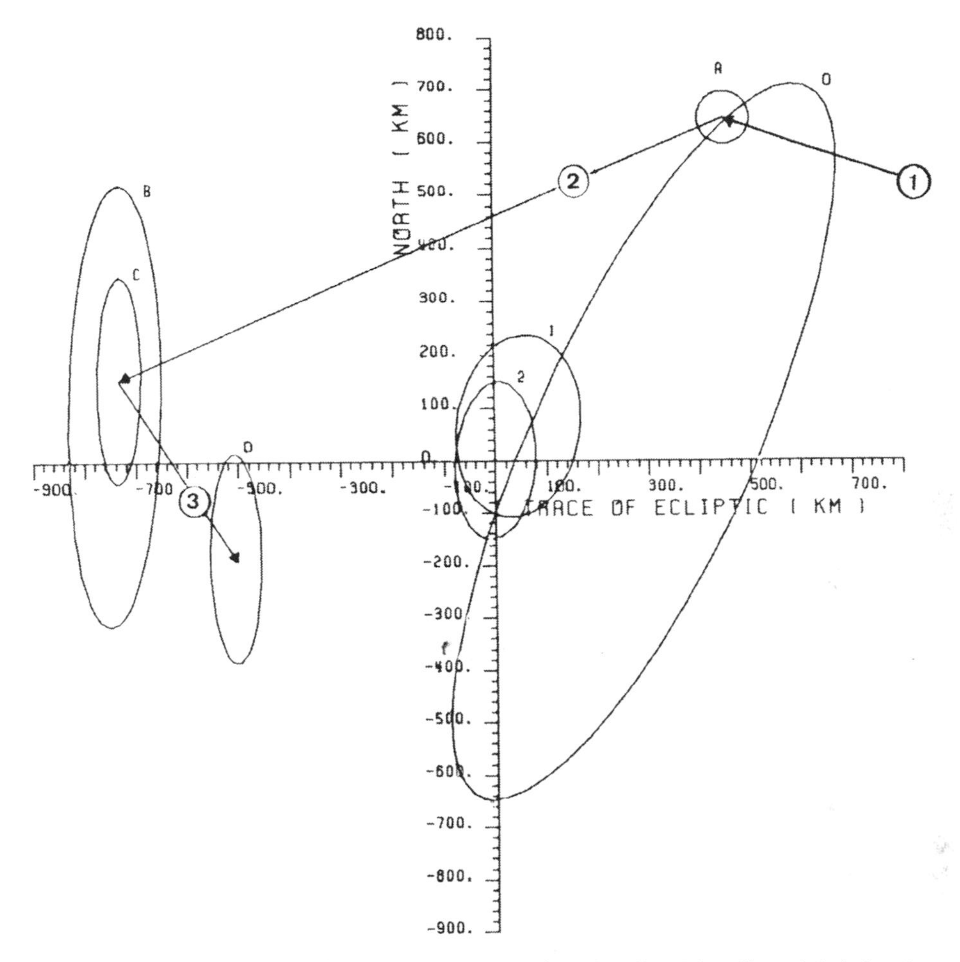

The relative positions of Giotto's closest approach and Halley. The ellipses labeled 0, 1 and 2 represent (respectively) the estimated positions of the comet's nucleus based on terrestrial observations, by Vega 1 and jointly by Vega 1 and Vega 2. The ellipses A, B, C, and D represent the possible positions of Giotto after each course correction and trajectory determination. The arrows 1, 2 and 3 represent the three Giotto course corrections. (ESA)

19:23 UTC the magnitude of the magnetic field started to increase; 10 minutes later it peaked and declined, probably marking the crossing of the bow shock at the predicted range of 1.15 million km. Another increase more than half an hour later could have been either a second bow wave crossing or fine structure within the cometary magnetosphere.[268] The ion mass spectrometer suggested that Giotto may have crossed the bow shock several times over an interval of a few minutes.[269] The Johnstone plasma analyzer did not find a discontinuity equivalent to the bow shock that develops upstream of a planetary magnetosphere, but did report several sharp transitions (the first of which was at a range of 1.13 million km and 125,000 km thick) over which the solar wind was progressively slowed and deflected. Other discontinuities consisted of oscillations in which the solar wind speed and density decreased and then recovered. To the plasma analyzer, therefore, Halley had only a weak shock wave with several precursors which inflated and deflated. The electron spectrometer section of the Rème plasma analyzer also noted the shock wave as an abrupt increase in the electron density. Beyond, there was a transitional zone that ran to within 550,000 km of the nucleus, inside which the electron density and temperature fluctuated wildly. In fact, to the Rème plasma analyzer the structure of Halley's comet closely mimicked that of Giacobini–Zinner, albeit on a much larger scale.[270,271]

Giotto activated its camera at 19:43, and performed the acquisition sequence to locate the comet. The first image was taken at 20:55 from a distance of 767,000 km – which is twice as far as the Moon is from Earth. Heading inbound over the next 3 hours it took a picture every 4 second as the spacecraft rotated, operating in the single-sensor mode in which only one CCD sensor was effectively working. This mode would be used so long as the angle between the spacecraft's spin axis and the relative velocity remained small – that is, while the comet was more or less directly ahead. Last minute commands were sent to the camera to ensure that if it lost track of the nucleus it would reacquire it.[272] This sequence, with a total of 2,043 images, was to study the innermost coma at low resolution. It showed a fan shaped jet that was expanding sunward from a bright spot near the nucleus into a sector spanning in excess of 70 degrees. At least seven fainter jets could be seen inside this broad fan, blending into the ambient glow not far from the nucleus. No jets could be seen on the night-side of the coma.[273] At a distance of 287,000 km, exactly 70 minutes before closest approach, the dust detector noted the first hit on the Whipple shield, with 30 more over the next half hour, none of which was sufficiently energetic to puncture the front shield. The nucleus was able to be identified in images of the bright coma taken within about 145,000 km, and at about 70,000 km it was finally resolved. All that could be discerned at this time were two bright blobs and a dark irregular shape, and most scientists instinctively interpreted the blobs as the bright nucleus and the dark as the shadow of the nucleus cast on the enveloping cometary haze. The truth was not recognized until later: the blobs marked the bases of two prominent dust jets situated 5.5 km apart, and the darkness was the night-side of an inactive region of the black nucleus. With the resolution increasing by 85 meters with every passing minute as the spacecraft closed in, a handful of fainter jets were detected, and a bright oval was seen on the night-side beyond the terminator which was suggestive of a hill or some other surface relief that was catching the light of the Sun.[274] At 23:58, when 25,000 km from the nucleus, the camera automatically switched to its multiple-sensor mode in which it used all four sections of its CCDs, took color and polarized images, and sent partial frames spanning 74 × 74 pixels. The magnetic field strength increased until it peaked 16,400 km from the nucleus. This marked the cometopause, within which there was only material of cometary origin. The solar wind and the magnetic field which it carried 'draped' around the cometopause and then shaped the cometary ion tail as it continued downstream. A little more than 3 minutes before closest approach, an impact punctured the front shield; only 10 more would do so during the entire encounter.[275] At about this time the baffle of the star mapper was perforated by dust, rendering the data from this device unusable and subjecting Giotto to its first large attitude disturbance. Shortly thereafter, the spacecraft closed within the 8,000 km range that marked the closest point of approach for the Vegas.

As Giotto continued to close in, the field of view of the camera showed an ever diminishing portion of the nucleus, with the pointing system firmly locked onto the brightest part. Although the resolution was now theoretically better than 50 meters, it was probably worse because the camera's mirror was being sandblasted by dust. When 4,660 km from the nucleus Giotto became the only member of the armada to completely transit the cometopause to enter the ionosphere of the comet, where the field strength was zero because the nucleus had no intrinsic magnetic field and the ions were cold. This region had been predicted from the similarity between how a comet interacted with the solar wind and how other non-magnetized bodies (like Venus) did so – and indeed with the 'artificial comet' created in Earth orbit in 1984 by the AMPTE (Active Magnetospheric Particle Tracer Explorer) experiment.[276] Meanwhile, the plasma analyzers reported dense clouds of hot plasma which were almost certainly produced as dust was vaporized on striking the spacecraft at very high speed. A 40-milligram particle that hit 44 seconds before closest approach was the heaviest recorded by Giotto. It triggered all three of the piezoelectric sensors on the front shield. About 28 seconds before closest approach and 2,050 km from the nucleus, the spacecraft crossed the terminator into the sunward hemisphere of the comet. At this moment all of the dust detectors reached their peak impact rates, just as the spacecraft's position matched a jet in terms of the line of sight from Earth.

Twelve seconds and 1,703 km from the nucleus the camera reset, terminating its observations. At the start of the encounter the camera tube had been facing directly ahead, but it was now angled 40 degrees to the side, making it a much larger target for the dust. At about this time the temperature of the camera's radiator increased, probably as a result of a dust impact that slowed the period of the spacecraft's spin from 3.998 to 4.010 seconds. Finally, at a range of 770 km what everyone dreaded finally happened: first the spacecraft switched to its backup transmitter, then the telemetry briefly fluctuated before being lost. This was due to an off-center hit on the shield 7.6 seconds from closest approach by what must have been a fairly large dust mote which caused the spacecraft to wobble sufficiently to swing the beam of the high-gain antenna away from Earth. At 00:03:02 UTC on 13 March it reached the minimum flyby distance of 596 km in front of the nucleus. Just as people were convincing themselves that the loss of signal must signify that Giotto had been destroyed, after 21.75 seconds of silence there was a burst of telemetry. There were further intermittent contacts with the signal growing stronger each time until, after about a minute, it possible to extract coherent data. This showed that the probe was alive and relatively healthy, but it was nutating with a period of 16 seconds and this was causing the beam of the high-gain antenna to point at Earth only intermittently. But as is standard for spin-stabilized spacecraft, Giotto was equipped with nutation dampers: long tubes filled with mercury inside which a small metal ball was free to float, and in nutation the ball's motion dissipated energy through friction with the

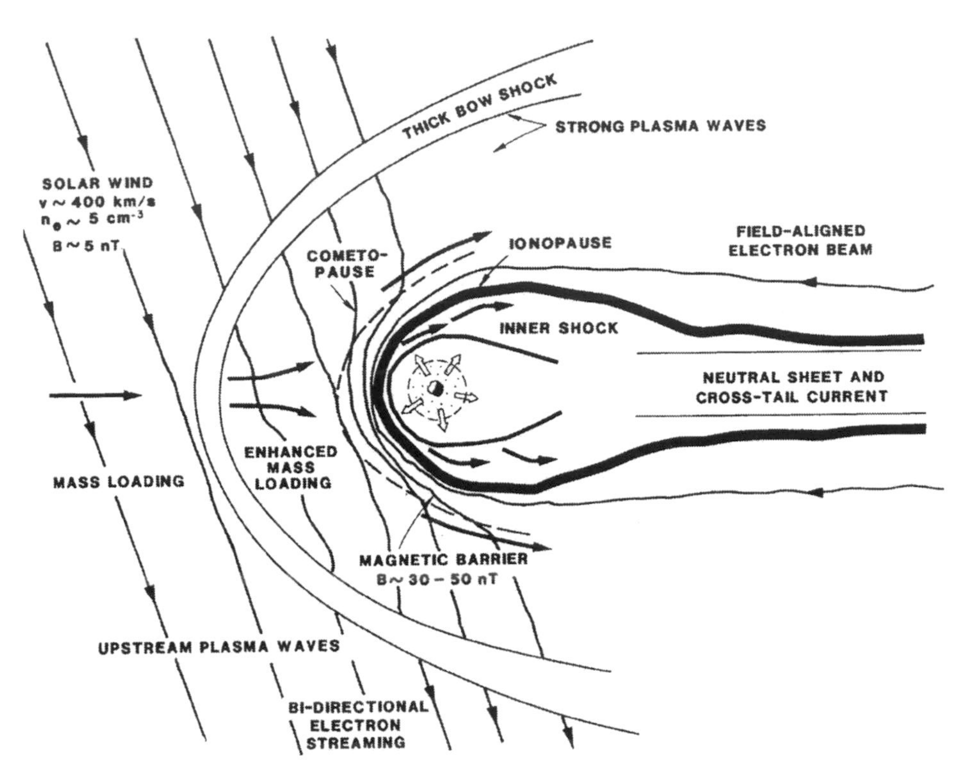

The model of the interactions between a comet and the solar wind as inferred from the Halley flybys (From Flammer, K.R., "The Global Interaction of Comets with the Solar Wind". In: Newburn, R.L., Neugebauer, M., Rahe, J., (eds.) "Comets in the Post-Halley Era", vol. 2, 1991; reprinted with permission of Springer Science and Business Media)

liquid until it restored the original orientation of the spin axis. That is exactly what happened, and 32 minutes after the encounter Giotto was again able to maintain its antenna pointing at Earth.[277] In the meantime, it had left the ionosphere at a range of 3,930 km and crossed the cometopause at 8,200 km. Luckily, the magnetometer was one of a few instruments to have its own data storage and these measurements were able to be replayed in the ensuing hours. The outbound bow shock crossing at about 700,000 km was not as clear as that inbound, the transition to the supersonic solar wind being much more gradual. In fact, even the magnetometer, which had detected a distinct shock inbound, found only a region of increased magnetic field fluctuation.

As the engineers studied the telemetry, they realized that in the minutes during which Giotto had been nutating, unprotected parts of its body had been sandblasted by cometary particles, and some of its instruments and systems had been damaged. In addition to the damage to the baffles of the star mapper and the camera, there was some degradation of the spacecraft's thermal control system. Remarkably, the solar cells lost less than 2 per cent of their power. The plasma clouds created by the vaporization of impacting dust corrupted software and interfered with the operation of some of the electronics. There was a dramatic rise in the temperature of the dust shield as it was sandblasted. So significant were the dust strikes that the spacecraft had been slowed by 23 cm/s. In fact, it would persist in wobbling by several tenths of degrees as a result of having lost some 600 grams of material from its periphery, although it had also acquired several grams of dust during its transit of the coma. Later tests showed the star mapper to be usable while it was on the dark side of the spacecraft, when sunlight was unable to enter the baffle. Alternative ways were devised for attitude determination. Although the star mapper was no longer able to locate Earth because this was on the sunward side, it was readily able to find Mars on the opposite side. And the antenna pointing was optimized utlizing a technique similar to the CONSCAN (Conical Scan) technique used by Pioneers 10 and 11 in which the spin axis was slowly rocked back and forth until the signal strength was maximized. A complete reset and reconfiguration sequence was performed on the camera but it could not find the comet, preventing outbound imaging. When further tests showed that the camera could not detect Jupiter either, it was inferred that the dust striking the tube must have smashed it, and the camera was no longer viewing the sky. In addition, the neutral mass spectrometer had been lost, one sensor of the ion mass spectrometer and one sensor of the impact detector were lost, one plasma analyzer ceased to function 1.5 hours after closest approach and the other one was damaged but still able to return data.

Giotto confirmed most of what had been inferred about the nucleus from the Vega flybys, namely that it is an elongated object of very low albedo, most of the surface being a layer of non-volatile material, and that only a small part was active. But it also eventually provided extraordinary new insights. Like every CCD, those mounted on Giotto would have required a cooling system to enable them to create images free of electronic noise but mass and power constraints ruled this out and so the camera was cooled by passive means. In fact the telemetry indicated that the temperature of the sensor was about 15°C warmer than predicted, which made the images noisy. It took months to calibrate and remove this noise, and it was several years before the definitive results were released. Whereas the pictures published in 1986 had been blurry and difficult to interpret (not least owing to the use of 'false' colors that made the live coverage a public relations fiasco) the final ones revealed the comet in astonishing detail.

First, the distant views were processed to highlight jets and fainter filaments, of which a total of 17 were counted, including one that seemed to be coming from the night-side of the nucleus – this could have been an effect of perspective, or perhaps there was sufficient thermal inertia for this site of activity to remain warm enough in darkness to produce jets. One of the jets appeared to point directly at Giotto, and may have been the one that issued the dust which disabled the camera. Because the night-side of the nucleus stood out against the glow of the coma, it was possible to determine its dimensions fairly accurately, confirming the impression gained from the Vega imagery that it is about 16 km long and 8 km wide. After almost 2 years of data processing, the Giotto scientists were able to produce a mosaic from images of varying resolution: since the camera was programmed to lock onto the brightest

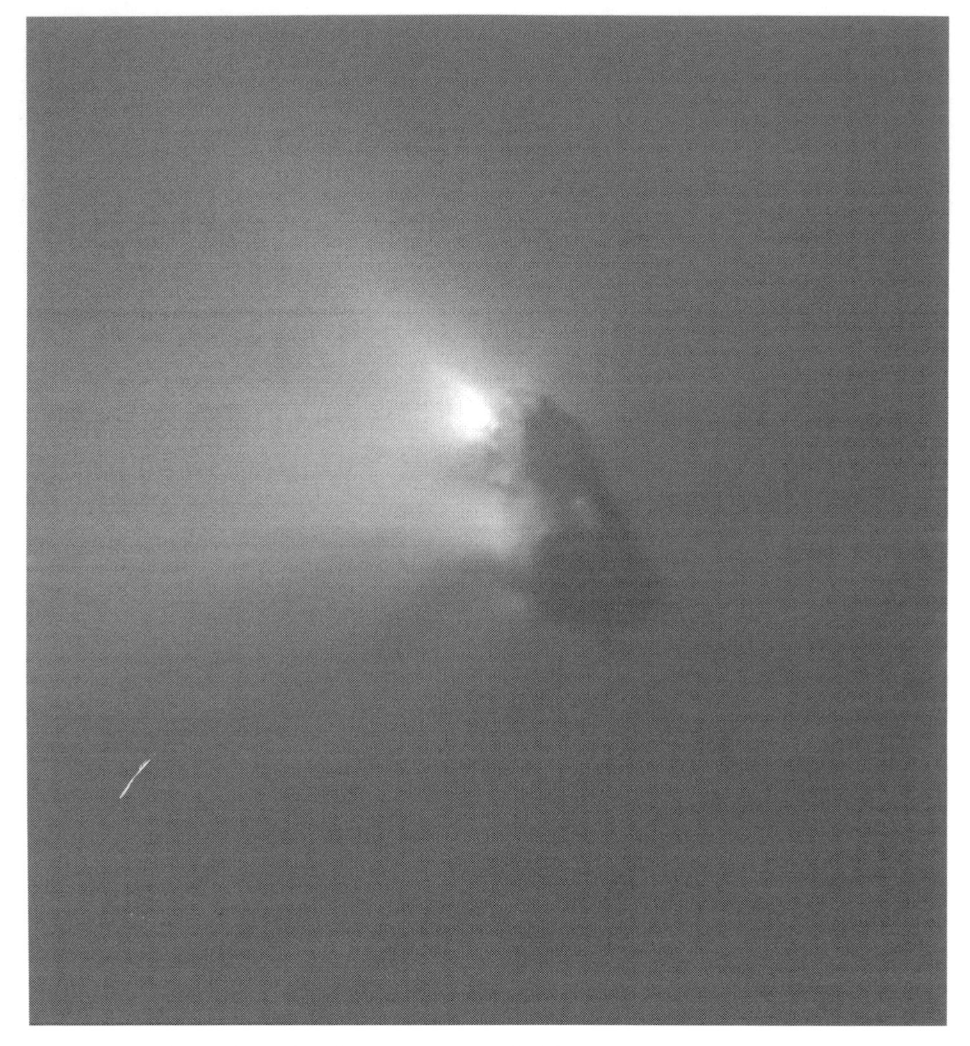

A view of the nucleus of Halley's comet compiled from 68 individual images of differing resolution by the Halley Multicolor Camera of the Giotto spacecraft. It shows the night-side of the elongated nucleus in silhouette against the background of dust reflecting sunlight. Jets of gas and dust can be seen originating from three regions on the nucleus. The moderately bright spot on the dark side of the nucleus is probably a hill at least 500 meters high which is catching the rays of the Sun. The circular feature between the two active regions could be a crater. (Courtesy of Dr. H.U. Keller; copyright 1986, the Max-Planck-Institut für Sonnensystemforschung, Lindau/Harz, Germany)

feature in its field of view, the highest resolution (some 60 meters per pixel) was of the areas nearer to the bright jets, and the resolution fell off to about 320 meters per pixel at the antipodes. About 75 per cent of the side of the peanut-shaped body that was visible to the camera was in darkness. The details on the terminator included: a roundish feature some 2 km wide at the limb that could have been a shallow crater distorted by the perspective, a depression, or an active region of some kind; a series of regularly spaced hills and a finger-like projection near the base of the brightest jet of dust; and a smooth depression adjacent to the crater that appeared to contain at least two sites of jet activity. Finally, the oval feature on the night-side that was prominent in the early images was revealed to be a mountain some 2.2 km beyond the terminator whose base covered an area of 1 × 2 km and whose summit was still illuminated because it rose at least 500 meters above its surroundings. Remarkably, the sources of the jets were a fair match for the active regions which were mapped by analyzing the pictures of the comet's 1910 apparition. In particular, it appeared that the largest emission areas occurred where the mapped sources intersected each other. Unfortunately, because the camera produced no post-encounter images, only one hemisphere of the nucleus was documented, and this, owing to the flyby geometry, was the night hemisphere![278,279,280]

No effects of rotation were seen during Giotto's 3 hours of observations, but it is difficult for flyby spacecraft to determine the rotational state of a body – not just its period of rotation but also the direction of its spin axis and the extent to which this may be precessing. According to Suisei and the Vegas, the nucleus appeared to be rotating with a period of about 52 hours, with an uncertainty of just a few hours. But terrestrial telescopes and the International Ultraviolet Explorer satellite in orbit around Earth found a periodicity for the brightness of the comet, for the production rates of some molecules and dust, and for its outbursts of about 7 days, which was about 5 days longer. These conflicting observations were reconciled by concluding that the rotational state superimposes rotational and precessional components. The rotational period does indeed appear to be 7.2 days but, in contrast to conventional wisdom, is aligned along the longest axis of the nucleus.[281,282]

None of the armada of spacecraft was able to measure the mass of the nucleus, in part because this is so small, but mainly because the high relative speed of their encounters meant that the comet had little opportunity to deflect their trajectories. However, its mass could be estimated from how the rocket-like effects of its jets altered its orbit, and when this was combined with an estimate of its volume from spacecraft images its density could be derived. This proved to be consistent with a mixture of solid water ice and dust. The neutral mass spectrometer confirmed that 80 per cent of the coma was water, some 16 tonnes being issued every second and leaving the nucleus at a speed of 900 m/s. The identity of the remaining 20 per cent was not immediately evident, as only in a few cases were the parent molecules observed directly; in all other cases it was necessary to develop chemical models of how these evolved with time and distance from the nucleus. Carbon monoxide and dioxide, methane, ammonia and cyanidric acid were found thus, as well as various hydrocarbons. In fact, two species of organic molecule – protonated formaldehyde and methanol – were not recognized until 5 years after the flyby and, remarkably, the former, despite its being a more complex molecule, was as abundant as carbon dioxide.[283] Water-based ions were abundant of course, but carbon, oxygen, sodium, sulfur and iron were also detected, with molecular sulfur and sulfhydric acid being the likely parent molecules. Strangely, the relative abundance of carbon monoxide appeared to increase with increasing distance, probably indicating that it was being released not only by the nucleus itself but also by the ejected dust grains. The water on Halley's comet (as on Earth and meteorites) contained more deuterated 'heavy water' than is typical of interstellar gas. Furthermore, the isotopic ratios of several elements also seemed to indicate that this comet formed in the solar system, rather than having formed elsewhere and been captured.

As the Vegas had discovered, small dust particles in the coma were much more abundant than models had predicted. In fact, most of the particles analyzed by the mass spectrometer were of the order of 10^{-16} grams. One surprise was that mixed in with the silicon, magnesium and iron in the dust which had traveled far from the nucleus, and had therefore been subjected to solar heating for a longer time, were lightweight elements such as carbon, hydrogen, oxygen and nitrogen which could reasonably be expected to evaporate. This probably meant that these elements were present as complex organic polymers which were stable in such conditions. From their composition, these polymers became known as the CHON molecules.[284] But the analyzer on Giotto and its counterparts on the Vegas failed to detect many other elements such as potassium and phosphorus which ought to have been present if it were true (as asserted by the panspermia hypothesis) that comets had 'seeded' the inner solar system with life. Nevertheless, the presence of complex organics meant that comets could have supplied the basic building blocks of life, if not actually life itself.[285] The dust detectors confirmed that at large distances from the nucleus small particles predominated, but as Giotto neared its point of closest approach there was a surge in activity and larger particles became dominant. A total of 12,000 impacts were recorded during the encounter, with the largest particles being more than one-thousand-billion times more massive than the smallest. One of the impact plasma sensors suffered a mishap that limited its productivity. As a late modification, each of the plasma detectors had been fitted with a removable thin cover as protection from the grains in the exhaust of the rocket motor that inserted the spacecraft into solar orbit. The covers on all the instruments were commanded to open in February 1986, but the signal to confirm the retraction of one was not received. The readings taken during the encounter with the comet indicated that this had indeed failed to open, which was a failure that had never happened in ground tests![286,287,288]

The instruments were turned off on 15 March, to draw the extremely successful encounter to a conclusion. Between 19 and 22 March Giotto performed a three-part course correction totaling 110 m/s that set up a close encounter with Earth in July 1990 (exactly 5 years after it departed) preparatory to a possible extended mission. After the new orbit was computed, it was refined by corrections on 1 and 2 April to ensure that the spacecraft would be conveniently positioned for radio-telescopes to pick it up again after its years in hibernation. Finally, early on 2 April, Giotto was commanded into an orientation in which its spin axis was aligned perpendicular to the plane of its orbit. Whilst this meant that the solar cells on the body of the drum-shaped craft would be illuminated throughout its orbit, it also meant that the high-gain link to Earth had to be terminated. This done, the spacecraft switched off all of its non-essential systems.[289]

Two NASA spacecraft had distant encounters with Halley as the comet receded from the Sun: on 20 March Pioneer 7 was 12 million km on its anti-sunward side; and on 28 March ICE was more than 30 million km on the sunward side. NASA proclaimed ICE to be the first spacecraft to visit two comets – as though a range of 0.2 AU constituted an encounter!

Having shed 400 million tonnes of water and dust during its perihelion passage, Halley retreated to the outer solar system and its dark crust slowly cooled towards its normal temperature of a few degrees on the absolute scale. However, the show was not quite over. In 1991, when it was halfway between the orbits of Saturn and Uranus, an outburst briefly renewed the coma. Perhaps the declining temperature caused crystalline ice to change its phase to amorphous ice, and this released some trapped gas and dust.[290] After 1992 observations continued of the bare nucleus as it sped towards aphelion. The pace at which the technology of astronomical detectors is improving suggests that it will be possible for the first time to monitor the comet through most, if not all, of its orbit. At the time of writing, the most recent pictures of were obtained in 2003 by the Very Large Telescope of the European Southern Observatory, one of the largest telescopes in the world. The heliocentric distance was comparable to the radius of Neptune's orbit (although in a different plane) and the nucleus was the faintest solar system body yet observed. In all likelihood there will be no attempt to observe it again until at least 2020.[291,292] It will reach aphelion in December 2023 and then fall back. Perhaps when it reaches perihelion in August 2061 a human crew will venture out to inspect it. One option would be to insert a robotic spacecraft into orbit around the nucleus to study this in detail as it resumes its dormant state. On its return in 2134 Halley will pass closer to Earth than it has for more than a millennium, and this will substantially alter its orbit; so much so, in fact, that it is not possible to accurately predict its future. However, we can say that in a few millennia the ice in the nucleus will be so thoroughly blanketed with the dark fluffy material that it will remain dormant even at perihelion, and will thereafter bear no resemblance to the celestial spectacle that either fascinated or terrified humanity throughout recorded history.[293]

EXTENDED MISSIONS

Since each of the spacecraft of the international armada survived its encounter with Halley in reasonable condition, the possibility of an extended mission was studied in each case.

The options for the Vegas, announced only after the Halley encounter, included a flyby of a small near-Earth asteroid, sampling meteoroid orbits and investigating a dormant comet's tail – such as the one which observations by the Pioneer Venus Orbiter suggested was trailing (2201) Oljato, whose orbit classified it as an Apollo-type asteroid but could actually be a dormant cometary nucleus.[294] In particular, it

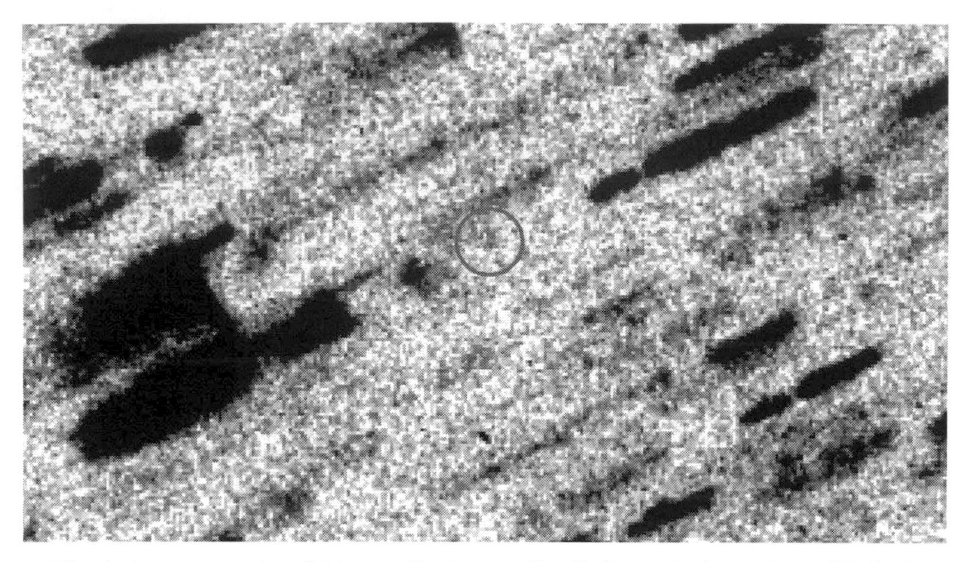

The dark spot at center of this negative image of 'trailed' stars is the nucleus of Halley's comet in September 2003. This is the most distant view if it ever taken. (ESO)

was planned that in 1987 Vega 2 should fly within 6 million km of (2101) Adonis, another Apollo-type asteroid suspected of being an old comet.[295] However, at such a range it would not be possible to undertake imaging and other remote sensing. In any case, it was soon found that there was insufficient fuel remaining to adopt the requisite trajectory.[296,297,298] A minimum extended mission was eventually carried out in which the Vegas returned data on dust streams as they crossed the orbits of comet Denning–Fujikawa, of the long-lost periodic comets Biela and Blanpain, and of Halley itself, in the latter case probably when their orbits took them back to the initial encounter point. Vega 1 ran out of attitude control propellant on 30 January 1987, and contact with Vega 2 continued until 24 March.[299] Although the Vegas marked the last deep-space use of the bus which was introduced by the 1971 Mars mission, it had proved reliable and so was adapted for other roles. In March 1983 a version without the course correction engine and propellant tanks was inserted into Earth orbit carrying an ultraviolet telescope and X-ray spectrometers to undertake an astronomical mission named Astron. A second one, named Granat, had X-ray and gamma-ray detectors. It was launched in December 1989 and operated until 27 November 1998.[300,301]

Although designed to last only 18 months, the twin Japanese spacecraft still had a lot of propellant remaining, and so were capable of undertaking more interesting extended missions. Both were in orbits which would return them to the vicinity of Earth in 1992. If maneuvers were made to reduce the 15-million-km flyby range, it would be possible send them on to Venus (Sakigake and Suisei) or Mars (Sakigake only). However, the instrument suite was not really appropriate to remote-sensing of Venus, and a Mars mission was impractical because the spacecraft had not been designed to operate that far from the Sun. Therefore, other solutions were studied.

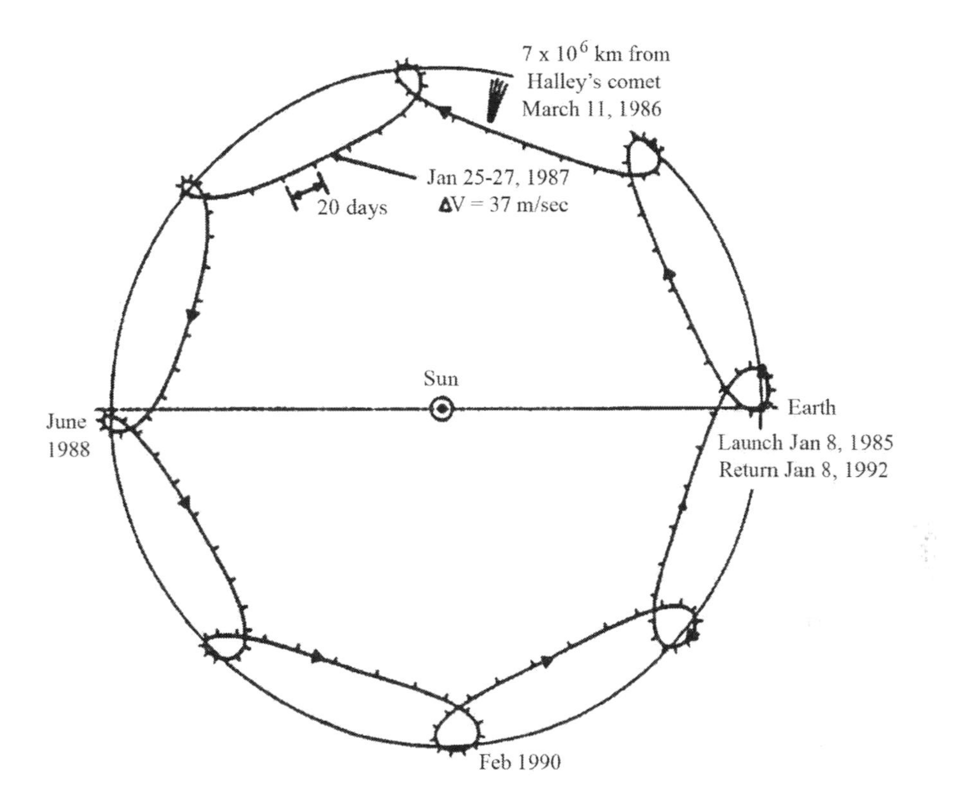

The trajectory of Sakigake from its launch to its first return to Earth in 1992, in a frame of reference which rotates with the Earth (i.e. relative to a fixed Sun–Earth line).

Sakigake made a complex series of maneuvers in January 1987 to set up its new mission. On reaching Earth it was to enter a heliocentric orbit that would enable it to make several passes through the tail of the Earth's magnetosphere, and thereby contribute to the International Solar–Terrestrial Physics program – for which ISAS also intended to launch the Geotail satellite in 1992. After four passes, Sakigake would have a 10,000-km encounter on 3 February 1996 with 45P/Honda–Mrkos–Pajdušáková, as this short-period comet flew within 29 million km of Earth.[302] On this occasion, Sakigake would make the first approach to a comet from its antisolar direction – starting off with the tail, then the coma and nuclear region and finally crossing the bow shock. Even without maneuvering, it would have a 14-million-km flyby of Giacobini–Zinner on 29 November 1998. In practise, Sakigake made its first flyby of Earth on 8 January 1992, almost exactly 7 years after launch, and this deflected it into an orbit ranging between 0.916 and 1.154 AU that sent it deep into the geotail, where its magnetometer and other instruments enabled it to make extensive observations. It returned to Earth on 14 June 1993, although this time the targeting errors were unusually large because its orbit was poorly defined, and then again on 28 October 1994. As there was very little hydrazine remaining, no more maneuvers were attempted. Telemetry was lost on 15 November 1995, but minimal contact was maintained using a beacon signal until it was switched off on 7 January 1999, on the 14th anniversary of its launch. As for Suisei, when it was near its maximum distance (1.9 AU) from Earth in April 1987 it maneuvered so that the Earth flyby would put the spacecraft on course to pass many millions of kilometers sunward of 55P/Tempel–Tuttle on 28 February 1998. This 33-year periodic comet was of great interest for being the parent of the Leonid meteor stream that in 1966 produced one of the most spectacular displays of the 20th century. It was hoped that Suisei would be able to go on to encounter Giacobini–Zinner on 24 November 1998. However, it ran out of hydrazine on 22 February 1991. Its orbit would have brought it within 900,000 km of Earth on 20 August 1992.[303,304,305,306]

One early proposal for Giotto was to reactivate it at solar conjunction in 1988, in order to use its radio to 'sound' the solar corona, but the Deep Space Network would not be available for tracking because its antennas would be in the process of being upgraded in preparation for Voyager 2's encounter with Neptune in 1989, so this proposal was discarded.

In fact, even before Giotto was launched, Martin Hechler, an orbit specialist at ESA, had noticed that it (or its hulk!) would return to the vicinity of Earth in July 1990.[307] The accuracy of the trajectory provided by the kick motor meant that the spacecraft did not need to make a course correction immediately after it set off, and would therefore have a significant reserve of propellant after the Halley encounter. In September 1985 the Giotto team invited Robert Farquhar to suggest comets that might be visited during an extended mission. One option was an encounter with the short-period comet 26P/Grigg–Skjellerup in July 1992.[308] This had been discovered by John Grigg in New Zealand on 23 July 1902, and recovered by John Francis Skjellerup in South Africa in 1922. It was later realized that an 1808 sighting of a comet that was then lost was the same object, and that in the intervening years its orbit was greatly changed by a series of approaches to Jupiter. Observations after 1922 established that it had a 5-year period. It had been observed at every return since. In 1982 its nucleus was detected by the Arecibo radio-telescope operating as a radar, and found to be only 400 meters across (making it at least 20 times smaller than Halley). With a water production rate of the order a few tens of kilograms per second (fully two orders of magnitude less than Halley and one order of magnitude less than Giacobini–Zinner) it would provide an interesting point of comparison for cometary scientists. And, just as for Giacobini–Zinner and Halley, it is known to be the parent of a meteor shower, albeit a weak one.[309,310,311,312] The relative speed of the encounter would be a mere 14 km/s, with the comet 12 days past its perihelion. Unfortunately, other geometric parameters would not be as favorable because (as related above) the spacecraft had been designed for the Halley encounter. The need to simultaneously maintain the antenna pointing at Earth and maximally expose the solar cells to the Sun meant that the dust shields would face almost 70 degrees off the direction in which the dust would approach – in effect, the spacecraft would be almost sideways on, and extensive damage to the unprotected solar cells could be expected. Moreover, the increased encounter distance from the Sun (1.1 rather than 0.9 AU) meant that the power would be barely sufficient to operate the spacecraft.

Although there were other options, including 79P/du Toit–Hartley, 103P/Hartley 2 and an encounter with Honda–Mrkos–Pajdušáková in 1996 several days before that planned for Sakigake, the Grigg–Skjellerup flyby had the operational advantage of being earliest and at an acceptable compromise in terms of range from the Sun for power and thermal control and Earth for communication.[313,314]

However, before a mission extension could be approved, the spacecraft's status had to be determined. A reactivation test was scheduled as it made its approach to Earth. The Deep Space Network antenna at Madrid in Spain uplinked commands, and at 15:06 UTC on 19 February 1990 a faint unmodulated carrier confirmed that the transmitter had activated after 4 years of silence. The first days of tracking were devoted to establishing the orbit and assessing the spin rate, and then the spacecraft was ordered to despin and point its high-gain antenna directly at Earth. After some initial problems and at least one telemetry failure, it did so. The full status readout showed that although Giotto had suffered hardware failures while hibernating, its overall health was good. The solar cells were working adequately, but further tests would be needed on the batteries. In addition to the damage to the Whipple shield, the thermal blankets were allowing the craft to overheat. The fact that the camera's baffle had been destroyed was confirmed by an analysis of the spacecraft's inertial characteristics, which suggested that it had lost some mass at its periphery, and by the solar cells, which showed that the baffle no longer cast a shadow. The camera was checked again, but it was unable to see anything. Suspecting that the viewing aperture was blocked by a piece of the shattered baffle, the tube was swiveled in an attempt to dislodge the obstruction, but in vain. The mass spectrometer and dust spectrometer also would not be usable in the Grigg–Skjellerup encounter, because these instruments had been tailored to collect material coming at a very high speed from a precise direction, and this geometry would no longer apply. Hence, only the plasma analyzer and the optical probe would be able to provide an indication of the composition of Grigg–Skjellerup's coma. And of course the dust detector would be much less sensitive, in part because for this encounter the particles would arrive at only one-fifth the speed as at Halley and also because they would strike the sensors at an almost grazing (rather than a perpendicular) angle. The magnetometer team were delighted that their instrument would no longer be disturbed by the magnetic field generated by the motor of the camera! The conclusion was that the spacecraft would be able to undertake a meaningful mission.

On 2 July 1990 Giotto became the first interplanetary spacecraft to return to the Earth's vicinity. As it flew by at an altitude of 22,731 km, it made observations of parts of the magnetosphere that had never previously been visited. The encounter changed the spacecraft's heliocentric orbit from one that was wholly interior to that of Earth to one which ranged between 0.994 and 1.165 AU. A maneuver 2 weeks later put it on a 'collision course' with the best available estimate of where Grigg–Skjellerup would be at the time of the encounter. There being no Vegas this time to act as pathfinders to refine the ephemeris, this was the best that could be done. An actual impact was most unlikely, owing to the large uncertainties in the orbit of the comet and the small size of its nucleus. On 23 July the spacecraft resumed its state of hibernation. With the health of the spacecraft verified, the scientific objectives of the

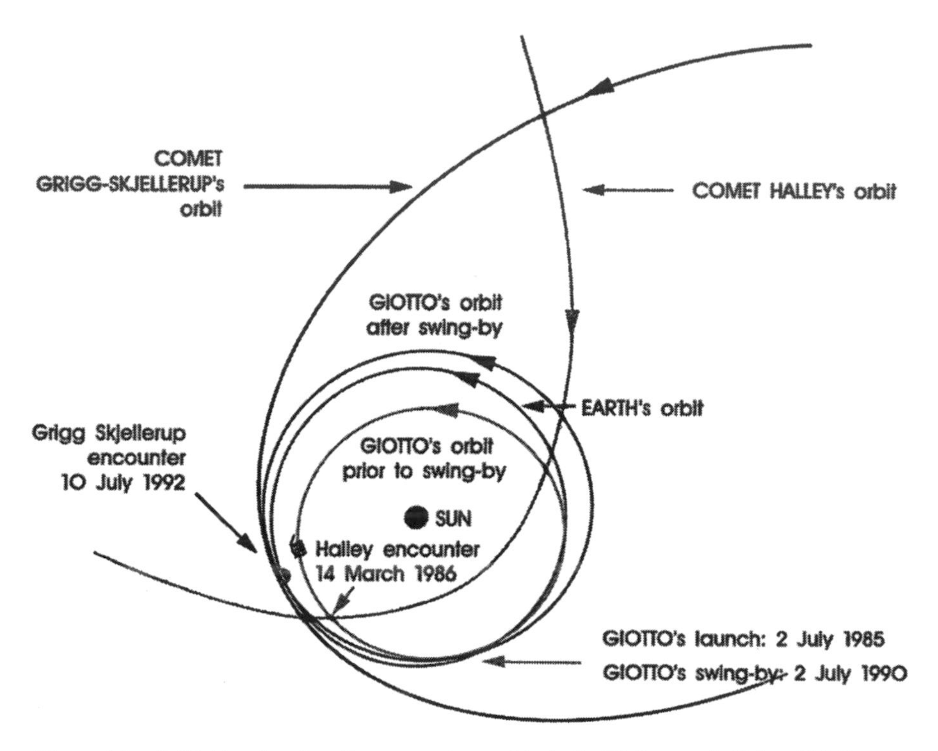

The orbit of Giotto showing its encounters with both Halley and Grigg–Skjellerup. (ESA)

encounter defined and the trajectory established, in June 1991 ESA finally agreed to devote a total of $14 million to the Giotto Extended Mission (GEM).[315]

Grigg–Skjellerup was recovered in September 1991 by astronomers at the Calar Alto Observatory in Spain, 10 months before perihelion. Giotto's second revival was initiated on 4 May 1992, and took 3 days. Once it had been confirmed that no problems had developed, a final test was made of the camera – this time ordering it to point directly at the Sun, but it saw only the soft glow of light being scattered by the ruined baffle. In fact, this came as something of a relief to the science planners, because even although the batteries were usable the minimal power margin would have made it difficult to operate the camera together with the other instruments. On the basis of observations by telescopes in the southern hemisphere, the spacecraft made a course correction on 8 July to adjust the aim point by a modest 145 km. An image taken by the European Southern Observatory in Chile 15 hours prior to the encounter showed that the comet had begun to develop a small dust tail.

On 9 July the instruments were activated, and the next day the first evidence of the comet's presence was detected at a range of 440,000 km, in the form of water-group ions which had been picked up and accelerated by the solar wind. This data indicated that the nucleus was producing a mere 68 kg of water per second. In the meantime, the magnetometer reported a wave field and other instruments saw the solar wind slowing and being deflected as it interacted with the comet. At a range of 19,900 km smooth variations in the magnetic field indicated a 'bow wave' rather than a bow shock. Some 4,600 km deeper into the coma the particle counts began to rise, and periodic peaks in the rates detected by the EPONA instrument on the inbound leg corresponded neatly to the cyclotron period of water-group ions in the prevailing magnetic field. As the range continued to reduce, the instruments began to detect heavier ions.[316] Although Giotto entered the coma at 50,000 km, as indicated by the optical probe detecting gas emissions, this saw no evidence of dust until the range had reduced to 17,000 km. Based on the brightness profile of the inner coma and magnetometer data, it was calculated that the flyby occurred at 15:18 UTC on 10 July at a range of just under 200 km. This time the spacecraft did not enter the ionosphere close to the nucleus, in which the magnetic field would disappear. The geometry of the encounter was difficult to determine, but it seems that Giotto flew over the 'evening' side of the nucleus, missing one of the most interesting (and still unexplored) regions of a comet where the cometopause stretches back on the night side and down the tail. Although the optical probe detected sunlight that was being scattered by dust, no impact was recorded until 12 seconds after closest approach, when the dust detector was hit by a 100-microgram particle (nicknamed "Big Mac" for principal investigator Anthony McDonnell in a moment of scientific euphoria!) which at least partially penetrated the shield. This was followed 3 seconds later by the 2-microgram "Barley" and 40 seconds after that by the final 20-microgram "Bretzel". Other impacts must have occurred, but owing to the encounter geometry they were not detected. In particular, a total velocity change was measured after the flyby which corresponded to the spacecraft collecting a coating of 39 milligrams of cometary dust. In addition, 5 seconds before the "Big Mac" strike, the antenna was probably hit by "Whopper", a particle having a mass not exceeding 50 micrograms, which set up a wobble that caused the antenna to move off-Earth for a few seconds and took 90 minutes to damp out. The fact that Grigg–Skjellerup was a relatively dust-free comet was confirmed by the absence of 'anomalous' plasma produced by the vaporization of impacting dust.[317,318]

A tantalizing observation was made by the optical probe less than a minute after the closest approach when it saw a second peak in the brightness of the coma. The most likely explanation is that Giotto passed within 50 km of a small companion nucleus that was located 1,000 km from the main nucleus, was 10 to 100 meters in size and was surrounded by its own coma. Other peaks could have been caused by particles impacting on the spacecraft's body, or by narrow jets erupting from the nucleus.[319] At 15:44, heading outbound, there was a step in the particle counts due to a distinct bow shock. The magnetometer detected this 4,600 km farther out, at a range of 25,400 km from the nucleus. Then, more than an hour after crossing the bow shock, back in the solar wind, the particle count rates unexpectedly surged for about 10 minutes and the magnetic field showed deviations from its average. The fact that this enhancement showed a very fine structure resembling that of Grigg–Skjellerup, complete with water-ion periodicities, suggested that the spacecraft had serendipitously encountered a very small companion 90,000 km from the primary nucleus. While Grigg–Skjellerup had been about 45,000 km across from bow wave to bow shock, this secondary structure was only about 9,200 km from side to side. Unfortunately, telescopic coverage was sparse at this time, and the existence of this secondary nucleus was unable to be confirmed.[320,321]

From beginning to end, the Grigg–Skjellerup encounter lasted less than 2 hours. Over the following days, Giotto switched off its instruments one by one, performed further engineering tests of the camera, and on 21 July adjusted its course to fly by Earth on 1 July 1999 at a range of 219,000 km. In the euphoria of the moment, the possibility of redirecting it to a third encounter around 2006 was suggested, but the remaining fuel (about 4 kg) would probably have been insufficient.[322] On 23 July the spacecraft was put into hibernation for the third time. It was hoped to be able to revive it in 1999, if not for a new mission involving the Earth's magnetosphere or even the Moon then at least to assess how it was being affected by its long time in deep space. However, tracking support was not forthcoming and it flew by Earth in silence. Although observers attempted to take images of the spacecraft, it was too far away for such a small object to be detected without exact foreknowledge of its orbit.[323,324,325]

LOW-COST MISSIONS: TAKE ONE

In 1980 JPL, in an effort to revitalize planetary exploration in America, initiated a study entitled LESS (Low-cost Exploration of the Solar System) that was meant to develop missions within a shorter time and at a lower cost. But this effort was short lived, partly because the laboratory had little experience of low-cost missions other than Mariner 5 which was a left-over, and Mariner 10 which used as many off-the-shelf systems as possible. The latter example did little to reassure the conservative engineers of the virtue of extreme cost-cutting, because it had endured a series of mishaps.[326,327] The concept was soon revived, however. The same year NASA had established the ad-hoc Solar System Exploration Committee (SSEC), whose tasks included proposing a strategy to overcome the crisis in the planetary exploration program. In its report in 1983 it recommended developing missions with realistic (i.e. cheaper) budgets, concentrating on the scientific requirements, avoiding costly new technologies, and using as much off-the-shelf hardware as possible (although the case of Seasat, lost when a 'standard' electrical component failed, indicated the need to tightly control supposedly proven hardware). Specifically, the report called for two classes of mission. The Planetary Observer class (initially referred to as the Pioneer class) would operate in the inner solar system, draw their technology from Earth-orbiting satellites and be cost-capped at $150 million (in 1982 dollars). The Mariner Mark II class would be for the outer solar system and would cost as much as $300 million each. The report recognized that 'flagship' missions in the style of Viking or Galileo, each of which cost around a billion dollars, would be unlikely to be approved, and this ruled out a mobile laboratory for Mars, returning a sample from that planet to Earth, and an ion-propelled tour of objects in the main asteroid belt.

The SSEC recommended that NASA devote a stable $300 million of its annual

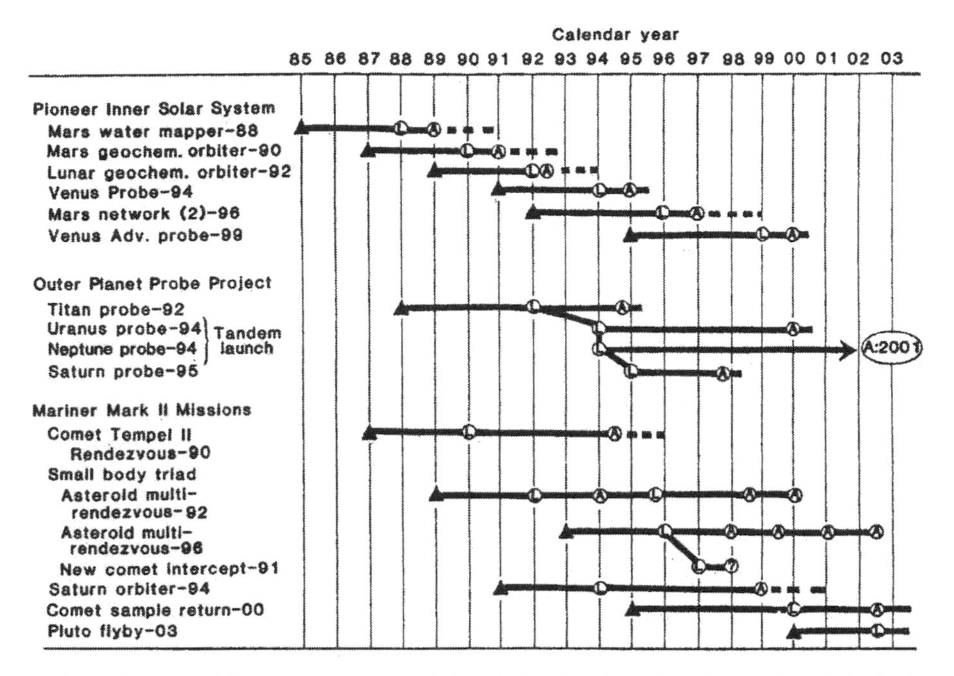

An early timetable proposed by the Solar System Exploration Committee. Note that Planetary Observer-class missions were still referred to as Pioneer missions. Circles with L indicate the time of launch and A the time of arrival. (From Waldrop, M.M., "Planetary Science in Extremis", Science, 214, 1981, 1322–1324; reprinted with permission of the AAAS)

budget to solar system exploration, to preclude the Congress having to assign each mission the 'new start' status that would make it vulnerable to political rivalries. It was also recognized that in response to missions becoming less frequent there had been a tendency to 'load up' those that flew with as many instruments as possible, thereby driving up both complexity and cost in a 'Christmas tree' effect. It was felt that more frequent flight opportunities (on lower cost missions) would reverse this trend. And of course a stable budget would enable NASA to start to plan missions to hitherto neglected targets such as comets and asteroids.[328]

The strategy for the Planetary Observer class was to modify the structures and systems of communications, meteorological and Earth-observation satellites to the deep-space environment of the inner solar system to facilitate missions to Venus and Mars (but not Mercury, for which the thermal issues were excessive) and such small bodies as might stray into this region. The architecture was to be sufficiently flexible for a mission to be mounted as a timely response to the discovery of a new object or phenomenon. All the major aerospace firms, including Hughes, General Electric and TRW, had satellite buses that could support this concept. In addition to inheritance of design, one of the key requirements of the Planetary Observer was to share the investment of developing the basic spacecraft over as many missions as possible. The potential of this approach had been demonstrated by the Explorer satellites that

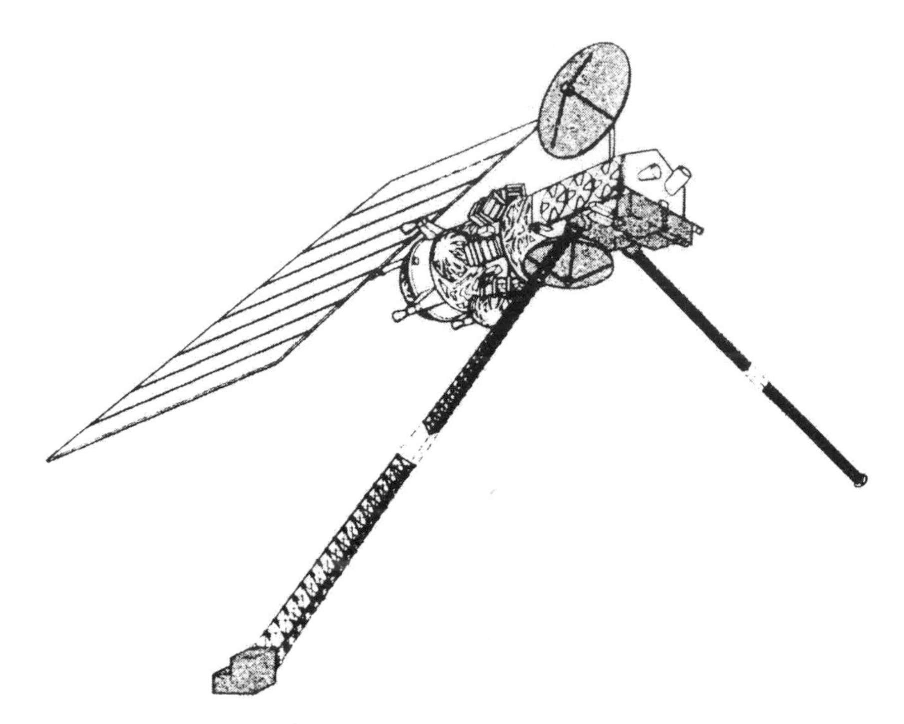

A sketch of the spacecraft for the Mars Geoscience/Climatology Orbiter Planetary Observer mission. Most of its components were to be reused from satellites such as the NOAA and Tiros meteorological families.

carried physics and astronomy payloads. Another tenet was that the program should require no new "enabling technology" to be developed in order to ensure success. However, a parallel program was envisaged that would enable new instruments and techniques to be developed to the point that they were sufficiently well understood to grant their inclusion in a scientific payload. The spacecraft was to be released in low orbit by the Space Shuttle, then use a low-cost kick stage for the escape maneuver. Of course, making Shuttle flights cheaper would be essential to the overall cost-effectiveness of this reinvigorated planetary exploration program. First would be the Mars Geoscience/Climatology Orbiter, to answer the questions arising from the Viking missions. It would be followed by the Lunar Geoscience Orbiter and by the Near-Earth Asteroid Rendezvous (NEAR) mission. In addition, a Venus Atmospheric Probe could address some of the issues raised by the Pioneer Venus Orbiter and Multiprobe missions. Other ideas included a Mars Surface Probe to deliver a number of penetrators, a similar mission to study the upper atmosphere of that planet, and a Comet Intercept and Sample Return (of which more later).[329]

As with the Planetary Observer, the Mariner Mark II relied upon the adoption of a standardized bus that could readily be reconfigured with systems and instruments for a specific mission. The spacecraft was to be capable of being adapted to a wide variety of missions, including Saturn orbiters, Uranus and Neptune flybys, Pluto encounters and comet/asteroid rendezvous. More advanced missions could sweep up gas and dust from comets and upon returning to the vicinity of Earth release an entry capsule to enable the material to be thoroughly analyzed. However, like HER, such sampling would do little to preserve the physical and chemical states of the material. An even more ambitious mission could land on the nucleus of a comet to collect and preserve a sample. A spacecraft could use a Mars slingshot to enter an orbit that would facilitate fast flybys of some main belt asteroids and close-in slow encounters with others. Another proposal was to configure the Mariner Mark II bus (for once powered by solar panels instead of RTGs) for the Planetary Observer-class Mars Geoscience/Climatology Orbiter mission. However, most Mariner Mark II missions would be devoted to the outer planets of the solar system, starting from Saturn. A Saturn orbiter could deliver a capsule to the atmosphere of either Saturn or Titan at the start of a 3-year orbital tour during which it would make numerous flybys of the

An early-1980s computer rendering of a Mariner Mark II orbiter for Saturn.

many satellites and map the hidden surface of Titan using a synthetic-aperture radar. Atmospheric probes based on that of the Galileo mission could also be delivered to Uranus and Neptune. There was no shortage of worthwhile missions!

The Mariner Mark II spacecraft would use a modular bus which incorporated as much heritage technology as possible. A central module would house the common electronics, and have standardized mechanical, electrical and electronic interfaces. Scientific instruments and other sensors would be carried externally, on booms and scan platforms. Stabilized imaging would facilitate the optical navigation required to refine encounters with comets and asteroids and to target the release of the Titan atmospheric probe. On the Saturn orbiter, a scan platform could carry the radar that would remain trained on Titan during each flyby of that moon. As far as possible, mission-specific electronics and scientific instruments would be dealt with in terms of 'black boxes' conforming to common interfaces and data formats. The launcher would be the Space Shuttle, with either a three-stage version of the solid-fuel IUS or the more powerful Centaur as an escape stage, in either case possibly augmented by an additional kick motor for missions that required a very high starting speed. It was expected that there would be four or five missions during the 1990s. Since most missions would require substantial propulsive capability to enter orbit around or to match speed with its target, the bus was to include a bipropellant engine. A study was conducted to determine whether a single standardized engine could be made to accommodate all of the requirements, or whether a different engine would be needed in each case. Atmospheric and Earth-return capsules would be mounted between the body of the spacecraft and the propellant tanks, so these would need to be jettisoned prior to releasing the capsule. Excluding propulsion, propellant and piggy-back probes, the Mariner Mark II was to be about 600 kg, of which 100 kg would be available to scientific instruments. Given the emphasis on remote sensing, the spacecraft would be 3-axis stabilized. In addition to the use of a standardized, reconfigurable bus, another key to major cost-savings would be the automation that would maximize the ability of the spacecraft to take care of itself, as opposed to a strategy of requiring constant monitoring and input from the ground. During the long cruise periods, the spacecraft would transmit a housekeeping signal to indicate whether it was content or was requesting intervention. Although Mariner Mark II was to exploit systems proven by the previous generation of deep-space missions, it was recognized that many technological advancements had been made during the hiatus in the planetary exploration program, and some of these were to be adopted – in particular modern electronics, communication systems, fiber-optic gyroscopes without moving parts, CCD star-cameras for attitude control, image compression algorithms and cutting operating costs by using the nascent Internet to disseminate data to participants.[330]

The Solar System Exploration Committee specifically recommended four new missions. One was a 'descoped' version of VOIR initially called the Venus Radar Mapper and later named Magellan. It would drastically reduce the cost of VOIR by using spare parts left over from Voyager and Galileo and by utilizing a series of hardware and software shortcuts to deliver most of the scientific results. Another mission was the Mars Geoscience/Climatology Orbiter. It was approved in 1984 as

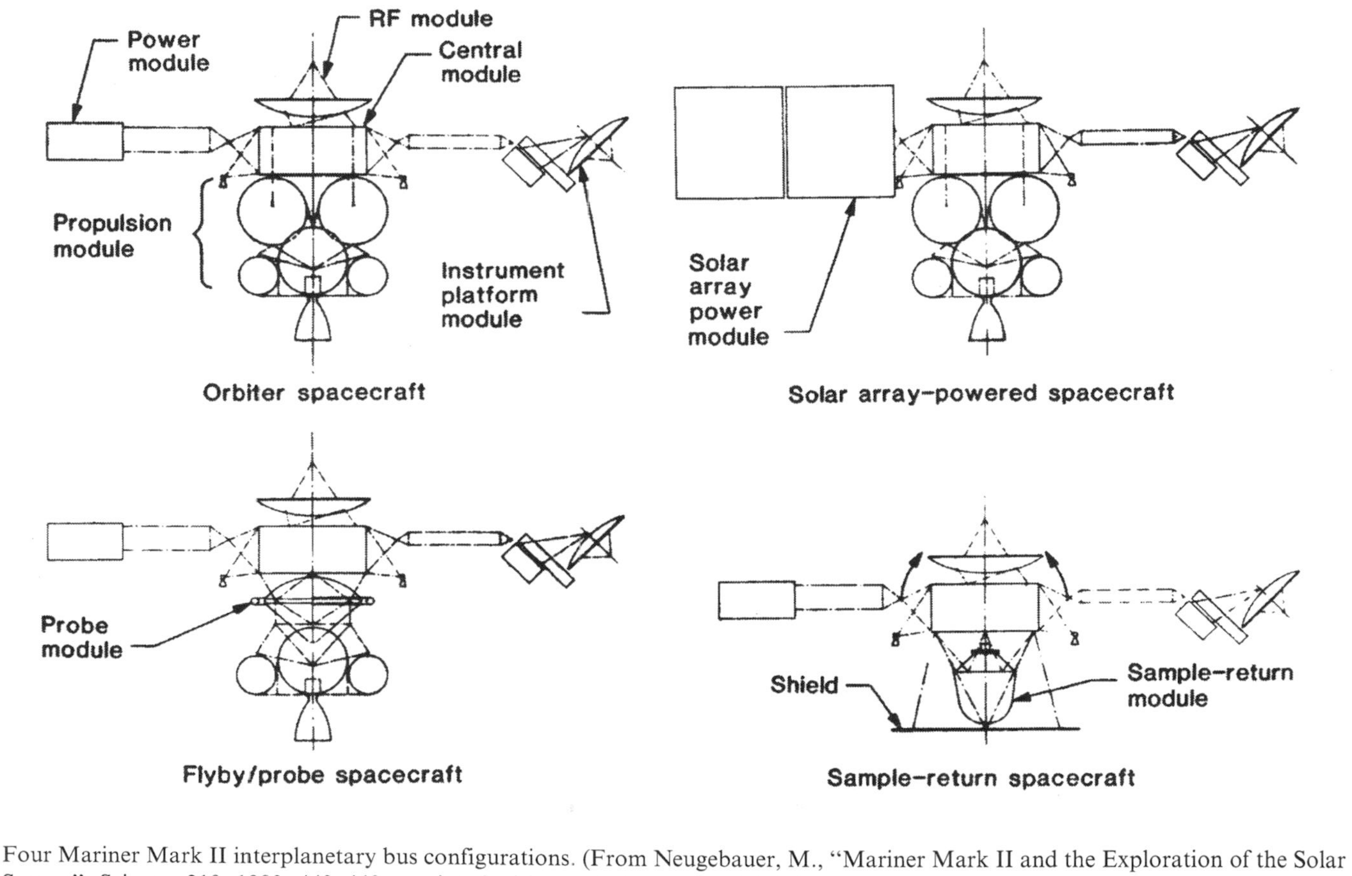

Four Mariner Mark II interplanetary bus configurations. (From Neugebauer, M., “Mariner Mark II and the Exploration of the Solar System”, Science, 219, 1983, 443–449; reprinted with permission of the AAAS)

the first of the Planetary Observer class, and later renamed Mars Observer. In the Mariner Mark II class were the Comet Rendezvous/Asteroid Flyby (CRAF) and an orbiter for Saturn, named in honor of the Italian–French astronomer Giovanni Domenico Cassini who devoted much of his life to observing the planet, its rings and many satellites, which would deliver a probe into Titan's atmosphere.[331,332]

In parallel with NASA's Solar System Exploration Committee study, in 1982 a Joint Working Group for US–European cooperation in planetary exploration was created under the auspices of the European Space Foundation and the US National Academy of Sciences. This recommended that a program of cooperative missions be conducted by the end of the century. Its highest priorities were a Saturn Orbiter and Titan Probe, a Multiple Asteroid Orbiter and a Mars rover.[333]

Unfortunately, even as the Mars Observer mission was being approved in 1984, Congress denied the Planetary Observer program the financial continuity that had been a key part of its rationale, and without start-up funding for the other missions the program collapsed. It was in any case doomed owing to its presumption that the Shuttle would fly frequently and cheaply. In response, the scientists resumed trying to load up the few spacecraft that were available with all the instruments that could be accommodated. In due course, Mars Observer would demonstrate the naivety of adapting Earth-orbiting satellites for use in deep-space. Nevertheless, after years of uncertainty there was now hope for America's exploration of the solar system, and in particular for JPL, which was now the only NASA center involved in planetary missions. As 1986 began, Voyager 2 was approaching Uranus on its Grand Tour, Galileo and Ulysses (the international out-of-ecliptic mission) were ready to be launched, the development of Magellan and Mars Observer was underway, and the international armada was closing in on comet Halley. The mood was so optimistic that 1986 was declared the "international year of space science". Unfortunately, on 28 January the Shuttle Challenger was lost, together with its seven astronauts, and with the remaining Shuttles grounded and 'expendable' rockets almost phased out the American space program in general – and planetary exploration in particular – was exposed to grave doubts. This prompted a complete reassessment of America's means of access to space. One early decision was that the Shuttle would not be able to serve as the National Space Transportation System. In particular, it must not be used to launch commercial satellites. The production lines for conventional rockets were restarted. If anything, this opened up possibilities for the planetary program. But the program was dealt a serious blow in mid-1986 by NASA's decision not to allow the Shuttle to carry the hydrogen-powered Centaur escape stage in its cargo bay. The Galileo, Ulysses and Mariner Mark II missions were all to have used the Centaur. In some cases these spacecraft were simply too heavy to be transferred to another launch vehicle. One way or another, they would require to be significantly redesigned, which would push up the overall cost of the missions. Yet funds would be particularly short because NASA would almost certainly raid its science budget in order to pay for the modifications to the Shuttle.

Meanwhile, the USSR, having scored a success with the Vegas at both Venus and Halley, announced that it was to resume missions to Mars with an ambitious program that would culminate in the return of a sample to Earth. Relations between the two Superpowers were finally beginning to thaw, in large part because the new Soviet leadership had abandoned the military build-up and paranoid confrontation of the Cold War and was increasingly devoting its attention to social and economic reforms. Exploiting this spirit of cooperation, Soviet and JPL engineers, managers and scientists established the first direct contacts for many years in order to discuss matters of mutual interest.[334]

COMET FRENZY

After the international armada of spacecraft provided exceptional and unexpected results at Halley which indirectly showed just how little we knew about comets, the various space agencies proposed a variety of new and more complex missions to comets. There were four major themes: (1) spacecraft which were to investigate a number of nuclei in order to establish a statistical database of their characteristics; (2) spacecraft to collect samples of dust and gas from the coma and return this to Earth; (3) spacecraft to rendezvous with a single comet and accompany it along its orbit to make a detailed study; and (4) spacecraft that would return to Earth with samples from the very surface (and interior) of the target nucleus.

The first type is represented by Vesta, which was a Soviet–French proposal that will be described in relation to asteroid missions (see below).

Cometary dust had been collected in the Earth's upper atmosphere and in Earth orbit. In the 1980s an interesting experiment to collect dust was undertaken jointly by French and Soviet scientists by installing extremely pure metallic targets on the exterior of Space Station Mir and exposing them to sample the most active meteor showers. Similar experiments were planned for flights by the Space Shuttle and for Space Station Freedom.[335] As related earlier, one of the objections against trying to return samples during a flyby of Halley in the style of a HER-like mission was that the material would be 'atomized' by the high-speed impact, with the result that the original chemistry would be lost. A sample-return mission should return a sample in a close approximation to its pristine state. Of course, the obvious advantage of a coma sample-return mission would be that the material would be analyzed not by onboard instruments of limited capabilities but in a laboratory using state of the art methods. And in contrast to complex sample-return missions involving landing on the nucleus, sampling the coma could be done with a less sophisticated (and hence cheaper) spacecraft. Interest in such a mission was revived by JPL researcher Peter Tsou, who showed in 1984 that 'underdense' materials like polymer foam or the exotic material aerogel, which is a silicon-based foam of an extremely low density, could halt a particle arriving at 10 km/s without imposing a thermal load on it that would modify its chemistry. This triggered a number of new studies of missions to return a coma sample.[336] On the basis of the work by Tsou, the US–European Joint Working Group devised an Atomized Sample Return mission in the style of HER to gain samples of dust from the coma of a comet by using either a metallic target or a low-density material that would capture the dust intact. The samples would be either delivered directly to the Earth or aerobraked into orbit for later recovery by a

Shuttle.[337] In 1984 a Giotto 2 mission was proposed as a joint NASA/ESA venture. ESA would provide the bus, which would use the basic structure as its predecessor but with a sample collection mechanism and entry capsule replacing the solid-fuel kick motor. The capsule would be based on the one developed by the US Air Force for its highly successful CORONA reconnaissance satellite. A camera would locate the comet early on and enable the spacecraft to refine its trajectory to make a flyby at a range of 80 km from the nucleus, where 'parent molecules' would be collected before they could be chemically modified by their environment. Several potential targets were identified for interception in the 1990s, but the agencies were so slow to consider the issue that ESA's 'comet interceptor' team dispersed, and in the end the mission did not advance beyond the study phase.[338]

Many of the scientists who supplied experiments for Giotto set out to build upon the Giotto 2 concept by proposing CAESAR, an acronym which, depending on the source, stood for Comet Atmosphere Encounter and Sample Return or for Comet Atmosphere and Earth Sample Return. The name referred to one of history's most famous comets – the one that appeared in July of 44 B.C. several months after the assassination of Julius Caesar. The spacecraft would present an array of deployable collectors of up to 5 m^2 in area to sample the gas and dust in the coma. As in the case of Giotto 2, the large particles would be brought to a halt by a soft medium such as polystyrene foam and remain relatively intact and unchanged. The smaller particles would be atomized on impact, and their remains recovered from the walls of the collector cells. Gases would be collected by chemically inert surfaces similar to the 'Swiss flag' of extremely pure aluminum foil deployed by Apollo astronauts on the lunar surface to trap solar wind ions then returned to Earth for analysis. No fewer than 38 sampling opportunities were identified in the 1990s. One option was for CAESAR to be launched by an Ariane 3 in December 1989 in order to sample comet 73P/Schwassmann–Wachmann 3 in May 1990. It would then return to Earth and release its capsule. In the initial design, this would use a retrorocket to enter orbit for later collection by a Shuttle, but as the design progressed the capsule was revised to make a direct atmospheric entry.[339,340]

Meanwhile, the success of the ICE mission prompted scientists of the Goddard Space Flight Center to devise a Multi-comet Sample Return mission which would make extensive use of Space Station Freedom. It would require a 770-kg spacecraft of the Planetary Observer class and a pair of 250-kg coma probes, which would be launched together by a Shuttle in 1992 but fly independently. They would observe the Sun during their interplanetary cruise. One of the probes would collect samples during a close flyby of the nucleus of comet d'Arrest in July 1995 and on returning to Earth would fire a retrorocket to enter an orbit from which it would be retrieved by an orbital tug that would take it to the space station. Meanwhile the other two spacecraft would fly by Honda–Mrkos–Pajdušáková, and the sample collected in its coma returned to Earth in a similar manner. The main spacecraft would then go on to encounter Giacobini–Zinner in 1998 and Tempel 2 in 1999. In the meantime, astronauts on the space station would equip the coma probes with clean collectors and then relaunch them to sample the latter two comets within a few weeks of the main spacecraft performing its encounters.[341] However, this ambitious idea (which was considered to be competitive with rather than complementary to JPL's CRAF) remained at the study stage. Undeterred, the Goddard team joined forces with their JPL counterparts and proposed the Comet Intercept and Sample Return as a Planetary Observer-class mission. As it had been recognized that a direct entry was the simplest and least expensive solution to returning a sample to Earth, this mission envisaged releasing a CORONA capsule. Comets Kopff, Honda–Mrkos–Pajdušáková and Giacobini–Zinner were all candidates for interception between 1995 and 1998.[342,343]

Some of the Americans who had worked on the HER and Comet Intercept and Sample Return proposals then teamed up with Japanese researchers from ISAS to propose the Sample of Comet Coma Earth Return (SOCCER) mission. This was to encounter one of the many Jupiter-family comets, the leading candidates including Finlay, Churyumov–Gerasimenko, Wirtanen, du Toit–Hartley, Kopff and Wild 2. The baseline scenario called for a launch in 2000, an encounter with 15P/Finlay in 2002 at a range of under 100 km (and possibly as small as 10 km in order to ensure that pristine 'parent molecules' would be collected) and a return to Earth precisely 4 years after it was launched. Finlay was discovered in 1896, and is a small comet in a 7-year orbit. Although its brightness dramatically decreased during the 20th century, as did its rates of gas and dust production, it was considered to be a good candidate for collecting cometary particles. On returning to Earth, SOCCER would fire its engine to enter an eccentric orbit that would be circularized by aerobraking, then be collected by a Shuttle. ISAS modified its MUSES-A (Hiten) engineering lunar spacecraft to test using a series of passes through the rarefied upper atmosphere to progressively lower the apogee of an orbit. To protect SOCCER during this phase of its mission, it would face a cone of refractory material in the forward direction. A Shuttle would detach the sample canister and abandon the spacecraft in orbit. As a rehearsal, ISAS planned for its Space Flyer Unit satellite to be launched in 1995 by a Japanese rocket and recovered by a Shuttle. SOCCER was seen from the start as a low-cost mission, and its appeal was further increased by the fact that even this would be shared. ISAS, exploiting Japan's experience in making digital imaging chips, was to provide the imager, various instruments and the bus. NASA would provide JPL's sample cell system and in-orbit retrieval. The spacecraft itself would be fairly mundane, consisting of an octagonal bus with the dust collection system on one end (and doubling as a dust shield during the encounter) and the high-gain antenna on the other end. Six solar panels would extend from the bus for the cruise, but be retracted during the comet encounter. A bipropellant engine would make various maneuvers, including course corrections during the cruise, two final comet targeting burns, a large deep-space maneuver, Earth-capture, aerobraking 'walk-in' and 'walk-out', and the final orbit circularization. The engine would therefore have to be rated for a major total change in velocity. Although SOCCER would be spin-stabilized most of the time, it would be able to adopt 3-axis stabilization when this was required. The CCD camera would be used both to obtain navigation images of the comet for precise targeting and for science – the latter including distant studies of the coma and close-in views of the nucleus at much better resolution and quality than had been possible in the case of Halley. The spacecraft would be launched by a

Japanese M-V (sometime also called the Mu-5; a rocket that was to replace the Mu-3SII and be capable of boosting about 500 kg to escape velocity). When it was realized that a start in 2000 would be impractical owing to constraints on ISAS's launch schedules, the launch was postponed to 2001. The plan was to make a flyby of 22P/Kopff in November 2002, with 81P/Wild 2 as a backup. Because it had also been selected for NASA's CRAF mission, Kopff was well observed. When it was realised that Kopff was beyond the scope of the M-V, the launch was switched to an American Delta.[344,345,346] The mission study continued until 1993, at which time ISAS opted instead for MUSES-C, which was intended to return a sample from a near-Earth asteroid.[347]

Meanwhile, building on the study for the unfunded International Comet Mission which had envisaged a Halley flyby and a Tempel 2 rendezvous, JPL had set out to recover some of the intended science with the Comet Rendezvous/Asteroid Flyby (CRAF; unofficially named Newton). This would require the spacecraft to orbit the nucleus of one of the Jupiter-family comets and deliver a penetrator hard-lander to its surface. By the end of 1983, a scientific working group had been established to define the objectives and nominal payload for this mission, which was to be the first of the Mariner Mark II program.

The original plan was for CRAF to be launched in 1990 and perform a flyby of asteroid (772) Tanete on its way to rendezvous with comet Kopff. When the start-up funding was denied, the launch was rescheduled to 1991, the flyby reassigned to asteroid (476) Hedwig and the rendezvous to comet Wild 2; but again funding was denied.[348,349,350,351] After a redesign, two options were available: the first setting off in September 1992 for 10P/Tempel 2, and the second in 1993 for 6P/d'Arrest. The Shuttle would deliver the spacecraft and its Centaur stage into low Earth orbit. In fact, a direct trajectory to Tempel 2 would be prohibitively expensive because the plane of the comet's orbit was inclined at 11 degrees to the ecliptic. The spacecraft was therefore to be inserted into a heliocentric orbit that would produce an Earth slingshot which would simultaneously increase the aphelion of CRAF's orbit and match its inclination to that of the comet. Meanwhile, in July 1985 NASA invited scientists to submit proposals to participate in the mission. The scientific objectives were to characterize the geology, morphology and composition of the nucleus and determine how it changed as a function of heliocentric distance; to investigate the chemistry of the coma; and to study the dynamics of the tail and its interaction with the solar wind. The trajectory to Tempel 2 would provide an opportunity to image and attempt to measure the mass of the main belt asteroid (45) Hestia. This was a reddish object about 214 km in size whose spectral characteristics resembled some carbonaceous meteorites. If the propellant margin allowed, a second flyby might be attempted, one candidate being 17-km-sized (1415) Malautra.[352] One hundred days before the rendezvous, the spacecraft's camera would start to search the sky for the nucleus. By this time, Tempel 2 would be at aphelion near the orbit of Jupiter, and, because its orbital period was only 5 years, about 1,000 days from perihelion. The uncertainties in the ephemeris for a nucleus at aphelion could easily add up to tens of thousands of kilometers, so it was vital to spot it in time to make the maneuver that would place the spacecraft 5,000 km sunward of the nucleus. As CRAF slowly

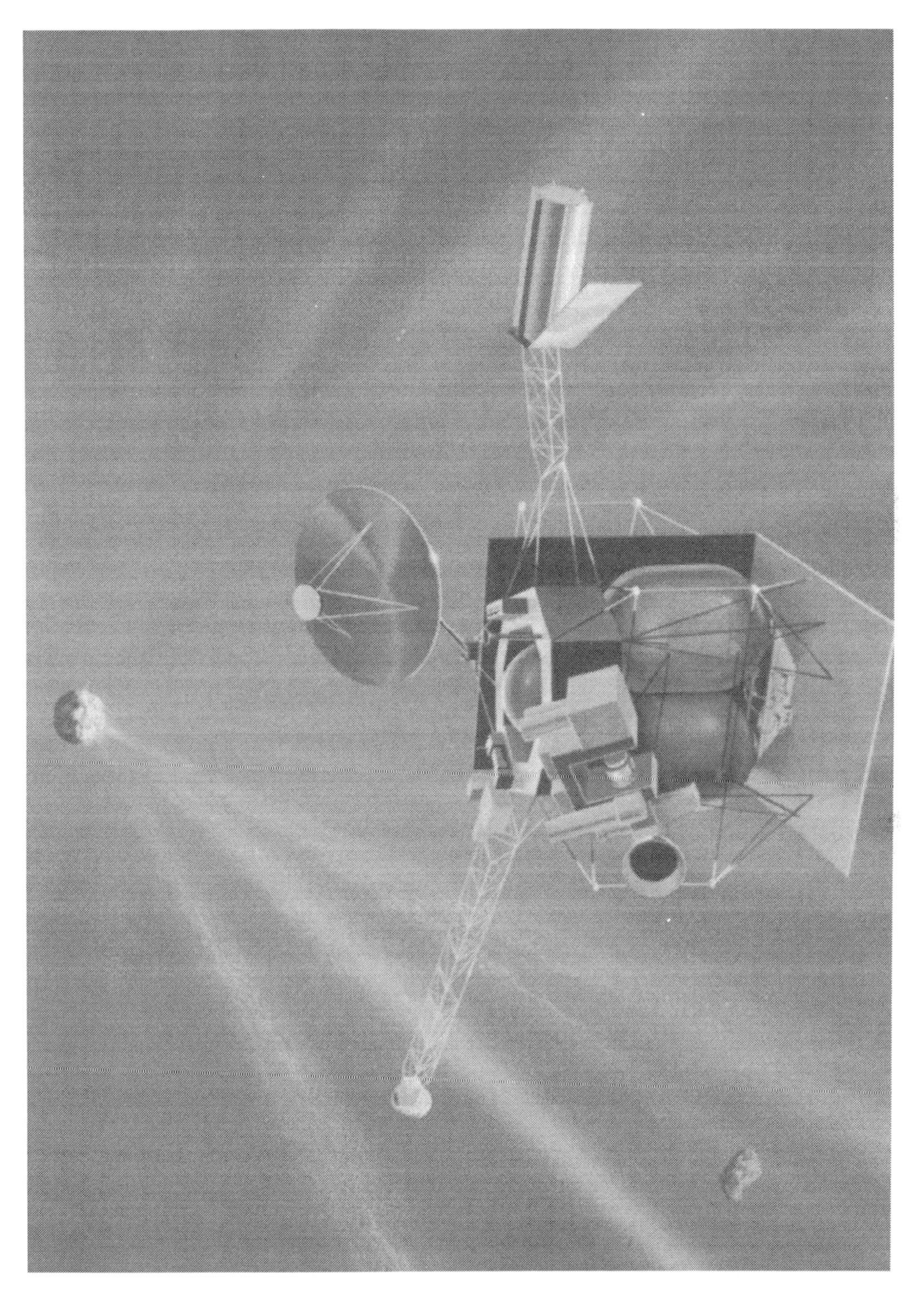

The initial concept of the CRAF spacecraft exploiting Voyager technology. (JPL/NASA/Caltech)

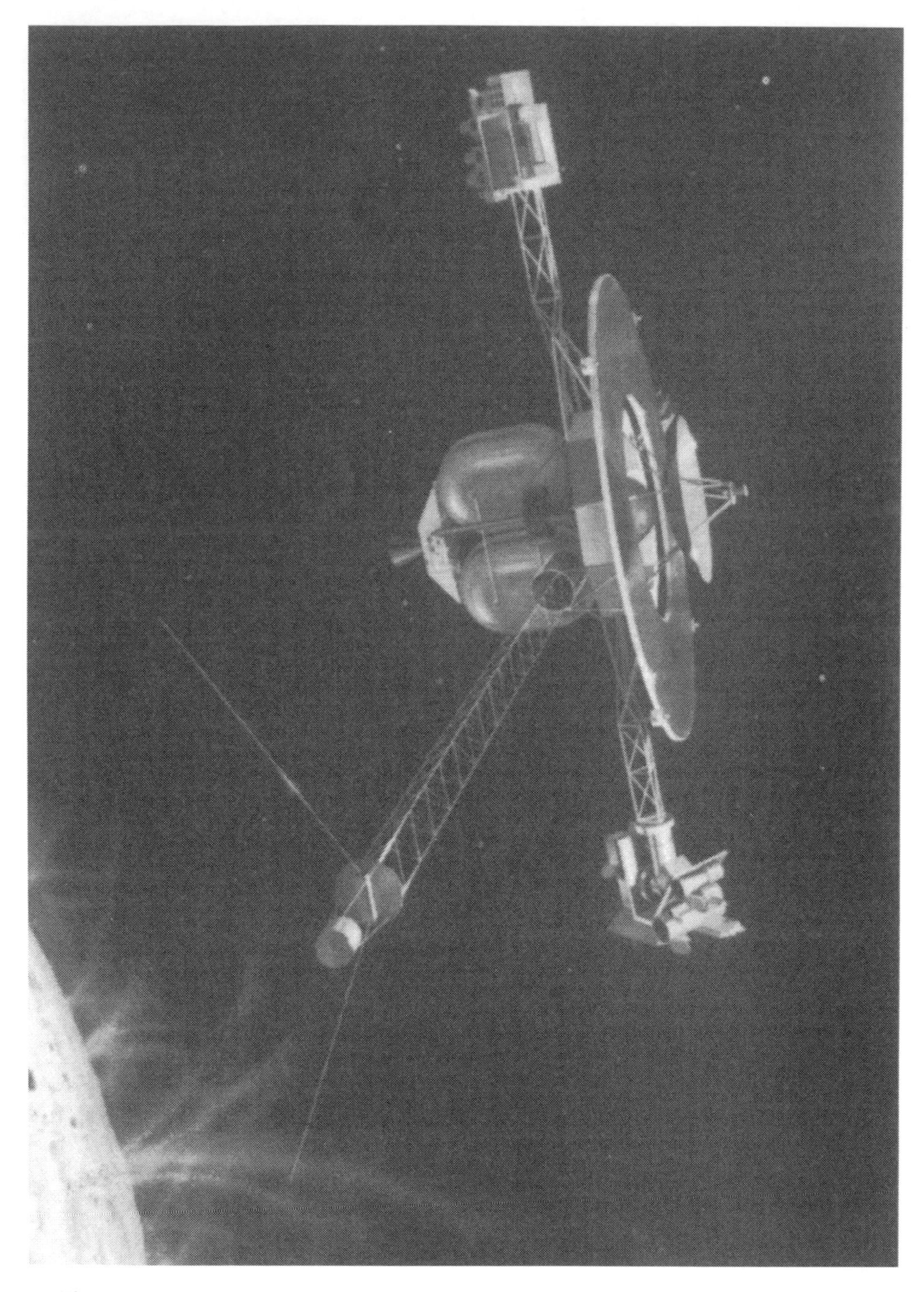

The Mariner Mark II version of the CRAF spacecraft as envisaged prior to the Challenger disaster. (JPL/NASA/Caltech)

The Mariner Mark II version of the CRAF spacecraft to be launched by a Titan IV, shown deploying the penetrator. (JPL/NASA/Caltech)

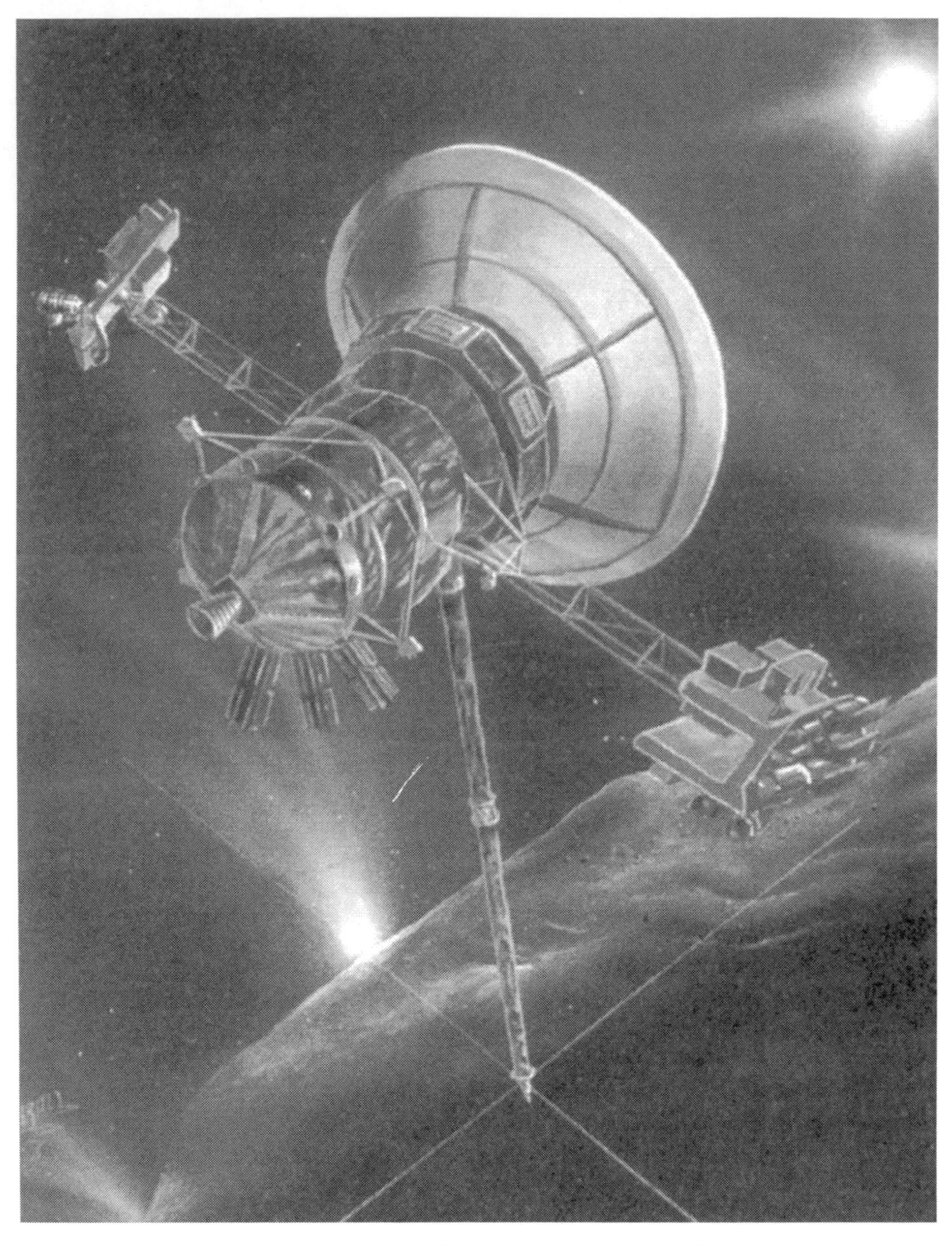

The final Mariner Mark II version of the CRAF spacecraft closely resembled how the Cassini Saturn orbiter was then intended to be built. (JPL/NASA/Caltech)

approached over the next 45 days, it would send the imagery and other data needed to plan the remainder of the mission. In addition to determining the size, shape and rotation of the nucleus, this early characterization phase would include a number of close passes to enable the mass of the nucleus to be estimated with some accuracy.

The ensuing 18 months would be spent orbiting the comet at altitudes as low as a few tens of kilometers, each circuit taking about a month. During this phase, the various instruments would fully document the state of the still-dormant nucleus. As the comet approached perihelion and started to develop a coma of gas and dust, the spacecraft would withdraw several thousand kilometers to safely observe the onset of activity on the nucleus and gain a sense of perspective of the coma, which would be studied at various wavelengths. If necessary, some of the instruments and other systems would close 'dustcaps' for protection. As the comet reached perihelion, which would be July 1999 in the case of Tempel 2, the spacecraft was to make a distant excursion into the tail to investigate its structure and the plasma phenomena within it. Although this would end the primary mission, there was the possibility of an extension in which the spacecraft would study how the activity subsided as the comet receded from perihelion, and determine the changes to the landscape of the nucleus. As the supply of propellant neared exhaustion, the spacecraft would very likely be steered to a slow-speed impact with the nucleus.

At the time of planning the CRAF mission, Mariner Mark II was to consist of a modified Voyager bus that would host most of the electronics and subsystems and support several booms to carry a magnetometer, RTGs and two scan platforms, one of which was for instruments that needed only low-accuracy pointing and the other for instruments requiring highly accurate pointing. In the case of CRAF, the power supply would be augmented by a circular solar panel which would take advantage of the fact that at certain times in the mission the Sun–comet–Earth angle would be quite narrow, meaning that when the high-gain antenna (itself inherited from the Viking orbiters) was pointed at Earth, the solar panel would be more or less facing the Sun. The spacecraft would have a dry mass of about 1,450 kg, and most of its structure would comprise a large propulsion module using a left-over bipropellant engine from Galileo and tankage for over 4 tonnes of monomethyl hydrazine fuel and nitrogen tetroxide oxidizer.

The wide-angle and narrow-angle CCD cameras were each to have a carousel of filters selected to enable the composition of the comet to be identified. The typical imaging resolution would be about 50 cm, but by allocating one carousel slot to a magnifying lens it would be possible to image small areas at a resolution of about 5 cm. It was also proposed that if the cameras could be operated continuously as the spacecraft sailed through the asteroid belt, they could obtain interesting data on the population of meter-sized objects there.[353] The other instruments were to include a visible/infrared spectrometer to map the composition of the surface of the nucleus in 320 different wavelengths, and an infrared radiometer with which to investigate its thermal structure. Another instrument would expose sample collection surfaces to the dust and then examine the grains using an electron microscope. A German dust analyzer of the type used by the Vegas and Giotto would be complemented by an instrument to determine the elemental chemistry of dust grains and the molecular

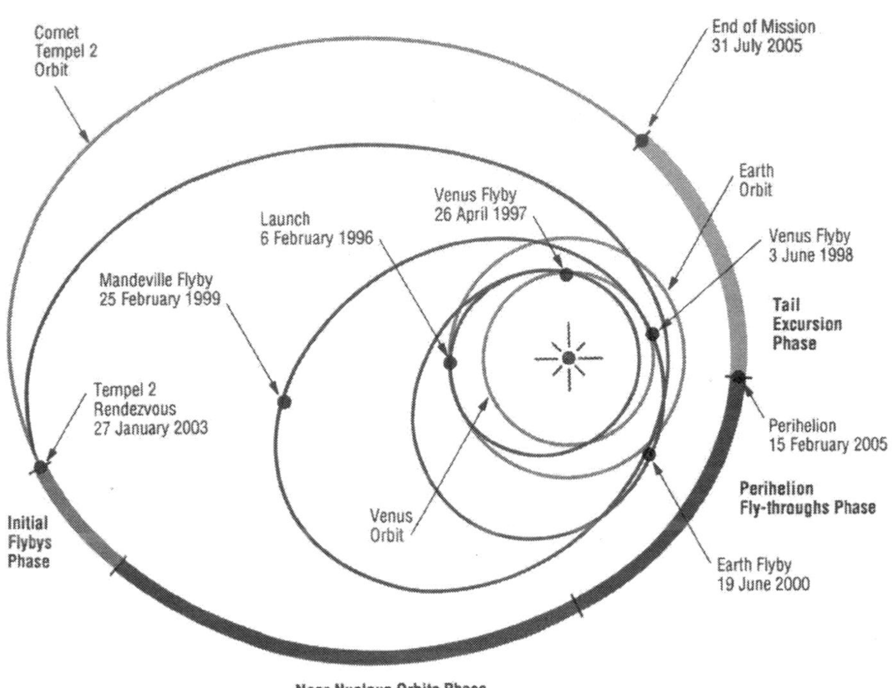

The trajectory of CRAF was to employ a Venus gravity-assist to reach Tempel 2, as intended just prior to the mission's cancellation in 1991. (JPL/NASA/Caltech)

composition of ice and gas. A dust sensor would monitor the flux of dust and the distribution of the masses, velocities and electric charges of the particles. It would also provide a timely hazard alert to close the covers on other instruments. Two ion spectrometers would measure the composition of neutral and ionized gases in the coma and ionosphere, while a plasma instrument measured how their fluxes varied with the activity of the nucleus. A magnetometer would characterize the ambient magnetic field, and a radio-science experiment would measure the electron density and temperature in the cometary 'atmosphere'. The CRAF results were expected to be excellent science.

The verification of the 'dirty snowball' model for Halley prompted modeling of how such an object might have developed. One result of this work, the 'primordial rubble pile' model, proposed that a cometary nucleus was a loose accumulation of smaller snowballs left over from the formation of the solar system which were held together primarily by mutual gravitation. This could easily explain why nuclei had a tendency to split, and it also predicted that some nuclei might be accompanied by detached fragments that remain gravitationally bound to the parent body. Another study was the 'icy glue' model, according to which a comet's nucleus was made of boulders of a porous and refractory material, held together by a matrix of dust and ice. The Giotto and Vega imagery showed that activity on Halley was confined to a small number of active regions, and it was argued that these marked where the icy matrix was exposed to the Sun.[354,355] An instrument developed by the University of Arizona's Lunar and Planetary Laboratory would enable CRAF to test the validity of these models. A spear-shaped penetrator was to impact a flat, relatively pristine-looking portion of the nucleus to determine the elemental chemistry of the material in situ using a gamma-ray spectrometer and to measure its thermal characteristics. Such data gained a particular importance after the infrared spectrometer of Vega 1 revealed that Halley's nucleus was considerably warmer than expected, suggesting there was a thin crust of black insulating material. In fact, the penetrator was to be able to apply heat to the ice and measure how this was diffused within the material. A protruding collector would ingest a small amount of material during the impact, and this would be analyzed by a calorimeter and a gas chromatograph. If possible, a camera would be carried to take pictures of the surface. Six accelerometers would monitor the impact to provide an indication of the surface structure. The penetrator would have a mass of 18 kg, be 1.18 meters in length and have a forebody 6 cm in diameter. The spacecraft would move to an altitude of just a few kilometers, spin up the penetrator for stability in flight and release it. The penetrator would use its own engine to ensure its emplacement in the ground. The engine was initially to be solid-fueled, but it was later changed to a liquid engine so that the duration of the burn could be tailored to the mass and density of the nucleus, as determined by the preliminary studies. Depending on the strength of the crust, it was expected to dig at least 30 cm into the underlying ice. Batteries would provide power for about a week. The spacecraft was to be able to accommodate two penetrators, with the second (if budgeted) serving initially as a backup, and if the first one succeeded the second would be sent to a riskier site such as a dust vent.[356,357]

When the Shuttle was declared 'operational' in 1982, the US Air Force was so concerned that the vehicle would never achieve the planned flight rate that in early 1984 the Department of Defense issued a Space Launch Strategy in which it called for the introduction of an expendable launcher capable of delivering a Shuttle-class payload to geostationary orbit.[358] It was decided to upgrade the successful Titan III as the Titan IV, and make it compatible with the version of the Centaur developed for the Shuttle. NASA's decision after the loss of Challenger not to use the Centaur meant that a number of deep-space missions had to be offloaded to the as-yet-unflown Titan IV–Centaur. To enable CRAF to fit inside the payload shroud of the new launcher, the long propulsion module was replaced by a squat unit which incorporated the engine and plumbing for a cluster of four tanks. Germany agreed to supply this system under a $75 million arrangement in exchange for the flight of a dust particle analyzer. Several other modifications were made to save mass. For example, the Voyager-era structural bus was replaced by the lighter one of Galileo. However, even after these modifications the spacecraft would be too heavy for the Titan IV–Centaur to dispatch it on the intended trajectory, and so the mission was revised to include an additional circuit of the Sun and a Venus slingshot to pick up energy. The flight would take longer, but it would still be able to rendezvous with Tempel 2 within days of the comet's aphelion. And as a precaution against NASA being banned from using plutonium-powered RTGs, a solar-powered version of the spacecraft using a large roll-out solar panel was also investigated.[359]

By 1988 CRAF and Cassini were sufficiently well advanced that NASA decided to combine them as a single budgetary item for 1990, the rationale being that the commonality of the spacecraft design, management and operations could reduce the overall cost by as much as $600 million in comparison to starting development separately. Finally, the two missions were financed! In this incarnation, CRAF was to set off in August 1995 and its target would be comet Kopff, which orbits the Sun every 6.4 years. The mission design now called for an Earth slingshot in mid-1997, a flyby of the 88-km main belt asteroid (449) Hamburga, and the rendezvous with Kopff in August 2000. There were even contacts between JPL and the McDonalds fast-food chain about the advertising opportunity of the encounter![360] In an effort to minimize the overall cost of CRAF and Cassini, the propulsion module was again redesigned, this time becoming a long cylindrical unit tailored to the requirements of CRAF (that for Cassini would be flown with its tanks only partially filled) and a common high-gain antenna was optimized to Cassini's requirements. Meanwhile, the launch slipped to 1996, and the new mission plan involved one Earth slingshot, two Venus slingshots and a flyby of 110-km asteroid (739) Mandeville on the way to a rendezvous with Tempel 2 in January 2003.[361] Not unreasonably, the scientists no longer cared which comet they studied, so long as the mission finally got off the ground!

When in 1991 the Congress drastically reduced the combined budget for CRAF and Cassini, both launches had to be postponed to 1997. In 1992 the White House said that as a result of the soaring costs of the 'low cost' Mariner Mark II (it had by now reached $1.85 billion) NASA must cancel one of the missions. It was decided to sacrifice CRAF. The German space agency, the only major international partner in the project, was content, as it too was anxious to cut its budget in the wake of the expensive reunification with its eastern sister.[362,363] The details of the decision were not publicized, but the Cassini mission probably benefited from heavy international involvement (ESA was to supply the atmospheric probe for Titan, and Italy part of the communication system). Another constraint on Cassini was that it could not be delayed any further, because it would require a Jupiter slingshot to reach Saturn.

The cancellation of CRAF also marked the end of the Mariner Mark II concept of 'low cost' deep-space missions. It had failed in every respect. In particular, the program had become so expensive that it would never have been able to fulfill the promise of mounting frequent missions to many targets. There were many reasons for this, not least the extensive redesign of the bus following the loss of Challenger. What emerged was the precise opposite of what had been intended: the 'flagship' concept, in which a single spacecraft would carry as many instruments as possible, and a schedule so sparse that the participants would devote much of their careers to a single mission.

Meanwhile, in 1984 a survey committee established by ESA delivered a report entitled 'Horizon 2000' recommending that four 'cornerstone' scientific missions be undertaken before the end of the century. One theme was asteroids and comets, and one option was to return to Earth a sample of cometary material believed to have remained unchanged since the formation of the solar system. Later that year ESA awarded Matra in France a contract to investigate such a mission, using either solar-electric or conventional propulsion.

NASA's Solar System Exploration Committee had also recommended a comet-sampling mission for the beginning of the new century. JPL was making a separate but parallel study of such a mission as a follow-on to its ambitious (and expensive) Mars Sample Return. One concept envisaged using solar-electric propulsion from the initial heliocentric orbit to make a slow-speed rendezvous as the comet neared perihelion, then the large solar panels used to power the electric engines would be retracted and the spacecraft would switch to chemical propulsion. It was not even considered necessary to land on the nucleus, because tethered drills could be reeled out and back as the spacecraft 'hovered' 100 meters above the surface.[364]

Ministers of the ESA member states met in Rome in January 1985 and endorsed the 'Horizon 2000' report. The fact that the Giotto mission was at that time being prepared for launch raised the profile of 'primitive bodies', and it was decided that one of the cornerstone missions should return a sample of a comet. But because it was clear that such a mission would be extremely expensive ($800 million at least) it was offered to NASA as a possible joint venture that would complement CRAF. In July 1986 a Comet Nucleus Sample Return workshop was held in Canterbury in England, and recommended establishing joint ESA/NASA teams on science and technology to investigate making the mission an international cooperative project. The ESA/NASA mission then became known as Rosetta after the 'Rosetta Stone', a tablet on which there were inscriptions which enabled the Egyptian hieroglyphic language to be deciphered. The analogy was the belief that a sample of cometary material would produce a comparable leap in understanding of the links between the interstellar medium and the origin of the solar system. Scheduled for launch in the early years of the new century, it was to sample one of the short-period comets of the Jupiter family. The initial candidate was 67P/Churyumov–Gerasimenko, but when a launch date in 2003 was chosen 73P/Schwassmann–Wachmann 3 became the target.

NASA would supply the Rosetta mission with a Mariner Mark II bus equipped for attitude control, navigation, communication and RTGs for power. The lander, sampling system and return capsule would be provided by ESA. The Deep Space Network would provide tracking and data retrieval. The Titan IV–Centaur launcher would put Rosetta into a heliocentric orbit that would return the spacecraft to Earth after 2 years, at which time a slingshot would stretch the eccentricity of the orbit sufficiently to reach its target – which at that time would be close to aphelion, out near the orbit of Jupiter. Rosetta would spend about 100 days in the vicinity of the comet. The distant approach phase would determine the main geometrical and dynamical characteristics of the nucleus: its size, spin rate, axial alignment etc. The ensuing orbital observation phase would assess the patterns of activity, in particular any out-gassing, to evaluate the risks of operating close to the nucleus. During this time the landing site would be selected. After a candidate had been chosen on the basis of its morphology, a radiometer would assess its texture and roughness. Then the data from all the instruments would be evaluated to determine whether the site would satisfy the scientific objectives. Prior to attempting to land, Rosetta was to

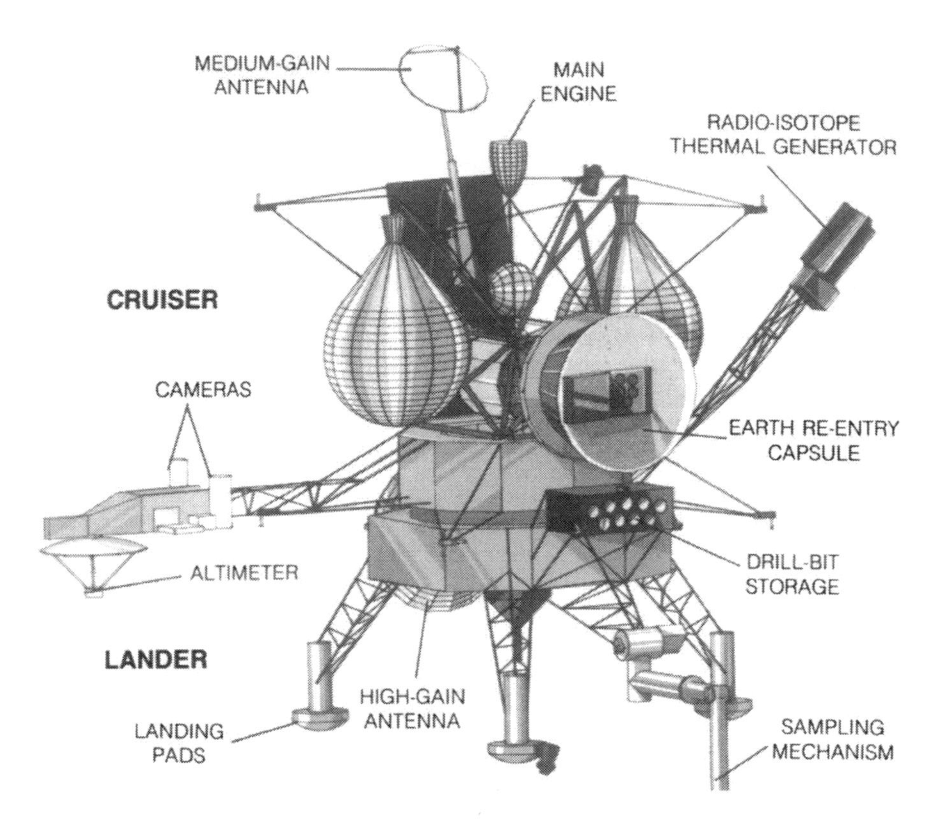

The joint ESA–NASA Comet Nucleus Sample Return spacecraft envisaged for the Rosetta mission. (ESA)

deploy a radio beacon to guide the descent. The lander would use a radar altimeter and a Doppler radar to control its approach. In view of the very weak gravitational attraction of the nucleus, on making contact with surface the lander would fire a harpoon-like device from each of its three foot pads to ensure that it would remain stable during the sampling operations. The sampling facility was to have a robotic manipulator with interchangeable 'end effectors'. After using a grasping tool to collect some surface samples, hopefully representing both volatile and non-volatile materials, it would switch to a coring drill to collect a stratigraphic sample as much as 3 meters deep. A drill that drew as little as 100 W of power and was capable of operating at –200°C was tested by the Italian firm Tecnospazio, one issue being to avoid heating the material.[365,366] After samples were placed in the low-temperature storage bay of the sample-return capsule, the NASA ascent stage would lift off and head for Earth. The ESA lander would probably be equipped with an autonomous power system to enable it to make in-situ observations as the comet approached its perihelion. Another possibility was to include some kind of 'sounder' with which to study the internal structure of the nucleus. As the return stage approached Earth it would jettison its RTG on a line that would pass by the planet, then release its capsule on an entry trajectory. In addition to providing a sample of the pristine material left over from the formation of the solar system some 4.6 billion years ago, it was hoped that the material might also contain dust and matter that *predated* the solar system. The distinctive elemental and isotopic composition of interstellar grains would yield insight into the history of nucleosynthesis in the galaxy. The complex carbon compounds expected to be found in a cometary sample would shed light on the processes that made complex chemistry out of simpler molecules. Scientists particularly wished to know whether comets might have 'seeded' planets with the building blocks of life.[367,368,369]

The cancellation of CRAF cast doubt on whether NASA would be able to give its support to Rosetta. In fact, ESA was facing financial difficulties, due in part to the cost of German reunification. Meanwhile, scientists in Europe were growing concerned that too much of the budget would be assigned to projects such as the Hermes spaceplane, which threatened ESA's science program in the same way that the development of the Space Shuttle had earlier ravaged NASA's program.[370] In response to all these pressures, a study was initiated of an ESA-only version of the Rosetta mission that eliminated the sample return activity (the most expensive part of the original concept to implement) and ended up looking remarkably similar to the CRAF. This time the project survived, and was brought to fruition at the start of the new century.

THE RISE OF THE VERMIN

The first asteroid was discovered on 1 January 1801. Others discoveries followed, and soon there were so many that astronomers dismissed them as "vermin of the sky". The possibility of sending spacecraft to asteroids (or 'minor planets' as they are more properly known) was discussed in the United States in the 1960s. In its 1966 *Planetary Flight Handbook* NASA even published trajectory data for flights to (1) Ceres and (4) Vesta, which are two of the largest.[371] Likewise, ESRO studied the possibility of a flyby while crossing the asteroid belt on the way to Jupiter.[372,373] Moreover, in reviewing NASA's planetary program the US National Academy of Sciences rated asteroids highly because they might provide insight into the earliest phases of planetary formation, and it was recommended that NASA start to plan an exploratory mission. One possibility was to send a spacecraft propelled by an ion engine and carrying 50 kg of scientific instruments on a flyby of (433) Eros, whose orbit brings it close to Earth; the journey would take a year.[374] Other studies in the 1970s included the possibility of landers based on Viking technology that would return samples of Eros or Vesta to Earth.[375] A preliminary study of the asteroid belt environment was conducted by Pioneers 10 and 11 in the early 1970s, on their way to Jupiter. A follow-up proposal was to send a solar-electric spacecraft carrying the 'Sisyphus' asteroid detector and the meteoroid puncture cells that were developed for the Pioneers, in order to more thoroughly survey the asteroid belt out as far as 3.5 AU from the Sun.[376]

Asteroid astronomy came of age in the 1970s and 1980s, as their importance in the study of the origin and evolution of the solar system grew increasingly evident. Part of this interest derived from the belief that, being so small, they could not have experienced significant thermal processing as a result of heat liberated by the decay of radioactive elements, and so should be primarily pristine material left over from the formation of the solar system. In 1975 a taxonomic scheme was introduced to classify asteroids spectroscopically by the (often vague) resemblance to meteoritic spectra. Asteroids of the S-class were taken to resemble stony meteorites, the dark C-class to carbon-rich (carbonaceous) chondrite meteorites and the M-class to iron-rich (metal) meteorites. In the ensuing years, more than a dozen other classes were added, with the V-class (which included Vesta, the brightest asteroid) resembling a rare class of volcanic meteorites – which in turn raised the intriguing prospect that at least some asteroids *had* been thermally processed! Then, in 1980 Luis Alvarez and his colleagues published the theory that the Cretaceous–Tertiary extinction of 65 million years ago, which marked the end of the age of dinosaurs, pterosaurs and large reptiles, was the result of a kilometer-sized asteroid impacting Earth. Also in the 1970s, the number of known 'near-Earth' asteroids dramatically increased, with new classifications being devised. The Aten asteroids, the prototype of which was discovered in 1976, all have their aphelia just beyond the Earth's orbit and orbital periods of less than a year.[377,378] For all of these reasons, and the fact that all of the major bodies in the solar system except Pluto had been (or would soon be) given at least a preliminary reconnaissance, the study of options to investigate the asteroids picked up in the early 1980s.

One of several new proposals submitted to ESA in 1979 was an asteroid mission called Asterex.[379] This was to be a 3-axis stabilized spacecraft with a bipropellant engine and two large solar panels. Its instrument suite, which was to be kept as simple as possible, had only a 'basic' camera, an infrared spectrometer and a radar altimeter. One mission plan called for launch by an Ariane 4 in April 1987, then a 3.5-year tour of the asteroid belt out to a heliocentric distance of 3.5 AU providing at least five encounters, one of which would be with Ceres, the largest.[380] European scientists recognized that it would be difficult to approach NASA to share the high cost of the mission because the lightweight payload would be difficult to divide satisfactorily between the two agencies.[381] The Asterex proposal was rejected by ESA, but revised and submitted 2 years later as AGORA (Asteroidal Gravity Optical and Radar Analysis). The baseline mission was to launch in 1990–1994, make at least two flybys of asteroids of differing sizes and taxonomic types at ranges of about 500 km and relative speeds of the order of 5 km/s, on the way to make a low-speed rendezvous with, and enter orbit around, a main belt asteroid having a diameter of at least 100 km – Vesta being the preferred option. During the encounters it would release 20-cm passive reflectors that would be tracked by the camera to determine the mass of the asteroid from the manner in which it attracted the reflectors. The mission would be launched by an Ariane 44L, the most powerful European rocket at the time. One option was to dispatch it on a trajectory to use a Mars slingshot to stretch its aphelion into the asteroid belt. The other possibility was to exploit the relatively small ion engines delivering a useful thrust that were finally becoming available in Europe and inject the spacecraft into an elliptical orbit with its

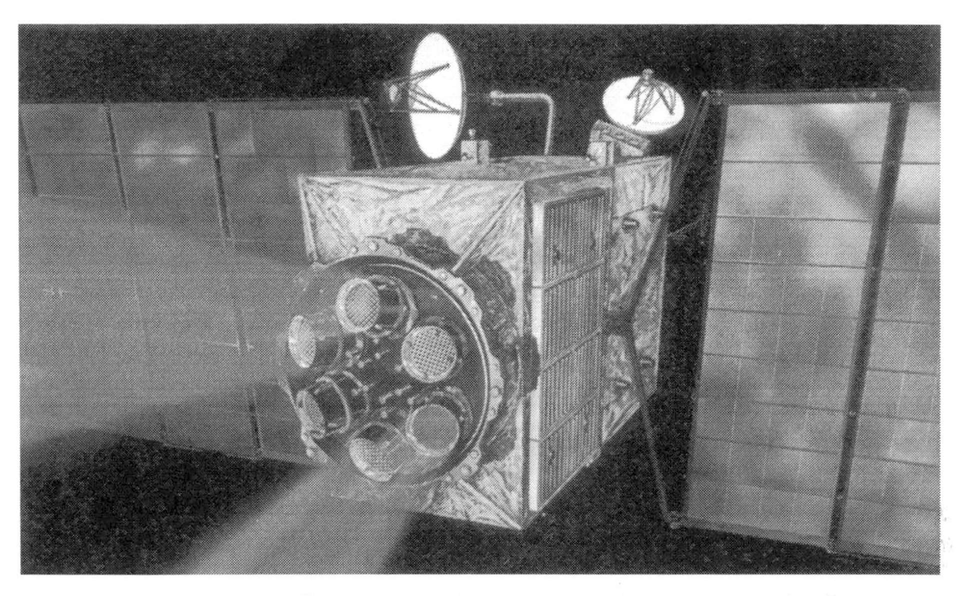

The ion-propelled European AGORA asteroid spacecraft. (ESA)

aphelion in the asteroid belt, then circularize this using the ion engine. In this case, however, very large (30-meter span) solar panels would be required, both because it would take about 4 kW to operate the engine at full thrust and because the energy of sunlight would be diminished in the region that the spacecraft was to undertake its most significant maneuvers.[382,383,384,385]

When AGORA was refused by ESA, the Joint Working Group on US–European cooperation in planetary exploration drew on the European experience in devising Asterex and AGORA and recommended the Multiple Asteroid Orbiter with Solar Electric Propulsion (MAOSEP). This called for a spacecraft with an ion engine that would visit asteroids representing the most common taxonomical classes (certainly C and S, and possibly also M, P and D). One mission scenario was to launch on an Ariane, make close flybys of four small asteroids and spend a few months in polar orbit around both Vesta and (17) Thetis. An alternative using a Shuttle–Centaur would orbit up to six asteroids. The time spent orbiting an asteroid would start with global reconnaissance in a high-altitude orbit whose period was several times longer than that of the asteroid's rotation, and would be followed by a detailed study in a low-altitude orbit.[386] But in early 1985 NASA said that it was not interested in an asteroid mission, and ESA initiated a technological study of the viability of mounting a 'test flight' of a European ion engine on a spacecraft which would perform a simplified asteroid rendezvous.[387,388]

National asteroid missions were also proposed in Europe in the 1980s. The West German branch of AMSAT, the Radio Amateur Satellite Corporation, proposed a 300-kg spacecraft that would be launched by an Ariane 4 and use an ion drive to visit small bodies within a heliocentric distance of 3 AU. The West German space agency made a feasibility study, and the Max Planck Institute of Lindau carried out an assessment of a possible payload. Like the satellites built by radio amateurs, this low-cost spacecraft would use 'home made' technologies, and its main instruments would be a camera left over from Giotto and an infrared spectrometer.[389,390]

The Italian CNUCE (Centro Nazionale Universitario di Calcolo Elettronico; National University Center for Electronic Computation) started a feasibility study in 1983 of the ECAM (Earth-Crossing Asteroid Mission) as a national mission to be launched by the Space Shuttle using a combination of the IRIS (Italian Research Interim Stage) and MAGE 1SB solid-fuel upper stages (the latter the same as used on Giotto). After the Italian scientific community and space industry showed some interest, the proposal was renamed Piazzi in honor of Giuseppe Piazzi, the Italian astronomer who was the first to discover an asteroid. After the loss of the Shuttle Challenger, the mission was redesigned to use an expendable launcher such as the European Ariane 4 or the American Atlas II, again using an IRIS as the kick stage. The baseline minimum payload was for four instruments: a multispectral camera, a reflectance spectrometer, an infrared radiometer and a radar altimeter. However, other instruments were considered, including even the possibility of a penetrator! Experience with the SIRIO experimental communication satellite would be applied in the design of the spin-stabilized spacecraft, whose launch mass would be in the range 400–900 kg. No fewer than 28 launch opportunities between 1996 and 2005 were identified targeting 11 Amor and one Apollo type of asteroid, many of them providing multiple opportunities. The Aten asteroids were out of reach because the intercept trajectories were too energetic for the spacecraft.[391,392,393,394,395] Backed by Aeritalia (the largest Italian aerospace firm), several astronomers, universities and national scientific institutes, it was put to the newly formed ASI (Agenzia Spaziale Italiana; Italian Space Agency), which gave it serious consideration but eventually refused funding – possibly owing to opposition from some in the Italian scientific community who desired to eliminate a source of potential competition for their existing programs.[396]

In parallel with investigating the possibility of mounting an asteroid mission in collaboration with the United States, ESA started a project called Vesta with CNES in France and the Soviet Interkosmos agency. The mission was first revealed by the Soviets in 1985, when they began to speak of missions that could use slingshots of Venus or Mars to make encounters with minor planets and comets during several loops of the Sun – in some cases as many as 20 encounters![397] A detailed design for the mission included two different spacecraft, one built by the French and the other by the Soviets, and they would be launched in pairs by a Proton and sent towards Venus.

On arriving at Venus the Soviet spacecraft would release atmospheric capsules, as the Vegas did, and then use a slingshot to adopt a trajectory that would intercept some near-Earth asteroids. The French spacecraft would use the Venus slingshot to return to Earth for another slingshot that would draw the aphelion out to the main asteroid belt. Several launch windows in 1991 and 1992 were identified that would facilitate flybys of the periodic comets 22P/Kopff and 78P/Gehrels as well as many asteroids on the way to the asteroid Vesta (hence the name of the mission). Whilst the Soviets would be able to use a modified version of either the old Venera bus or

Italy's small Piazzi spacecraft approaching an asteroid. (Courtesy of Luciano Anselmo)

the new UMVL bus, the French spacecraft would be a completely new design with two solar panels spanning a total of 20 meters, a high-gain antenna and a platform for scientific instruments and a radar altimeter. The instruments included cameras and an infrared radiometer. If the total mass of 870 kg permitted it, a Soviet-built penetrator would also be carried.[398,399,400] When the French (facing a funding crisis) asked for the mission to be delayed to 1994, the Soviets exploited this to insert it into their new program of planetary exploration, which was to shift the focus from Venus, where it had had tremendous success, back to Mars, where it had had little success. The redesigned Proton stack now included the French spacecraft, a Soviet

The Vesta spacecraft considered as a joint Soviet–European mission. (ESA)

module carrying two penetrators, a Soviet Mars descent module, and the UMVL flyby bus. The French mission plan had to be revised to allow for the fact that there would be no Venus flyby. Instead, it would use multiple slingshots of Mars to visit at least three small asteroids in addition to Vesta.[401,402] In the redesign, the mass of the French spacecraft was increased to 1,500 kg, and the 500-kg Soviet penetrator module was added. It would pass targets at distances ranging from 500 to 2,000 km and at relative speeds of 2 to 15 km/s. On approaching one target, it would spin up the penetrator module in order to stabilize it and release it. This would then use its own propulsion system to slow its speed relative to the target, locate it by imaging, and release two penetrators that would hit at locations some 10 to 20 km apart and use a 4-kg instrument suite to report the chemical composition, the magnetic and thermal characteristics, and the seismic environment.[403,404] The revised mission was put to ESA in 1988, by which time its launch date had slipped to 1996. But it was up against a gamma-ray observatory, an ultraviolet telescope, a space-based radio-telescope and participation in the Cassini mission – and the decision was to supply the atmospheric probe that Cassini was to deliver to Titan.[405] In the mid-1990s the Russians revived the concept as Mars–Aster, which was to use a Mars slingshot to intercept some comets, whose dust would be sampled by a new mass spectrometer, and asteroids, the largest of which would be investigated using penetrators. It was

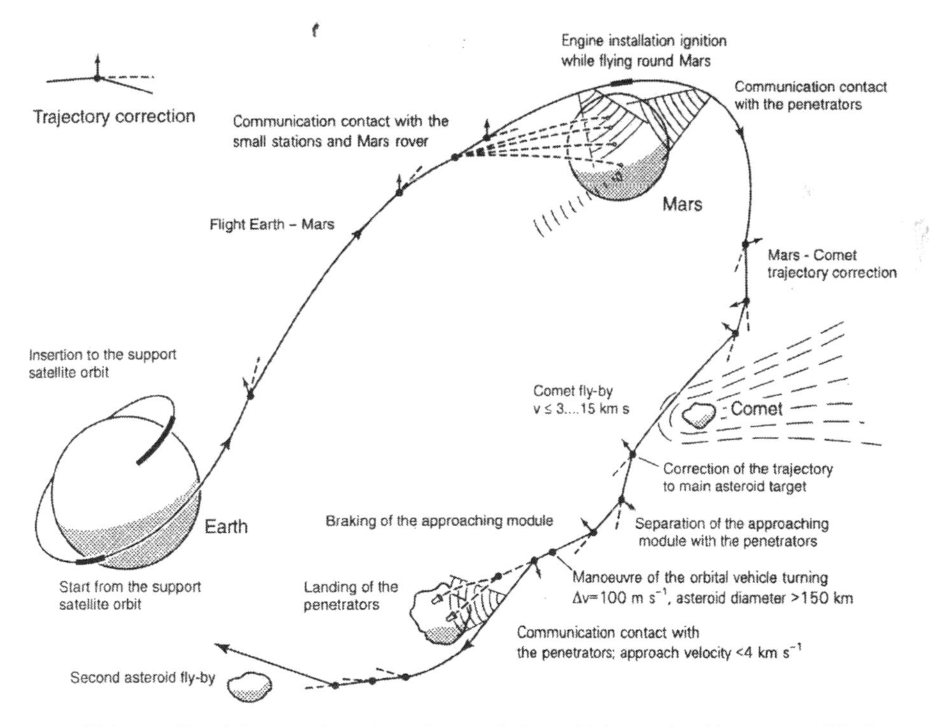

The flight profile of the Russian Mars–Aster mission which remained in a state of limbo during most of the 1990s. (Reprinted from Surkov, Yu. A., "Exploration of Terrestrial Planets from Spacecraft", Chichester, Wiley–Praxis, 1997)

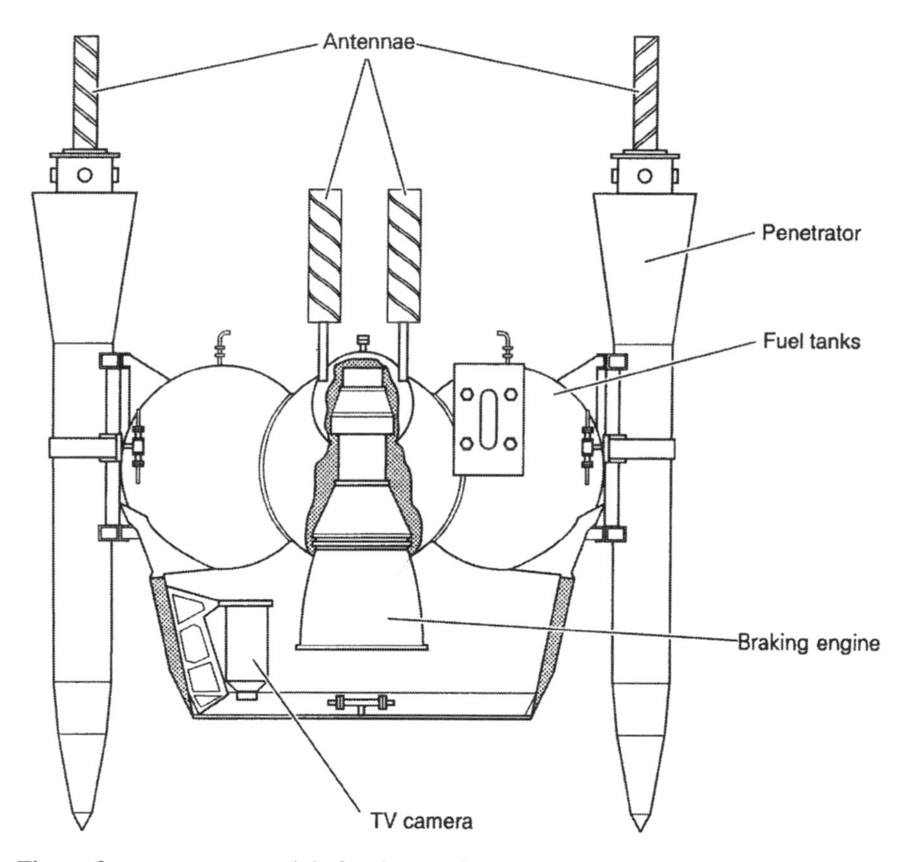

The surface penetrator module for the Russian Mars–Aster mission. (Reprinted from Surkov, Yu. A., "Exploration of Terrestrial Planets from Spacecraft", Chichester, Wiley–Praxis, 1997)

intended that this mission would test targeting techniques for interceptors designed to protect Earth from hazardous asteroids.[406,407]

There were also extensive studies of asteroid missions in the United States. For example, multiple asteroid reconnaissance missions were investigated using either ballistic trajectories (with and without planetary slingshots) or ion propulsion, with trajectories being identified in the latter case for low-speed rendezvous encounters with more than half-a-dozen asteroids. Furthermore, asteroid flybys were included in the architecture of planetary missions to the outer solar system, such as Galileo, Cassini and CRAF, and in 1983 the Solar System Exploration Committee proposed that NASA consider developing the Near-Earth Asteroid Rendezvous (NEAR) as a dedicated asteroid mission in the Planetary Observer class. After launch on a Space Shuttle, an IUS would send the spacecraft to a near-Earth asteroid, where it would maneuver to match orbits with its target. Depending on the mass of the asteroid, the spacecraft could either enter orbit around it or, if the gravitational attraction was too weak, fly alongside it and maneuver to perform repeated approaches. The shape, mass, topography and composition of the asteroid would be determined by a four-instrument baseline payload that included a CCD camera, X-ray and gamma-ray spectrometers, and a multispectral infrared mapper. In all likelihood, it would also have an altimeter and a magnetometer. A number of targets were identified among the Apollo and Amor-type asteroids, including Eros, the second largest near-Earth asteroid. As studies of the mission were being carried out, astronomers discovered a small object, initially designated 1982DB and then named (4660) Nereus, that proved to be the easiest known solar system object to reach beyond the Moon, and thus the one from which it would be energetically easiest to return a sample to Earth. In the end, the baseline mission settled on a 1994 Shuttle launch with the IUS replaced by a Transfer Orbit Stage, and asteroid (3361) Orpheus as the target. But then the Planetary Observer program was canceled soon after its first mission was authorized (this being Mars Observer) and the prospective NEAR mission was left without a spacecraft.[408,409,410,411] Fortunately, the proposal was kept alive ...

AN ARROW TO THE SUN

As will be related later, in the early 1970s ESRO (later ESA) interested NASA in a joint mission to observe the Sun at high latitudes. This would involve maneuvering a spacecraft into an orbit whose plane was steeply inclined to the Sun's equator (or, which is almost the same, to the plane of the ecliptic). Although ESRO's internal studies initially favored using an ion drive to gradually increase the inclination of the initial heliocentric orbit, when the project was adopted as a joint European–US program it was decided instead to achieve this by using a polar slingshot of Jupiter. At this point in its development the mission became associated with one which had been under discussion since 1958, when the Simpson Committee of the National Academy of Sciences, in recommending a scientific program for the newly formed NASA, had called for a spacecraft to penetrate the solar corona. Although this was not done, scientific interest remained high.[412] With the approval of the out-of-ecliptic mission, and a wave of enthusiasm for audacious deep-space missions, the idea of a small-perihelion mission resurfaced in 1975. The Italian space dynamics expert Giuseppe Colombo ('father' of Giotto and many other scientific missions) was a leading proponent of this Solar Probe. Although one of the originators of the idea of using Jupiter to 'set up' an out-of-ecliptic mission, he was a formidable, if isolated, adversary of the proposal – arguing that it should not be flown unless it carried the Solar Probe, which, as a result of the Jupiter flyby, would be robbed of almost all its heliocentric orbital velocity and literally 'fall' towards the Sun. In fact, his ideal was for this probe to dive straight into the Sun, as what he referred to as a 'solar plunger' or, more poetically, as an arrow to the Sun. As an alternative, a 'solar grazer' with its perihelion at about 4 solar radii was also considered. When ESA scientists showed an interest, a preliminary feasibility study for a small Solar Probe was prepared. It would require to be powered by an RTG since, paradoxically, although it would approach very close to the Sun, it would not be able to use solar panels because the cells would become too hot to operate efficiently. The fact that an RTG was needed meant that NASA would have to be invited to participate.

The scientific objectives of the Solar Probe were: to measure the distribution, density, velocity and temperature of solar plasma; to measure the magnetic field, which was thought to contain most of the plasma's energy; to detect plasma waves and energetic particles; to detect 'heavy' ions; and to explore the region in which solar plasma was heated and accelerated to supersonic speed in order to create the solar wind, the precise origin of which was still a mystery. A coronal light detector (similar in concept to the photopolarimeter used by the Pioneer missions at Jupiter and Saturn) could build up 3-dimensional presentations of the distribution of light, temperature and density within the corona. A primary objective, however, would be to measure some poorly determined gravitational and relativistic parameters, for which a Sun-grazing orbit would be particularly suited. As could be expected, ESA paid particular attention to the thermal design of the spacecraft, which proved to be a less difficult issue than initially feared. In fact, a theoretical analysis showed that a two-stage circular heat shield of aluminized graphite, and careful design of the lattice to connect it to the spacecraft, could ensure that the temperature of the main body did not exceed 40°C at a heliocentric distance of 0.02 AU (at that time, the well-insulated front shield would be 2,400°C). If approved, the Solar Probe could be launched in 1982, flyby Jupiter in July 1983 and reach perihelion in June 1985. Even if it were not traveling on a 'solar plunger' trajectory, it was not expected to survive beyond its first perihelion owing to thermal and configuration constraints which obliged the RTG to be jettisoned on the inbound leg. Despite the optimistic feasibility study, ESA's top managers were unconvinced that it would be proper to combine the out-of-ecliptic mission with the Solar Probe, primarily since to do so would delay the former, whose development was well advanced, and therefore the European Solar Probe was put aside.[413]

Thanks to Colombo's extensive connections with American scientists (it was he who suggested to JPL that Mariner 10 be put into an orbit which would enable it to return to Mercury periodically) JPL made an in-house preliminary assessment of such a mission, confirming that a simple lightweight thermal shield would keep the vulnerable systems cool. Because ESA had adopted a two-stage planar heat shield, JPL investigated solutions such as 'roof', inclined ellipse and conical shields. The scientific objectives of the RTG-powered JPL spacecraft would be similar to those of the European mission, but given JPL's expertise with imaging systems it would have a camera to study the structure and behavior of the corona, photosphere and chromosphere at high resolution. Its proponents stressed that an in-situ exploration of the near-solar environment would provide a leap in understanding comparable to that which resulted from the Mariner 4 flyby of Mars in 1965.

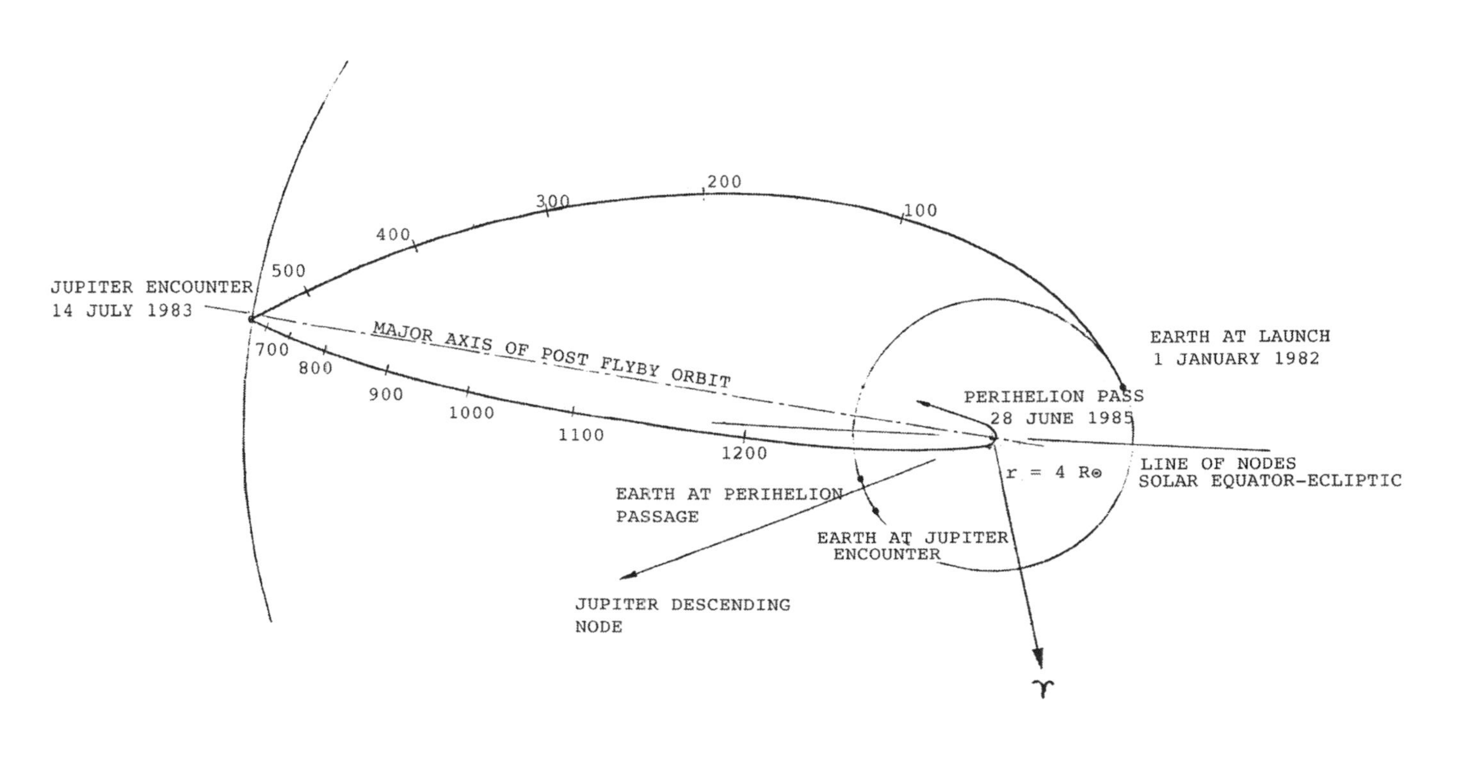

The trajectory of ESA's close-perihelion Solar Probe, including the orbit-shaping flyby of Jupiter. The US and Soviet probes would fly very similar orbits. (ESA)

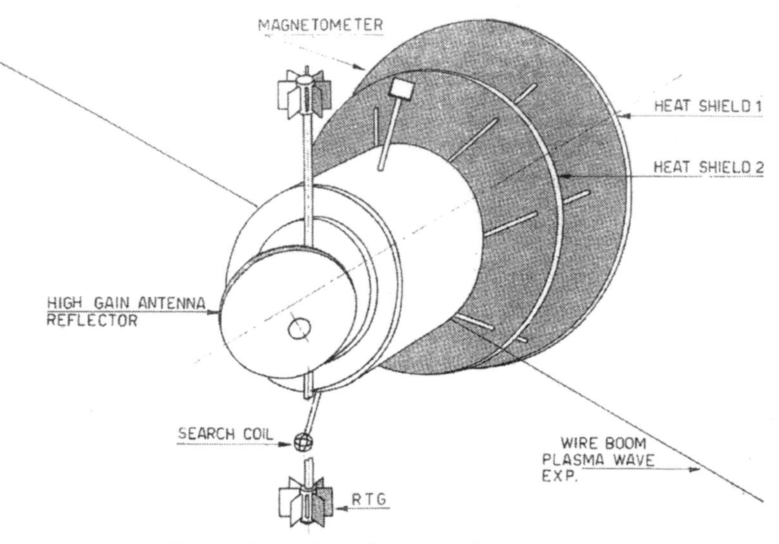

The configuration of ESA's Solar Probe. (ESA)

In recognition of its high scientific value the JPL study was adopted by NASA's Office of Space Science and given the dramatic name Starprobe. If it had flown, it would probably have been renamed in honor of Colombo, who died in 1984.[414] The 1,200-kg spacecraft was dominated by the large conical carbon–carbon heat shield whose shadow protected the RTGs, secondary heat shields, antennas, instruments, subsystems, etc. In close to perihelion, at most a few wires or appendages would be exposed to the full thermal stress. A 1-meter steerable antenna would be constantly pointed at Earth to return some 200 images of the Sun in real-time at perihelion. At that time, however, interference from the corona and the presence of the Sun in the field of view of the terrestrial antennas would limit the data rate to one-tenth of that early in the mission. Most of the data would have to have been recorded on board and replayed after the spacecraft had moved clear of the solar disk as viewed from Earth. Such a mission would be impossible to launch directly from Earth, owing to the energy to cancel the inherited heliocentric velocity motion. "In essence", said an article with Colombo as one of its authors, "such trajectories require [...] to stop the world while we get off". So, like the European Sun Probe, JPL's Starprobe would use a Jovian slingshot to reach the Sun. A Space Shuttle would deploy it and a Centaur stage would dispatch it towards Jupiter. The Jovian flyby would deflect the spacecraft back towards the Sun and twist the plane of its orbit perpendicular to the solar equator. The dramatic perihelion passage would see the spacecraft sweep from pole to pole in less than 14 hours![415,416,417,418] The development continued, but in 1982 the scientific objectives were revised with the gravity and relativity studies being deleted, leaving the mission dedicated to in-situ measurements of the solar corona.[419] After the loss of the Shuttle Challenger in 1986, the Starprobe was again redesigned to ride the Titan IV–Centaur, and in the process the imaging objectives were deleted. Despite its importance being reasserted many times and scientific boards rating it as

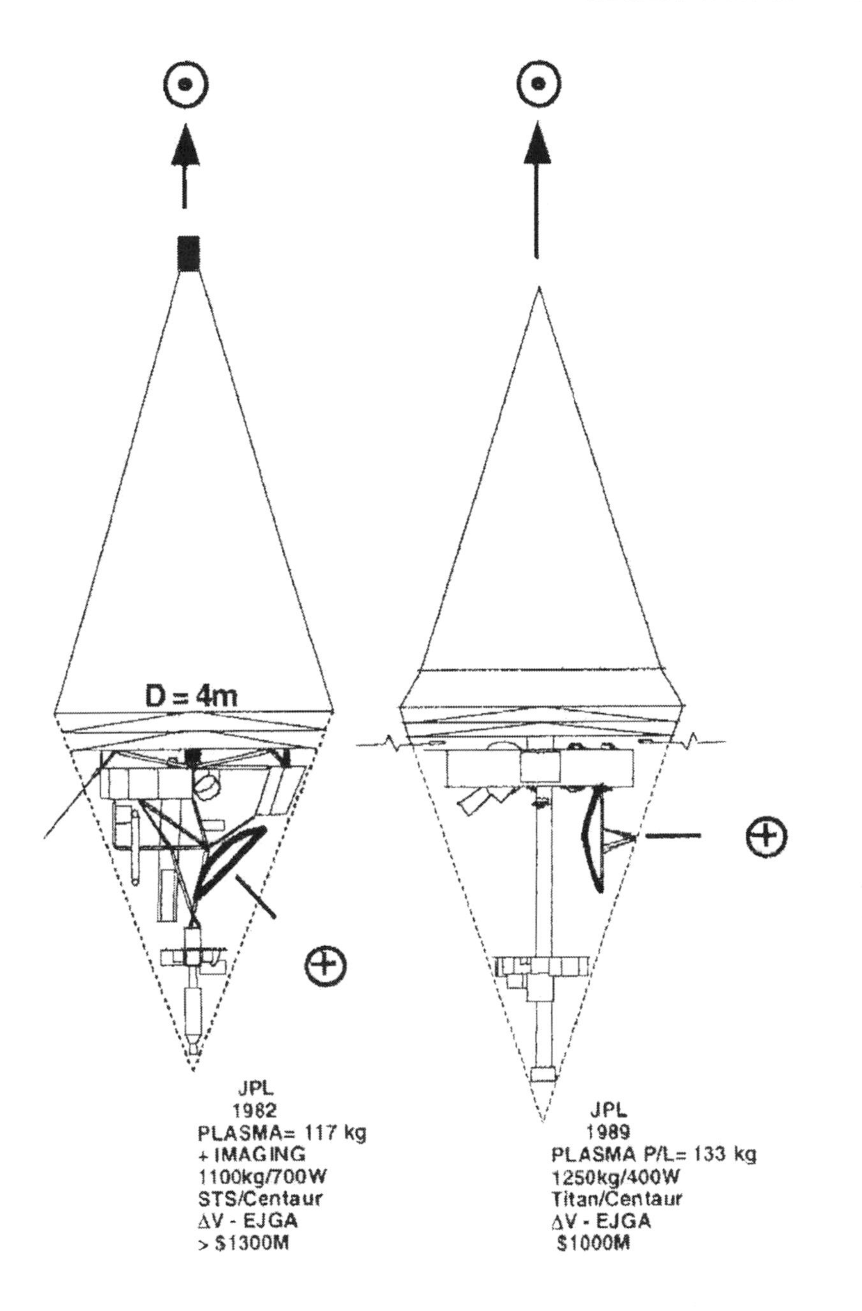

Two configurations of JPL's Starprobe close-perihelion spacecraft, as the design evolved during the 1980s. Note the large conical sunward thermal shield, and the high-gain antenna pointing Earthward.

one of the top priority missions, the small-perihelion solar probe idea languished for the remainder of the 1980s and was never flown – in part because its projected cost had risen to exceed $1 billion.

In the 1980s the Soviets also started their first serious studies of missions to the giant planets and to the vicinity of the Sun. The idea was that the missions should provide a leap in knowledge not only with respect to NASA's Pioneer and Voyager reconnaissance, but also the follow-up work of the Galileo and out-of-ecliptic missions. Soviet scientists and engineers identified three baseline mission profiles to address these requirements. In the first profile, a spacecraft would fly within 100 km of Jupiter's volcanic moon Io and then use a slingshot of the planet to increase the inclination of its heliocentric orbit and send it back to pass within 3 or 4 million km of the Sun after a flight lasting just over 3 years. However, it would be difficult to make observations of Io because the relative velocity of the encounter would be greater than 40 km/s, and at such close range this would produce an angular rate exceeding the very demanding one degree per second which the Vegas had faced in attempting to track the nucleus of Halley at their closest approach. An alternative was to limit the mission to the Jovian system and land a capsule on Io. Another possibility was to use Jupiter to reach Saturn, where the spacecraft would enter orbit around the planet and deliver a Titan probe. If this could be done ahead of the joint US–European Cassini mission, then all the better. However, the most thoroughly studied option was the YuS (Yupiter–Solntsye; Jupiter–Sun) mission. It received the support of the Academy of Sciences; and many institutions, including some of the leading technical and scientific universities, agreed to participate.

The YuS spacecraft would weigh 1,200 kg at launch and would incorporate the UMVL propulsion unit, a trajectory control module and a mission module. In fact, the trajectory control module was essentially the spacecraft's bus. It was to be built around a 320-cm-diameter parabolic antenna designed to communicate at distances in excess of 6 billion km from Earth. As the principal structural element, it would support tanks and (depending on the mission) up to six RTGs. Beneath the antenna was a conical equipment section, with attachments for the mission module. In the case of YuS, this would be a 460-kg spin-stabilized Solnechnii Zond (solar probe) shaped like a 'flying saucer' and powered by a chemical battery. At its center was a thermally insulated spherical module housing most of the systems and electronics. Communications would be through a parabolic antenna mounted flush against the anti-Sun side of the saucer. An alternative design was for a conical mission module with 60 kg of scientific instruments. The trajectory module would carry most of the instruments for the Jupiter flyby, and would be discarded about 10 days prior to the solar encounter, the active part of which would last only 16 hours and involve both directly sampling the corona and making high-resolution observations in the X-ray, ultraviolet and visible ranges. For non-solar missions, the trajectory module would carry landers, atmospheric probes etc.

The YuS project was named Tsiolkovskii after the early-20th-century Russian astronautics pioneer. On the schedule announced in 1987, the first mission, named Corona, was to be launched in 1995. At an international meeting about the solar–terrestrial relationship, it was suggested that Corona and Starprobe should be flown

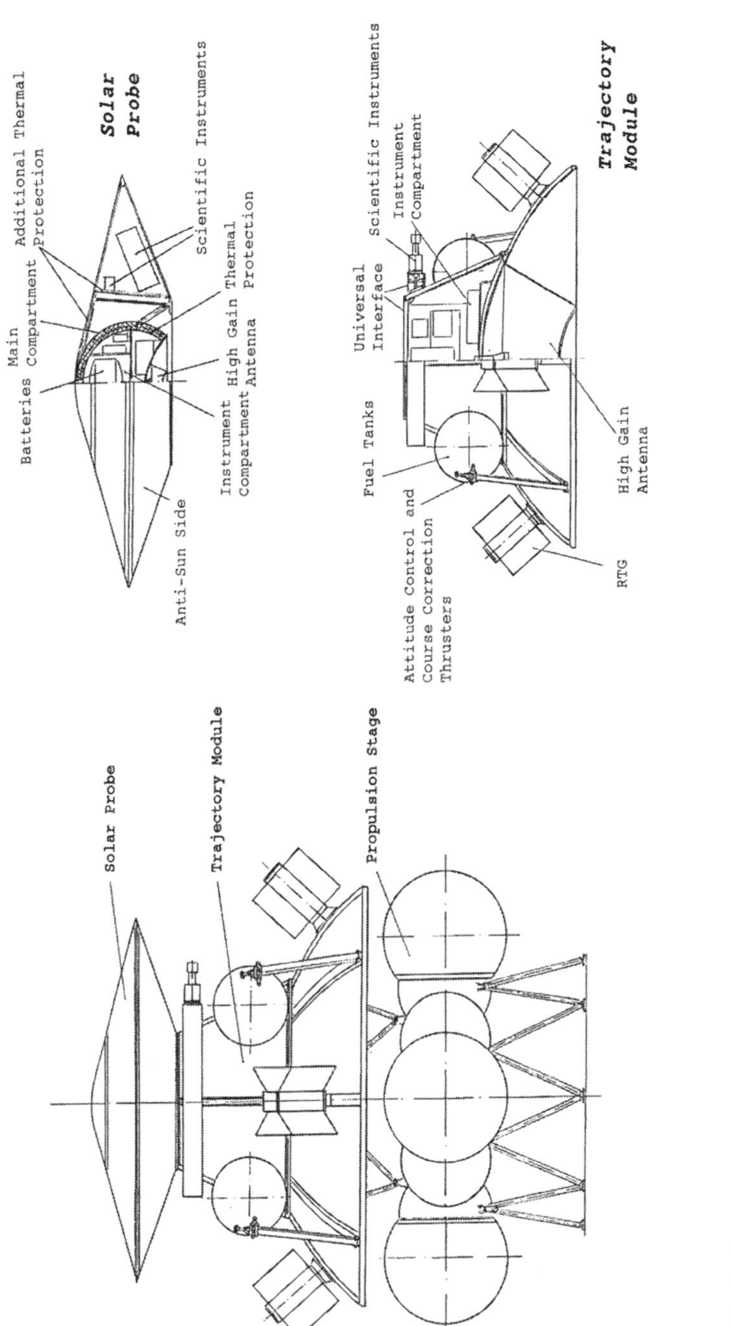

The Soviet YuS spacecraft comprised a propulsion unit, a trajectory module and a mission module. The envisaged missions included dropping a probe into Jupiter's atmosphere, and employing a Jovian gravity-assist to go on to Saturn with a probe for Titan or to enter an orbit with a small perihelion to release a solar probe.

at the same time in order to make complementary studies. In 1999 missions would be launched that would release probes into Jupiter's atmosphere and use that planet to reach Saturn carrying a Titan probe. The entry capsule for Jupiter would weigh 500 kg, and be able to endure a 1,500-g deceleration. The Lavochkin bureau built a special centrifuge to test it. One concept for the Titan probe was to use a balloon as the decelerator, and that after this had released the surface package it would rise to an altitude of about 10 km and report conditions for several days.[420,421,422,423,424,425] Unfortunately, owing to the financial crisis that followed the collapse of the Soviet Union the Tsiolkovskii project never left the drawing board, but Russian engineers and scientists continued to study advanced missions, in particular close-perihelion solar probes.

INTO THE INFINITE

Remarkably, in the 1980s JPL was studying even more amazing missions than the solar-grazing Starprobe. It had begun to investigate the possibility of an interstellar mission in the 1960s, and in August 1976 hosted a symposium entitled 'Missions Beyond the Solar System'. Although it was readily acknowledged that it was (and indeed still is) far beyond our capability to send a spacecraft to another star, it was deemed worthwhile to send a spacecraft to investigate the environment beyond the solar system. In November 1976, JPL started a small in-house study to draw up the scientific objectives and evaluate the requirements and mission architectures of this Interstellar Precursor Mission. The scientific goals included in-situ sampling of the heliopause (the boundary of the heliosphere beyond which the solar wind cannot reach) and characterizing the interstellar medium, in particular its composition and magnetic field, in order to put constraints on a variety of phenomena, including the Big Bang model of the universe. Because the solar wind modulates galactic cosmic rays and at some energies even prevents them from penetrating the heliosphere, once the spacecraft was beyond the heliopause it would be able to report their true fluxes. Another intriguing prospect was to play the data of the departing probe 'in reverse' in order to assess the scientific observations that a spacecraft might make on approaching another star. The Interstellar Precursor might carry a telescope to exploit a baseline of several hundred AU (i.e. between Earth and the spacecraft) to accurately determine the parallaxes (and hence distances) of many stars. This could also observe distant galactic and extragalactic objects. A secondary objective of the mission might be a reconnaissance of Pluto, if this had not already been done. Incidentally, in the period of interest for the mission (the early 2000s), Pluto would be crossing the solar apex, the direction in which the Sun is traveling and therefore the direction in which the heliopause could be expected to be both nearest and best defined.

A number of drive technologies were proposed for the Interstellar Precursor (the aim being to reach the heliopause as soon as possible) including both feasible but unproven ones like solar sails and powered flybys, and ones out of science fiction, including laser-sails, fusion motors and antimatter drives. A fairly conventional but untested nuclear-electric propulsion (NEP) system was selected. This would use a small fission reactor to power a cluster of ion engines which would run for years to reach a speed of 50 to 100 km/s. Although the US tested a fission reactor in space in the 1960s, this was a one-off. However, the Soviet Union has powered its ocean-surveillance radar satellites using reactors, so the technology existed. Nevertheless, despite many years in development, ion engines were at that time still regarded as experimental. The NEP booster would account for most of the 90-tonne mass. The spacecraft would probably have to be assembled in orbit, with the segments ferried up by Space Shuttle flights. A paper describing the mission optimistically observed that "1977 figures for [Space Shuttle] launch capabilities will be only of historical interest by 2000". It is certainly true that the 1977 projections now seem naive, but almost certainly not in the sense meant by the author! As the Interstellar Precursor approached solar system escape speed, heading towards Pluto, it would release a 1,500-kg piggyback craft based on Galileo that would use a large propulsive stage to slow down to enter orbit around Pluto. The Interstellar Precursor would have a 40-W transmitter and a 15-meter-diameter parabolic antenna that would unfurl on the spacecraft's axis (where the reactor, engines, tanks for mercury fuel, radiators and other systems would also be mounted) equipped with an annular feed to direct transmissions around the stream of high-energy particles from the engines. Such a communication system would be able to provide a significant data rate at up to 500 AU, a distance that would be reached within 50 years. Two booms would carry the platforms for instruments and telescopes and an auxiliary shorter-range antenna.[426]

The Interstellar Precursor would be the fastest object ever dispatched into space, but it would be limited to the local interstellar medium. Studies at JPL in the 1980s showed that advanced fission reactors and electric propulsion systems could reach a nearby star with a flight of several centuries; but whilst theoretically possible this is still far beyond present day technology.[427,428]

In the 1980s JPL's director Lew Allen made the Interstellar Precursor mission a pet program and gave it some of his discretionary funds in an effort to reinvigorate space exploration and advance our reach beyond the solar system. With respect to the earlier study, technologies were finally becoming available. In fact, in 1983 the US Department of Energy, Department of Defense and NASA began to develop a 100-kW nuclear reactor for use in space – although mainly for use by the Strategic Defense Initiative ('Star Wars') effort. This SP-100 reactor was to be followed by a multi-megawatt reactor. JPL studied using this technology for the TAU (Thousand Astronomical Units) mission which would reach a heliocentric range of 1,000 AU by a 50-year flight in a straight line. The spacecraft would be a 100-meter truss that would be assembled in Earth orbit, with an SP-100 reactor and an ion propulsion module on one end and a communication bus and two separate scientific spacecraft on the other end. After a decade of thrusting, during which it would first attain and then exceed solar escape speed, it would release the two spacecraft. One of these spacecraft would have an astronomy payload whose objectives would include the parallaxes of bright stars in our galaxy and the Magellanic Clouds (which are small companion galaxies) and a range of radio-astronomy studies. The other spacecraft would be spin-stabilized to investigate the magnetic fields, particles, dust, gas and

JPL's Interstellar Precursor spacecraft flying in formation with the Galileo-based Pluto orbiter. The nuclear reactor and ion engines of the Interstellar Precursor are on the opposite end of the body from to the antenna and instrument booms. The scale can be inferred from the fact that the umbrella antenna is 15 meters in diameter. (JPL/NASA/Caltech)

plasma of the heliopause and the interstellar medium beyond. Of course, although JPL established a 'TAU thinkshop', the mission was never anything more than a fascinating paper study. As regards stellar parallaxes, there was no need to set up a baseline of 100 AU or more, as significant progress could be made simply by using the diameter of the Earth's orbit around the Sun, and in 1989 ESA launched the HIPPARCOS satellite, which measured the parallaxes of more than 120,000 stars. This will be followed up in the 2010s by Gaia. Moreover, a pathfinder interstellar mission has since been flown by the Voyagers.

A number of other (more realistic) deep-space missions using NEP technology were also proposed, including orbiters for the outer planets; in particular a Saturn orbiter that could use its engines to 'hover' above the plane of the ring system in order to conduct a detailed study of their composition, structure and dynamics. But despite a vast amount of money being spent on it, the development of the SP-100 made little progress, and was canceled (together with the multi-megawatt version) in the early 1990s, by which time the collapse of the Soviet Union had prompted the US military to revise its priorities in space, and the United States and Russia began a joint project to refine the Russian Topaz-2/Yenisey reactor. Although the Ballistic Missile Defense Organization (the successor to the Strategic Defense Initiative) proposed testing the system by a flight to Saturn, NASA expressed no interest. In effect, the Topaz-2 is a solution awaiting a suitable problem.[429,430,431,432]

EUROPE TRIES HARDER

The Europeans started to formulate proposals for deep-space missions in the early-1960s, but none were undertaken. One reason was that until the introduction of the Ariane launcher in 1979, Europe had to use American rockets. Another reason was that space scientists in Europe were primarily interested in astrophysics, the fields and particles in space and solar physics, and so had little interest in exploring the solar system. This focus is illustrated by the fact that although the out-of-ecliptic mission (which was the first deep-space mission that Europe funded) used a Jovian slingshot, it was not really a planetary mission since its main focus was particles, cosmic rays and solar physics.[433]

One field in which European scientists excelled was cometary astronomy, which explains why missions to the comets figured so prominently in the early proposals, and why the Giotto and the Comet Nucleus Sample Return studies were so readily accepted. But despite Giotto's success, ESA was reluctant to pursue solar system exploration. All proposals for asteroid missions (another field in which Europeans excelled) were rejected, as were planetary missions. In the late 1970s ESA studied a small Polar Orbiting Lunar Observatory as a modest debut into the field of true planetary exploration, and immediately recognized that it could form the basis of a standardized, moderately priced planetary orbiter. Whereas in the 1960s there were few European planetary scientists, by now a new generation had grown up having worked in secondary roles on American deep-space missions, and as a result there was widespread interest in a planetary mission. A number of Mars missions were proposed, some as joint ventures with NASA, which would supply and operate the orbiter while the Europeans provided a surface rover.[434,435]

In 1981 ESA issued a call for mission proposals to the European space science community, and chose four of the responses for further study. Magellan (not to be confused with NASA's radar-mapper for Venus) was a far-ultraviolet astronomy mission. DISCO was a solar observatory. Asterex was an asteroid flyby mission. Another planetary mission was Kepler, named in honor of the early-17th-century astronomer who inferred the laws of planetary motion from the path of Mars in the sky. The plan was to enter an orbit around the planet with a low periapsis in order: to survey the gravity field and learn about the internal structure of the planet; to study the global temperature field and composition of its atmosphere; to address the still-open issue of whether the planet possessed a magnetic field; and to study how the solar wind interacted with the planet. Moreover, Kepler offered the option of cooperation with NASA, with two spacecraft orbiting the planet simultaneously (the American contribution being Mars Observer). The addition of a direct radio link between them would facilitate radio-occultation measurements of the atmosphere over a wider range of latitudes and local times than would be feasible with a single spacecraft.[436] In the words of the ESA Solar System Working Group, Kepler was "an opportunity for European scientists to face planetological problems by means of a relatively low cost mission", in that "the objectives of Kepler have not been covered by the previous US and USSR missions to Mars". One problem that had hitherto impeded acceptance of a European deep-space mission was that scientists in Europe had not established a common agenda which complemented rather than overlapped the ambitions of the two major players. If approved, Kepler would be launched in July 1988 by an Ariane 3 (or an Ariane 2 with an added escape stage) and would enter orbit around Mars in January 1989. The baseline spacecraft was a 2.8-meter-diameter 3.3-meter-tall spin-stabilized drum. More than half of the 800-kg 'wet' mass would be propellant for the orbit-insertion maneuver. The scientific payload was 46 kg.[437,438] The total cost was estimated at 188 million 'accounting units' (an artificial European currency used by ESA prior to the introduction of the Euro), or approximately $500 million.

Kepler, DISCO and Magellan were recommended for further studies, with the prospect of one being pursued as a scientific mission. Later, two other candidates were added. These were the Infrared Space Observatory (ISO) and the X-80 X-ray astronomy mission. ISO obtained the general support of the scientific community and was chosen in 1983. Kepler was left in limbo; then put in competition with the Cluster magnetospheric satellite constellation and SOHO (Solar and Heliospheric Observatory, a descendant of DISCO), both of which were more costly; and finally killed in 1985 (by which time its launch had slipped to 1990). A member of ESA's Space Science Advisory Committee lamented on this occasion that "there had been a failure on the side of ESRO not to go into the planetary field. The same thing [has happened] with Kepler". Also under study at ESA at the time was a mission involving multiple Venus orbiters, called Venture.[439]

The planet Mercury had long been of interest to European planetary scientists. A flyby of Mercury had been proposed to ESRO in 1969, and an orbiter was under

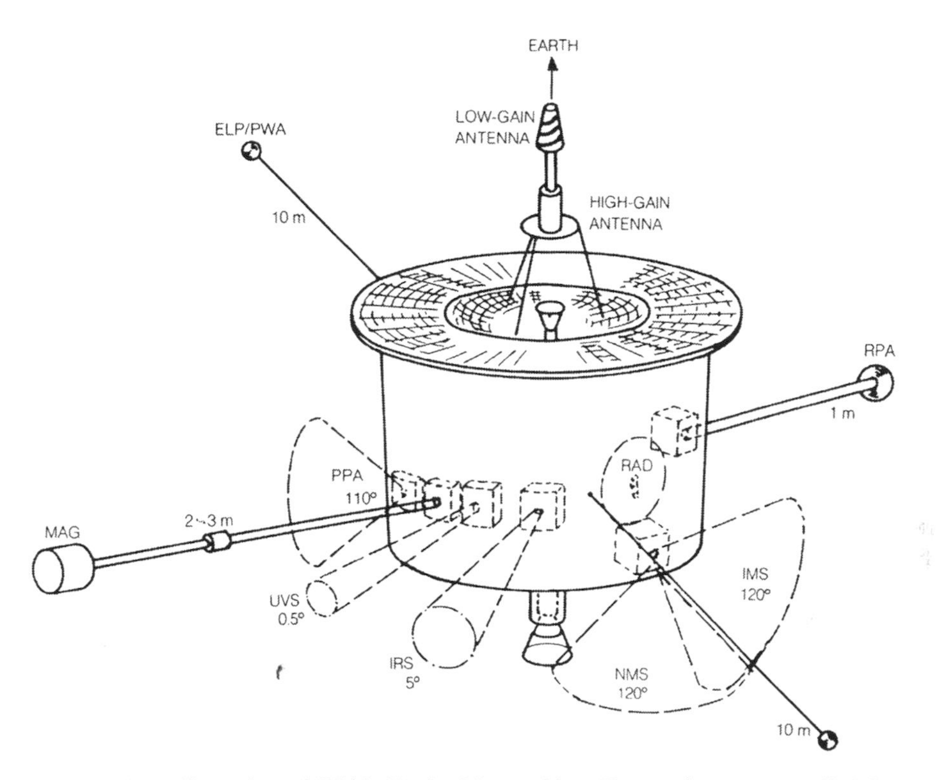

A possible configuration of ESA's Kepler Mars orbiter. Two options were studied for communications. In one case the spacecraft would include a de-spun high-gain antenna that would be pointed at Earth, as in the case of Giotto. Alternatively, the high-gain antenna would be fixed, and the spacecraft's spin axis aligned to aim the antenna at Earth – the option depicted here. Although it eliminated the need for a de-spin system, it complicated instrument pointing and power generation by solar cells. (ESA)

study during the entire 1970s. Studies focused on the fact that so little was actually known of this small planet. Owing to the repeating nature of its flyby geometry, Mariner 10 had been able to map only just over half of the surface, and experience with the early Mars flybys should have prompted scientists to refrain from drawing general conclusions. The origin and extent of Mercury's magnetic field, the most unexpected discovery by Mariner 10, remained a particular mystery. And despite centuries of telescopic study and the amusing 'moon accident' of Mariner 10, no one can say for certain that the planet has no moons. In fact, a paper published in 1986 theoretically proved that it may have retained a small satellite in a retrograde orbit 250,000 km out.[440]

In November 1985 a team of 52 European scientists submitted a proposal to ESA for Mercury Polar Orbiter (MPO, sometimes called Hermes even though this was the name of ESA's proposed manned spaceplane). The spacecraft was to enter a highly elliptical orbit with a 100-km periapsis, this altitude being a compromise between the remote orbit preferred by the particles and fields researchers and the proximity required by the imaging and remote-sensing scientists. A large suite of instruments would provide a complete and in-depth coverage of all aspects of the planet. A camera with a resolution of 10 meters and a laser altimeter were to chart the topography, and a multispectral infrared instrument, a gamma-ray and an X-ray spectrometer would analyze the surface composition. An ultraviolet spectrometer would analyze the amazing exosphere found by Mariner 10, a dust detector would place constraints on the flux of micrometeoroids, and a magnetometer would chart the magnetic field. Gravitational studies were also a priority, not only for insight into the planet's interior (which is dominated by an iron core) but also in relation to General Relativity (the first verification of Einstein's theory was the explanation of the 'anomalous' advancement of Mercury's perihelion). If the additional mass could be accommodated, the payload would include a 35-kg penetrator carrying an accelerometer to record the dynamics of the impact, heat flux meters to investigate the structure of the regolith, an alpha–proton backscatter instrument to analyze the chemistry of the soil and a seismometer to study the interior of the planet. Various ballistic options were identified for the flight. A small (367-kg) MPO could fly as early as 1994 on the most powerful version of the Ariane 4 launcher. A number of flybys with near-Earth asteroids were possible on the way, including two that have their perihelia inside Mercury's orbit: (1566) Icarus and (3200) Phaethon, the latter being the first asteroid to be discovered by a spacecraft, namely the IRAS satellite in 1983. However, because a ballistic mission profile would probably result in too small a probe, various other options were considered, including an ion propulsion system similar to that planned for the AGORA asteroid mission. The proposal was well received by ESA's Solar System Working Group, but judged to be too costly. As with many 'expensive' proposals, it was suggested that it be reworked as a joint European–US mission.[441] After several years in limbo, the proposal was revived in 1992 when ESA issued a request for proposals from the scientific community for a medium-cost mission. Three deep-space proposals made it to the final shortlist. In addition to several astronomy missions, these were the Moon Orbiting Observatory MORO, the INTERMARSNET international network of stations to be distributed across the surface of Mars (described later) and the renamed Mercury Orbiter.

The relatively conventional spin-stabilized Mercury Orbiter was to be based on the Cluster scientific satellites. It would be launched by a dedicated Ariane 5, and the journey would take 4 four years and involve a number of slingshots with Venus and Mercury. As previously, the spacecraft would enter an eccentric orbit, this time ranging between 400 km and 16,800 km to accommodate both the remote-sensing and particles and fields requirements. The primary mission in orbit around Mercury would last 3 local years (about 9 months) and provide global multispectral imaging at a resolution as good as 45 meters, a map of the surface composition, a magnetic field survey, and investigations of the origin and dynamics of the exosphere, of the structure of the magnetosphere, of the planet's interior, and of the interplanetary medium at Mercury's distance from the Sun. If approved, it could be launched in 2004 and reach its target in April 2008.[442] But again, the Mercury mission was not selected. In fact, none of the three deep-space proposals was chosen. Nevertheless, ESA's Space Science Advisory Committee recommended that a Mercury mission

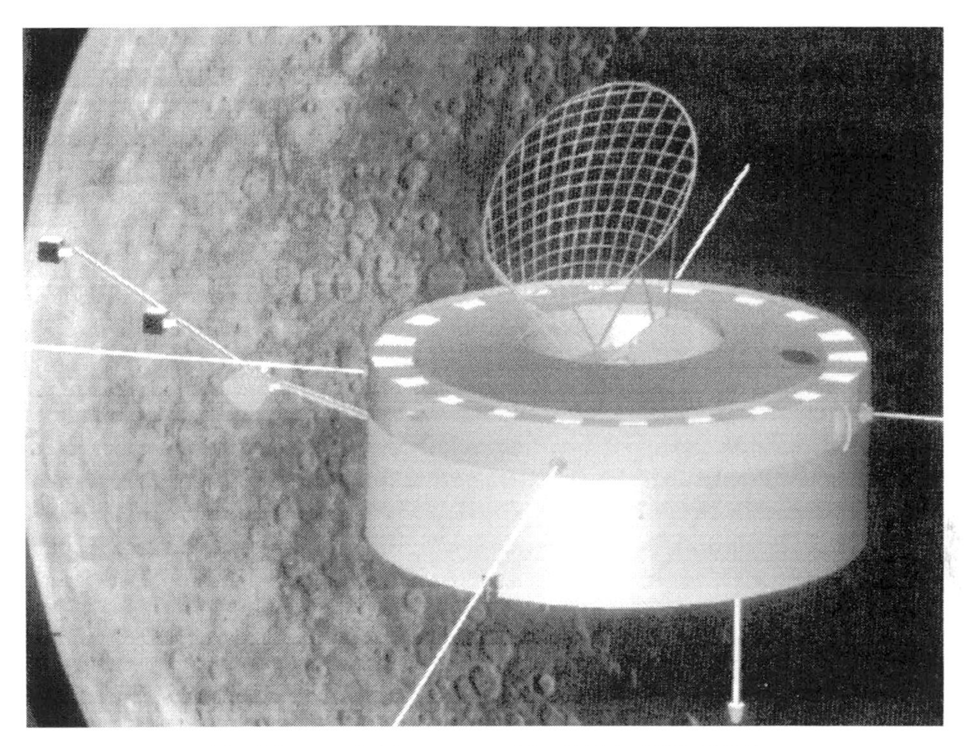

The Mercury Orbiter proposed by ESA in 1992. (ESA)

and one to detect low-frequency gravitational waves be pursued as 'cornerstones' of the Horizon 2000-Plus program that was proposed to member states in 1995.

In fact, Europe was not alone in planning a Mercury mission in the early 1990s. India had begun its space program in the 1960s, and throughout the 1970s had tried to launch a domestic satellite using its indigenously developed Scout-class Satellite Launch Vehicle (SLV). It finally succeeded in 1980, becoming the 7th nation to do so; or the 8th if one counts Europe. During the next decade India used the SLV and its augmented ASLV to launch a number of small satellites, some of which carried scientific and astronomical payloads. In the late 1980s it approved the development of a completely new launcher called the Polar Satellite Launch Vehicle (PSLV). This was to be capable of being augmented by a cryogenic upper stage to form the Geostationary Satellite Launch Vehicle (GSLV), whose debut was expected in the second half of the 1990s.[443] Because the performance of these launchers would be comparable to the American Delta II and the European Ariane, they could serve a national deep-space program. The Indian Space Research Organization undertook preliminary studies of missions to Mars and Venus but decided to focus on Mercury, probably because it was a relatively neglected target. A spacecraft with a dry mass of 250 kg could be launched by the GSLV around the end of the century, make a Venus flyby and then encounter Mercury. Flyby and orbiter profiles were studied. The payload of an orbiter would include particles and fields sensors, a magnetometer, a

CCD camera with a resolution of 50 meters, an ultraviolet spectrometer, gamma-ray, X-ray and infrared spectrometers, radiometers etc.[444,445,446] The plans were shelved, awaiting the introduction of the new launcher. In the meantime, India would have to develop a true planetary science community and develop a means of communicating with spacecraft in deep space.

REFERENCES

1 Friedman-1980
2 Westwick-2007a
3 Kerr-1979
4 For the 'War of the Worlds' see Part 1, page 170
5 For Venera missions see Part 1, pages 209–216 and 284–289
6 Elachi-1980
7 Westwick-2007b
8 Adams-1981
9 Muenger-1985
10 Westwick-2007c
11 For the Pioneer Venus missions see Part 1, pages 262–284
12 AWST-1980
13 James-1982
14 NASA-1980
15 Smith-1982
16 Butrica-1996a
17 Perminov-2004
18 Wilson-1987a
19 Siddiqi-2002a
20 Lardier-1992a
21 Murray-1989a
22 Burke-1984
23 Oertel-1984
24 Perminov-2004
25 Kelly Beatty-1984
26 Burke-1984
27 Oertel-1984
28 Kelly Beatty-1984
29 Perminov-2004
30 Burke-1984
31 Bockstein-1988
32 Blamont-1987a
33 Burke-1984
34 Kelly Beatty-1984
35 Ivanov-1988
36 Slyuta-1988
37 Kotelnikov-1984
38 Petropoulos-1993
39 Basilevsky-1988
40 Schaber-1986
41 Ivanov-1990
42 Alexandrov-1989
43 Alekseev-1986
44 NSSDC-2004
45 Kelly Beatty-1985a
46 Siddiqi-2002a
47 Maffei-1987a
48 For Edmund Halley and Halley's comet see Part 1, page liv
49 Michielsen-1968
50 Friedlander-1971
51 Friedman-1980
52 Ulivi-2006
53 Tsander-1924
54 Saunders-1951
55 Powell-1959
56 Schaefer-2007
57 For Purple Pigeons see Part 1, pages 256–261
58 Blamont-1987b
59 Spaceflight-1977
60 Time-1977
61 Murray-1989b
62 Friedman-1988
63 McInnes-2003
64 Logsdon-1989
65 Kronk-1984a
66 ESA-1979a
67 Hughes-1980
68 Kumar-1978
69 ESA-1979b
70 Friedman-1980
71 Logsdon-1989
72 Calder-1992a
73 AWST-1979

74 Covault-1979
75 Harvey-2000a
76 Kraemer-2000a
77 Akiba-1980
78 Wilson-1987b
79 Hirao-1986
80 Hirao-1984
81 Hirao-1987
82 ESA-1979c
83 Olson-1979
84 Calder-1992b
85 Reinhard-1986a
86 Dale-1986
87 Logsdon-1989
88 ST-1946
89 Whipple-1966
90 Lundquist-2008
91 For Whipple and the 'dirty snowball' theory see Part 1, pages liv–lv
92 Janin-1984
93 Calder-1992c
94 Wiison-1987c
95 Dale-1986
96 Jenkins-2002
97 Keller-1986
98 Calder-1992d
99 Barbieri-1985
100 Reinhard-1986b
101 Bonnet-2002
102 Calder-1992e
103 Kissel-1986a
104 Reinhard-1986b
105 Calder-1992f
106 Calder-1992g
107 Sagdeev-1994a
108 Friedman-1980
109 Blamont-1987c
110 Vekshin-1999
111 Perminov-2006
112 Perminov-2005
113 Blamont-1987d
114 Sagdeev-1986a
115 Simpson-1986
116 Sagdeev-1994b
117 Covault-1985a
118 Kissel-1986b
119 Gringauz-1986
120 Somogyi-1986
121 Riedler-1986
122 Grard-1986
123 Klimov-1986
124 Blamont-1987e
125 Blamont-1987f
126 Sagdeev-1986b
127 Perminov-2006
128 For a description of the standard Venera lander see Part 1, pages 209–210 and 285
129 Bertaux-1986
130 Moshkin-1986
131 Wilson-1987d
132 Perminov-2005
133 Wilson-1987d
134 Kremnev-1986a
135 Kremnev-1986b
136 Blamont-1987g
137 Harvey-2007a
138 Friedman-1980
139 Murray-1989c
140 Wood-1981
141 Farquhar-1999
142 Murray-1989d
143 Logsdon-1989
144 Waldrop-1981a
145 Friedman-1980
146 NASA-1986
147 Davies-1988a
148 Geenty-2005
149 Blamont-1987h
150 IAUC-3737
151 Maffei-1987b
152 Blamont-1987i
153 Sagdeev-1994c
154 Vekshin-1999
155 Blamont-1987j
156 Blamont-1987k
157 Basilevsky-1992
158 Blamont-1987l
159 Preston-1986
160 Sagdeev-1986c
161 Basilevsky-1992
162 Linkin-1986
163 Linkin-1986
164 Moshkin-1986
165 Zhulanov-1986
166 Gel'man-1986
167 Surkov-1986a
168 Andreichikov-1986

169 Surkov-1986b
170 Bertaux-1986
171 Surkov-1986c
172 Surkov-1986d
173 Preston-1986
174 Sagdeev-1986c
175 Perminov-2005
176 Sagdeev-1986d
177 Sagdeev-1994d
178 Dornheim-1985
179 Blamont-1987m
180 Klaes-1993
181 Wilson-1987e
182 Farquhar-1976
183 Farquhar-2001
184 Kronk-1984b
185 Kronk-1988a
186 Sekanina-1985
187 For Robert Farquhar and early comet exploration proposals see Part 1, pages 206–208
188 Kerr-1984
189 Farquhar-1983
190 Kuznik-1985
191 Farquhar-2001
192 Eberhart-1985
193 IAUC-3937
194 Farquhar-2001
195 Maran-1985
196 Von Rosenvinge-1986
197 Scarf-1986
198 Smith-1986
199 Oglivie-1986
200 Kelly Beatty-1985b
201 Kerr-1985
202 Cowley-1985
203 Covault-1985b
204 Mudgway-2001a
205 Williams-2005
206 Farquhar-2001
207 Wilson-1987b
208 Hirao-1986
209 Calder-1992h
210 Jenkins-2002
211 Münch-1986
212 Wilson-1987b
213 Calder-1992i
214 Calder-1992j
215 Dale-1986
216 Hirao-1986
217 Bonnet-2002
218 Calder-1992k
219 Calder-1992l
220 Kiseleva-2007
221 West-1986
222 Sekanina-1986
223 Gore-1986
224 Kronk-1999
225 Somogyi-1986
226 Grard-1986
227 Gringauz-1986
228 Somogyi-1986
229 Simpson-1986
230 Keppler-1986
231 Gringauz-1986
232 Sagdeev-1986b
233 Perminov-2006
234 Sagdeev-1986a
235 Kerr-1986
236 Kaneda-1986
237 Uesugi-1986
238 West-1986
239 Hirao-1986
240 Hirao-1987
241 Wilson-1987b
242 Simpson-1986
243 Vaisberg-1986
244 Gringauz-1986
245 Riedler-1986
246 Grard-1986
247 Perminov-2006
248 Lenorovitz-1986a
249 Sagdeev-1986a
250 Kerr-1986
251 Lenorovitz-1986a
252 Combes-1986
253 Krasnopolsky-1986
254 Moreels-1986
255 Vaisberg-1986
256 Mazets-1986
257 Simpson-1986
258 Sagdeev-1986b
259 Kissel-1986b
260 Savich-1986
261 Canby-1986
262 Hirao-1986
263 Saito-1986
264 Hirao-1987

265 Wilson-1987b
266 Festou-1986
267 Calder-1993m
268 Neubauer-1986
269 Balsiger-1986
270 Johnstone-1986
271 Rème-1986
272 Calder-1992n
273 Keller-1986
274 Calder-1992o
275 McDonnell-1986
276 Neubauer-1986
277 Jenkins-2002
278 Keller-1986
279 Keller-1988
280 Sekanina-1986
281 Sekanina-1987
282 Smith-1987a
283 Calder-1992p
284 Kissel-1986a
285 Calder-1992q
286 McDonnell-1987
287 McDonnell-1986
288 Balsiger-1988
289 Wilkins-1986
290 Prialnik-1992
291 Hainaut-2004
292 Hainaut-2007
293 Ferrin-1988
294 McFadden-1993
295 Ostro-1985
296 Sagdeev-1986b
297 AWST-1986a
298 Spaceflight-1992a
299 Perminov-2006
300 Davies-1988b
301 Verigin-1999
302 Bortle-1996
303 Uesugi-1988
304 Dunham-1990
305 Farquhar-1999
306 Siddiqi-2002b
307 Calder-1992r
308 Farquhar-1999
309 Kamoun-1982
310 Kronk-1984c
311 Kronk-1998b
312 Kresak-1987
313 Dunham-1990
314 Calder-1992s
315 Calder-1992t
316 McKenna-Lawlor-2002
317 McDonnell-1993
318 Bond-1993
319 McBride-1997
320 McKenna-Lawlor-2002
321 IAUC-7243
322 Flight-1992a
323 Schwehm-1992
324 Calder-1992u
325 Bond-1993
326 For Mariner 10's tribulations see Part 1, pages 172–196
327 Westwick-2007d
328 Westwick-2007e
329 Blume-1984
330 Neugebauer-1983
331 Waldrop-1981b
332 Waldrop-1982
333 JWG-1986a
334 Westwick-2007f
335 Borg-1994
336 Farquhar-1999
337 JWG-1986b
338 Tsou-1985a
339 Wilson-1986a
340 Eberhardt-1986
341 Covault-1985c
342 Tsou-1985b
343 Blume-1984
344 Albee-1994
345 Uesugi-1995
346 Kronk-1984d
347 Brownlee-2003
348 Wilson-1986a
349 AWST-1985a
350 Wilson-1985
351 AWST-1985b
352 Cunningham-1988a
353 Cunningham-1988b
354 Whipple-1987
355 Houpis-1986
356 Collins-1986
357 AWST-1989a
358 Weinberger-1984
359 Wilson-1987f
360 Westwick-2007g
361 JPL-1991

362 NRC-1998a
363 Spaceflight-1992b
364 Stuhlinger-1986
365 JP4-1992
366 Elfving-1993
367 Atzei-1989
368 Sedbon-1989
369 Wilson-1986b
370 Carlier-1993
371 NASA-1966
372 Ulivi-2006
373 Alfvèn-1970
374 Stuhlinger-1970
375 Cunningham-1988c
376 Schwaiger-1971
377 Cunningham-1988d
378 Alvarez-1997
379 Russo-2000a
380 Thomson-1982a
381 ESA-1980
382 Cunningham-1983
383 Langevin-1983
384 Balogh-1984
385 Cunningham-1988e
386 JWG-1986c
387 Cunningham-1985
388 Cunningham-1988f
389 Cosmovici-1983
390 Cunningham-1988g
391 Anselmo-1987a
392 Anselmo-1987b
393 Pardini-1990
394 Anselmo-1990
395 Anselmo-1991
396 Anselmo-2007
397 Cunningham-1988f
398 Covault-1985d
399 Lenorovitz-1985
400 Kelly Beatty-1985a
401 Furniss-1987a
402 Lenorovitz-1986b
403 Grard-1988
404 Cunningham-1989
405 Flight-1988
406 Surkov-1997a
407 Kovtunenko-1995
408 Farquhar-1995
409 McLaughlin-1985
410 Maehl-1983
411 Cunningham-1988h
412 McComas-2006
413 Ulivi-2008
414 Friedman-1994
415 Anderson-1977
416 Randolph-1978
417 Bender-1978
418 McLaughlin-1984
419 Anderson-1994
420 Sukhanov-1985
421 Kovtunenko-1990
422 Galeev-1990
423 Furniss-1987a
424 Furniss-1987b
425 Zak-2004
426 Jaffe-1980
427 Aston-1986
428 Forward-1986
429 Nock-1987
430 Etchegaray-1987
431 Day-2006
432 Westwick-2007h
433 For early European deep space exploration planning see Part 1, pages 88–89
434 ESA-1979d
435 ESA-1979e
436 JWG-1986d
437 Grard-1982
438 Thomson-1982b
439 Russo-2000b
440 Rawal-1986
441 Wilson-1987g
442 Grard-1994
443 Harvey-2000b
444 Flight-1993
445 Spaceflight-1992c
446 Mama-1993

5

The era of flagships

THE FINAL SOVIET DEBACLE

In the decade after NASA's highly successful Viking missions to Mars, which saw orbiters dispatch landers to the surface, a variety of ambitious follow-up projects were elaborated in the United States, the European Space Agency came close to approving its Kepler Mars orbiter, and the Soviet Union was planning to resume its exploration of the planet with a series of increasingly complex missions.[1]

In the late 1970s the Soviets decided to start with a mission to explore one of Mars's two moons, and attention soon focused on Phobos, the larger and innermost one. The initial plan was for the spacecraft to land on the moon, but because this would be extremely complicated in that body's weak yet irregular gravity field, the mission designers investigated having the spacecraft maneuver to within 20 meters of Phobos to collect samples by using harpoons. On deeming this operation to be too risky, the designers settled on remote sensing, active sampling using lasers and particle cannons and deploying a number of small landers.[2] Why the Soviets were so interested in exploring Phobos is not clear, and in fact the project was somewhat scornfully viewed by some American planetary scientist as a "poor man's asteroid mission" – their argument being based on the speculation that Mars's moons are captured asteroids.[3] The Fobos mission was formally approved only in early 1985, and publicly announced in March of that year. The mission would involve many of the same scientific and industrial players as devised the Vega mission, whose twin spacecraft were at that time heading for Venus. However, whereas the extremely successful Vega program was effectively managed by the IKI science team, control over Fobos was transferred back to the industrial contractor, the Lavochkin design bureau. The initial intention was to use the particularly advantageous 1988 launch window, but the project soon fell behind schedule; so much so, in fact, that meetings of the international scientific teams were organized to coincide with those of Vega in an effort to keep the program on track. At the same time, the Soviets created the new space agency, Glavcosmos, to coordinate its civilian scientific space program and to streamline its international contacts. Moreover, this agency would interface between the two institutes of the Academy of Sciences involved in space exploration – namely the IKI and the Vernadsky Institute of Geochemistry and Applied Chemistry – and the space industry that was controlled by the military, which in the case of planetary exploration was mainly the Lavochkin design bureau.[4,5]

Many countries from both sides of the European divide took part in the Fobos mission, including Austria, Bulgaria, Czechoslovakia, Finland, France, East and West Germany, Hungary, Ireland, Poland, Switzerland and Sweden, as well as the European Space Agency itself. NASA provided tracking support through its Deep Space Network and American researchers had secondary roles on some of the investigations. Owing to its long-standing involvement in the Soviet deep-space program, France was the most important foreign partner in the project; cooperating on many instruments and experiments, as well as in the process of navigation. On this occasion, moreover, Soviet and American scientists exchanged data in support of their respective missions. In particular, Americans planning the Magellan Venus radar mapper received tapes of raw data from Veneras 15 and 16, and in return the Soviets were given information and images of Phobos obtained by Mariner 9 and the Vikings in order to assist in computing the tiny moon's ephemeris.[6]

As regards the spacecraft hardware, after a full decade during which it had been assigned only a low priority, the development of the new UMVL bus was finally completed. It was hoped that this bus would facilitate a number of new missions in the 1990s which would lead eventually to the return to Earth of a sample of Mars's surface, the Vesta flight to asteroids and comets, continued Venus exploration, and a return to the Moon – a body which had been neglected by the Soviets since the mid-1970s. The UMVL bus was built around a propulsion module which served as its primary structural component. This had a toroidal fuel tank and four outrigger spherical tanks with propellant for attitude control and small course corrections. It had 28 maneuvering thrusters: 24 with a thrust of 50 N, and the others of 10 N. There were in addition 12 hydrazine monopropellant thrusters delivering 0.5 N of thrust for attitude control. Several electronic modules completed the basic bus. To this could be added solar panels, instruments, other equipment and (if required for landing missions) footpads. Stabilization could be either over three axes to ensure the requisite pointing of remote-sensing instruments, or in a 'drift mode' with the axis normal to the solar panels maintained pointing at the Sun and the spacecraft free to rotate around this at slow angular rate. The attitude control system used star and Sun sensors, gyroscopic and accelerometry platforms, thrusters for control and a triply redundant computer. In the case of the 1F spacecraft (the UMVL variant designed for the Fobos mission) two square solar panels were carried, which had at their tips thrusters for attitude control and omnidirectional antennas. Beneath the panels were mounted the antennas of the radar altimeter, the Doppler radar and the low-altitude radar, all of which were to be used in approaching Phobos. A ground-penetrating scientific radar (described elsewhere) could also be carried. On top of the 1F was a cylindrical tower containing the chemical batteries and the electronics for the attitude control system, transmitters and receivers, housekeeping systems and the scientific instruments. The tower was topped by an articulated high-gain antenna with two degrees of freedom. Interestingly, in some respects the 1F bus was a retrogressive step in comparison to Vega, in that the remote-sensing instruments were body-mounted instead of being installed on a scan platform (as in the case of Vega), which would require the spacecraft to break contact with Earth while taking data, and the data rate of the relatively small high-gain antenna was only between 4 and 20 kbps, whereas Vega was capable of transmitting at a rate of 64 kbps. For this reason, a data recorder was mounted with a capacity of 30 million bits to serve as a buffer.

The UMVL bus incorporated a 3,600-kg ADU (Avtonomnaya Dvigatel'naya Ustanovka; autonomous engine unit) stage which the Lavochkin design bureau had obtained by extensively redesigning the old common propulsion stage of the E-8 heavy lunar probes. For most of the missions envisaged for the new spacecraft, the ADU (also known as Fregat) would act as a fifth stage of the Proton launcher and provide in-flight propulsion. For the Mars missions it consisted of four 102-cm and four 73-cm spherical tanks with a capacity of 3,000 kg of unsymmetrical dimethylhydrazine (UDMH) and nitrogen tetroxide, and a single central engine mounted on a special suspension and steering system to provide control in pitch and yaw; other configurations were envisaged for Venus or lunar missions. The single Isayev KTDU-425A (Korrektiruyushaya Tormoznaya Dvigatelnaya Ustanovka; course correction and braking engine) could deliver a thrust of between 9.8 and 18.6 kN, had a total burn duration of 560 seconds, and could be restarted as many as seven times. Including the Fregat stage and 500 kg of instruments, the Fobos spacecraft had a total mass at launch of about 6,220 kg.[7]

The Fobos mission profile was in many ways unusual and untried; so much so that NASA engineers privately considered the plan to be a bit too ambitious for the Soviets. Firstly, since the Proton could dispatch only about 4.5 tonnes to Mars, the plan was for the final burn of its stage D not to inject the 1F into a heliocentric orbit but into an eccentric Earth orbit with an apogee of the order of 130,500 km.[8] In due course, the Fregat would burn for 142 seconds to head towards Mars. It would perform course corrections as necessary, brake into an initial orbit around Mars and subsequently shape this orbit to one nearly matching that of Phobos, and then be jettisoned. After a slow approach and several intermediate maneuvers, the 1F would start a very close reconnaissance of the moon at a distance of 35 km. All of the subsequent 'active' phases of the mission would have to be scheduled for when the spacecraft was within line of sight of Earth for communications, and when the surface of the moon was in sunlight for visibility. The radar altimeter would lock onto the surface at a range of 2 km, and the spacecraft would slowly maneuver to make a fly by of Phobos at an altitude of about 50 meters and at a relative speed of 2–5 m/s. During the 20-minute flyby, the spacecraft would undertake analyses by remote sensing, take pictures with a resolution of about 6 cm and release one or two landers. Given the roughness of the surface, pockmarked by craters, grooves and crevices, robust artificial intelligence software would have been necessary to stabilize and orient the spacecraft during this low-altitude pass, but with hindsight (especially in view of the unremarkable performance of the computers which were carried) it is doubtful that the flyby could have been successfully executed. After the flyby, a two-

The Soviet Fobos mission was the first use of the new UMVL planetary bus. It is shown here in Mars orbit, after the Fregat stage has been jettisoned.

impulse burn would return the spacecraft to a stable Martian orbit to conduct a detailed study of the planet, its atmosphere and interactions with the solar wind, as well as monitoring the Sun at a variety of wavelengths. Two almost identical spacecraft would be launched under a $480 million program. The whole mission would last about 460 days from launch. There were suggestions that if the first flyby were to be successful, the second spacecraft could be retargeted towards Deimos, the smaller of the planet's moons.[9,10]

The Soviets adopted the 'Christmas tree' approach of their contemporaries in America by equipping Fobos with as many scientific and engineering instruments as possible. In fact, with almost 30 instruments to study Phobos, Mars, the solar wind and interplanetary space, it was the most heavily instrumented interplanetary craft ever flown.

The part of the payload which most attracted the interest of the press at the time was a mobile robot for Phobos. Work on this started in 1983 and lasted 4 years. It was the responsibility of the VNII Transmash institute of Leningrad, which had developed all previous Soviet planetary rovers as well as several soil penetrometry experiments. To enable the robot to overcome obstacles which were substantially larger than itself, a 'leaping' locomotion system was chosen. Although it massed 50 kg, the PrOP-F (Pribori Otchenki Prokhodimosti-Fobos; instrument to evaluate cross-country characteristics on Phobos) 'hopper', or 'frog', was designed to hop around in the extremely weak gravity of Phobos, which is 2,000 times weaker than that of Earth. It would be carried on a side of the spacecraft's central 'tower' and be released during the low-altitude flyby, with the separation device controlling its initial direction and speed. Approximate parameters would be a horizontal speed of 3 m/s and a vertical speed of 0.45 m/s. A truss-shaped damper would ensure that the lander settled onto the surface without excessive bouncing and also prevent it from rolling downslope. Once the lander had come to a stop, it would release the damper. The robot would then right itself using four 'whiskers', and power on its scientific instruments. The PrOP-F was a hemisphere of 50-cm diameter connected to a downward-facing truncated cone with scientific instruments on its base. An X-ray fluorescence spectrometer would use radioactive iron and cadmium sources to irradiate the surface material in order to analyze its composition. It would also be equipped with a magnetometer; magnetic susceptibility probes; an accelerometer; a gravimeter; an electric resistance probe; a dynamic penetrometer which would drive a wedge into the uppermost few centimeters of the surface to measure its bearing, compressibility and shear strength; thermistors to make contact with the surface to directly measure its temperature; and radiometers to measure its thermal flux. Once the observations were made, the 'whiskers' would be flexed in order to propel the robot on a 20-meter hop to repeat the analyses at a different place. This cycle would continue until the batteries exhausted their 4-hour operating life. All the lander's data would be transmitted directly to the mothership, which, by the time that the robot fell silent, would be about 300 km away.[11,12,13] Helicopter drop-tests were made in Turkmenistan and Kamchatka, and the ability of the system to operate in very low-gravity was tested on a modified Ilyushin Il-76 aircraft usually used to give cosmonauts experience of weightlessness.

The PrOP-F hopper for the Martian moon Phobos, shown here during integration. (VNII Transmash)

The Fobos spacecraft was also to deploy a second, stationary lander. This DAS (Dolgozhivushaya Avtonomnaya Stanziya; long-duration autonomous station) was the responsibility of the Lavochkin bureau itself. It was stowed on the top of the 1F bus, close by the 'tower', with minimal clearance with respect to other instruments and packages, and a pair of arms to move it up and out for release. Once free, cold gas thrusters would propel the squat, roughly hexagonal lander towards the moon's surface, whilst simultaneously setting it spinning for stabilization. Sensors on the underbody would detect contact with the surface and trigger both the firing of downward-acting solid-fuel engines and of a harpoon connected by a lanyard to hold the lander in position – depending on the hardness of the surficial material, the harpoon could be expected to penetrate between 1 and 10 meters. The DAS would then wait 10 minutes to allow the dust to settle, literally, after which it would extend three legs to raise its instrument platform 80 cm above the surface, and deploy three solar panels. Also mounted on its top shelf were transmitter and receiver antennas to communicate directly with Earth, and Sun sensors to enable the solar panels to be appropriately oriented. The designers recognised that at some point in the mission the shadow cast by the Earth-facing antenna might fall on the solar panels and had hoped to develop software to address this issue, but the tight schedule did not allow time. Similarly, two alternative data compression algorithms were developed by Soviet and French engineers but there was insufficient time to compare them and choose the best. Since the processor did not have the capacity to install both, it was decided that the DAS on one of the Fobos spacecraft would use the Soviet algorithm and the DAS on the other would have the French algorithm. The French supplied a CCD camera to take high-resolution 'ground level' pictures of the surface of Phobos. Because the data rate would be 4 bps at the start of the lander's mission, surging to 16 bps for the main, 3-month phase, and down to 8 bps at its end, even with an efficient data compression algorithm it would require three or four communication sessions to transmit the data for a single image frame! The French also supplied the Stenopee optical experiment that would track the position of the Sun in order to monitor the libration motions of the moon. West Germany supplied an X-ray and alpha-particle backscattering spectrometer to determine the chemical composition of the surficial material. A VLBI transponder was carried to enable an international network of radio-telescopes to precisely track the lander (in the same manner as the Vega balloons were tracked in the Venusian atmosphere) in order to refine the moon's ephemeris to a very high accuracy.[14] All of the other instruments on the DAS lander were supplied by the Soviets and in addition to a seismometer included temperature and acceleration sensors on the harpoon to enable this to be used as a penetrometer. The seismometer would be sufficiently sensitive to detect the PrOP-F hopping around. If the second Fobos spacecraft were not redirected to Deimos, it was to set its DAS down about 5 km from the first, and it was expected that the seismometer of the first lander would record the touchdown and harpoon-firing of the new arrival. The dual-processor main computer for the DAS was developed in Hungary, with input from the IKI.[15,16]

The orbiter's instruments were dedicated to three different themes of research: solar physics and astrophysics; Martian and interplanetary particles and fields; and remote sensing of Mars and Phobos.

The main solar physics instrument was a Soviet–Czechoslovak solar telescope consisting of a coronagraph (a telescope equipped with an occulting disk to mask the photosphere in order to observe the fainter corona) and two parallel objectives for different wavelengths in the X-ray band, with all of the optics mated to image intensifiers and CCD arrays. Together with similar telescopes on Earth, this would hopefully give scientists a full 360-degree image of the Sun.[17] In addition, a 3-channel photometer was to precisely measure the solar irradiance to detect oscillations on the Sun caused by pressure and gravity vibration modes. Developed by the Swiss Observatory of Davos with assistance from France, ESA, the Soviet Crimean Observatory and Hungary, it was the first instrument dedicated to helioseismology (as the study of our star's proper vibrations is called) to fly in space, and was to provide extensive data during the cruise to Mars. An ultraviolet radiometer provided by the Soviets would monitor the flux of solar radiation. A Soviet–Czechoslovak photometer consisting of five identical gas-discharge cells would monitor the full

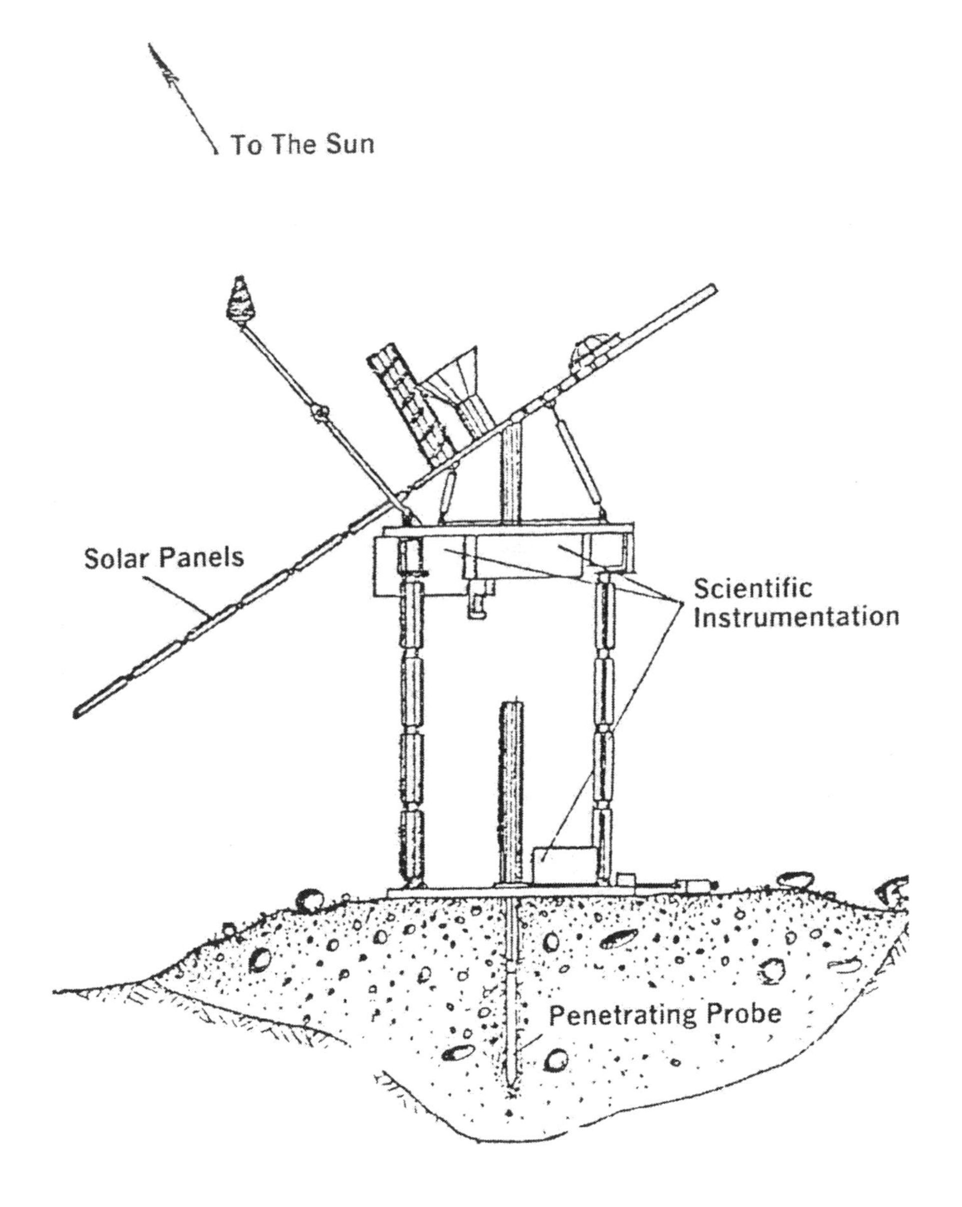

The DAS long-duration lander for Phobos.

disk of the Sun at soft X-ray wavelengths. American satellites of the GOES series carried detectors with similar characteristics, and combining their data with that returned by the Fobos spacecraft would enable different hemispheres of the Sun to be monitored.[18] Two instruments were carried to augment the network of spacecraft in interplanetary space that were monitoring gamma-ray bursts.[19] The French provided one instrument dedicated to low-energy bursts, and a Soviet sensor fitted with French electronics was for high-energy bursts. At Mars, the latter would also serve as a rudimentary spectrometer to study the composition of the surface. Both instruments were mounted on the tip of one of the solar panels.[20]

Unlike NASA missions – which, since Mariner 4 had suggested that Mars had no intrinsic magnetic field, had emphasized remote sensing over the measurement of particles and fields – the Fobos missions (as with previous Soviet Mars probes) would also study space near Mars as well as the interplanetary environment. Two separate 3-axis fluxgate magnetometers were onboard: one a joint project between the Soviets and Austria which was derived from the Venera and Vega instruments, and the other a collaboration between the Soviets and West Germany. They were on a 3.5-meter boom with the Austrian one at the tip and the German one mounted 1 meter closer in.[21] A plasma-wave system was developed by ESA in cooperation with the French CNRS (Centre National de la Recherche Scientifique; national scientific research center), the American University of California in Los Angeles, the IKI and the Polish Space Electronics Laboratory. It included a dipole antenna consisting of a Langmuir probe to measure electron fluxes, and two 10-cm spheres set 1.45 meters from each other to receive signals from plasma instabilities and electromagnetic waves. It was also to monitor any electric charge picked up by the spacecraft while in close proximity to Phobos.[22] Another European instrument was a cooperation between ESA, the IKI, the German Max Planck Institute of Lindau and Hungary's Central Research Institute for Physics. This low-energy telescope could detect atomic nuclei with masses ranging from hydrogen to iron, and was to measure the flux, spectra and composition of the solar wind and cosmic rays. The plan was for it to complement a similar instrument on ESA's Ulysses out-of-ecliptic mission, but the loss of Shuttle Challenger in January 1986 delayed that launch to 1990.[23] ASPERA (Automatic Space Plasma Experiment with a Rotating Analyzer) was a Swedish–Soviet–Finnish instrument consisting of a scanning platform and two spectrometers to measure the composition, energy and distribution of ions and electrons in an almost complete sphere around the spacecraft.[24] A West German–Soviet–Hungarian–Austrian experiment would measure separately the energy and angular spectra of hydrogen and helium ions, as well as the heavy ions which were expected to be present in the near-Mars environment.[25] A joint Soviet–Hungarian electrostatic analyzer would simultaneously detect electrons and ions coming from eight directions in the vicinity of Mars. American scientists from the University of Michigan participated in this study.[26] The particles and fields payload was rounded out by a Soviet array of three gas-discharge counters to monitor charged particles in interplanetary space, a Soviet–Austrian–West German energy, mass and charge spectrometer, and a Soviet–Irish–Hungarian–West German twin-telescope particle detector.[27]

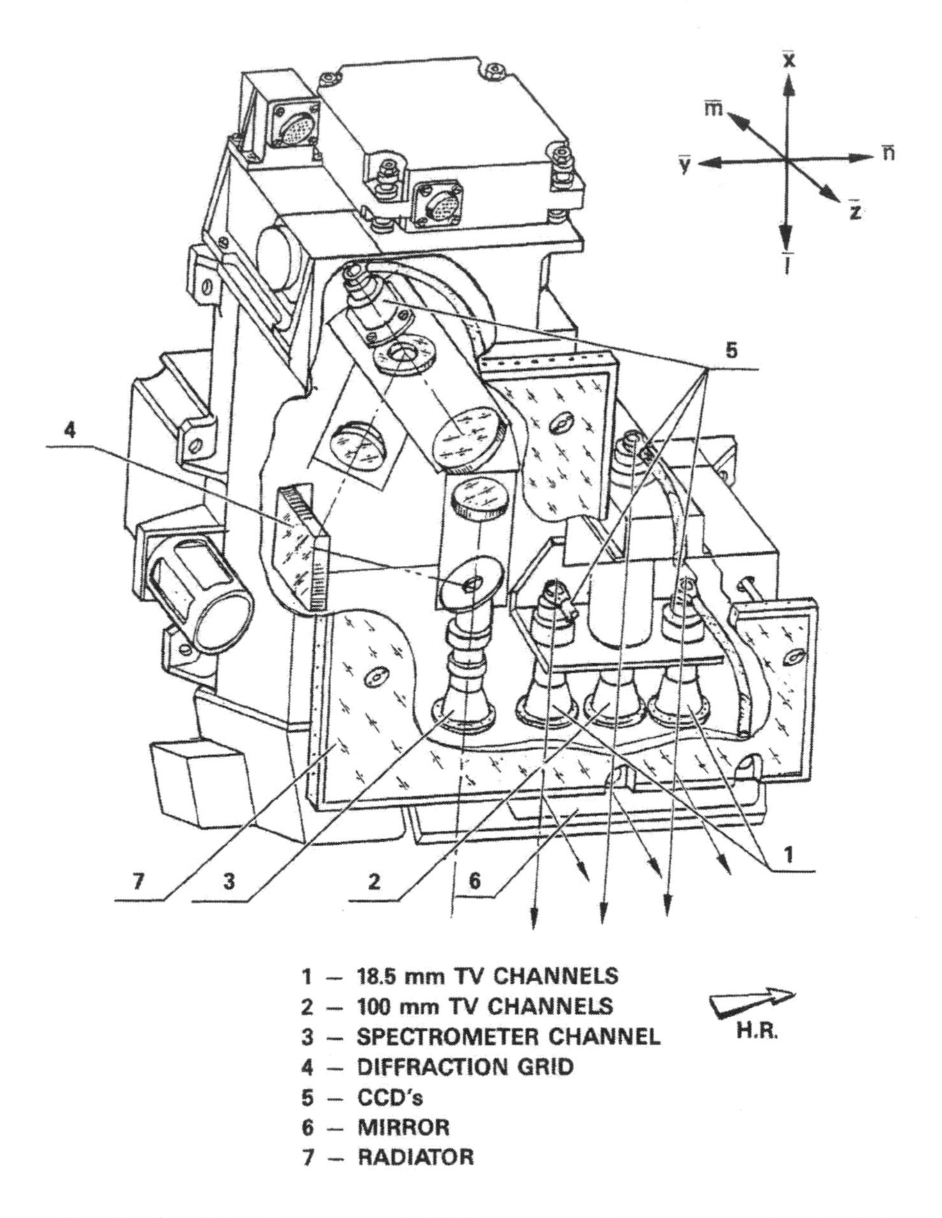

The Russian–Bulgarian integrated CCD camera and spectrometer for the Fobos mission. (Astronomy & Astrophysics, reprinted with permission)

The remote-sensing suite for Mars and Phobos was just as extensive. The main science and navigation camera consisted of wide-angle and narrow-angle optics for CCD detectors with arrays of 288 × 505 pixels, and an integrated spectrometer. East Germany developed a memory for this instrument that was capable of storing over 1,000 images, and Bulgaria developed all the electronics and undertook the final assembling and testing. At various stages of the development, however, the teams received assistance from colleagues in France, the US and Finland.[28] A combined radiometer and photometer for infrared, visible and near-ultraviolet wavelengths was a Soviet–French joint venture, and was to measure the thermal and reflectivity properties of the outermost layer of regolith on Phobos and Mars, the temperature of the Martian stratosphere, and the optical characteristics of atmospheric aerosol particles.[29] Another French–Soviet cooperation was an imaging spectrometer (the first ever flown on a planetary spacecraft) operating in the near-infrared. It was to take spectra of single pixels of the Martian surface to provide information on the 'column depth' of carbon dioxide (and hence altimetry of the site viewed) and on the chemical composition and mineralogy of the surfaces of Mars and Phobos. The single-pixel 'scan' would track across the ground as the spacecraft flew. The data from the initial eccentric orbit would form a track 1,600 km long at a resolution of about 5 km, but once the spacecraft had maneuvered into a Phobos-like orbit the resolution of the planet would be 30 km.[30] One of the most interesting instruments was the Termoskan multispectral imager (which was also the first to be flown on a planetary mission). This all-Soviet instrument consisted of photodetectors cooled by a closed-cycle cryogenic device using liquid nitrogen as coolant, and it operated in two spectral bands in the visible and near-infrared. By using a scanning mirror that oscillated in a direction perpendicular to the spacecraft's motion, the 'camera' could produce a swath about 650 km wide and indefinitely long with a resolution of about 1.8 km. Its multispectral data could be used to produce images which presented information on the temperature, thermal inertia and texture of the terrain. Remarkably, despite the interest in its investigations, Termoskan was a last-minute addition to the suite.[31] The French–Soviet Auguste experiment applied techniques developed to study the Earth's atmosphere from space by measuring the absorption of solar ultraviolet and infrared in viewing the limb at orbital sunrise and sunset. It consisted of a Sun-pointing mirror and a Cassegrain telescope which illuminated a beam splitter to feed one spectrograph to detect ozone in the Martian atmosphere, a second spectrograph to detect carbon dioxide, water and deuterated water, and an interferometer to detect atmospheric oxygen and water.[32,33,34]

A gas-scintillation neutron detector was to identify areas of enhanced humidity which could be further investigated for evidence of biology, as well as identifying basaltic provinces and rocks possessing a very high iron content.[35] A gamma-ray spectrometer was on the outer edge of one of the solar panels, 3 meters from the spacecraft whose natural radioactivity would add a background signal to the data. During the initial eccentric orbit this instrument was to map the composition of the rocks exposed at the surface of the planet in a narrow belt along its equator. It may be remembered that despite the extensive exploration of Mars in the 1960s and 1970s, the composition of the surface had been determined only at the two Viking landing sites and a limited number of areas remotely sensed by the Soviet Mars 5 orbiter.[36] Gamma-ray spectra would also be taken while making the low pass over Phobos.[37,38]

Three instruments were carried specifically for the Phobos flyby. A laser mass spectrometer was to analyze the composition of the outermost few micrometers of the moon's surface. When the altimeter readings indicated that the spacecraft was sufficiently low, the instrument would fire laser pulses at a spot several millimeters in diameter to evaporate the material, and some of the ions thereby released would be collected and analyzed electrostatically as the spacecraft passed over. This was a joint effort by Soviet, Austrian, Bulgarian, Finnish, German (both East and West) and Czechoslovak scientists, with British scientists assisting with data analysis. At 70 kg, it was the most massive instrument of the entire payload. In fact, there were some concerns because although it was clearly not a prototype for a military space-based laser weapon, its performance made it similar to the kinds of laser required for some strategic defense applications.[39] At the same time another mass spectrometer would irradiate the surface of Phobos with a beam of krypton ions and record the scattered ions. The instrument was barely powerful enough to sputter ions from the uppermost few nanometers of the surface, but this would include the layers which were most 'weathered', and so would provide information on the surface processes that have occurred over the ages. It was a cooperation between the Soviet Union, Austria, Finland and France, with the latter providing the ion gun. Altogether, the two 'active' instruments were to analyze about 100 distinct sites on the satellite.[40] During the close flyby of Phobos, the multi-wavelength Grunt (Soil) radar, whose antenna was mounted beneath one of the solar panels, would 'sound' the surface to a depth of at least several meters to determine the structure and dielectric characteristics of any layering.[41] And of course, the data gained from tracking the spacecraft during the flyby would refine knowledge of the moon's orbit and mass. At one point, the engineers at JPL investigated the possibility of linking the Fobos orbiters to their own Mars Observer, as they had planned to do with ESA's Kepler orbiter, but the communication systems would have required substantial hardware modifications and this was not pursued.[42,43]

The first 1F spacecraft was launched as Fobos 1 on 7 July 1988, and followed by Fobos 2 on 12 July. Both successfully performed the innovative Fregat escape maneuver. The degree of openness exceeded even that of the Vega flights. Western journalists and officials (including the former Apollo 11 astronaut Michael Collins and a US Air Force delegation) were allowed to visit the building in which the Proton launchers were integrated, and to watch the preparation and launch. Indeed, the Fobos 2 rocket was probably the first Soviet launcher to carry advertisements on its body; these representing two Italian and Austrian steel companies which had been supplying the USSR for some time.[44] The two spacecraft made their first course corrections on 16 July of 8.9 m/s and 21 July of 9.3 m/s respectively to encounter Mars in late January 1989. The intention was for Fobos 1 to make its Phobos flyby on 7 April 1989, with Fobos 2 doing so on 13 June. The first data from the mission came from the European plasma-wave experiment, which documented the crossing of the Earth's magnetospheric bow shock, some 200,000 km out. Soon thereafter, the gamma-ray spectrometer was powered on for calibration.[45] In August the solar telescope on Fobos 1 returned 140 X-ray images of the Sun which, in addition to recording the fine detail of the photosphere, caught a solar flare.[46]

The first indication that the mission had not been prepared as thoroughly as it ought to have been came on 2 September, when there was no response to a routine call to Fobos 1. In fact, this spacecraft was not heard from again. It was discovered that during the previous communication session on 29 August an engineer had sent an erroneous command which, instead of activating the gamma-ray spectrometer, had started a program resident in the read-only firmware that had been used to test the attitude control system on the ground and ought to have been cleared prior to launch but had not been for lack of time. This program concluded by switching off the attitude control thrusters, leaving the spacecraft to slowly drift out of alignment into the most dynamically stable attitude, which unfortunately was with the rear of the solar panels facing the Sun. By the next attempt to contact the spacecraft, it had drained its battery and was unable to respond. Efforts to re-establish contact were made throughout September and October in the hope that the vehicle might have fortuitously faced the 'front' of its solar panels towards the Sun for long enough to recharge its batteries, but there was no response, and on 3 November the Soviets said that Fobos 1 had been written off. It is not clear how the erroneous command escaped detection by controllers prior to its being transmitted to the spacecraft. By one account there was a dispute between the mission control centers in Moscow and Yevpatoria regarding who was in charge. It was decided that Moscow would effectively command the mission from 18 August, with Yevpatoria testing all commands on the ground simulator prior to transmitting them to the spacecraft. On 29 August the simulator was out of commission but it was decided, irresponsibly as it turned out, to send the command anyway. To further complicate the situation, the spacecraft's programming had not been written to reject such a 'suicidal' order. In the words of Roald Kremnev, the deputy director of Lavochkin, the engineer who sent the false command (and who also subsequently identified the error) "was not [allowed] to participate in the later operation [of Fobos 2]".[47]

This was yet another case of a Soviet spacecraft being lost on its way to Mars. European astronomers at the La Silla Observatory in Chile reportedly tried to identify Fobos 1 on photographs taken using their powerful telescopes, but were unable to find it.[48,49,50,51] It was a pity that this was the first occasion on which the Soviets had sent a pair of non-identical spacecraft. While only Fobos 2 carried the PrOP-F hopper, only Fobos 1 had the X-ray solar telescope and the science radar. Moreover, the loss of one spacecraft meant that it would not be possible to attempt the active seismic experiment in which the first long-lived lander on Phobos would monitor the arrival of its counterpart from the second spacecraft. As a result of this accident, the schedule was revised to bring forward the Phobos flyby by Fobos 2 to early April.

Meanwhile, Fobos 2 continued to collect data during the interplanetary cruise. For example, its two gamma-ray burst instruments detected some 200 events in all, 400 solar flares, 10 'hard' solar flares and 10 'soft' gamma repeaters. The fine time structure of bursts was monitored, in addition to their extremely variable spectra, and this data provided hints as to the origin of these outbursts. The solar oscillation experiment on Fobos 2 also returned a great deal of high-quality data between July 1988 and January 1989, even though it suffered from large and unexpected pointing errors.[52]

The outlook for Fobos 2 on arrival at Mars was not particularly good. Although the Soviets confirmed rumors that it had suffered problems, they insisted that these would not significantly impair the mission. But amid reports of failed instruments and of the spacecraft having switched from a high-rate data return to a lower rate, a meeting of the international team of scientists involved was initially deferred and then canceled without being rescheduled.[53] The problem was that one of the triply redundant attitude control system computers had failed completely and a second was malfunctioning. If the malfunctioning computer were to be lost, then the third computer, which was functional, would be unable to out-vote its two siblings. This would end the mission. In addition, the main transmitter had malfunctioned, and the less powerful backup transmitter was being used, which reduced the data rate. On 23 January 1989 the inert Fobos 1 flew by Mars. That same day, Fobos 2 made its second course correction, this time of 20.8 m/s, and then on 28 January it fired its engine to slow down by 815.1 m/s in order to enter a preliminary orbit of Mars ranging between an apoapsis of 81,301 km and a periapsis altitude of 864 km with a period of 76.5 hours. It had an inclination relative to the equator of just 1 degree, which was comparable to that of Phobos.

On the early revolutions, ESA's plasma-wave instrument detected the crossings of the planetary bow shock, magnetosheath, magnetopause and magnetosphere (but they were not always identifiable), as well as a forerunner 'foot' which appeared to flap to and fro with respect to the planet. In the case of a non-magnetic or weakly magnetic planet like Mars, these phenomena arose from the interaction between the solar wind and the planet's ionosphere; so much so that the bow shock was less than half a Mars radius sunward of the planet (whereas it is in excess of 10 radii sunward of Earth and 100 radii sunward of Jupiter, both of which planets are strongly magnetized).[54] In fact, Mars provided a rare chance to observe how the solar wind directly interacted with the plasmas around a planet. The results showed there to be a transition between the two plasmas, with the deceleration of the solar wind often occurring via a succession of stepped shock-like features.[55] Also during these early orbits Fobos 2 observed plasma and magnetic field perturbations when it crossed the orbit of Phobos, which researchers suggested may have been due to the presence of a ring of gas or dust along the moon's orbit. In fact, the existence of such a ring had been proposed in 1971. Despite extensive theoretical work on the subject, however, these 'Phobos events' were not definitively explained. One additional event was observed on 1 February between Phobos and Deimos which may have been due to Deimos presenting an obstacle to the solar wind.[56]

As the orbit was gradually lowered and circularized, the periapsis passes were used to conduct remote sensing of the equatorial regions of Mars. On 11 February the first Termoskan imaging data was obtained along a 120-degree swath including Elysium, Aeolis, Amazonis, Memnonia and Tharsis. Near the end of the swath, Pavonis Mons entered the camera's field of view. Owing to the elliptical orbit and

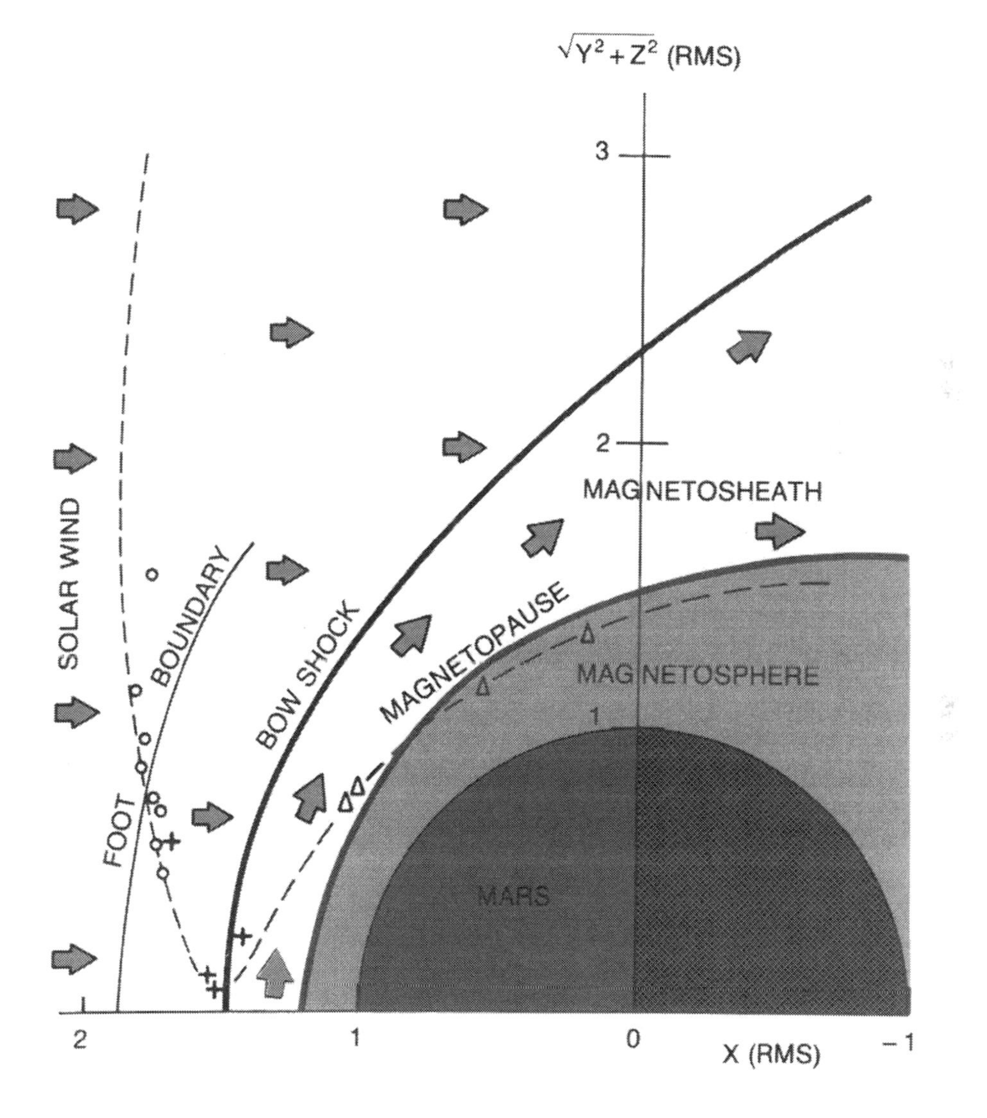

The locations of the Martian magnetospheric boundaries as observed by ESA's plasma-wave instrument on the Fobos 2 spacecraft during its first four orbits. Distances are measured in Martian radii of 3,393 km. (ESA)

altitude of periapsis, the scanned area was 120 km wide at its narrowest point with a resolution of 300 meters.[57] Most of the gamma-ray spectra were obtained during these early orbits, with 4 hours of data being taken at each periapsis and 1 hour of calibration and ambient background readings at apoapsis. The remote-sensing data indicated the surface composition along a wide strip in the equatorial zone to be remarkably similar to that measured at ground level by the Viking landers at two locations in the northern hemisphere. This probably meant that most of the planet was covered by wind-deposited dust, making a homogeneous surficial layer. It was also noted, however, that a contribution may be present from a bedrock whose best analog appeared to be the kind of basalt found on oceanic islands on Earth.[58,59] The orbit was lowered on the fourth periapsis, and then thrice more, concluding on 18 February with the final 722-m/s burn to circularize the orbit at an average altitude of about 6,270 km (which was only a few hundred kilometers higher than Phobos) with a period of 7.66 hours matching that of Phobos. Having successfully achieved this 'observation orbit', the spacecraft jettisoned its Fregat stage to clear the field of view of most of the instruments. Meanwhile, on 17 February the Deep Space Network had obtained its first cache of interferometric data to precisely locate the spacecraft in preparation for the rendezvous with Phobos.[60] The circular orbit was well suited to studying the Martian environment, because it passed through the interface between the ionosphere and the inbound solar wind. The data from the particles and fields instruments over many orbits showed that the position of the bow shock ranged between 0.45 and 0.75 Mars radii, and mapped the underlying transitional layer (dubbed a 'planetopause' by analogy with a comet's cometopause) at which large electron plasma densities were recorded.[61,62] On 9 March the spacecraft noted the passage of an interplanetary shock produced by a solar flare that erupted 3 days earlier. This is one of only a few cases in which a solar energetic particle eruption has been observed by spacecraft located near both Earth and Mars, with the shock reaching Mars about 26 hours after it swept by Earth.[63]

Meanwhile, on 21 February Fobos 2 flew by Phobos for the first time, imaging it at ranges between 860 and 1,130 km, primarily to refine the moon's ephemeris. Remarkably, although the uncertainty in the moon's position was initially several hundred kilometers, it was located close to the center of the first image. As a result of these and similar images over ensuing days, by early March the moon's position was known to within several tens of kilometers, and as the observations continued it was pin-pointed to within just a few kilometers. In addition, on 27 February Deimos was imaged at a range of 30,000 km crossing a field of stars in Taurus which also included Jupiter. A number of narrow-angle images of Jupiter were taken in order to calibrate the pointing system, and then wide-angle images were taken of Jupiter, Deimos, Aldebaran and other stars in order to refine the orbital parameters of this satellite.[64] Fobos 2 used its thrusters to make small orbital adjustments on 7 and 15 March, and on 21 March it reached a station-keeping orbit relative to Phobos in which slight differences in eccentricity and of inclination with respect to Mars had the effect of making the moon appear to trace out ellipses on the anti-sunward side of the spacecraft at ranges varying between 200 and 600 km, to keep it in view. The next steps would establish an orbit that approached closer to the moon in order to arrange the flyby on which the spacecraft would release its two landers.[65]

By the third week of March, Fobos 2 had taken good data on Phobos and Mars, including high-quality images and thermal scans of both. Over the entire mission it took a total of 37 pictures of Phobos, including 13 higher resolution ones using the narrow-angle camera from ranges as small as 190 km and with resolutions of about 40 meters. These complemented the imagery from NASA's Mariner 9 and Viking orbiters. In the end, Fobos 2 viewed about 80 per cent of the surface. In particular, an area to the west of the crater Stickney (which is the largest crater on the moon, and had previously been seen only at very oblique angles at 100-meter resolution) was mapped in greater detail and found to be remarkably free of the grooves that appear to radiate out from Stickney; evidently either grooves were not created in this area or they were and have since been buried by regolith. It was also possible to observe the moon at low phase angles; i.e. with the Sun positioned behind the spacecraft. Although the absence of shadows 'washed out' topographic detail (as it does with the 'full' Moon viewed from Earth) it yielded photometric information and revealed that many fresh craters and grooves had bright material on their rims. After analyzing these observations, Soviet scientists chose the area over which the spacecraft would make its low pass, and possible target sites for the long-duration static lander and the hopper.

Precise tracking of Fobos 2 provided a quite accurate determination of Phobos's mass which, together with an estimate of its volume derived from images, gave a value for the moon's density that was significantly less than that from the Viking data, suggesting either that the interior of the moon was porous or that it contained significant quantities of ice.[66]

Taken on 28 February 1989, this picture of Phobos hovering over Mars is one of the most famous images from the Fobos 2 mission.

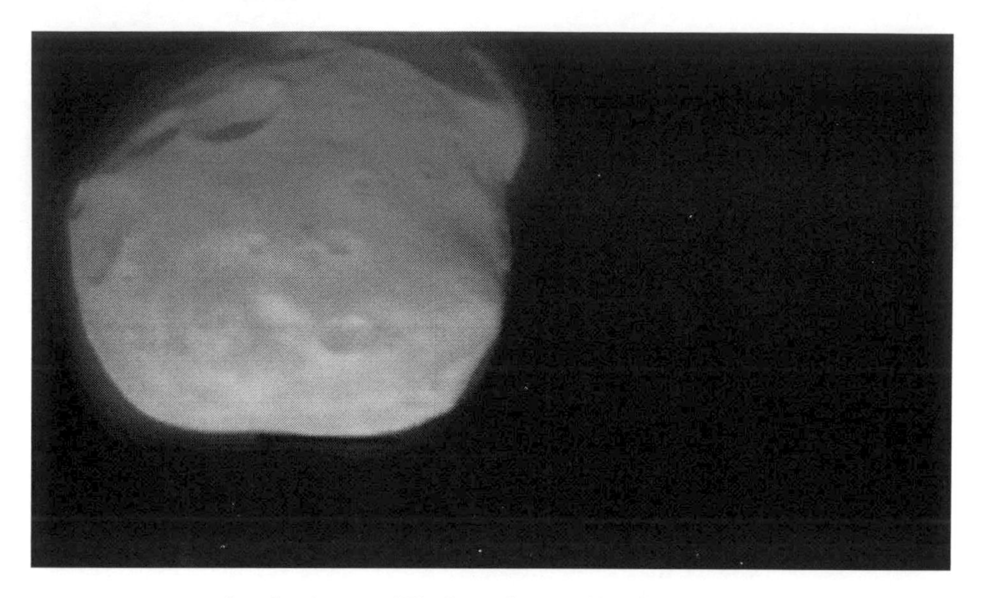

Another image of Phobos taken on 28 February 1989.

While in the initial eccentric orbits the French imaging spectrometer had taken two high-resolution swaths over the Tharsis region: the first one crossing Pavonis Mons and the second crossing the Biblis and Ulysses Paterae. While the spacecraft was in its Phobos-matching orbit, the instrument made a further nine observations. In addition to Arabia Terra, Isidis Planitia and the dark Syrtis Major, it was able to map over 25 per cent of Tharsis and Valles Marineris, gaining at total of in excess of 36,000 spectra. Precise altimetry of Pavonis Mons provided a detailed profile of its caldera. The data also provided a map of hydrated minerals in Valles Marineris, on Tharsis and at other sites. In particular, the flanks of the volcanoes atop Tharsis appeared to be richer in hydrated minerals than the surrounding plateau. Two scans of Phobos were obtained on 25 March from a distance of about 200 km and with a surface resolution of 700 meters. In these spectra the moon appeared more akin to carbonaceous chondrite meteorites than to a more water-rich chondrite. Although the degree of mineral hydration appeared to vary by up to 10 per cent from place to place on Phobos, the moon was distinctly drier than Mars.[67] Once the spacecraft was in a circular orbit, the resolution of the Termoskan along and perpendicular to the ground track were matched, providing images with little distortion. Three more imaging sessions of Mars were held in this orbit: one on 1 March and two on 26 March; one in daylight and the other in the evening. These scans covered most of the terrain types (only the characteristic polar terrains were unobservable) and included a wide variety of geological features, in particular the entire Valles Marineris.[68,69]

When Fobos 2 was close to Phobos, the magnetometers yielded intriguing data which suggested that the moon may have a weak intrinsic magnetic field – as some asteroids have since been found to possess.[70,71] The question of whether the planet has a global magnetic field remained unanswered: although when the spacecraft was

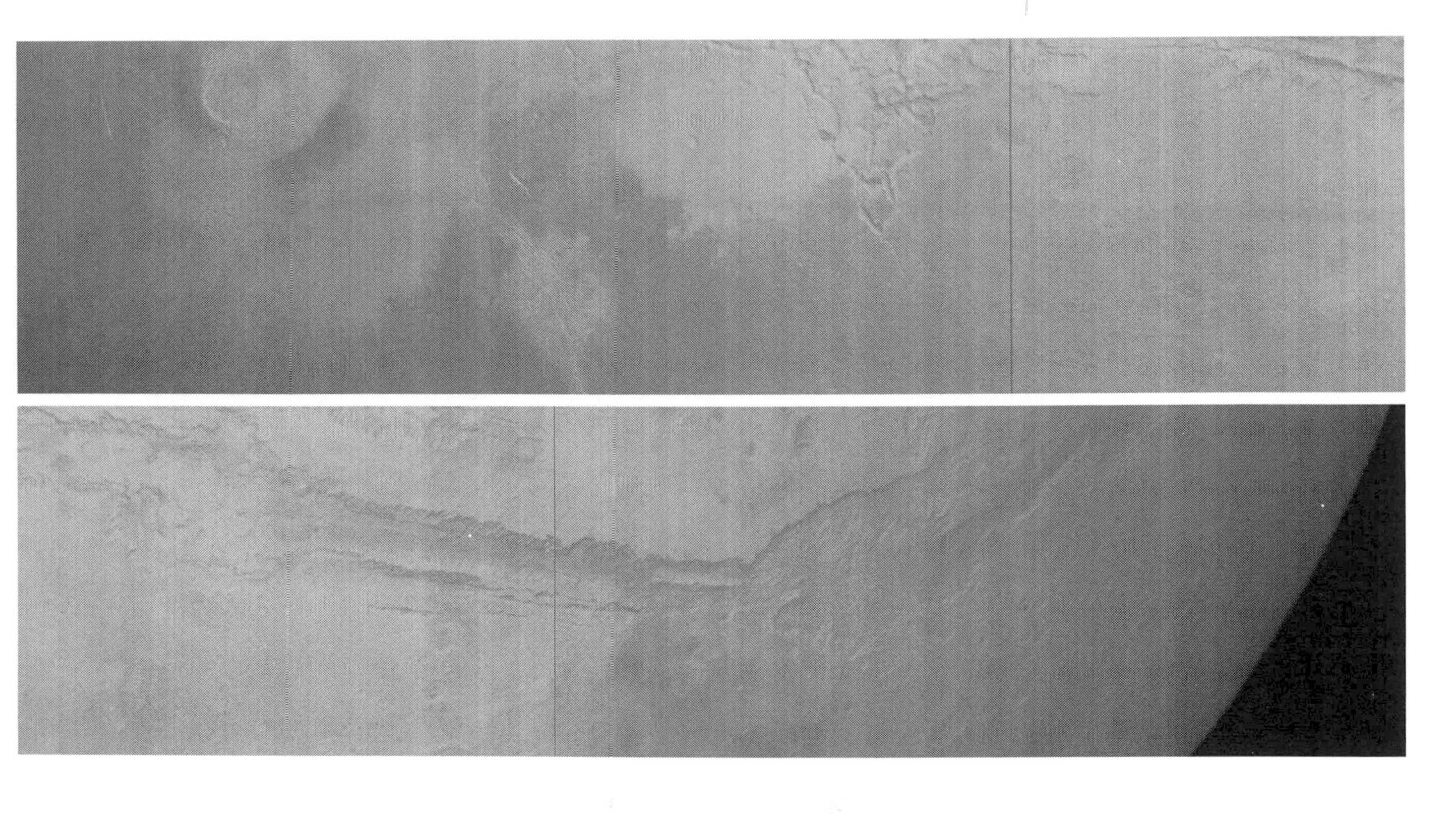

A full Termoskan swath obtained by Fobos 2 on 26 March 1989. It features Arsia Mons (top left), Noctis Labyrinthus on the summit of the Tharsis rise (top centre) and part of the Valles Marineris complex (lower frame).

in an elliptical orbit any field at large distances from the planet appeared to be induced by the solar wind, once it was in a lower circular orbit the possibility of 12-hour and 24-hour field periodicities hinted at the existence of an intrinsic field. Some of the most amazing particles and fields results were those provided by the Swedish ASPERA plasma analyzer and the German ion spectrometer. When the spacecraft was in the planet's ionosphere these instruments detected a surprisingly high flux of oxygen ions that were either energized by solar radiation or 'picked up' by the solar wind. Mars would appear to be losing of the order of a kilogram of oxygen per second; which could explain how it came to lose much of its initially much denser atmosphere, as well as its water. Although this mass loss is similar to that measured in the Earth's magnetic tail, for our planet's dense atmosphere this rate of loss is negligibly small. For Mars however, it translates to the planet having lost over the history of the solar system an amount of water equivalent to a global ocean with a depth of 1 or 2 meters.[72,73] Despite suffering difficulties with its Sun-tracking system, the French Auguste instrument found atmospheric ozone at high altitude, made measurements of the atmospheric water content, provided vertical temperature profiles, measured the opacity of dust in the atmosphere, and found a spectral feature that may indicated the presence of formaldehyde – a chemical with potential implications for the possibility of life. A total of 32 solar occultations were observed while crossing the limb into the planet's shadow, but due to stabilization difficulties while in eclipse only one at egress.[74,75,76,77,78,79]

Despite these successes, Fobos 2 was suffering serious difficulties. In particular, the second of three attitude control computers was still malfunctioning, and now so too was the only remaining backup transmitter. On 25 March a second series of interferometric tracking observations were made by the Deep Space Network. Two

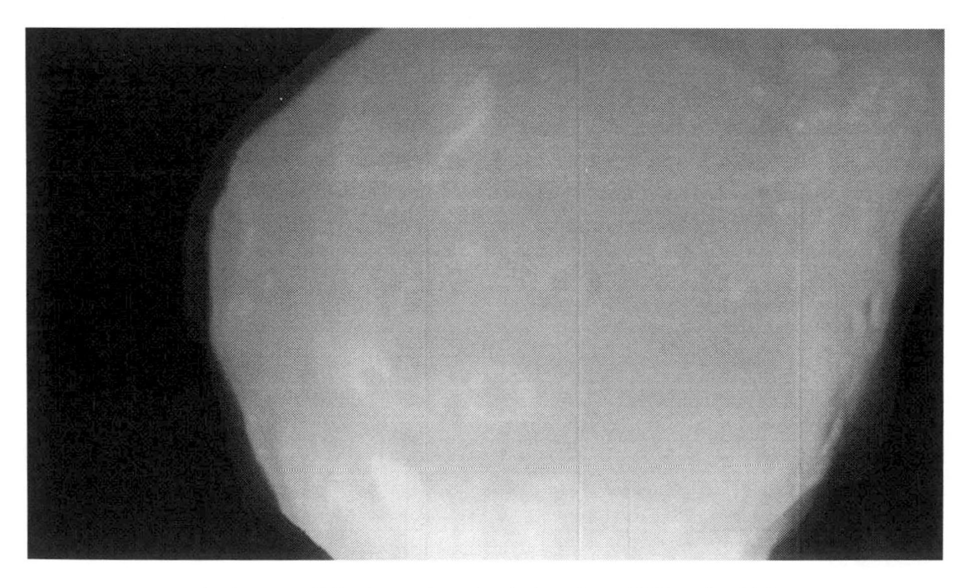

An image of Phobos taken by Fobos 2 on 25 March 1989.

A portion of the final Termoskan swath from Fobos 2. Since Phobos was orbiting Mars at almost the same speed as the spacecraft, the manner in which the image was built up represented the moon's shadow on the surface as a dark cigar-shaped feature.

A Fobos mission press meeting. Standing is Vyacheslav Kovtunenko, director of the Lavochkin bureau. Roald Sagdeev, director of IKI, is behind the microphones. On the extreme right is Yuri Koptev, a member of Lavochkin who would head the Russian Space Agency in the early 1990s.

days later, on 27 March, the spacecraft rotated once again to take another series of navigation images of Phobos in order to prepare for the slow flyby on 9 or 10 April on which the landers would be released. For this television session the spacecraft was to switch off its transmitter, turn to take the pictures of Phobos, and then turn back and re-establish communication with Earth. All that was heard from it was a weak signal some 4 hours after contact had been severed, which lasted 13 minutes. When this was finally decrypted, it was realized that the spacecraft was spinning on an unintended axis, and that unless it soon managed to face its solar panels to the Sun it would exhaust its battery. Unfortunately, nothing more was heard from it.

The first, simplistic and self-absolving explanation provided by the Lavochkin engineers was that the spacecraft had been disabled by an impact with material that occupied the moon's orbit, but all models (made in the Soviet Union by the IKI and in the United States by JPL) showed that the risk of collision with such debris was so insignificant as not to be a viable failure mode. In all likelihood, the second attitude control computer had finally failed, and the sole functioning computer was unable to out-vote the two dead ones. A contributing factor may have been that the battery could have run low during the television session, and Lavochkin had elected not to include a program (available to previous spacecraft) which would switch off all non-critical systems in the event of power starvation in order to extend the time available to the recovery process. In the spirit of 'perestroika', sources at the IKI criticized the poor planning and preparedness of the mission. The 3 years between approval and launch were evidently insufficient time for Soviet industry to develop robustness in their systems and software.[80,81]

It is worth mentioning a rather fantastic 'explanation' that has been offered for the loss of Fobos 2. Because the spacecraft was in nearly the same orbit as Phobos and was moving at a similar speed, Termoskan images of Mars often included the shadow of the moon on the planet's surface as a dark oblong cigar-shaped feature. Some particularly naive UFO enthusiasts suggested that the shadow was proof that a giant alien spaceship had approached the Soviet probe and shot it down![82]

Although on the one hand Fobos 2 could be classified as a successful mission because it managed to provide fresh and unprecedented data of Mars and Phobos – more data, in fact, than the total from all the previous Soviet missions to Mars – on the other hand it failed its primary Phobos exploration objectives, never managing to conduct a remote-sensing chemical analysis of the moon's surface or to deliver the landers. There was a proposal to launch a backup spacecraft as Fobos 3 as early as 1992 to recover the lost science, but there probably would not have been time to implement the changes dictated by the poor performance of the originals, and this plan was not pursued. However, Soviet scientists and engineers continued to think about exploring Phobos.

MAPPING HELL

Immediately after VOIR was canceled, JPL engineers and the scientists involved in Venus radar astronomy set out to investigate how the scientific objectives might be recovered by a less costly mission that would be more politically acceptable. It was estimated that the cost could be halved by deleting all of the instruments apart from the radar, reusing as much hardware as possible and simplifying the mission. The revised proposal became known as the Venus Radar Mapper (VRM).

The cost-saving measures were many. First, VOIR's high-resolution radar mode was deleted, thereby limiting the radar mapping to an equivalent optical resolution of 200–500 meters (nevertheless, this represented a considerable improvement over the Soviet Veneras) and greatly reducing the scientific data rate. Also deleted were all the other scientific objectives involving the atmosphere and ionosphere. The only instruments to be carried were the synthetic-aperture radar and the radar altimeter. The radar altimeter was to provide vertical profiles of areas ranging in size from 2 to 20 km attaining a relative accuracy of 5 meters, but when orbital uncertainties were taken into account the overall result would be a vertical resolution of between 30 and 50 meters. Passive microwave emission observations would also be able to be made by using the radar antenna and receiver as a radiometer. Radio tracking of the spacecraft during limb crossings would provide information on the atmosphere, and long-term tracking would provide data on the planet's gravity field. The use of an elliptical rather than a circular mapping orbit would also reduce costs, because the spacecraft would not require to perform the complex (and essentially untested) aerobraking maneuvers required for circularization, and it would not need to have an orbit-shaping engine. Moreover, whereas the original concept had required two antennas, one to collect radar data and the other to transmit it to Earth in real-time, if the spacecraft used an eccentric orbit a single antenna could collect radar data at periapsis and this would be taped for transmission at apoapsis. Although operating the radar as the spacecraft first approached periapsis and then drew away from the planet would vary both the altitude and speed (as the range varied between 300 and 3,000 km its speed relative to the planet would vary between 8.4 and 6.4 km/s) and result in degraded and varying resolution, new developments in computing meant that advanced digital synthetic-aperture radar processing (in contrast to the analog systems of Seasat and VOIR) would compensate for these effects. Nevertheless, to optimize the performance of the radar to take account of the varying distance and speed, parameters such as the pulse repetition frequency would have to be changed hundreds of times during a periapsis passage. But if the price to be paid to be able to fly the mission was the need to overcome such complications, then so be it. The point was to recover the science.

Although VRM would not be part of either the Planetary Observer or Mariner Mark II programs, it still would be built to their philosophy, in particular the use of hardware developed for other programs. The bus and the dual-role radar and high-gain antenna were spares from Voyager; the power, attitude control and command systems and tape recorders were from Galileo; the medium- and low-gain antennas were from Viking and the Mariners; the thrusters were from Voyager and Skylab; the fuel valves and filters were from Voyager; the radio amplifiers were from Ulysses; a tank from a Space Shuttle auxiliary power unit was used for hydrazine; and so on. By narrowing the scientific focus, switching to an elliptical orbit and by vigorously pursuing the strategy of reusing existing hardware, the estimated price tag of $300 million was about half that of VOIR. The mission was one of the four recommended by the Solar System Exploration Committee, and in 1984 it was approved by NASA – the first new-start planetary mission in many years. Martin Marietta of Denver was awarded a $120 million contract to design, assemble and test the spacecraft, and to assist with launch and in-flight operations. Hughes Space and Communications, having made the synthetic-aperture radar for the Pioneer Venus Orbiter, would provide the new radar system.[83,84,85,86]

In 1985, in line with NASA's new policy of naming planetary spacecraft after people, VRM was named Magellan in honor of the Portuguese navigator Fernão de Magalhães, who from 1519 until his death in 1521 led the first circumnavigation of the Earth and provided the first global appreciation of our world, just as the orbital radar would hopefully do in the case of Venus.

Magellan was to be launched in April 1988 by a Shuttle and the escape stage would be a Centaur G, the shorter of the two versions of the Centaur that had been adapted for the Shuttle. It would enter orbit around Venus in October 1988, and its primary mission would last 243 days, which is the length of time the planet takes to rotate once on its axis. The orbit would range between about 250 and 8,000 km, with a period of 189 minutes and an inclination of about 86 degrees to the equator. The periapsis would be in the northern hemisphere in order to obtain excellent data on the high northern latitudes, where some intriguing topographical features were known to be situated. The almost perfect spherical shape of Venus meant that the

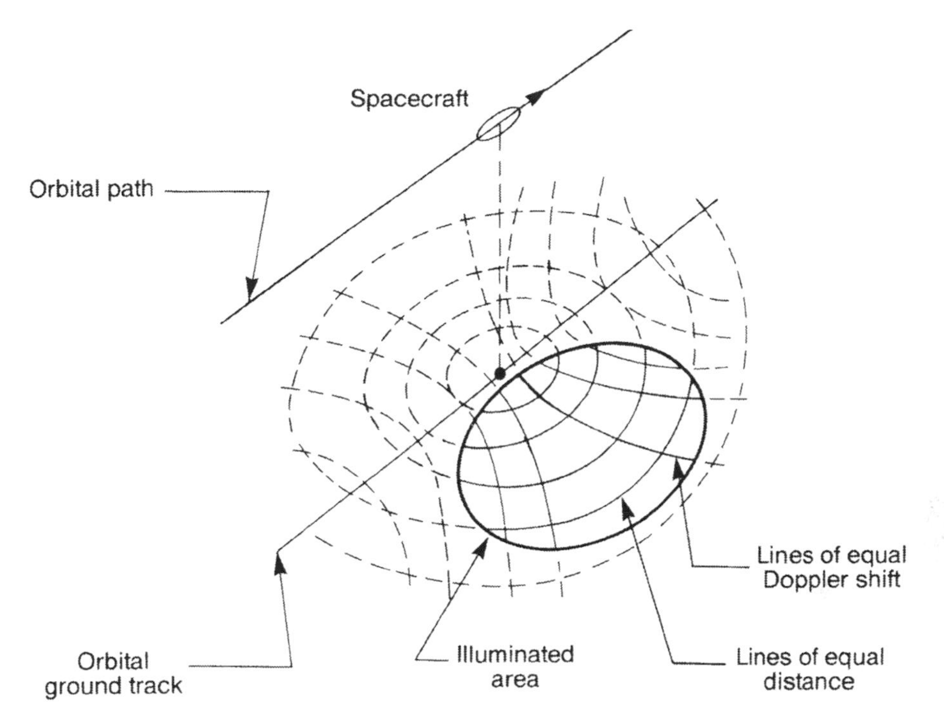

The observing geometry of the Magellan synthetic-aperture radar. The resolution on the surface across the orbital ground track was obtained from the time delay or distance coordinate, whilst the resolution along the track came from the Doppler shift coordinate. The radar beam illuminated an area off to one side of the ground track, for otherwise it would be impossible to discriminate between echoes which came from the left and those from the right.

orientation of the orbit would be so stable that at successive periapses the ground track at the equator would be displaced by only about 20 km, with the result that the 25 × 15,000-km radar swaths (dubbed 'noodles') would conveniently overlap. Moreover, to preclude excessive overlap of swaths at the north pole successive mapping passes would alternate from between the north pole and an intermediate southern latitude, and between a temperate northern latitude and a greater southern latitude. During each swath, the spacecraft would slew around in order to view the surface at a steadily varying angle, while maintaining the antenna 35 degrees off to one side of the ground track (as opposed to 10 degrees in the case of the Veneras) to better distinguish terrains having rough surfaces. Only 37 minutes of each orbit would be devoted to taking radar data. During the remainder, the spacecraft was to turn to point its antenna towards Earth, transmit the recorded data, and undertake engineering tasks such as recalibrating the attitude control system to ensure that successive swaths would be correctly positioned and that the Earth-pointing was accurate in order to enable the data to be transmitted at the highest possible speed. During its primary mission, Magellan was expected to map over 70 per cent of the

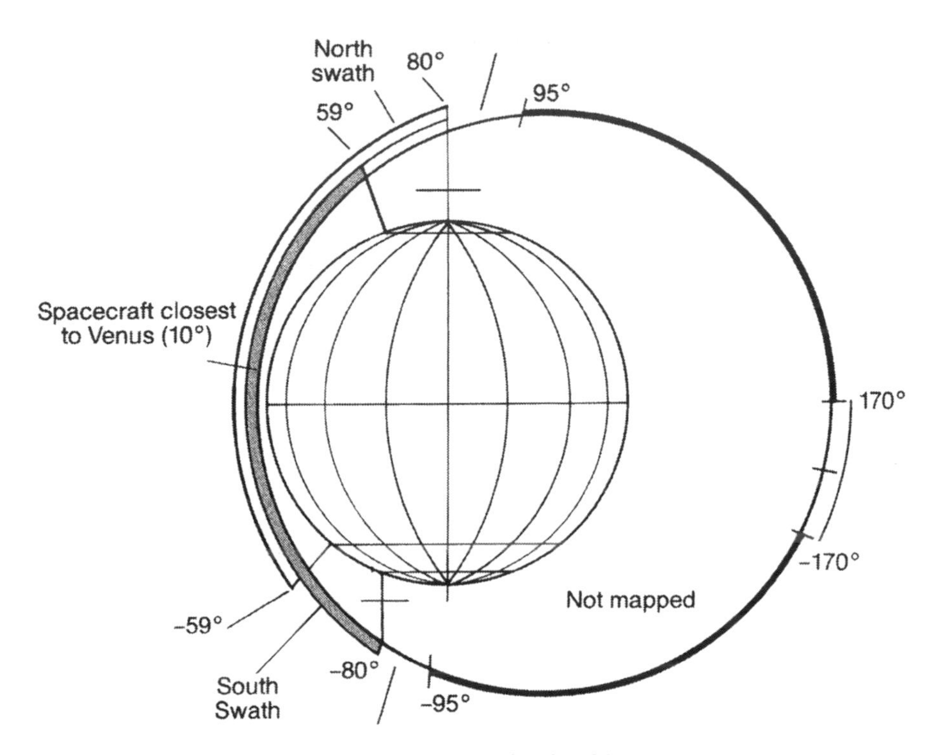

When the Magellan spacecraft was at the periapsis of its eccentric orbit of Venus it would alternate between radar swaths covering the high northern or high southern latitudes. The remainder of the time, it would point its antenna at Earth to transmit the radar data.

planet's surface. If the mission were to be extended, it would be able to fill in any gaps (known as 'gores'), image interesting features from different perspectives for stereoscopic analysis, and increase its coverage of the southern hemisphere. At the conclusion of the mission, owing to the fact that our planet is mostly shrouded by water, scientists expected to know more about the surface of Venus than they knew about the Earth.

Although the Challenger disaster affected the Magellan mission, the impact was less traumatic than for the other American planetary missions then in the pipeline. The cancellation of the Centaur G meant that Magellan had to be revised to use the IUS instead. Because it was not known whether the Shuttle would have resumed flying by the April 1988 launch window (it did not) it was decided to postpone the launch. The next window, in November 1989, would have required Magellan to set off within days of Galileo, which was to initiate its long voyage to Jupiter with a flyby of Venus. Rather than rely on back-to-back Shuttle missions, JPL devised an unprecedented 'fourth type' of trajectory in which Magellan would be launched in a 1-month window that extended from late April to late May 1989 and travel more than one and a half times around the Sun to reach Venus, instead of the usual half

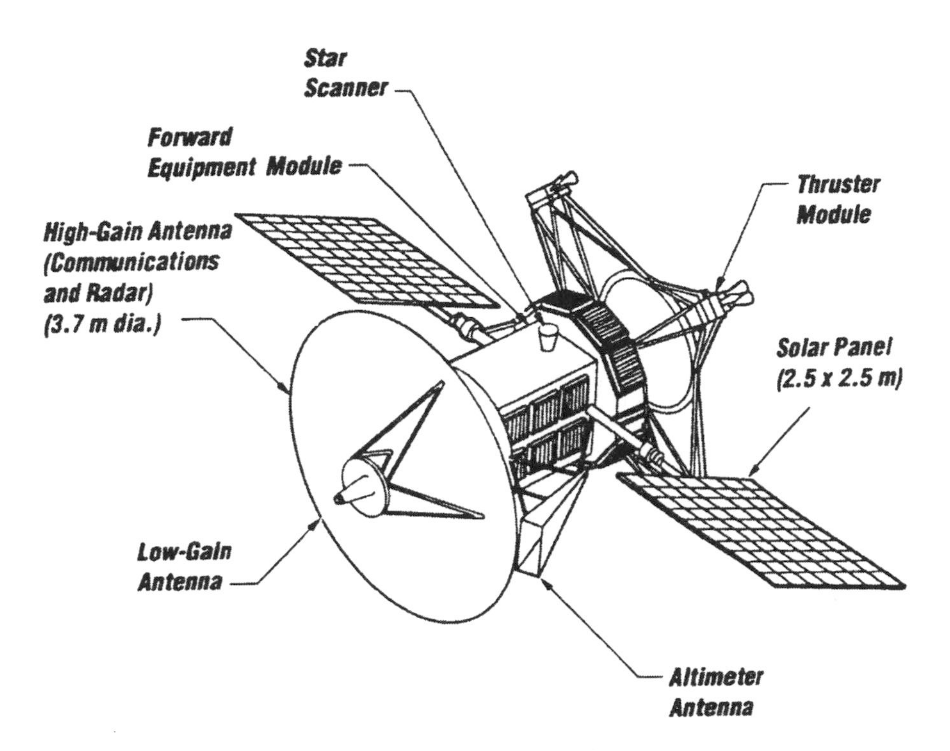

The Magellan spacecraft.

circuit. Although this trajectory had some advantages in terms of the energy of the departure from Earth and on arrival at Venus, it imposed a 15-month cruise and an 18-day hiatus for solar conjunction midway through the primary mission. If this launch window were to be missed, the next opportunity for a conventional transfer to Venus would not open until May 1991.

The 3,453-kg Magellan spacecraft was dominated by the Voyager antenna that was to be used for both radar data collection and communications. At 3.7 meters in diameter, this spanned the payload bay of the Shuttle. To provide communications at times when neither the high-gain nor the medium-gain antennas could be aimed at Earth, there was a low-gain antenna from Mariner 9 on the feed of the big dish. The high-gain antenna would transmit at either 268.8 kbps or 115 kbps, depending on the size of the Deep Space Network antenna that was used. The radar data from a periapsis pass would be stored on two redundant Galileo-heritage tape recorders, and replayed over two 57-minute sessions when Magellan was near apoapsis. The 1.5-meter-long horn of the radar altimeter was installed alongside the main antenna and offset 25 degrees from its axis so that when the synthetic-aperture radar looked off to the side the other would view the surface more or less vertically. Behind the antennas was a boxy module 1.7 × 1 × 1.3 meters in size which contained all of the radar electronics, some of the communication systems (including the medium-gain antenna protruding from one side, which was to be used primarily during the cruise to Venus and during the orbit-insertion maneuver), reaction wheels for the attitude control system, a star mapper for attitude determination and batteries. The exterior of this equipment module was covered with thermal control louvers to maintain the radar electronics, which consumed 200 W of power, inside its allowed temperature range. Next was the 10-sided Voyager bus, which was 2 meters across and 42.4 cm tall and housed the main computer, tape recorders and a number of other systems. At the center of the bus was a tank for 132.5 kg of hydrazine, which was sufficient for many years of operations in orbit of Venus. On each of two sides of the bus was a solar panel that was hinged against the frame of the spacecraft for launch and in its deployed state could be tilted in order to track the Sun. Each solar panel was 2.5 meters on a side, and at Venus's distance from the Sun would generate 1,200 W of power. At the rear was a propulsion module featuring a cross-shaped truss, at each tip of which was a pod housing a pair of 445-N thrusters, one 22-N thruster and three 0.9-N attitude control thrusters that were also to be used to trim the orbit around Venus. Also mounted on this truss was a small tank of helium to maintain the pressure in the hydrazine tank. The propulsion module also provided attachment points for the structure that would connect the spacecraft with the IUS at launch, and for the 2,146-kg STAR-48B solid-fuel motor that was to be used for the orbit-insertion maneuver at Venus (the same kind of motor as was often used as a kick-stage for terrestrial geostationary satellites). Magellan was 4.6 meters from the tip of the feed on the high-gain antenna to the propulsion module, and with the solar panels deployed it spanned 10 meters.[87,88]

Despite being promoted as a relatively inexpensive mission, by the time that it was ready for launch Magellan's price tag had almost doubled to $550 million; due in part to the delays caused by the Challenger disaster, but also as a result of cost overruns – mostly concerning the development of the radar (over which JPL had at one point to regain control from the contractor) and to redesigns and improvements which made the spacecraft more capable but also more expensive. In fact, as early as 1984 JPL had identified numerous "half-million-dollar items" for improving the spacecraft and its performance, some of which were implemented during the years that the Shuttle was grounded. One improvement that was not funded was to put a 30-cm aluminum skirt around the circumference of the antenna in order to boost its radar performance. Another source of overruns was that on the revised schedule Magellan would be launched before Galileo, rather than after it, which meant that some of the spare parts that would otherwise have become available from Galileo had to be retained by that program in case they should be required, with the result that new parts had to be purchased.[89] The manufacturer delivered the spacecraft to Cape Canaveral in early October 1988, but on 17 October it suffered an unusual accident when a technician put a connector into the wrong socket of a test battery, causing a short circuit. This started a fire, but it was readily extinguished and the spacecraft suffered only minor damage. At that time, the radar, high-gain antenna and communications electronics had not yet been fitted, and because the 'remove before flight' covers were still in place over most of the installed components they were protected. Nevertheless, it took several days to clean the soot and grease from the spacecraft. Overall, the incident cost the mission $80,000.[90,91]

The Magellan/IUS stack leaves its cradle in the cargo bay of Space Shuttle Atlantis.

The first attempt to launch Space Shuttle Atlantis on mission STS-30 was made on 28 April 1989, and was aborted 31 seconds before ignition owing to a problem with a pump on the Shuttle. It eventually lifted off on the third attempt, on 4 May, after a 1-hour hold for the weather. It was the first US planetary launch since the Pioneer Venus Multiprobe and International Cometary Explorer were sent aloft in 1978, and also the first planetary mission to be dispatched by the Shuttle, which had imposed such delays on the scientific and planetary exploration programs that most scientists "[wanted] nothing to do with [it]".

Five revolutions after launch, Atlantis released the 'stack' comprising the two-stage IUS, the motor for the Venus orbit-insertion maneuver and the spacecraft at an altitude of 296 km above a point 1,000 km southwest of Los Angeles. Shortly thereafter, the astronauts visually confirmed that the solar panels had hinged out correctly. Although the deployed panels would act like flexible appendages during the escape maneuver and introduce anomalous dynamics which the attitude control system of the IUS would have to overcome, it had been decided to deploy them in advance in order to protect them from the efflux from the roll-control thrusters of the IUS, which were close to their ends. Sixty minutes after it was deployed from the Shuttle, the IUS fired its two stages in sequence and injected Magellan into the planned heliocentric orbit ranging between 1.011 × 0.699 AU. Atlantis set down at Edwards Air Force Base in California 4 days later, having conducted a variety of microgravity experiments.[92]

On 7 October Magellan crossed inside the orbit of Venus and reached its first perihelion, but of course the planet was nowhere near. It then reached aphelion at Earth's orbit in early March 1990, and started back inward. Three corrections were performed during the cruise on 21 May 1989, 13 March and 25 July 1990 to refine the approach to Venus. A number of minor problems were encountered during the cruise, including spurious signals in the star tracker caused by solar protons hitting its detectors. Software patches were written to reject most of the false signals and so restore the tracker's capabilities, which would be essential to ensuring that the attitude of the spacecraft would be able to be updated to within a fraction of a degree on each orbit of Venus. High temperatures were recorded on the main bus, and also on the thrusters – where the heat induced gas bubbles to form in the hydrazine lines. The overheating in the bus was remedied by turning the spacecraft to cast the shadow of the high-gain antenna on the affected bays.[93] Two radar rehearsals were also conducted during the cruise: in December 1989 to test its basic working, and then in May 1990 to simulate a full mapping run over an interval equivalent to 20 orbits.[94]

As Magellan was cruising, two researchers at Brown University, Rhode Island, proposed a controversial theory which argued that Venus had undergone a process of plate tectonics similar to that which occurs on Earth. In particular, they said that in the low-resolution radar imagery from the Pioneer Venus Orbiter the equatorial Aphrodite Terra resembled the mid-Atlantic spreading zone where new lithosphere is continuously created. Moreover, Ovda Regio, a 3,500-km ellipsoid in Aphrodite, bore a striking resemblance to Iceland, which is a plateau on the spreading ridge. The researchers predicted that the higher resolution imagery from Magellan would reveal fine details which would prove their theory.[95,96]

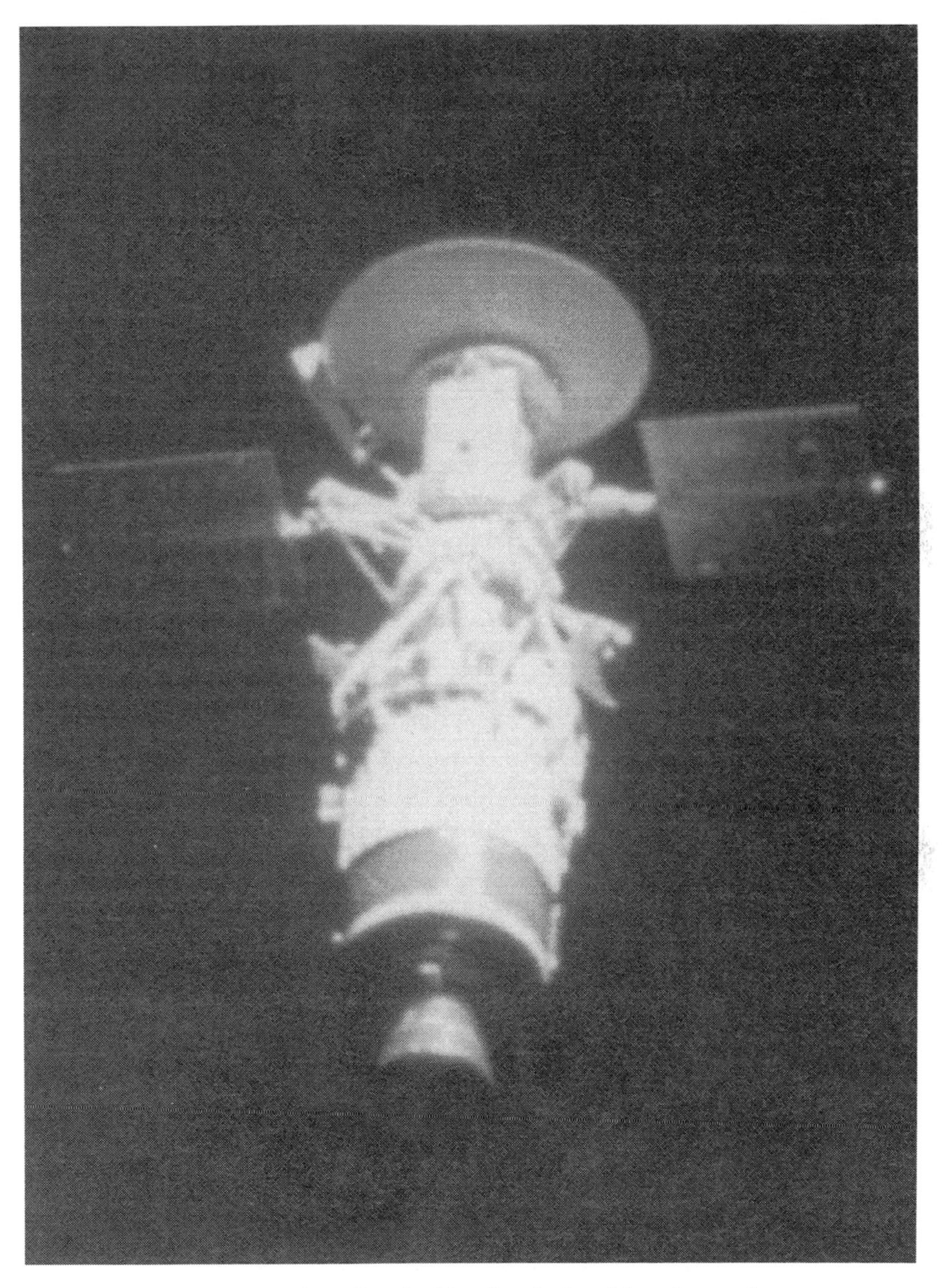

Once clear of the Shuttle, Magellan deployed its solar panels.

On 10 August 1990, Magellan fired its solid rocket to enter orbit around Venus: the aim point was within 100 km of that planned, and the burn was so precise that an orbit ranging between 289 and 8,458 km with a period of 3.26 hours at an inclination of 85.5 degrees to the equator and a periapsis at 10°N was achieved directly from heliocentric orbit, so no refinements would be required to achieve the mapping orbit. Magellan was only the second US spacecraft to enter orbit around the planet. In fact, the first, the Pioneer Venus Orbiter, which had arrived in 1978, was still operating, and it attempted a fascinating engineering experiment in which its photopolarimeter tried to image the plume from the retrorocket of the new arrival crossing the dark hemisphere, but it was too faint. The perfect orbit was extremely welcome, as the propellant saved would be able to be used to extend the mission to fill gaps in the map and also to change the viewing geometry of the spacecraft and its radar because radar echoes (and the images created from them) depend on the angle at which the beam strikes the surface – features that could not be seen at one angle of incidence may stand out at another, or when the polarization of the beam is changed. Magellan suffered several more glitches at this time: about a minute after the insertion burn one of the four gyroscopes of the attitude control system malfunctioned and was automatically turned off, and a backup memory became corrupted several seconds after the pyrotechnic bolts were detonated to jettison the spent motor casing.[97]

After several days of checks, the radar was powered up on 15 August and the engineers set about 'tuning' it to obtain the best possible imagery. If all went well, the plan was to make test mapping runs during the next few days and initiate routine mapping on 29 August. But as the scientists were celebrating the "stunning clarity" of the images produced from initial test data, the Deep Space Network reported not receiving telemetry from Magellan after it performed the star sightings required to update its attitude at apoapsis. A faint signal was detected 14.5 hours later, only to disappear and recur at intervals of 2 hours, which suggested that the spacecraft was slowly spinning and sweeping the beam of its antenna past Earth once per rotation. Ten hours later, Magellan was instructed to stabilize itself in an attitude that would allow it to point its medium-gain antenna at Earth. When the recorded telemetry was recovered the engineers deduced that the attitude control computer had lost its health-monitoring 'heartbeat', and this had prompted the spacecraft to enter a 'safe mode' that terminated all operations and switched communications to the low-gain and medium-gain systems. In this case, a slew should have been made to align the antenna to Earth and the solar panels to the Sun, but the two reference stars which it was to use to make this maneuver had either been missed or mistaken, and the spacecraft had ended up in an unexpected attitude and unable to communicate with Earth. Some hours later, a series of faults in the attitude control system had caused it to switch over to a simpler 'primitive' orientation control program in a 1 kilobyte Read-Only Memory that used thrusters instead of reaction wheels to maneuver. At that point, as the spacecraft began to cone (as it should) its antenna swept by Earth and control was able to be regained.[98]

But success in recovering Magellan was short-lived, because a few days later, as the checks were continuing, it again drifted out of attitude and contact was lost.

After 4 hours of waiting for the spacecraft to re-establish contact, the controllers at JPL, fearing either that it might misalign the solar panels and prevent the batteries from charging or that it might attempt unplanned Earth search modes which would waste precious fuel, decided to have the Deep Space Network send commands in the blind. After another 4 hours without a signal, a command was issued to disable the software that monitored the 'heartbeat', and control of the ailing spacecraft was regained. However, the engineers were at odds to explain what was happening to their spacecraft. Although several kilograms of hydrazine had been wasted, enough remained to support many years of orbital operations. While the engineers tried to determine the problem, the scientists presented a preliminary mosaic of the 34-km-diameter crater Golubkina produced from the early test data. It had a resolution as good as 120 meters, which was an order of magnitude better than that provided by the Veneras, and revealed for the first time details of its central peak, terraced wall and surrounding ejecta. Everyone was eager to complete the tests and start routine mapping.

A fortnight after the initial problem, attitude control was returned to the normal system. On 12 September the high-gain antenna was pointed towards Earth for the first time in almost a month, and the resumption of the maximum data rate enabled all the engineering data that had been stored on board since the first anomaly to be downloaded, allowing the engineers to conduct a detailed analysis of the problem. Meanwhile, the spacecraft continued to suffer attitude control problems as a result of taking star sightings at apoapsis.

Nevertheless, the radar was reactivated on 14 September and regular mapping began the following day.[99,100] Operations had to be suspended from 26 October to 10 November while Venus was on the far side of the Sun as viewed from Earth, at superior conjunction. The gap in the coverage would be filled in one local day later, in June and July 1991. There continued to be glitches, including overheating of components and excessive vibrations of the solar panels caused by oscillations in the attitude control system. The worst problem occurred in December, 3 months into the mapping mission, when one of the two tape recorders developed a rapidly worsening rate of errors and had to be switched off.[101] The first cycle of mapping concluded on 15 May 1991, having covered 83.7 per cent of the surface. Magellan had exceeded by a large margin the data from all the previous planetary missions combined. But as the overheating problem had become progressively worse, more and more of each mapping pass (at worst 55 minutes of it) was devoted to turning the spacecraft so that the high-gain antenna would shade the electronics. This was particularly critical when the orbital geometry was such that the propulsion module would be exposed to the Sun. Moreover, on 10 May Magellan suffered the fifth and longest loss of signal, this time being out of contact for 32 hours. But this was the last such outage, because in July the cause of the problem was finally found. In certain conditions, a software shortcut was putting the attitude control computer in a logical loop. The problem was exacerbated by the section of memory which had been corrupted at orbit-insertion.[102]

The second mapping cycle followed on immediately after the conclusion of the first, and imaged a swath to the right of the ground track rather than to the left in

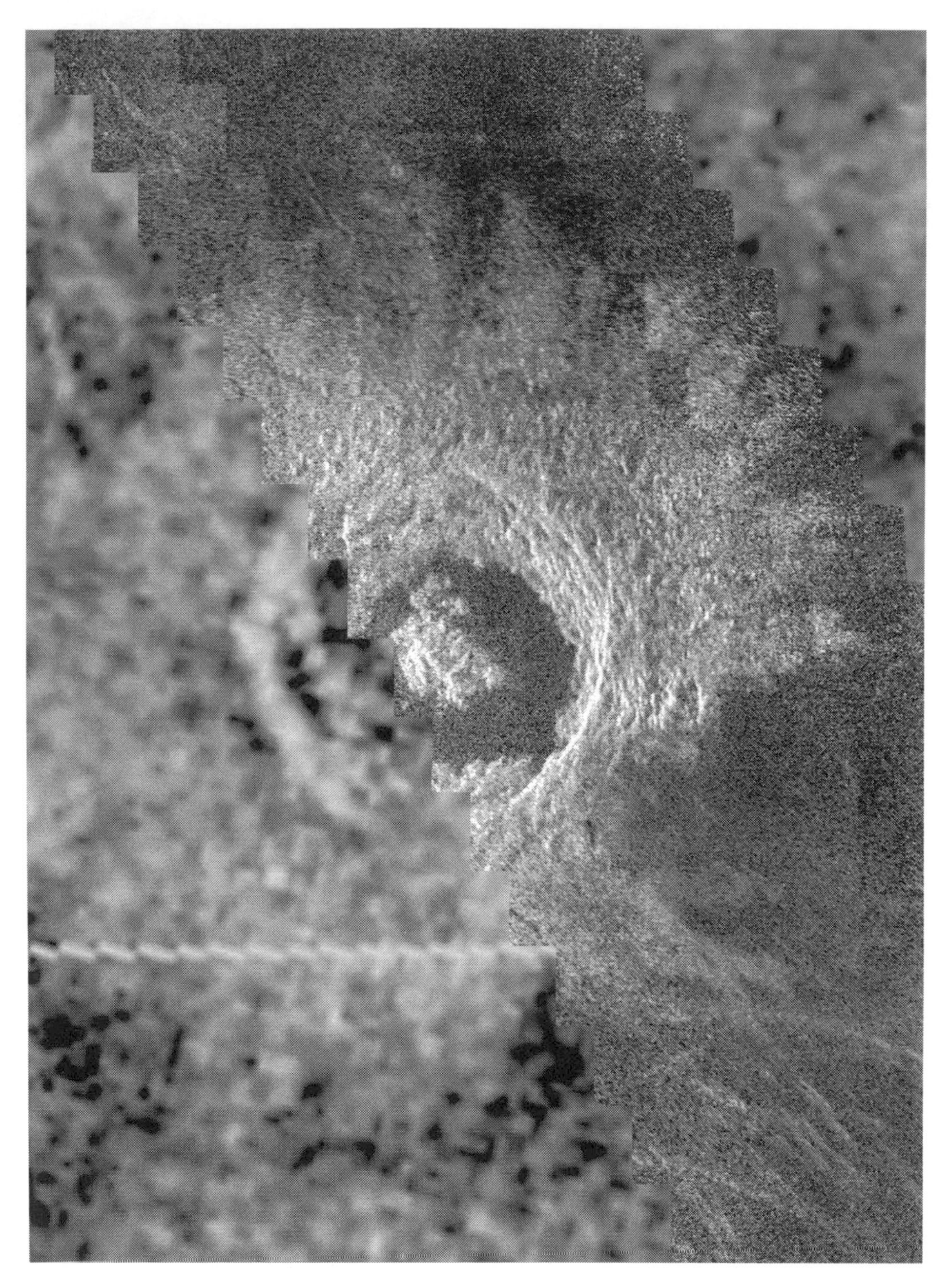

A Magellan image of the 340-km-diameter impact crater Golubkina taken while testing the mapping radar, set against a Venera image of the same landscape. The improvement in resolution is striking. (JPL/NASA/Caltech)

order to observe the terrain at a different angle of illumination. This cycle would fill in the gaps left by the communication outages, overheating and tape recorder problems, and expand the coverage to high southern latitudes.

Although Magellan and its radio system were optimized for radar observations, it was possible to 'sound' the atmosphere during radio occultations using the dual-frequency technique by which a signal transmitted from Earth first passed through the planet's atmosphere to reach the spacecraft and was then returned, in this case with the signal being transmitted at both of the spacecraft's operating frequencies. The experiment was conducted only once during the mission, on three consecutive orbits of the second cycle between 5 and 6 October 1991, and the two frequencies allowed the absorptivity of the atmosphere to be probed down to an altitude of 34 km. The polarization of the signal was also measured to detect the effects of clouds and lightning, and although no such effects were seen, the experiment did provide detailed profiles of the abundances of various gases in the planet's atmosphere, in particular of gaseous sulfuric acid.[103]

On 4 January 1992 the main high-gain antenna transmitter failed, and ceased to return data. Its backup was already known to be defective, since it drew excessive power, overheated and as a consequence introduced 'noise' to the transmission that impaired the data. When it was turned on as a test, it successfully returned data but it overheated and had to be switched off after just 25 minutes. If neither transmitter could be recovered, this would mark the end of the mapping mission, which had by now covered 96 per cent of the planet's surface. Even so, it would still be possible to obtain gravity data using the medium-gain antenna. The solution implemented later that month was to operate the backup transmitter at the lower data rate of 115 kbps to overcome the heating problem. Radar imaging resumed at the start of the third Venusian day, and the objective was to fill in the few remaining gaps. The week in which the radar could not be used was not wasted, because the Doppler-shift on the transmission was monitored in order to collect preliminary data for the gravity survey.[104,105]

A number of special radar experiments were performed during the first cycles. Five orbits on 12 July 1991 were allocated to a high-resolution test which required Magellan to maintain its beam at a constant angle of incidence with respect to the surface of about 25 degrees, rather than allowing it to vary from about 15 degrees over the poles to 45 degrees at periapsis. This effectively doubled the resolution in the 'along the track' direction to 60 meters. And starting on 24 July several orbits were used to collect swaths with the radar looking first to the left of the track and then to the right in order to make stereoscopic views of some southern hemisphere regions. The resulting 'stereo pair' images allowed the heights of single features to be determined to an accuracy of 70–100 meters. This test was so successful that the third cycle was replanned to concentrate on collecting stereoscopic data. On 13 December 1991 several orbits were dedicated to polarimetric observations of Rhea and Theia Mons. The spacecraft was rotated 90 degrees around its roll axis so that vertically polarized radio echoes could be collected to complement the horizontally polarized ones taken earlier. By subtracting the two radar echoes, it was possible to measure the dielectric constant of the surface. Finally, stereoscopic imaging of the Maxwell

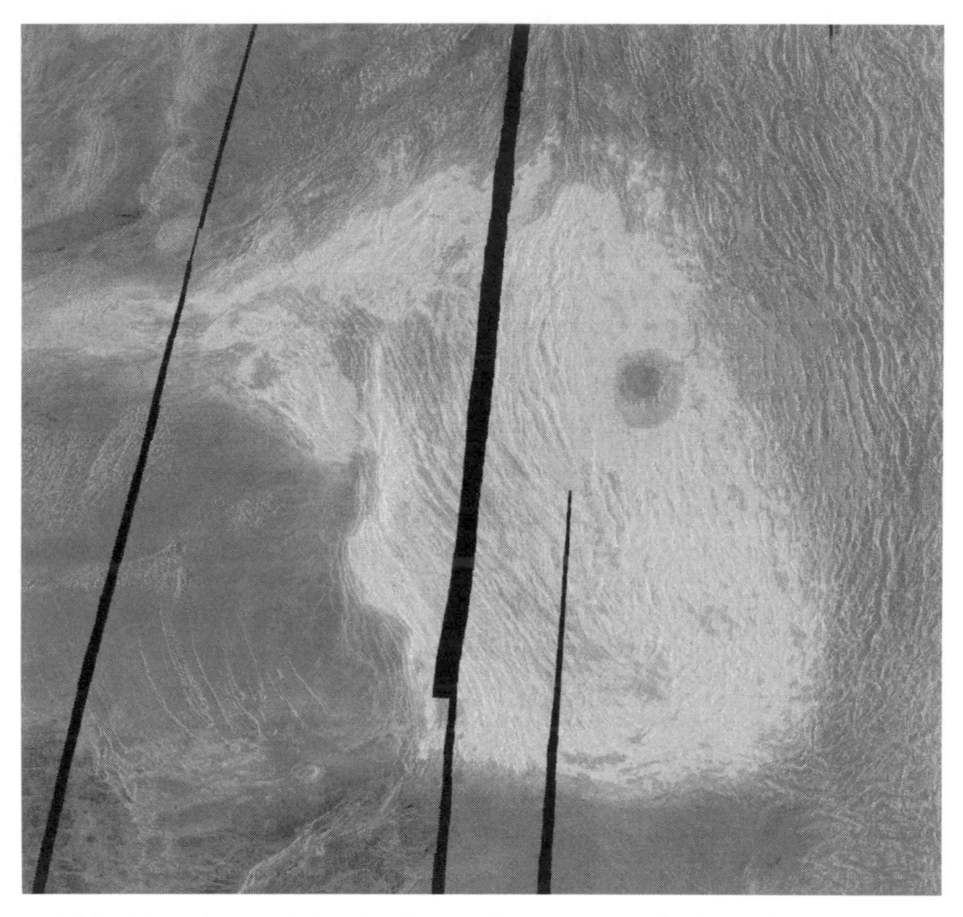

A Magellan radar swath showing the complex structure of the Maxwell Montes. The unusual brightness of the elevated terrain is probably due to a surface coating of a material that is strongly reflective at the radar's wavelength. The double-ring of the Cleopatra crater is clearly seen. The black streaks are the result of missing data. (JPL/NASA/Caltech)

Montes, Sif Mons and Gula Mons was scheduled between 24 January and 7 February 1992. In particular, scientists hoped to analyze the cliff that forms one side of Maxwell, which was evidently so steep that the altimeter was unable to resolve it. However, the observations could only be made at a time when the spacecraft's apoapsis would be occulted by Venus, which would limit the data that could be returned. Moreover, the opportunity occurred just weeks after the backup transmitter was brought into operation at its lower rate, which further reduced the data return. Although the observation was made, the Maxwell cliff was not able to be imaged owing to the fact that when it was in the field of view the recorder had to change track and reverse the tape direction – the tape had four tracks, alternating between forward- and backward-recorded ones.[106]

The mapping had to be curtailed when the apoapsis of Magellan's orbit was on the opposite side of Venus, limiting communications, but with the periapsis on the Earth-facing side of the planet conditions were ideal for making the gravity survey. During these orbits, the high-gain antenna was held pointing at Earth in order to receive and return a signal. When the relative motions of Earth and Venus, and of the spacecraft's motion around Venus were subtracted, the residual Doppler shift served to measure the varying gravitational pull from a patch of surface about as wide as the altitude of the spacecraft's orbit. Because an acceleration of less than 1 mm/s^2 would impart a measurable Doppler shift, the technique was remarkably sensitive. The first collection of data between 22 April and 16 May 1992 occurred over Artemis Chasma, one of the many rifts of Aphrodite Terra. At the same time,

A hemispherical view of the Magellan data centered at a longitude of 180 degrees. (JPL/NASA/Caltech)

the venerable Pioneer Venus Orbiter was providing gravity data over the southern hemisphere, although unfortunately in October it entered the atmosphere and was destroyed.[107,108]

After Magellan's backup transmitter once again overheated in July 1992 it was switched off to save it for the campaign late in the third cycle to fill the last major gap in the mapping coverage. By the conclusion of this cycle on 13 September, the radar had mapped 98 per cent of the surface. Unfortunately, it was not possible to implement an interferometry experiment in which radar data collected at the same geometry on different cycles would allow very fine surface detail to be resolved.[109] The fourth and future cycles were to be dedicated to the gravity survey, for which the slow rate over the high-gain antenna was more than adequate.

Each of Magellan's mapping swaths yielded about 100 megabits of imaging and related data – which was almost as much as the combined total from the radars on Veneras 15 and 16. The inflow of data was so prodigious that the computers at JPL used to process it were often running several weeks behind. Another challenge was to provide at least a preliminary inspection of the resulting images; so much so that the project scientists invited colleagues to bring in their post-doctoral and graduate students to assist. The mission also fostered a new era in the relationship between Soviet (later Russian) and American scientists, when three geologists who had 'cut their teeth' interpreting the Venera data at the Vernadsky Institute in Moscow were recruited to assist in analyzing the data.

Magellan confirmed Venus to be a world dominated by volcanism, with related terrains covering at least 80 per cent of its surface. There were structures ranging in size from domes several kilometers across to shields spanning at least 100 km. Almost the entire surface was dotted by thousands of small volcanoes concentrated in clusters. The shields (individually similar to their largest terrestrial counterparts) were located atop large regional rises which might have formed above upwelling plumes deep inside the planet. They could still be erupting. In addition, there were lava plains covering tens of thousands of square kilometers. These lavas appear to have been of low viscosity, and to have flowed in the same manner as the lunar maria. But unlike the lunar maria and the sand deserts of Mars, the plains of Venus had electrical characteristics indicating that, on average, they comprised hard soil. A more viscous and sluggish lava made pancake-like structures in the form of flat circular domes over 50 km across, having steep sides which rose no more than 1 km. These tended to occur more or less in chains of overlapping structures. As lava withdrew from the pancakes, it left troughs and pits on the upper surface. Although similar domes exist on Earth, those on Venus are at least 50 times larger, perhaps because the ambient temperature at the surface of Venus is so great that lava was able to flow further before cooling and solidifying. Remarkably, we may know the composition of a pancake, since there is such a feature within the target ellipse of Venera 8, the lander which performed the first gamma-ray spectrometer analysis of the surface. Another remarkable class of feature were the 'arachnoids' first seen in the Venera radar imagery. These networks of radial ridges in patterns resembling spider's webs (hence the name) are unique to Venus, and were probably formed by a combination of volcanism and tectonism.[110]

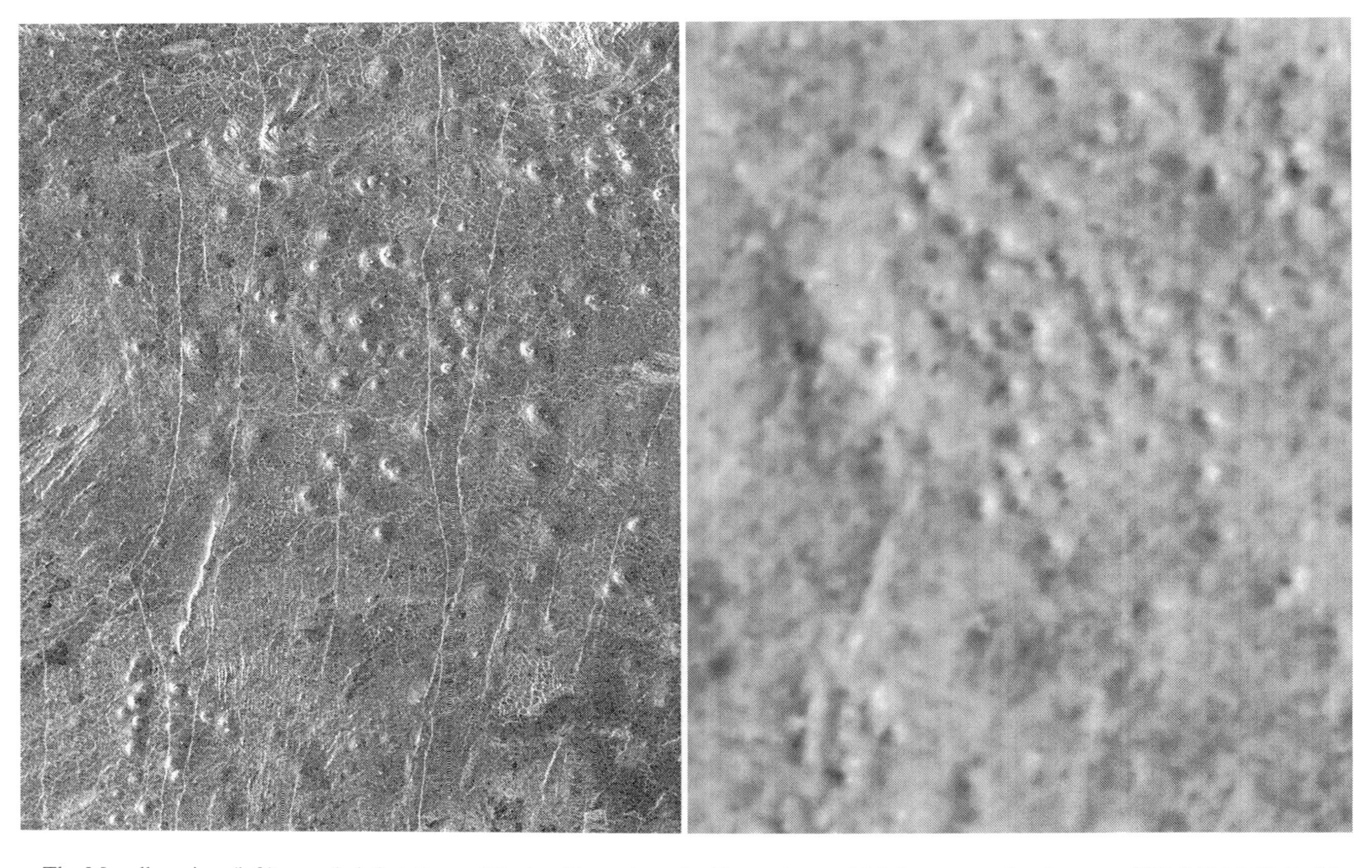

The Magellan view (left) revealed that the multitude of 'spots' on the Venera image (right) were small volcanoes. (JPL/NASA/Caltech)

A Magellan image of the Navka region covering most of the uncertainty ellipse of the Venera 8 landing site. As the circular structure (top right) is a pancake volcano, it is possible that the lander analyzed such lavas. (JPL/NASA/Caltech)

Another volcanic landform peculiar to Venus were meandering rivers made not by water but by an extremely low-viscosity lava such as carbonatite (rich in molten carbonates and salts) which, being just barely warmer than planet's surface, would have flowed for long distances and eroded deep channels. Similar but much shorter lava channels exist on the Moon, and one, Rima Hadley, was explored in 1971 by the Apollo 15 astronauts. The largest of these Venusian rivers (which the scientists jokingly dubbed 'canali') is Baltis Vallis, part of which was evident in the lower resolution imagery from the Veneras. Although only a few kilometers wide, it runs for 6,800 km, and is believed to be the solar system's longest channel. If it were on Earth it would run from New York to Rome! Although the Venusian rivers all lack lakes and tributaries, some sport other river-like features, including delta estuaries. They must be relatively old, since in some place they run uphill where the surface has since bulged upward.[111]

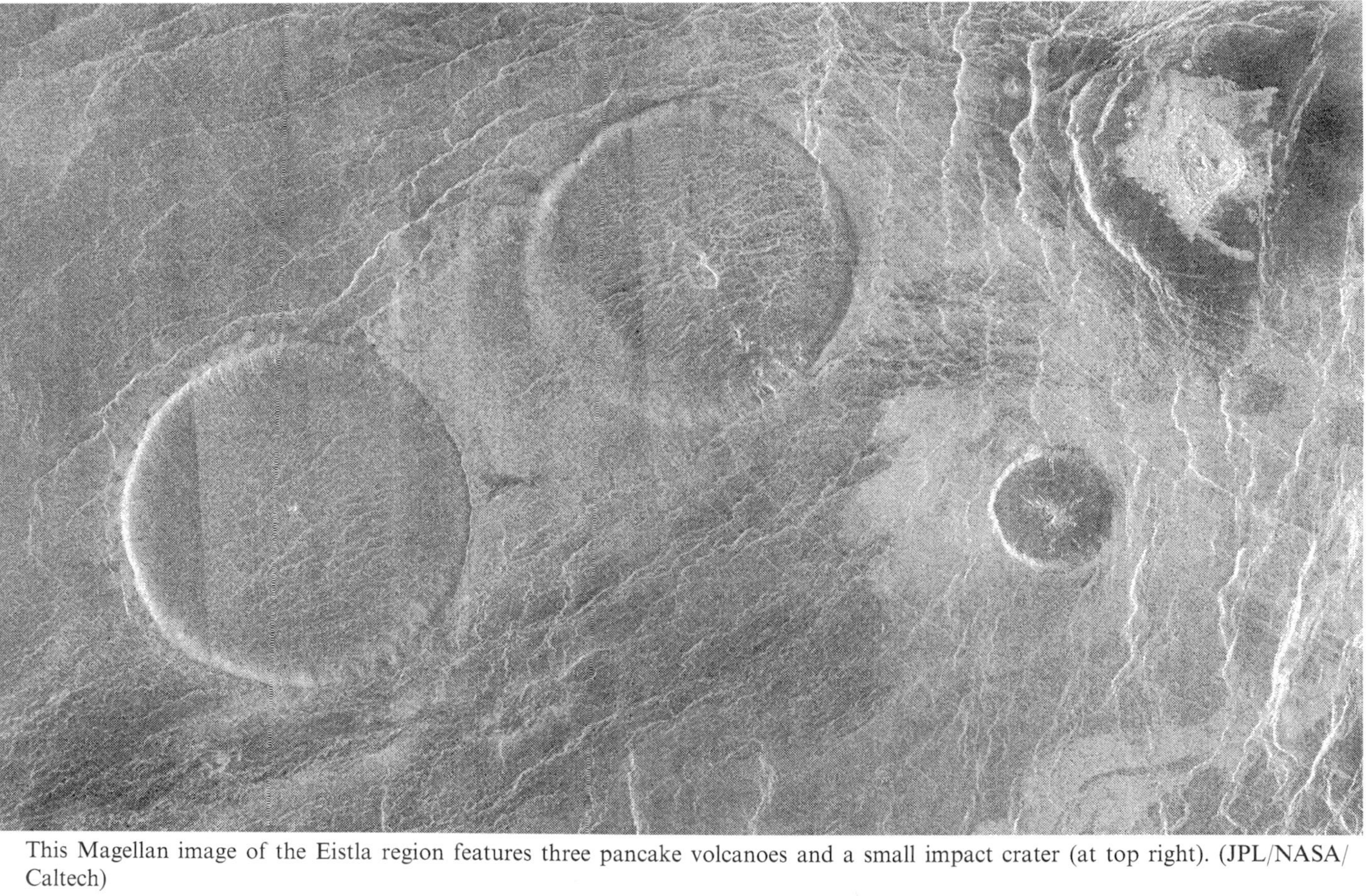

This Magellan image of the Eistla region features three pancake volcanoes and a small impact crater (at top right). (JPL/NASA/Caltech)

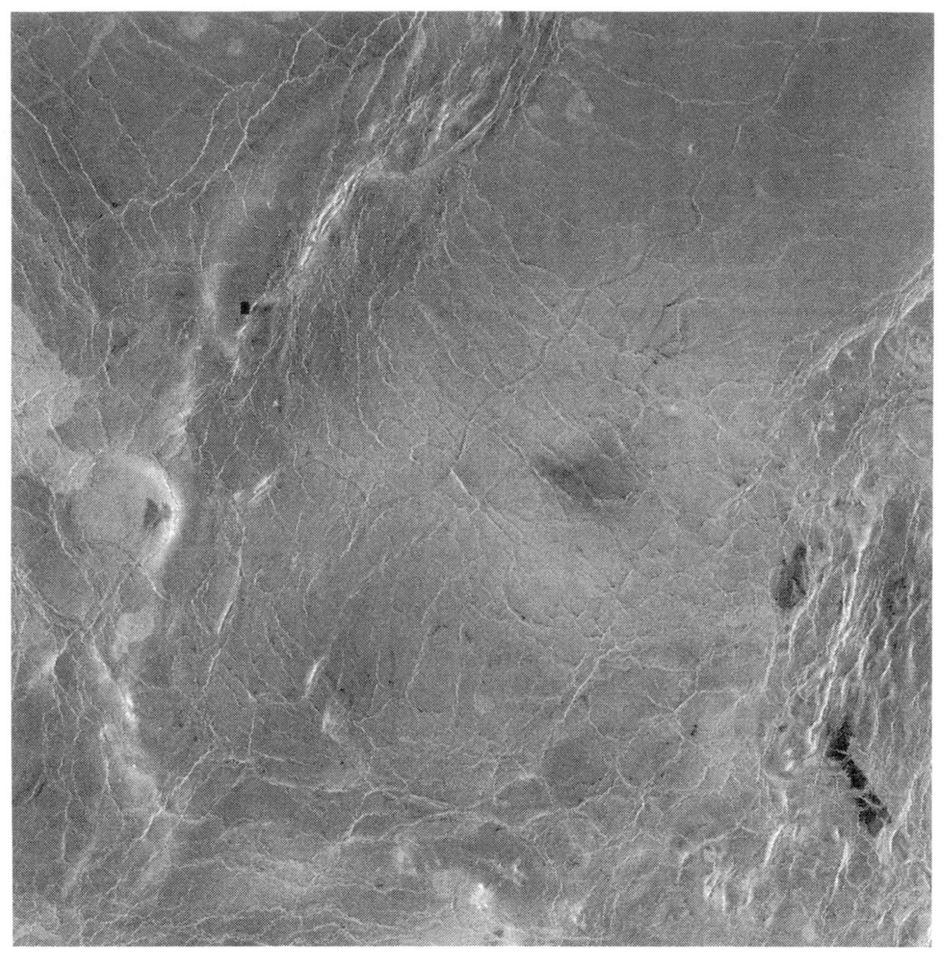

Meandering obliquely through the center of this Magellan image is a portion of Baltis Vallis, which is the solar system's longest channel. (JPL/NASA/Caltech)

Magellan's view of the Maxwell Montes revealed that contrary to the Soviet conclusion from the Venera data, Cleopatra is a double-ringed impact crater about 100 km across, not a volcanic caldera, therefore the designation Cleopatra Patera was withdrawn. The main evidence for this was the surrounding blanket of ejecta. However, as Venera 16 showed, it also has volcanic characteristics, most notably a flow extending out from a breach in its northeastern rim.[112] One theory was that the mountains of Venus were supported isostatically, as blocks of low-density rock 'floating' on a denser mantle material. Another theory involved the active uplift of mantle plumes. However, the fact that the slopes on the western side of Maxwell proved to be as steep as 35 degrees hinted that the origin of this highest of Venus's mountains might remain to be discovered. As researchers had discovered in the 1970s, the highest mountains have an anomalously high radar reflectivity. Data produced in parallel with Magellan's radar imagery revealed that this was not due to the landscape being particularly rough but to the dielectric characteristics of the surface, which appeared to be coated by a metallic material.[113] Although pyrite was proposed as a candidate this proved to be unstable in that environment, and metal salts involving chlorine, fluorine and sulfur and other metal-rich minerals are now believed to be more likely. There is a strong thermal gradient in which the surface temperature decreases with increasing elevation, and while such materials would evaporate on the low-lying terrain they might well condense on terrain over 3.5 km high and create a veneer of radar-reflective material a millimeter (possibly even a centimeter) in thickness.[114]

A network of ridges and rift zones was seen along most of the equator, running from the continent-like Aphrodite Terra to Beta Regio, and through the highlands of Atla Regio. These (and other) features were clear evidence that the surface has been subjected to tectonic stresses. While there were no distinct spreading regions such as the terrestrial mid-Atlantic ridge or East African rift system, or subduction trenches like those of the Pacific Ocean's 'ring of fire', there were features which bore a certain resemblance. For example, the two long canyons of Diana and Dali Chasmata situated in Aphrodite Terra had a raised rim on one side, a cliff several kilometers tall and a downward-sloping other side which hinted that they marked where slabs of crust might be in collision, with one slab sliding beneath the other and pushing it up to form a steep cliff. Moreover, some of the 'continents' appeared to have large volcanoes and canyons at their center which could mark where the crust was spreading apart. But the fact that volcanoes did not occur in chains (and even along the rifts volcanism was confined to individual volcanoes in the chasmata rather than on the associated ridgeline) suggested that the process of plate tectonics was not active in the terrestrial sense. The reason for this is debated. Perhaps it has something to do with the presence of water on Earth, which is drawn down into the subduction zones and stimulates volcanic activity; or with the greater plasticity of the 'hot' Venusian crust; or with the way in which the interior convects to release the heat of the core.

The coronae were another type of structure that had no a terrestrial counterpart. These were circular regions hundreds of kilometers across, with narrow concentric ridges at their periphery. The largest, Artemis, was 2,100 km across and edged by a 120-km-wide belt of ridges. There are about 350 of them, and they might mark where the surface was bulged upward by a rising plume of magma. Their centers often contain radial fractures, volcanic domes, and flows where it would seem that eruptions occurred repeatedly. When the uplift was withdrawn, the corona relaxed and the ridges formed around its periphery. Magellan documented these features at various stages of their evolution. Like the large volcanoes, they would appear to be associated with geological structures in the equatorial regions of Atla, Beta and Themis. The parquet terrain found by the Veneras could mark where downwelling plumes deformed, fractured and piled up the surface. However, this interpretation is disputed, and another proposal is that upwelling plumes were involved. Another type of structure that might be related to vertical movements in the mantle were the parallel and evenly spaced wrinkle ridges which would seem to be the result of the

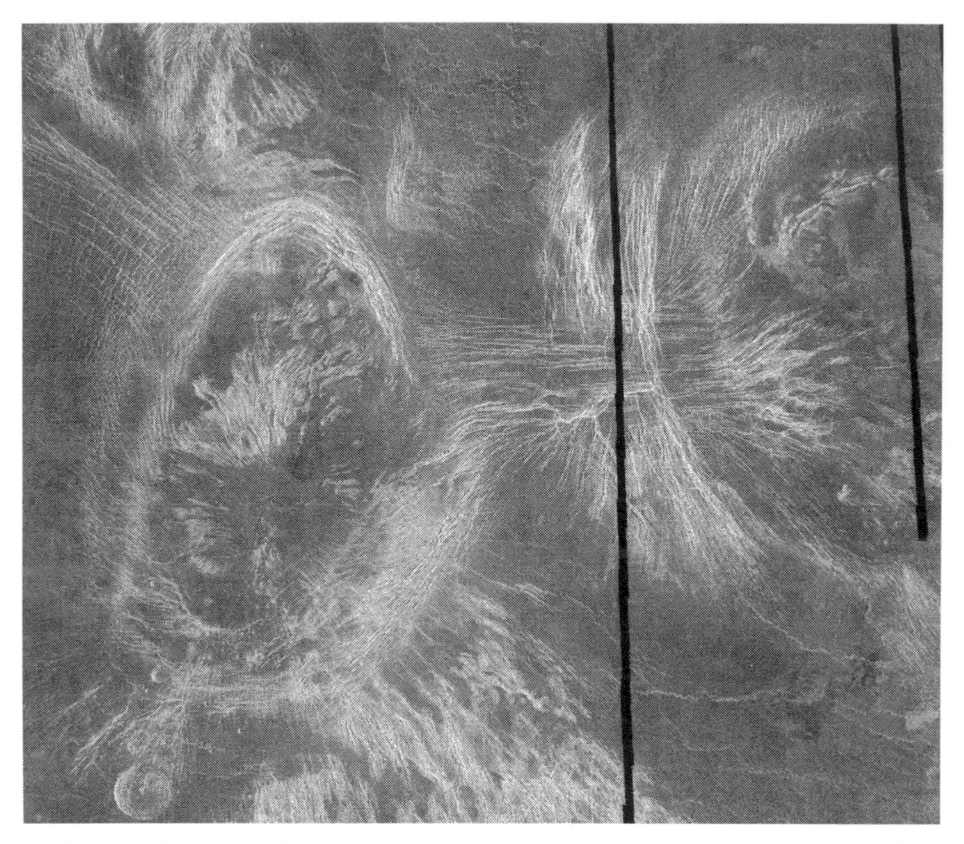

This Magellan image of the Fortuna region shows two coronae: Bahet (left) is about 230 km long and 150 km wide, and Onatah (right) is over 350 km across. (JPL/NASA/Caltech)

compression stress imposed on the crust by the formation of other features. In fact, these wrinkles were the most common feature on the planet, and most appeared to be aligned with the Aphrodite Terra continent.

Although plate tectonics in the terrestrial sense is not operating on Venus, it is clearly a tectonically active planet. Its form of tectonism has been nicknamed 'blob tectonics' since the crust is mobile in a vertical rather than horizontal manner.[115] But the inferences about the upwelling and downwelling motions in the mantle are tentative, to say the least, because we have no data. If a seismometer network such as was left on the Moon by the Apollo astronauts could be deployed on Venus we would gain some insight into the interior, but the hellish surface environment poses a severe technological challenge. The main objective of the Soviet proposal for the DZhVS long-duration lander was seismometry, but this mission was not flown and the task remains to be attempted.[116,117,118,119,120]

Magellan found thousands of impact craters, and because the final 'raw' image resolution was 120–300 meters per 'pixel' (depending on the surface elevation) it confirmed the conclusion from the Veneras that there were no craters smaller than about a kilometer in diameter. The Soviet scientists had argued that the objects that would make smaller craters were unable to reach the surface, since they burned up in the planet's inordinately dense atmosphere. A similar effect on Earth precludes craters less than 100 meters in diameter. Many of the craters on Venus appeared to be asymmetric, which probably meant that the atmosphere had managed to break the projectile into pieces during the descent. A surprisingly large number of craters possessed a flat floor and a central peak. The craters appeared to be surrounded by smooth patches that appear dark in the radar images, and might be rock which was ground to dust by the atmospheric shock wave that preceded the descending body. Circular patches of smooth terrain without a central crater might indicate where the shock wave reached the ground but the object responsible for it exploded in the air. The bright center on some of these 'splotches' could mark where associated debris showered the surface. The radar-bright (and hence rough) terrain surrounding other craters suggested fluid flows that could be either impact melt or lava erupted when the impact triggered local volcanism. Radar-dark (and hence flat) U- or parabolic-shaped tails which extended to the west of some craters appeared to be formed by an interaction between the meteor's shock wave, the atmosphere and the terrain.[121] The broad radar-bright fans associated with some craters could have been made by ejecta that attained sufficient altitude for the super-rotating airflow to carry it many kilometers before it fell back. Other craters appeared to have long radar-bright tails of windblown dust. Although the surface winds measured by the Venera landers were mild, the fact that the atmosphere was so dense meant they would be capable of mobilizing dust and sand. As in the case of Mars, the orientations of these 'wind streaks' enabled the air circulation on regional and global scales to be studied. In other places dune fields were found, and around the planet's largest impact crater, the 280-km-diameter multiple-ringed Mead located in Aphrodite Terra, there were yardangs, which are corridors of rock carved by windblown sand. As yardangs on Earth can form only in very friable landforms, the terrain surrounding Mead must be similarly weak.

The geographic distribution of the impact craters and the statistics of their sizes and shapes revealed that not only were small craters absent owing to the screening effect of the atmosphere, but large ones were also lacking. This (and the relatively small number of craters in comparison to Mercury or Earth's Moon) suggested that some process had destroyed the oldest and largest craters, and that those which are present were excavated in 'recent' times – estimates of this 'horizon' vary between 500 and 800 million years. Moreover, in marked contrast to Mars, which has a mix of older and more cratered terrains and younger and fairly pristine ones, the impact craters on Venus were essentially uniformly distributed. Another clue was the fact that very few of the craters on Venus appeared to have been covered and degraded by either volcanoes or folding of the crust. As had happened on many occasions in the history of science, in attempting to explain these characteristics the geologists were divided into catastrophists and uniformitarians. The catastrophists postulated that a combination of volcanism and other geological activity totally resurfaced Venus 500 million years ago, since which time the planet has been largely inactive.

Dark circular patches in Magellan radar images, such as here in the Lakshmi region, are believed to mark where the ground was subjected to the shock wave of a meteoroid exploding in the dense Venusian atmosphere. (JPL/NASA/Caltech)

The uniformitarians argued that Venus has always been active to some degree, but resurfacing is extremely effective, erasing all but the most recently made craters. The currently prevailing view is that volcanism on a global scale did indeed make a clean slate of the surface around half a billion years ago, since which activity has continued at a greatly reduced level. The nub of the issue is that some structures are so large that they could only have been produced by volcanism on a vast scale, and yet ongoing activity would appear to be necessary for the maintenance of the 'runaway greenhouse' effect of the atmosphere. However, the data from the three mapping cycles did not reveal any changes in the landscape to prove that Venus is currently geologically active. (There was a 'false alarm' in August 1991 when what seemed to be a recent landslide was found in images taken one cycle apart, but it was just an artifact of viewing the same feature at different perspectives.) Nevertheless, the fact that areas such as a large region around Sappho Patera are completely lacking in impact craters strongly argues that some degree of volcanic activity must still be ongoing.

Something which Magellan could not reveal about Venus, was what the planet was like prior to the near-global resurfacing event. Many lines of evidence suggest that early in its history the planet was a more benign environment, that liquid water could have existed on its surface, and that it could have sustained life that either developed indigenously or was inherited as a result of 'fertilization' by meteorites from Earth or possibly Mars (if this planet has ever hosted life). It is possible that when the surface became inhospitable, micro-organisms migrated to a more suitable environment high in the atmosphere, because at an altitude of 50 km the temperature is modest and stable, there is sufficient water to sustain simple life as we know it, and by shortening the night the super-rotating wind increases the scope for photosynthesis-like reactions. One way to prove that there is life in the atmosphere would be to collect samples at a certain altitude, either for immediate analysis or for return to Earth. But perhaps the best (and maybe even the only) place to look for a record of the early history of Venus, including the development of life, would be our Moon. This is because the lunar surface must have collected debris from all over the solar system, and it has been estimated that in addition to 20,000 kg of material from the Earth and 180 kg from Mars, over the history of the solar system an area 10 km on a side could have collected up to 30 kg of material from Venus.[122,123]

By the end of its third cycle, Magellan had effectively finished radar mapping, which released the spacecraft to pursue other scientific objectives, in particular the gravity survey, to which the fourth cycle would be devoted, and on 12 September 1992 the periapsis was lowered to 180 km for this purpose.

But because meaningful gravity data could be taken in this eccentric orbit only when Magellan was within 30 degrees of periapsis, located at 10°N, it would omit many interesting features such as the Maxwell Montes. It would clearly be better to lower the apoapsis and circularize the orbit in order to obtain gravity data on a global basis.[124,125] However, to do this using a propulsive maneuver would require about 900 kg of propellant and only 94 kg was available. It was therefore decided to use the aerobraking technique by which VOIR would have circularized its orbit in preparation for radar mapping. First, Magellan would fire its engine to lower its periapsis to skim the fringe of the atmosphere in order that the friction encountered on successive passes slowed the spacecraft and lowered the apoapsis to the altitude required for the final circular orbit, then the engine would be fired again to lift the periapsis. There was concern that electronic components of the solar arrays might be damaged and that some of the solar cells might crack or even become detached, but because the mission had achieved its main scientific objectives it was decided to accept the risk. Even if Magellan were to be damaged, aerobraking would yield indirect measurements of the conditions (mainly density and temperature) of the upper atmosphere. And the attempt would serve as an engineering experiment that would assist in planning future planetary missions. However, the 1993 NASA budget that had canceled the CRAF cometary mission had severely reduced funding for Magellan, and to further extend the mission its 1992 allocation had to be phased to cover operations through to mid-1993. Aerobraking had been under consideration for Magellan since the late 1980s, and plans for it had been developed in 1991, but JPL had initially estimated that it would take at least $40 million and a team of 200

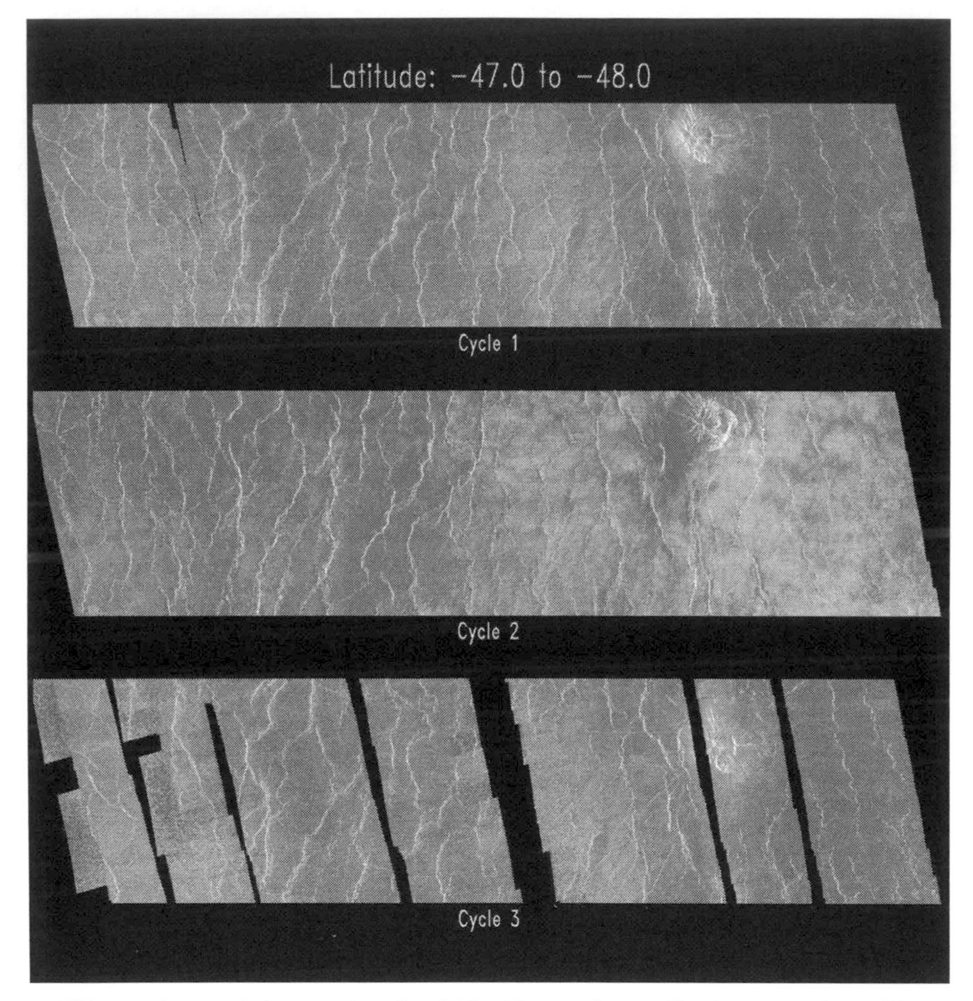

This terrain containing a series of wrinkle ridges and a small volcano was imaged by Magellan over three cycles one Venusian day apart. Although the terrain to the right of the volcano changes its appearance between the first and second cycles, the fact that it reverts in the third cycle proves that the difference in appearance is due to the changing geometry of the swaths, because on the first and third cycles the radar was viewing to the left side of the nadir point, whereas on the second it was viewing to the right – in effect the radar was illuminating the surface differently. (JPL/NASA/Caltech)

people to circularize the orbit and conduct two full cycles of gravity mapping. This was much more than NASA could afford, and so JPL developed a proposal which would cost only $8.2 million and involve 30 people. If the agency had not come up with the money, Magellan would have become the first US planetary mission to be switched off in flight. In fact, apart from the Pioneer spacecraft in the inner solar system whose funding had simply been discontinued after three decades, all deep-space missions had been tracked until they failed or exhausted their consumables. It is possible that Magellan benefited from the failure of Mars Observer in August 1993, because the funds allocated for its operations were able to be 'scavenged' by other missions.[126]

Aerobraking began on 25 May 1993, at the end of the fourth cycle, with the so-called 'walk-in' maneuver to dip the periapsis into the atmosphere. Magellan then turned so that the rear of its solar panels faced the direction of travel, the high-gain antenna trailed behind to provide passive aerodynamic 'weathercock stabilization', and the medium-gain antenna pointed more or less at Earth. The first aerobraking pass through the upper atmosphere lasted for about 4 minutes and subjected the spacecraft to an aerodynamic drag of 9 N (approximately the same as the weight of a 1-kg mass on Earth). The engineers had decided on a dynamic pressure limit of 0.32 Pa. A significantly lower pressure would make the aerobraking inefficient and achieving the desired change of orbit would take much longer than intended, and a greater pressure might generate sufficient heat to melt the solder of the solar panels or debond the honeycomb composite of the high-gain antenna. After re-emerging, the spacecraft slewed its antenna back to Earth and reported that apart from a slow rocking to and fro it had been aerodynamic stable, and the recorded temperatures were well within safety limits. As the aerobraking continued, every few orbits JPL calculated propulsive maneuvers to overcome the periapsis drifting either too low or too high. A total of 12 such adjustments were performed during the aerobraking phase. If ever the data were to show that the spacecraft was in immediate danger, it would be commanded to perform an emergency maneuver at the next apoapsis to lift its periapsis. All these maneuvers required a high degree of adaptability, as the orbit parameters were changing in an unpredictable manner. Fortunately, extensive data on the gravitational field and the outer atmosphere and had been collected by the Pioneer Venus Orbiter during its long orbital mission and terminal atmospheric entry. In July the orbit crossed the terminator and the aerobraking began to occur on the night-side, where the atmospheric density was different. Each pass reduced the period of the spacecraft's orbit by between 5 and 12 seconds, and the apoapsis by between 6 and 15 km. On 3 August the apoapsis was down to 540 km, which was deemed adequate to provide high-resolution gravity data, and the 'walk-out' was performed to lift the periapsis to the more stable altitude of 180 km, just above the atmosphere. The aerobraking was a great success, not only because it enabled the apoapsis to be lowered by 7,927 km at the cost of only 37.8 kg of propellant, but also because the spacecraft had survived 730 atmospheric passes without damage – in fact, its condition had improved because the contaminant on its surface that had earlier caused overheating issues had been removed![127,128] After the experiments in the Earth's atmosphere by the Atmospheric Explorer C and by the Japanese Hiten moon probe, Magellan became the first spacecraft to demonstrate the feasibility of aerobraking as an operational technique, thereby enabling mission planners to use it to reduce propellant loads, operational complexity and mission cost, in particular on missions to Mars.

The fifth and sixth cycles were almost wholly dedicated to the gravity survey in the near-circular orbit. The result was a map of the varying gravitational intensity with a resolution of about 200 km which showed a very high correlation between gravity and topography, with elevated terrain producing the greatest field strength and vice versa. From this it was concluded that the 'continent' of Aphrodite Terra was isostatically supported, and that the highlands of Atla and Beta Regio must be dynamically supported since the required depth of isostatic compensation would be implausibly great. Scientists were particularly eager to map the gravitational field of Mead, which at 280 km wide and 1 km deep was the largest crater and the only opportunity to study the structure of the lithosphere. The crater was detected on at least 10 consecutive orbits, and the results showed neither it nor the surrounding terrain to be isostatically supported. The low latitudes had been mapped during the fourth cycle in the original elliptical orbit, but there was a gap at higher latitudes that included Artemis, Atalanta, Tethus, Aino Planitia and Lada Terra, and which, owing to celestial mechanics constraints, could be filled only if Magellan were to continue until March 1995, but by now it had lost entire strings of cells off its solar panels, and it was questionable whether it would survive to that date.[129,130,131,132]

A number of other results were obtained as a by-product of operating in the low orbit, including an improvement of the planet's ephemeris and the orientation of its spin axis. During March and April 1994, several thruster firings further modified the

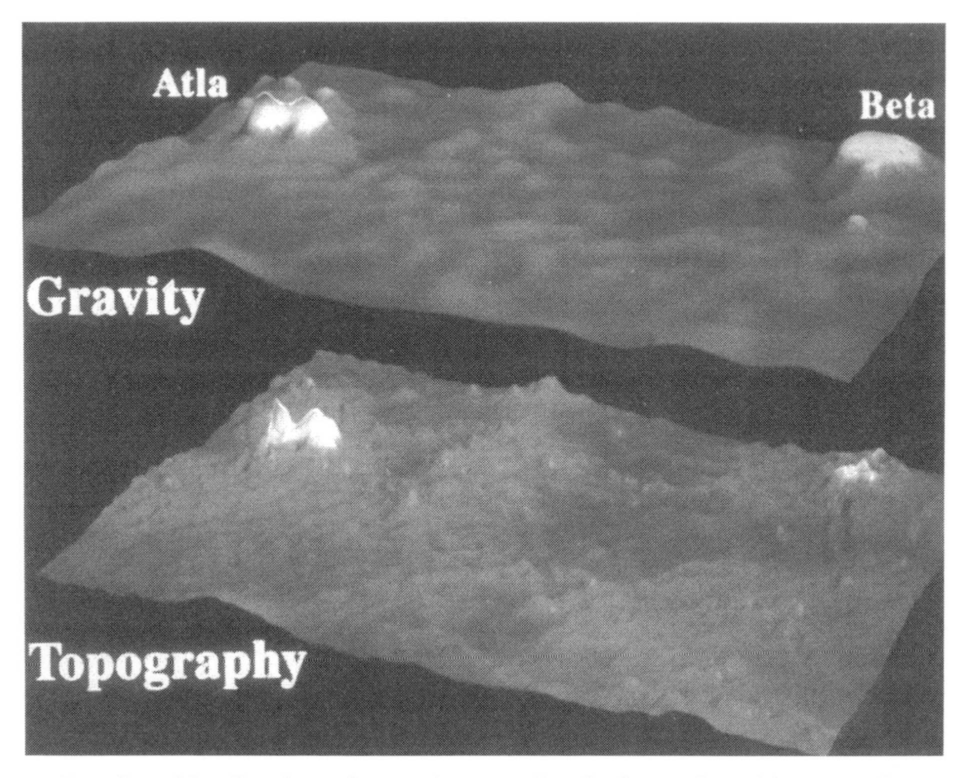

Data from Magellan shows that gravity anomalies closely correlate with topography.

parameters of Magellan's orbit, in particular to reduce the apoapsis to 350 km. In addition to the gravity survey, several experiments were performed during the sixth cycle. When Venus was at superior conjunction in January 1994, Magellan's radio signal was used to 'sound' the structure of the solar corona and the electron density in the vicinity of the Sun.[133] Then, bistatic radar observations were made in which the spacecraft's radio carriers were aimed at Venus and their reflection was recorded on Earth, in particular to study how the polarization of the waves changed upon reflection by the various terrains.[134] In another experiment, the periapsis was lowered to 172 km to once again skim the upper atmosphere. This time, however, the solar panels were tilted at opposite angles, like the blades of a propeller (hence the 'windmill experiment') and the reaction wheels were used to prevent Magellan from spinning. The telemetry from the reaction wheels would serve to measure the torque that the atmosphere imposed on the vehicle, and thereby enable important aerodynamic and thermodynamic parameters to be calculated which could assist in planning future aerobraking missions. The first windmill experiment was made on 30 August 1994, and there was a 'campaign' in September. In all, 13 passes were made, two each with the solar panels angled at 10, 20, 30, 45 and 70 degrees and three at 90 degrees. Because the tape recorder was unusable, the high-gain antenna was aimed at Earth to provide telemetry in real-time. On 28 September Magellan maneuvered to further lower its periapsis, and then again on 11 October, this time to 138 km in order to make a final series of windmilling passes. All non-essential systems were switched off, as were all the spacecraft's safeguards. The first passes were performed with the solar panels angled at 30 degrees, and then the angle was increased to 75 degrees – in which orientation the panels provided little power and the batteries would soon be drained by the reaction wheels. Although the periapsis remained more or less unchanged by these passes, the apoapsis fell dramatically with each consecutive orbit, finally reaching 280 km. Contact was lost at 10:02 UTC on 12 October 1994, during the 15,032th orbit, as a result of power starvation. It was estimated that within two days Magellan would burn up. The final resting place of the remains of this highly successful spacecraft is therefore unknown.[135]

The Magellan mission marked the end of an astronomical discipline which had flourished since the early 1960s – Earth-based Venus radar astronomy. Although there remained some studies that could be done from Earth (including trying to find rain on Venus) mapping was no longer productive and radar astronomers had to seek other targets, eventually identifying the radar detection and mapping of near-Earth asteroids and comets as a worthy venture.[136] Another outcome of Magellan was the false impression that everything knowable about Venus was known. But the interior remained completely unknown, which impaired a full understanding of the radar data. Also, in developing Magellan in the aftermath of the cancellation of VOIR, the instruments to study the atmosphere were deleted and these scientific observations remained to be made. Finally, it was unclear whether the planet was still geologically active. Nevertheless, other than a few flybys by missions heading elsewhere the planet remained largely ignored for more than a decade.

THE RELUCTANT FLAGSHIP

Building upon the success of its early-1970s Pioneer Jupiter missions, and on over a decade of preliminary studies, NASA's Ames Research Center began to plan the next step in the exploration of Jupiter, namely the placing of a spacecraft into orbit around the planet and dropping a probe into its atmosphere. This Pioneer Jupiter Orbiter with Probe (JOP), as the mission was called, was to make extensive use of existing hardware. In particular, the orbiter would be an upgraded Pioneer Jupiter spin-stabilized bus fitted with an engine for orbit-insertion and a frame to carry the probe, which would itself be a derivative of the capsule of the Pioneer Venus Multiprobe mission. In 1975 Ames was authorized to initiate development of the mission, which was to be launched by the Space Shuttle in 1982, and ESA offered to provide the propulsion module for the orbit-insertion maneuver at Jupiter. But a few months later NASA headquarters transferred responsibility for the project to JPL, which, with the Vikings heading for Mars, the Voyagers a year from launch and no further missions approved, feared that it would soon be out of the planetary exploration business. In the new arrangement, JPL would provide the orbiter and Ames would provide the entry capsule.

Whereas the Ames orbiter would have carried mainly instruments to investigate particles and fields, for which a spinning platform was ideally suited, and would have had only rudimentary imagers, photometers, radiometers and spectrometers, JPL envisaged a 3-axis-stabilized Mariner Jupiter Orbiter that would be equipped with a high-resolution camera and other remote-sensing instruments that would require accurate pointing. Because a fully stabilized spacecraft would be less effective for particles and fields work, JPL suggested that once the spacecraft had attained orbit around Jupiter it should release a spinning subsatellite with a suite of particles and fields instruments. ESA was invited to provide the subsatellite, but declined on the basis that such a contribution presented little challenge. Nevertheless, a significant West German contribution to the mission was agreed.[137,138] JPL eventually reached a compromise between the two 'souls' of the mission, in which the orbiter would comprise a spinning section on which the particles and fields instruments would be mounted, and of a 3-axis-stabilized 'despun' section for everything else. Although this made the vehicle considerably more complex, it did not exceed the expected capacity of the Space Shuttle. On the other hand, the dual-spin arrangement was an innovation, and the technical development required to implement it would drive up the mission costs. So much so, in fact, that it was realised early on that the budget ruled out the usual procedure of dispatching a pair of spacecraft in case one were to fail. And to rely upon a single spacecraft would require a degree of reliability so great that this in itself would further escalate costs. But by now JPL was confident that it had sufficient experience to run a single-spacecraft mission successfully.[139]

JOP was proposed as a NASA 'new start' in August 1976 and endorsed first by the Ford administration's Office of Management and Budget and then the Carter administration, and when cancellation seemed possible less than a year later it was supported in part by an unprecedented public campaign (including a convention of *Star Trek* fans) before finally receiving House approval in July 1977. Another 'new start' was approved that year: the Space Telescope, which would share many of the travails of the Jupiter orbiter owing to their reliance on the Space Shuttle. In recognition of the importance of the Space Telescope to the scientific community, and of how placing the Grand Tour in direct competition with it in the early 1970s had killed off this planetary mission, the Jupiter orbiter team chose not to pitch the two programs against one another; not least because the Space Telescope had a stronger support base in Congress. In part, this strategy led to the mission cost, as presented for approval, being substantially underestimated at $270 million.[140]

At the beginning of 1978 the Jupiter Orbiter with Probe was renamed 'Galileo' in tribute to the Italian astronomer who, in January 1610, made the first telescopic study of the planet and discovered its four major moons. It was to be mated with a 3-stage all-solid Interim (later Inertial) Upper Stage and launched by a Shuttle in 1982. Although at that time the IUS being built by Boeing for the Air Force was the most powerful payload stage under development for the Shuttle, it would not be able to send Galileo directly to Jupiter, so a Mars flyby was planned that would deliver the spacecraft to Jupiter in 1985. When it became evident that the date of the Shuttle's inaugural flight would have to be slipped by at least a year, the launch of Galileo was postponed to 1984. Unfortunately, this time the gravity-assist at Mars would be marginal. In addition, the projected payload of the Shuttle was diminishing. There was no way that NASA could switch Galileo to the Titan IIIE–Centaur, as used by the Vikings and Voyagers, because heavy payloads such as this were touted as one of the *raisons d'être* of the Shuttle. Moreover, such a solution would have required purchasing rockets from the Air Force and modifying the Titan III pad at Cape Canaveral to accommodate this particular spacecraft, which would further increase the cost of the mission. As a result of this combination of factors, JPL faced having to decide whether it would be better to reduce the propellant load and delete some of the scientific instruments, or to split the mission between two spacecraft which would be dispatched separately, one to go into orbit around Jupiter and the other to deliver the atmospheric probe.

Fortunately, at about this time NASA's Lewis Research Center set out to adapt the Centaur cryogenic upper stage for carriage by the Shuttle, and it was decided to cancel both the three- and two-stage versions of the IUS. Because the Centaur was almost 50 per cent more powerful, it would be able to dispatch the original Galileo spacecraft directly to Jupiter. However, because the Centaur would not be ready in 1984, the Galileo launch was slipped a year. The Shuttle had its inaugural flight in April 1981, and after its fourth flight was declared operational the following year. But by then the Reagan administration was giving serious thought to canceling the Galileo mission. It survived because: (1) its development was so far advanced that only a minor amount would be saved by its cancellation; (2) because it had strong backing by both the scientific community and the public; and (3) because JPL had presented it as of significance to the nation's military because it would be a highly autonomous vehicle, as a military satellite would require to be in time of war, and the shielding that would protect the spacecraft's electronic systems from the high-energy particles in the Jovian environment would provide information on how well a satellite would operate in a nuclear war. Just as the mission's prospects appeared to

An early-1980s rendition of the Galileo Jupiter orbiter and probe.

be secure, lobbying by rival NASA centers and aerospace companies succeeded in having the decision to cancel the IUS reversed, and Galileo was transferred back to this. Unfortunately, by this time the launch had slipped so far that a Mars flyby was not possible, and so JPL came up with a revised plan by which the IUS would place Galileo into an elliptical heliocentric orbit, and at its aphelion the spacecraft would maneuver to adopt a trajectory which would result in a flyby of Earth two years after launch to gain the gravity-assist needed to reach Jupiter. This circuitous route would not only take many more months, but the deep-space maneuver would use so much propellant that the spacecraft would be able to conduct only a reduced tour of the Jovian system. On the other hand, Galileo would serendipitously have flown within 30 million km of comet Halley. However, common sense prevailed and in July 1982 the Centaur was reinstated; not just to dispatch planetary missions but also to try to recapture some of the communication satellites market away from the European Ariane launcher. Galileo was transferred back to the Centaur, and the direct route to Jupiter. On the downside, the launch had to be slipped to May 1986 to allow for the delay in the development of the Centaur. In fact, two forms of the Centaur were

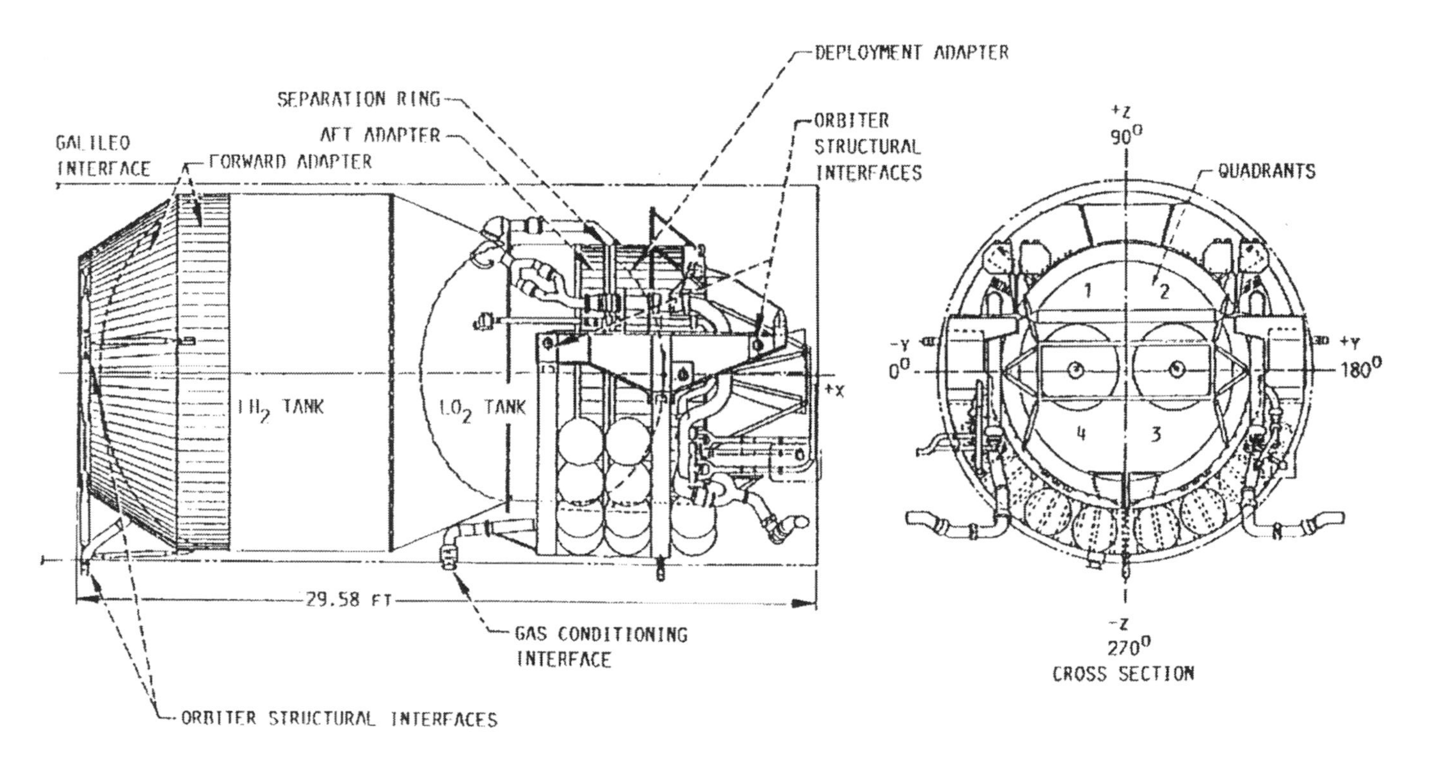

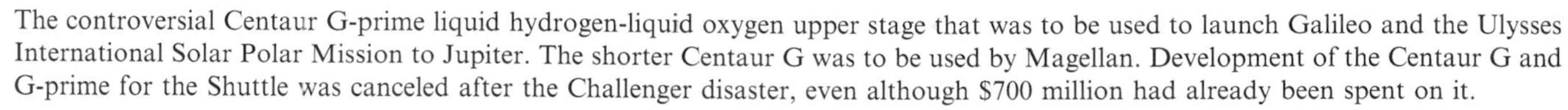
The controversial Centaur G-prime liquid hydrogen-liquid oxygen upper stage that was to be used to launch Galileo and the Ulysses International Solar Polar Mission to Jupiter. The shorter Centaur G was to be used by Magellan. Development of the Centaur G and G-prime for the Shuttle was canceled after the Challenger disaster, even although $700 million had already been spent on it.

intended, both of which would exploit the full width of the Shuttle's payload bay. The Centaur G would be shorter to accommodate the large spacecraft of the Department of Defense, which the stage would insert into geosynchronous transfer orbit. The Centaur G-prime (which Galileo would need) would be longer to carry more propellant, and its performance would be more than thrice that of the Atlas–Centaur.[141,142]

As the original 1982 launch date for Galileo came and went, it was realized that if it had been dispatched on time the spacecraft would have suffered so much from radiation on its interplanetary cruise and tour of the Jovian system as to thoroughly disrupt its science operations. In 1983, therefore, it was decided to replace the most delicate electronics with versions 'hardened' to withstand the radiation in close to Jupiter that had almost disabled Pioneer 10 and severely affected Voyager 1.

The Galileo spacecraft comprised a spun section which rotated around the main axis at 3 rpm, typically, and a despun section which rotated in the opposite sense at precisely the same rate in order to 'stand still'. For particularly important events, such as major propulsive maneuvers and the separation of the atmospheric probe, the sections could be locked together and the entire spacecraft spun at a higher rate for enhanced stability. The spun section was much larger than the despun section. At the top was the dual-band high-gain antenna. Because at 4.8 meters in diameter this was wider than the payload bay of the Shuttle, it was designed as a gold-plated molybdenum wire mesh affixed to 18 graphite-epoxy ribs designed to unfold in the manner of an umbrella once the spacecraft was on its way. Although based on the design proposed for the canceled Grand Tour mission, the mechanism was proved by the TDRS (Tracking and Data Relay Satellite) launched by the Shuttle in 1983. The antenna was to provide a data rate of 134 kbps from Jupiter (equivalent to one frame per minute from the main camera) in addition to other science, engineering and 'housekeeping' data. The tower which accommodated the feed for the antenna and the retaining system for its ribs also held sensors for some of the experiments and, at its tip, a low-gain antenna to transmit science and engineering data at rates of up to 7.68 kbps.

Beneath the antenna was a Voyager-heritage compartment to house electronics and other systems. Affixed to this compartment were two 5-meter trusses, each of which had a single RTG at its tip, together providing 570 W at launch (expected to decline to about 485 W by the close of the mission). These units could be moved slightly up or down to balance the spin of the spacecraft – in particular in response to changes in the moment of inertia as propellant was used. From another side of this compartment protruded an 11-meter boom which carried most of the instruments of the spun section, including two magnetometers, one half way along and the other at the tip, and two plasma-wave 'whisker' antennas which projected outwards in a perpendicular sense from the tip. Next in line was the propulsion module. As a result of the early discussions of possible European participation, this was built by MBB in West Germany as that country's principal contribution to the mission – the Germans also supplied some of the instruments and scientific teams. It comprised four identical titanium tanks, two for monomethyl hydrazine and two for nitrogen tetroxide, plus tanks of helium for pressurization. This module gave Galileo its distinctive 'bulbous waist' appearance. A total of 925 kg of propellant was carried. There were two clusters of six 10-N thrusters for making small course corrections and controlling the vehicle's attitude and spin, set at the ends of two short booms fitted with curved shields to protect the spacecraft from combustion efflux. As the nozzle of the 400-N main engine would be blocked by the presence of the atmospheric probe until this was released a few months from Jupiter, it would be fired only three times: (1) in a brief calibration burn, (2) at orbit insertion and (3) to raise the periapsis. Tests late during the development of the propulsion module revealed that the thrusters overheated so badly that instead of being able to fire continuously for up to 8 minutes they could sustain only a few seconds. However, NASA determined that the mission would be able to be achieved by firing the thrusters in short bursts, even though this would make maneuvers more complex to plan, slower to execute and increase the overall propellant consumption.[143]

At the base of the propulsion module was the spin-bearing assembly to interface to the short despun section. This assembly comprised the motor to impart the spin, the optical encoders to monitor the rate of spin, slip rings to carry power across the interface and rotary transformers to transfer signals. It took extensive development to ensure that data transmission through the rotary transformers was not marred by 'noise' from the remainder of the spacecraft or from the space environment.

Two platforms were mounted on the despun section. One was the scan platform carrying the imaging, spectrometric and radiometric instruments. Azimuth pointing

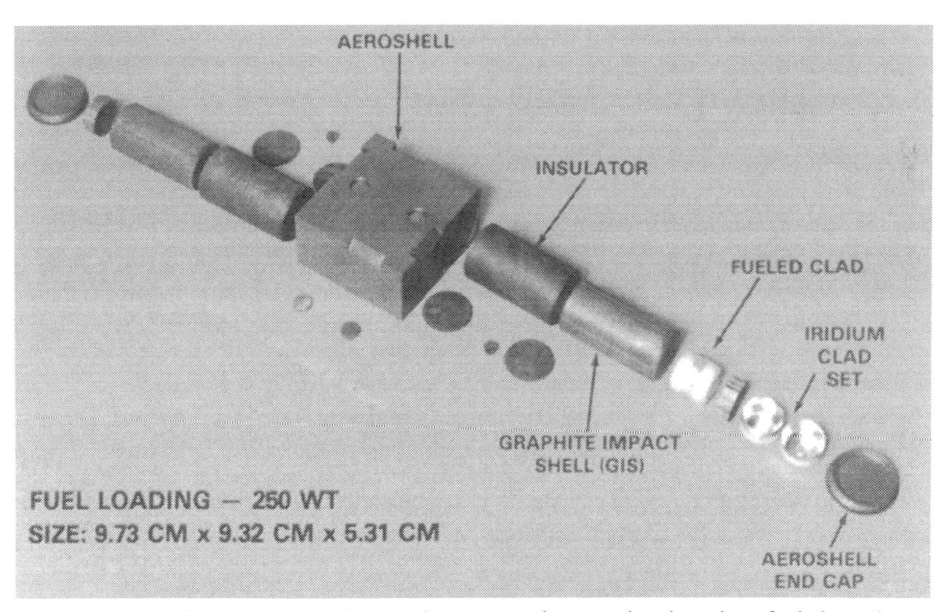

The 'General-Purpose Heat Source' incorporated not only plutonium fuel, but also protection in case of explosion or accidental re-entry. Eighteen such blocks were stacked inside a converter unit to form an RTG of the type used by the Galileo and Ulysses spacecraft.

Galileo's high-gain antenna in its deployed configuration during a ground test. The failure of the 'umbrella' to unfurl in flight severely reduced the amount of data that the mission could return.

was controlled by the despin motor, and the platform itself selected the elevation. The other platform held a 1.1-meter-diameter parabolic antenna to track the probe as it entered Jupiter's atmosphere. Finally, at the bottom of the despun section was the ring attachment flange for the probe, which also doubled as a skirt for the main engine.

Galileo's attitude control system was one of the most complex ever flown on a planetary spacecraft. It utilized Sun and star sensors; accelerometers to monitor the rate of spin and the performance of the main engine, particularly during the orbit-insertion maneuver; a gyroscopic platform to monitor the high-precision pointing of the scan platform; optical encoders to monitor the relative positions of the spun and despun sections; and so on. In addition to extensive black-carbon blanketing that gave both thermal insulation and micrometeorite protection, the spacecraft had small electrical heaters and 120 isotopic heaters strategically positioned to keep its electronics warm against the chill of space. In consideration of the flux of charged particles circulating in the Jovian magnetosphere, the blankets were grounded onto the body of the orbiter to preclude electrostatic discharges.

The launch mass of the orbiter in its fully loaded state was 2,233 kg. Affixed to its base was the 339-kg atmospheric probe that was based on the large probe of the Pioneer Venus Multiprobe and, like that probe, was built by Hughes in partnership with General Electric under contract to Ames. It was 1.24 meters across and 86 cm tall, and consisted of three parts: a conical 152-kg heat shield, a dome-shaped nylon aft cover, and the 121-kg spherical descent module. The heat shield consisted of an aluminum main structure over which was laid an extensive layer of carbon phenolic

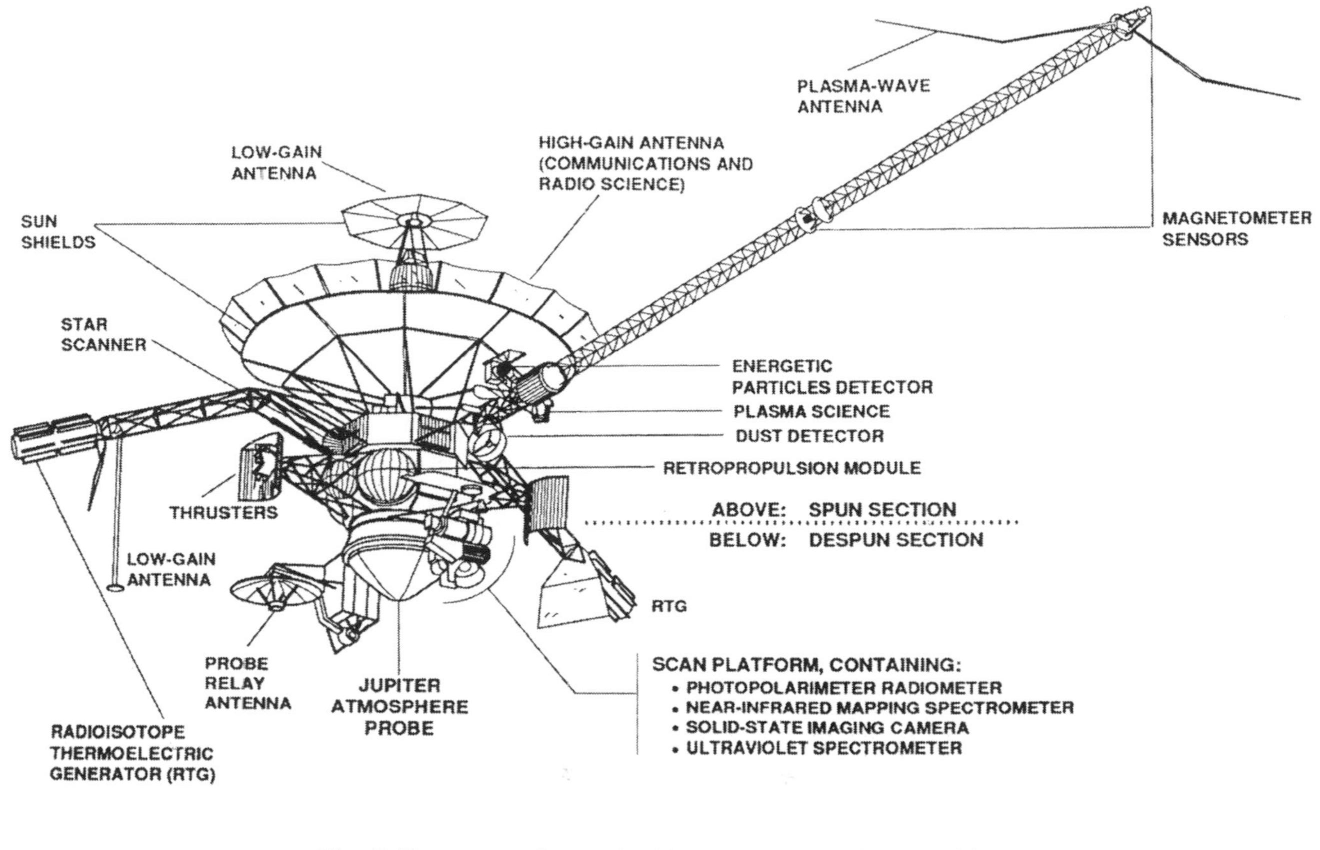

The Galileo spacecraft as revised following the Challenger accident.

ablative. On entering the Jovian atmosphere at 47 km/s, the shield was expected to be subjected to such heating that almost half of its original mass would evaporate. In fact, the capsule would be targeted to enter near the equator at the dusk limb, so that the 10 km/s rotation of the upper atmosphere would reduce the relative speed by one-sixth. Had it been targeted to enter over the dawn limb, against the rotation of the planet, then its speed would have been 70 km/s, which would have made the heat flux almost 3 times greater. Unlike Pioneer Venus, this probe was not made pressure-tight, and it actually included a 'chimney' to equalize the internal and external pressures. This decision was partly to save mass, and also because the maximum pressure in which the probe was to operate was not as great as on Venus. Only those instruments and systems which required it were placed in hermetical housings. Extensive thermal blanketing was to maintain the electronics within the acceptable temperature range until the probe reached below the 10 hPa level of the atmosphere. For redundancy, it had two independent radio systems to transmit data. The Galileo orbiter had a 600-meter tape recorder with a 900-megabit capacity to store this data. This was a contingency against a failure of the despin system obliging the spacecraft to break contact with Earth in order to slew to track its probe. The probe was to remain dormant during the interplanetary cruise. It would be released with only its main timer operating. The timer would initiate a sequence of operations 6 hours prior to the predicted time of arrival, but the main entry and descent phases of the mission would be initiated only by accelerometers sensing deceleration. The probe would draw power from the orbiter for periodic checks during the cruise, but once free it would draw from 39 lithium/sulfur-dioxide batteries that were expected to sustain operations in the atmosphere for between 60 and 75 minutes.

After enduring a peak deceleration of 250 g (in fact, the probe was designed to withstand 400 g), and having slowed to about 0.9 Mach in the local environment, an adaptive algorithm would fire a mortar through the aft cover in order to deploy the pilot parachute, which would in turn draw off the cover and pull out the main 2.4-meter-diameter Dacron parachute. Three slightly canted vanes were to make the capsule spin at between 0.25 and 40 rpm as it descended, otherwise its scientific measurements would be somewhat degraded or the Doppler shift on the link to the orbiter would be excessive.[144] The operation of the probe and its parachute system were tested in 1982 and 1983 by airdrops from an altitude of almost 30 km above Roswell, New Mexico (where the parachute systems for the Voyager and Viking Mars landers were tested during the 1960s and 1970s) and then in wind tunnels to eliminate some of the shortcomings.[145]

The capsule's scientific objectives included sampling the Jovian environment at decreasing distances from Jupiter during the final approach, and then, on entering the atmosphere, to thoroughly characterize this in terms of chemical composition, the sizes of the cloud particles, the layering of the clouds, energy fluxes, and so on. The composition of the atmosphere was expected to have a percentage of helium similar to, but smaller than, that of the Sun (which had also been measured by the Voyager flybys) and higher percentages of carbon, nitrogen, sulfur and oxygen. The concentrations of other noble gases were unknown, as also were their isotopic ratios. Determining these values would provide information on the evolution of the

The capsule for the Galileo atmospheric probe was based on those of the Pioneer Venus mission.

The capsule for the Galileo atmospheric probe being tested in a wind tunnel.

The parachute of the Galileo atmospheric probe being tested in a wind tunnel at NASA's Langley Research Center.

atmosphere. The probe would also provide in-situ measurements of the profiles of the pressure and temperature in the atmosphere, for which the flybys had provided conflicting results. It was expected to encounter three layers of cloud. Ammonia at the 600-hPa pressure level (i.e. 0.6 bars) was known to constitute the visible surface. Models showed that this should be followed by a layer of icy crystals of ammonium hydrosulfide (NH_4SH) at 1,500 to 2,000 hPa and a layer of water vapor at 4,000 to 5,000 hPa.

The probe had seven instruments. A neutral mass spectrometer fed by two inlets in the forebody would measure the chemical and isotopic composition at various depths. In addition to the most common atomic species, this instrument would seek noble gases such as argon, krypton and xenon, and attempt to confirm whether (as hypothesized on the basis of the Voyager remote sensing) methane and ammonia reacted with lightning to create more complex organic molecules. An atmospheric structure instrument would measure the temperature, pressure, density and average molecular weight over a broad range of altitudes. In addition to being transmitted to the orbiter, its data would be used in situ as a reference for the other instruments. A nephelometer would determine the size, density, shape, etc, of cloud droplets by firing an infrared laser across a gap to a short arm carrying five mirrors that would reflect the beam back to the instrument. A net-flux instrument consisting of two radiometers, one looking upward the other downward, would compare the energy that Jupiter receives from the Sun with that radiated from its interior to determine its heat budget. This instrument would also provide data on the cloud structure and layering. Three instruments were built by or in cooperation with West Germany. A helium detector would precisely measure the helium-to-hydrogen ratio, which was believed to have remained unchanged since the formation of the planet. This value was to be used to calibrate the cosmological Big Bang theory. A lightning detector located on the probe's aft shelf consisted of an aerial and two photodiodes, each of which viewed through a fisheye lens. It would monitor how the ambient brightness varied with depth, and detect transient optical flashes from nearby thunderstorms and electric disturbances from distant ones. There being no solid surface, lightning discharges would be between one cloud and another. (In fact, in terrestrial terms this is the commonest yet least understood lightning.) The aerial would also serve as a detector of background radio noise and magnetic fields. Sharing the same electrical system as the lightning detector was the energetic-particle instrument. The latter was the only instrument that would collect data in the final hours of the approach to the planet (specifically, from the orbit of Io down to the top of the atmosphere) rather than while on the parachute. It was to analyze the particles which were sufficiently energetic to penetrate through the heat shield while crossing the torus of plasma in Io's orbit, the planet's radiation belts, and the zone occupied by the ring and small inner moons. In addition, the radio to the orbiter would measure the attenuation of the signal as it crossed the atmosphere and ionosphere, and an ultrastable oscillator would enable an analysis of the Doppler shift to reconstruct the trajectory followed by the probe within the atmosphere, and thereby the wind speeds; in particular to determine whether the belts and zones visible at the surface extended deep into the atmosphere, as this would have implications for the energy source and mechanism driving the jetstream winds.[146]

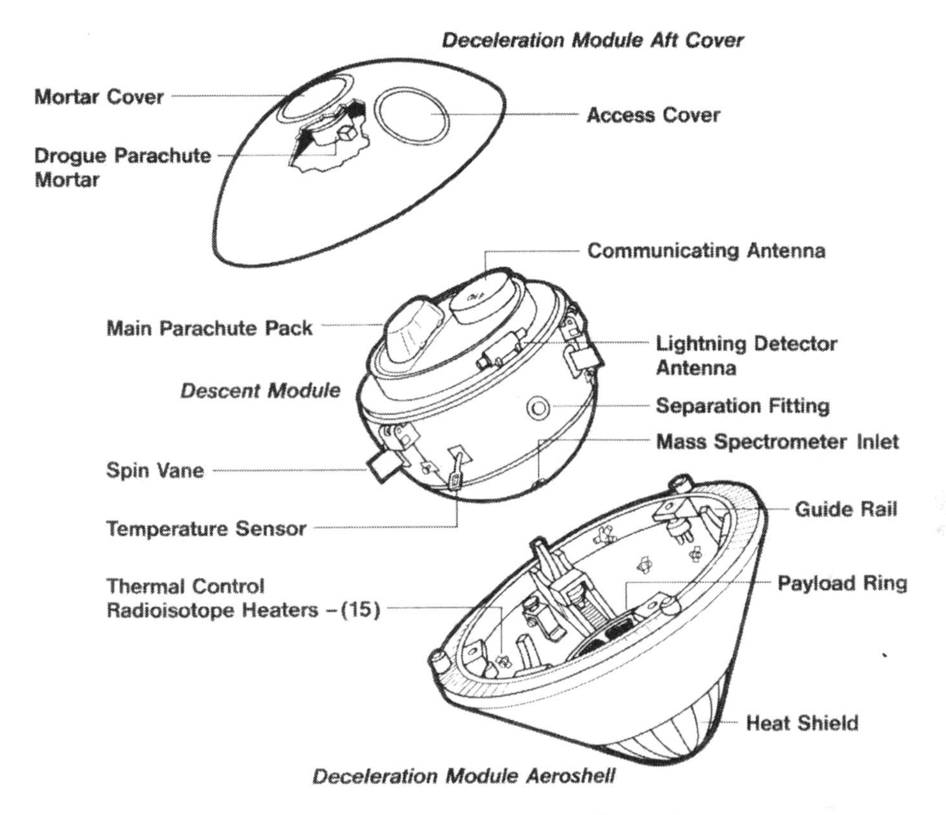

The main components of the Galileo probe.

The orbiter had nine instruments: five mounted on the spun section for particles and fields, and four on the scan platform of the despun section for remote-sensing of the planet and its satellites.

A dust detector would collect and characterize dust during the cruise and in the Jovian system. Like similar instruments on the Vega, Giotto and Ulysses out-of-ecliptic missions, this was built and managed by German scientists and based on a sensor which was flown on ESA's HEOS 2 (Highly Eccentric Orbit Satellite). An energetic-particle detector would measure electrons and ions with atomic masses ranging from hydrogen to iron in the Jovian magnetosphere, and study how these were lost and replaced. In a redesign after the Challenger tragedy, its capabilities were increased by adding a time-of-flight experiment. The magnetometer utilized two sets of three sensors, with the set for strong fields 7 meters along the 11-meter magnetometer boom and that for weak fields at its tip. This would not only map the Jovian magnetosphere, but also attempt to find out whether any of the satellites had its own magnetic field. As Pioneer 10 revealed, the solar wind draws the tail of the Jovian magnetosphere all the way out to the orbit of Saturn, and the plan called for Galileo's tour to include at least one highly eccentric orbit that had its apoapsis positioned far down the tail. In addition, two plasma analyzers were to characterize low-energy electrons and ions and measure their composition, energy, temperature, motion and 3-dimensional distribution in space. The whisker antennas at the tip of the magnetometer boom would monitor electrostatic and electromagnetic waves in the plasma of the Jovian magnetosphere. As usual, the radio science team would study celestial mechanics, relativity, the interplanetary medium, radio occultations, and so on. Following the Challenger tragedy, a heavy-ion counter was added as an engineering instrument to determine the radiation hazard from heavy ions in deep space and in the Jovian magnetosphere.[147,148]

Of the remote-sensing instruments, the near-infrared mapping spectrometer was an optical instrument to analyze the chemical composition of the surfaces of airless bodies such as asteroids and the Galilean moons by their reflection spectra, and of atmospheres by their absorption spectra. In effect, it produced a map in which each pixel was a spectrum of that position. Its spatial resolution was low, however. The gaseous molecules which it could detect included ammonia, phosphine, water and methane. A photopolarimeter would analyze the clouds and hazes of the planet to determine the size and shape of the particles and droplets. The instrument included a photometer and a radiometer. It would undertake polarimetric, radiometric and photometric observations of the satellites to determine the nature of their surfaces. One of the main objectives of the instrument was to precisely measure the planet's heat budget. The Voyagers had been equipped with photopolarimeters, but these had suffered a number of problems and had returned little useful data from Jupiter. Because Galileo's instrument was sensitive far into the infrared, it could measure thermal energy from Jupiter's interior, with different wavelengths corresponding to emissions from different depths. An ultraviolet instrument would be able to detect some complex hydrocarbon molecules in the atmosphere, and study processes such as auroras and airglow in the upper atmosphere. It would also monitor the Io torus and search for gaseous envelopes around the satellites, in particular atoms and ions of nitrogen, sulfur, atomic hydrogen and oxygen. This instrument was in two parts, having an ultraviolet spectrometer on the scan platform and an extreme-ultraviolet spectrometer on the spun section – the latter a spare from Voyager that was added during the hiatus which followed the Challenger tragedy.

As usual for a JPL mission, the main (and at 30 kg the heaviest) instrument was the solid-state imager. This comprised a 176-mm-diameter 1,500-mm-focal-length Cassegrain telescope mated to a Texas Instruments CCD with an 800 × 800 pixel array. Soon after CCDs were invented in the late 1960s, JPL noted that their small size and low power would make them ideal for deep-space imagers. In particular, they suffered none of the shortcomings of vidicons: namely an uneven photometric response which always made parts of a picture brighter that the rest, and geometric distortions that obliged scientists to etch 'reseau' references on the faceplate of the sensor as the basis for later correction. Although Galileo was the first mission to be assigned CCDs by JPL, its launch was so delayed that by then this technology had already flown on several missions – including the Soviet Vegas and the European Giotto. The sensitive chip was 'armored' on all sides apart from its optical window by a 1-cm-thick layer of tantalum that would absorb all but the most intense Jovian

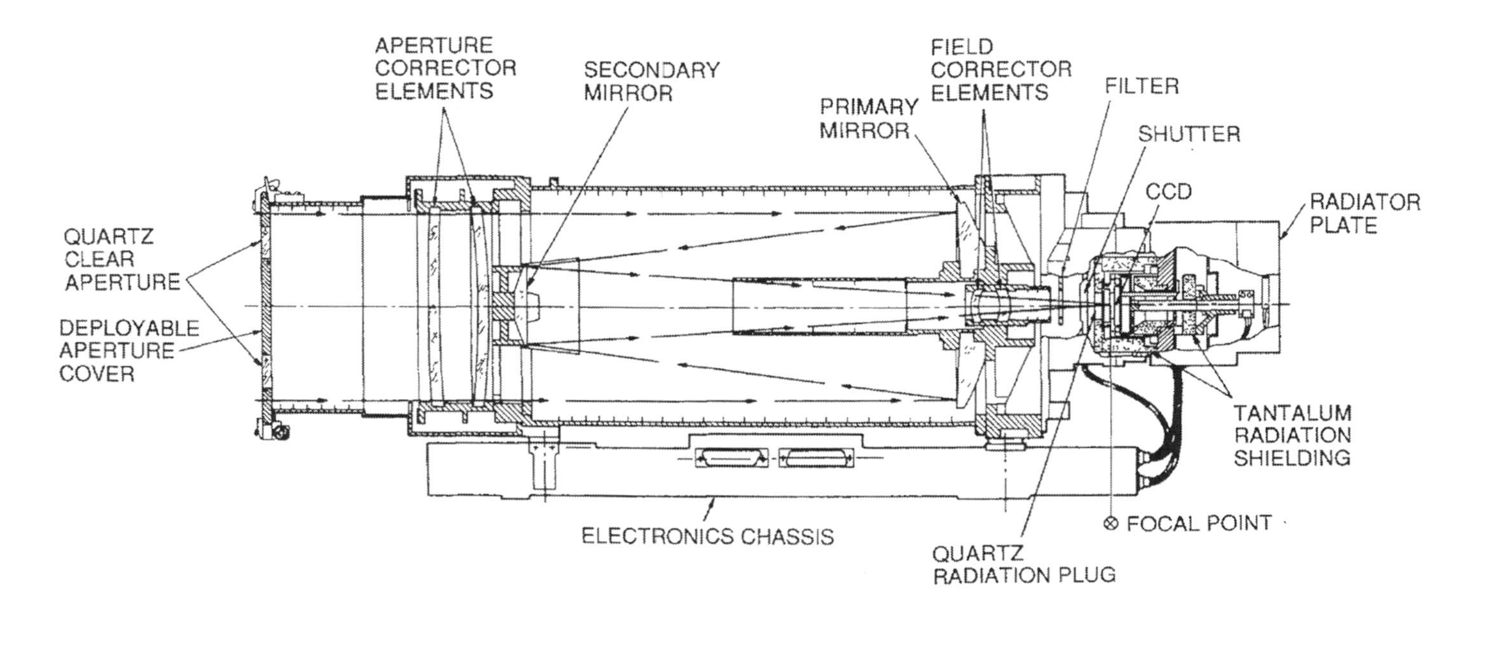

Although the solid-state imager developed for the Galileo mission marked the first time that a CCD detector was used on a planetary spacecraft, the protracted delay in its launch meant many such imagers flew before it. Unlike most spacecraft, Galileo carried only a narrow-angle camera. (JPL/NASA/Caltech)

radiation. An 8-position carousel carried filters selected to optimize various types of investigation and also to enable color pictures to be reconstructed. In particular, two near-infrared filters were centered on wavelengths at which gaseous methane was respectively a moderate and a strong absorber, and their data would enable the depths of features in the Jovian atmosphere to be inferred. Most of the imaging of the planet, however, would be done using two filters at wavelengths between the methane absorption bands.[149] During their brief flybys the Voyagers had returned several thousand images of Jupiter and its satellites, but Galileo was to enter orbit and make a thorough study that would yield at least 50,000 images, each of higher quality than could have been attained by a vidicon. Also, the more sensitive CCD was better able to observe the very faint Jovian ring, and to search for auroras and lightning on the night-side of the planet (which had been barely feasible using a vidicon). And, of course, by flying closer to the satellites Galileo would be able to obtain images at a resolution 100 times better than the best from the Voyagers.

The schedule for Galileo was the most ambitious ever planned for the Shuttle, in that it was to be launched during the same May 1986 Jupiter launch window as the US/European Ulysses out-of-ecliptic mission: Challenger was to be launched as STS-61F on 15 May with Ulysses, and return to Earth on 19 May. The next day Atlantis would be launched as STS-61G with Galileo. In addition to concern about the hectic schedule, which would have placed unprecedented strain on the ground support teams, the presence of the Centaur stage led to the missions being dubbed 'Death Star' flights. This was due to concerns about the structural integrity of the Centaur G and G-prime tankage when subjected to the stresses of a Shuttle launch. Another issue was how to dump the Centaur's cryogenic propellents if the Shuttle had to undertake an emergency landing following a launch abort. There were also some concerns about the Shuttle itself, since even when running its main engines at 109 per cent of their nominal thrust, with the heavy Centaur aboard it would be able to achieve only a very low 169 km orbit.[150,151]

Trusting that the Shuttle would safely send Galileo on its way, astrodynamicist Giuseppe Colombo prompted the celestial navigators at JPL to analyze the orbits of the thousands of asteroids on file for possible 'targets of opportunity' during the interplanetary cruise. This revealed that a minor revision of the course would allow a 10,000-km flyby of (29) Amphitrite on 6 December 1986 in return for delaying the spacecraft's arrival at Jupiter by 3 months to December 1988. Amphitrite was discovered on 1 March 1854 by Albert Marth, and is an S-type body (suggestive of stony meteorites) about 200 km across with a relatively rapid rotational period of 5.4 hours. Two other candidates were (1219) Britta and (1972) Yi Xing but both were much smaller and almost nothing was known about them. In December 1984 NASA approved the $15-million Amphitrite 'detour' in order to gain humanity's first look at one of the 'vermin of the sky'.[152,153,154] Looking ahead to when Galileo reached Jupiter, the mission designers devised a 'mini-Grand Tour' in which the spacecraft would use flybys of the Galilean satellites as an almost free means of constantly reshaping its orbit to investigate different aspects of the system during its 2-year nominal mission.

In December 1985 Galileo was taken by truck to Florida for fueling, installation of the RTGs, checkout and mating with the Centaur. On 28 January 1986 it was in the final round of checks, with no 'show stoppers' threatening its launch schedule, when Challenger, lifting off to deploy the second TDRS satellite and make a study of comet Halley at perihelion, exploded 72 seconds into its flight, killing the seven astronauts onboard. Although it was evident that neither Ulysses or Galileo would be able to be launched in 1986, work continued to prepare them for flight, in order to have them ready in case it proved possible to launch them in the 1987 window. When NASA ruled that the Shuttle would not be allowed to run its engines at 109 per cent, JPL devised a plan to launch a half-fueled Centaur that would put Galileo into a resonant orbit which would return the spacecraft to Earth after 24 months in order to use an Earth flyby to gain the energy required to reach Jupiter, thus adding 2 years to the cruise. Unfortunately, in June NASA canceled the Shuttle version of the Centaur, deeming a cryogenic stage too risky to be carried on a piloted spacecraft. To date $700 million had been spent in developing the new Centaur, modifying the Shuttle to carry it, and creating the infrastructure to prepare two Centaurs and their payloads in parallel. This cancellation of the Centaur left JPL with no option but to redesign the Galileo mission to use the IUS – hopefully in such a manner as not to require the spacecraft to be split in two.

In August 1986 the celestial navigators devised a trajectory that would allow an integrated spacecraft to reach Jupiter in 6 years. It would be launched in October 1989 on the fifteenth mission after the Shuttle's return to service. The plan was to build up its energy by a series of gravity-assists in the inner solar system. The IUS was to put it on course for a Venus flyby, and after two Earth flybys 2 years apart it would reach Jupiter in December 1995 – 11 years later than envisaged when the mission was conceived, by which time Jupiter would have made almost a circuit of the Sun. Unlike other similar solutions, this one did not require any major deep-space maneuvers. This had the advantage of saving most of the propellant for the tour of the Jovian system. If it looked as if Galileo would miss the 1989 window, NASA intended to ask the Department of Defense to make available one of its new Titan IV heavy lifters and dispatch Galileo in May 1991. If the Shuttle managed to launch Galileo on time, the Titan IV would backup Ulysses and become the prime launcher for the CRAF cometary mission.[155]

One can only wonder how much simpler, and probably cheaper, Galileo would have been if, like the Voyagers, it had been designed from the start to be launched by a Titan IIIE–Centaur; all the more so, in fact, because its troubles did not end with launch...[156,157] The new mission profile would cause it to travel as close to the Sun as the orbit of Venus, an environment for which it had not been designed. To prevent the heat from damaging the high-gain antenna, a Sun shade was affixed to the tip of the furled umbrella. And since this would require the spacecraft to adopt an attitude in which its main axis was pointed at the Sun, a larger shade was added beneath the antenna mount to protect the body. Instead of being unfurled soon after launch, the antenna would now remain furled until Galileo had made its first Earth flyby and entered the environment for which it was designed. Until then, it would be limited to the low-gain antenna. However, because the spacecraft would have to maintain a Sun-pointing attitude whilst inside the Earth's orbit, an articulated low-gain antenna

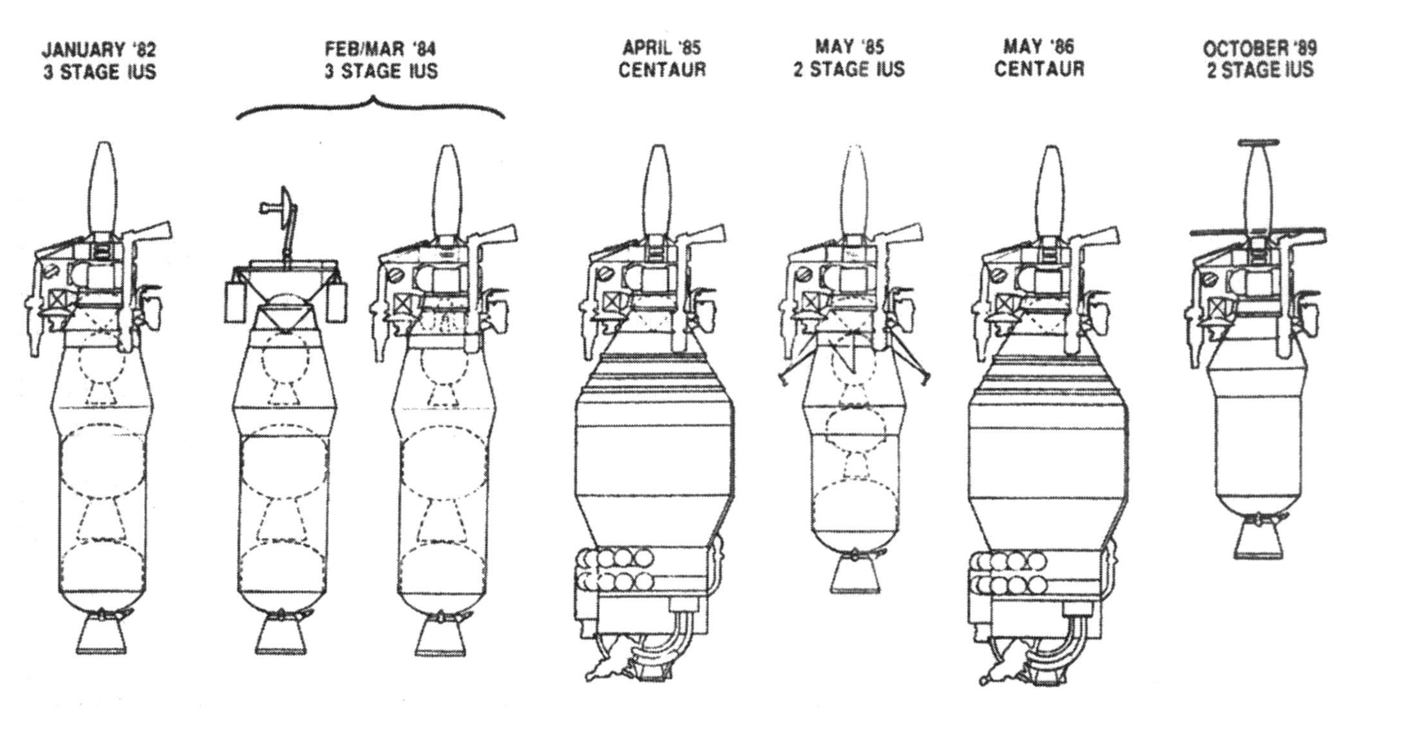

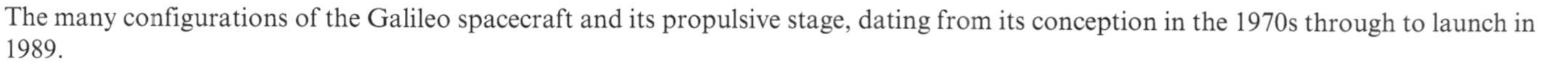

The many configurations of the Galileo spacecraft and its propulsive stage, dating from its conception in the 1970s through to launch in 1989.

had to be placed on one of the RTG booms in order to accommodate the varying angle to Earth. It was a case of one complication leading to another.

On the other hand, the new trajectory, provided new opportunities for science in the cruise phase. First, Galileo would be able to study Venus several months prior to the arrival of the Magellan radar mapper; such observations providing a useful calibration of its instruments. Then it would make two passes through the asteroid belt, one between the two Earth flybys and the second while outbound for Jupiter. An investigation found that the spacecraft could readily be retargeted to encounter an asteroid on each pass through the belt: (951) Gaspra in October 1991 and (243) Ida in August 1993. A flyby of 92-km-sized (63) Ausonia in April 1992 was also a possibility, but was apparently not seriously considered.[158] Named after a Crimean resort on the Black Sea, Gaspra was discovered in July 1916 by G.N. Neujmin at the Simeis astronomical observatory in Crimea. Ida was named after the nymph of Crete who cared for young Jupiter while he was hidden from his father, Saturn, and also for the mountain on Crete on which the young Jupiter lived. It was discovered in

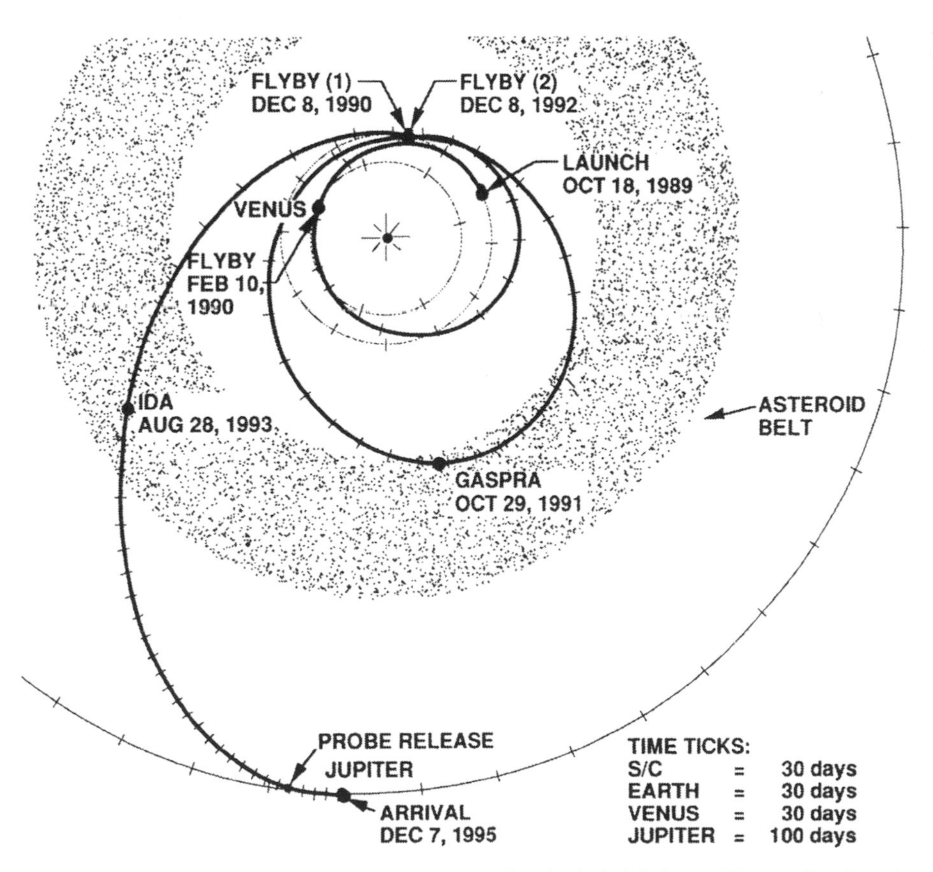

The circuitous journey taken by Galileo to Jupiter included flybys of Venus, Earth and two asteroids.

Galileo is prepared for mating with its IUS stage ready for launch on Space Shuttle Atlantis in 1989.

September 1884 by Johann Palisa in Vienna. Very little was known about either apart from the fact that they were both S-type and infrared measurements implied average diameters of 15.5 and 32.5 km respectively.

Meanwhile, the radiation leakage caused by the nuclear disaster at Chernobyl in the Ukraine had an impact on the Galileo mission when concerns were expressed about launching a spacecraft equipped with RTGs. If the Shuttle carrying it were to be lost in a Challenger-style accident then, critics said, the explosion would release lethal plutonium dioxide into the atmosphere. But after an incident in 1964 in which an upper-stage failure caused a Department of Defense satellite to fall back into the atmosphere and release the plutonium of its SNAP power source into the upper atmosphere, the design was improved to survive not only an explosion but also re-entry. In 1968 two such power units were successfully recovered from the sea floor after a NASA satellite was lost in a launch accident. And when the RTG carried by the Apollo 13 lunar module fell into the Pacific Ocean in 1970 there was no sign of contamination. The official position was therefore that RTGs were safe, not least because the plutonium isotope used released most of its energy in the form of heat, not radiation, which is precisely why it was used, and even if the casing were to be breached the contents were ceramic pellets that were unlikely to shatter and release dust that could contaminate the environment. Despite a last-minute lawsuit by one of the campaigners, the White House considered the matter as it must whenever an American nuclear-powered spacecraft is to take off, and agreed to the launch.[159]

ASTEROIDS INTO MINOR PLANETS

The window for the Galileo launch started on 12 October and ran to 21 November 1989, and in the first days was open for as few as 10 minutes, one of the tightest of the Shuttle program. An additional delay of almost a year in returning the Shuttle to service meant that it would be only the sixth post-Challenger flight. Meanwhile, the spacecraft had modifications and refurbishments which added $220 million to its price tag, with some of the components being rebuilt or replaced, including the atmospheric probe's parachute and batteries. (In fact, the cost, including a nominal two years of operations in Jovian orbit, would eventually reach $1.36 billion.) The orbiter and probe arrived separately at the Kennedy Space Center in June.

On the day of launch, two US Air Force aircraft loaded with cooling equipment for the RTGs were to stand by to fly to either Morocco or Spain in the event of the Shuttle having to perform a transatlantic abort. Another plane would be chartered by the Department of Energy to fly personnel to any potential site of a plutonium release. The attempt to launch on 12 October was canceled before the crew could board the Shuttle, owing to a problem with one of the main engines which imposed a 5-day delay. The second attempt was ruined by bad weather in Florida and by an earthquake that rocked the IUS control center in California. But at 16:54 UTC on 18 October, Atlantis successfully lifted off for the STS-34 mission. The checkout of the payload began 2 hours later. The radio system of the IUS was exercised and the range of motion of the gimbal of the first stage was tested under the watchful eye of a television camera. The cradle holding the 19-tonne IUS–Galileo stack was first tilted 30 degrees for the final checks, and finally to the 58-degree deployment position, thereby releasing the umbilicals to the orbiter. Six hours 21 minutes after launch, the annular cradle released its grip of the stack and springs gently eased it away at 15 cm/s. The operation was recorded by IMAX movie cameras. One hour later, with the Shuttle having withdrawn 80 km, the first stage of the IUS fired and was then jettisoned. Two minutes later the second stage fired. The overall effect of the maneuver was to reduce the heliocentric velocity by 3.1 km/s to enter an orbit ranging between 0.67 and 1.00 AU. The IUS adopted the correct attitude to release Galileo and then spun up to 2.9 rpm for stability. After being released 8 hours 12 minutes into the mission, the spacecraft coasted with the Sun shade protecting its high-gain antenna. Atlantis flew its 5-day mission, conducting various medical and scientific experiments, including studying the 'ozone hole' in the stratosphere, and then returned to Earth.[160] Meanwhile, the Galileo spacecraft was checked out. The instruments were switched on one by one for calibration and testing, with the data being stored on tape for later downloading via the low-gain antenna. In particular, after allowing time for trapped gases and humidity to leak from the spacecraft, the covers of the imaging infrared instrument were opened and a thermal 'image' was taken to assess the extent to which the various booms and appendages obscured the field of view. This fuzzy self-portrait was almost certainly the first ever taken by an interplanetary mission in cruise.[161,162] Although radio tracking by the Deep Space Network indicated that the IUS had done its job extremely well, it was necessary to fire the 10-N thrusters to calibrate their performance. On 9 November 1989, a 2 m/s burn was executed by repeatedly pulsing the thrusters. This established that the thrusters were delivering 102 per cent of their rated thrust, which was good news because it meant that they would consume propellant at a slower rate. A course correction on 22 December refined the Venus flyby so precisely that an optional third correction was deemed unnecessary.[163]

After cruising for 4 months on its 'fast' trajectory, on 10 February 1990 Galileo flew by Venus at a range of 16,106 km from the planet's center, and within 5 km of the aim point. The slingshot deflected the spacecraft's heliocentric velocity and increased its magnitude by 2.3 km/s, to place it into a 0.70 $\times$ 1.29-AU orbit which would yield the first Earth flyby 10 months later. Although not one of the 'design drivers' of the mission, the Venus encounter provided a 'target of opportunity' to calibrate and test the scientific instruments and, because those instruments were so capable, to obtain valuable data. The inbound trajectory skimmed the downstream flank of the planet's bow shock, which was analyzed by the magnetometer, particle and plasma instruments. About 17 hours prior to closest approach, Venus became observable by the scan platform (it had previously been occulted by the high-gain antenna's Sun shade). Recent telescopic studies of the night-side had revealed that in certain sections of the near-infrared spectrum the cool upper atmosphere passed thermal radiation from the hot atmosphere below and indeed from the surface. The spacecraft was therefore programmed to make two scans of the night-side using its near-infrared spectrometer during the approach phase. Radio bursts from lightning were sought by the plasma-wave experiment, and about 45 minutes before closest

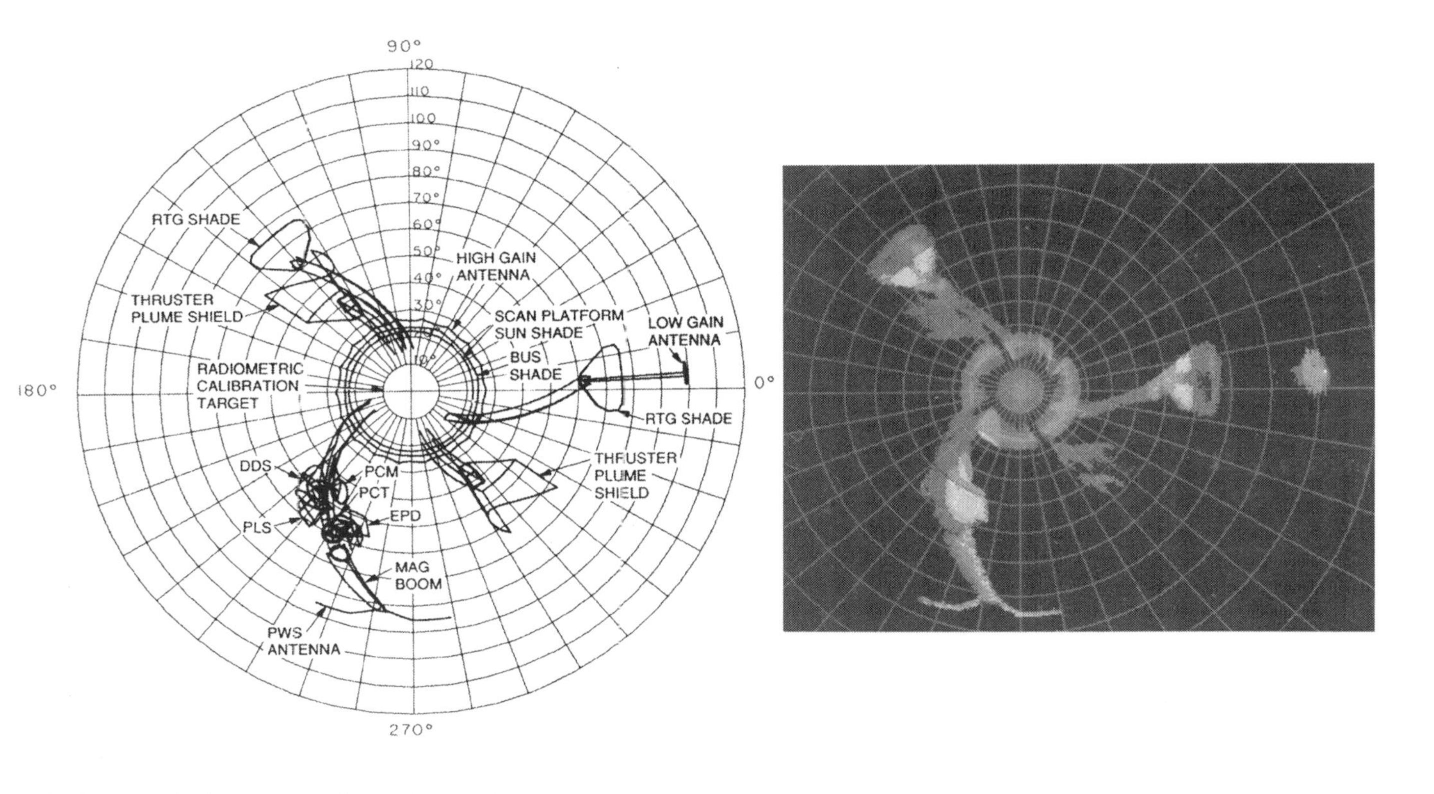

Galileo was the first spacecraft to take a 'self picture' in flight. However, in this view by the near-infrared mapping spectrometer the spinning spacecraft appears distorted. (Courtesy of Robert Carlson, reproduced from Carlson, R.W., et al., "Near-Infrared Mapping Spectrometer Experiment on Galileo", Space Science Reviews, 60, 1992, 457–502, with permission of Springer Science and Business Media)

approach the camera took visible-light images in the hope of detecting flashes. The photopolarimeter and ultraviolet spectrometer also made global scans of the night-side. The camera conducted a study of haze on the limb of the planet after closest approach, but observations on the outbound leg were mostly dedicated to particles and fields.[164] A total of 81 images were taken by the CCD camera during a 7-day period, of which 77 were eventually deemed useful. Since the high-gain antenna was still furled, the science data was recorded on board. However, three images were 'trickled' back through the low-gain antenna in order to assess their quality. It was discovered that a programming error had caused the camera to take an additional 452 pictures, but these were not stored on the tape. The full data was to be retained until the spacecraft was closer to Earth, allowing the low-gain antenna to operate at an increased data rate. When Galileo reached its 0.70-AU perihelion 2 weeks after the Venus flyby it experienced no temperature problems. On its way back to Earth, it made ultraviolet observations of the hydrogen clouds surrounding the two bright long-period comets Austin and Levy.[165,166]

By mid-November, the data from the Venus encounter had been transmitted to Earth. Infrared measurements that were taken during the approach phase provided some of the best snapshots of Venusian meteorology ever. In fact, by observing at two infrared wavelengths it was possible not only to identify the cloud structures at differing altitudes but also, if the contribution from the atmosphere was subtracted, the images revealed a number of 'hot spots' that correlated quite well with known surface features. By confirming that the atmosphere is indeed transparent at some wavelengths, Galileo literally opened a window for subsequent missions to use in studying the planet's surface processes. The visual camera had taken 10 pictures of the night-side in search of lightning flashes, but although the exposures captured stars in the background no lightning was detected. Over the day-side, images taken using violet, clear and near-infrared filters penetrated the atmosphere to differing depths. While the violet images showed the well-known features of the cloud tops, including the Y-shaped dark marking, bright polar collars, etc, the near-infrared showed other patterns which had never been seen before. A spiral pattern that was only faintly visible in the ultraviolet at intermediate latitudes was clearly evident in the near-infrared. This implied that the meridional (i.e. along the meridians) wind speeds at greater depths were slower than at the cloud tops. In the near-infrared, the polar collars were dark. A strong brightness discontinuity was seen running north to south across the equator and over the subsolar point, very likely marking the 'deep root' of the Y-marking. A comparison of the near-infrared and ultraviolet images confirmed that the upper atmosphere is 'super rotating', with the winds circulating tens of times faster than the planet itself spins, gradually slowing with decreasing altitude until stagnant at the surface. The plasma-wave detector reported nine extremely faint radio bursts for which lightning was the most plausible source, thus vindicating the decade-old Venera observations.[167,168,169,170]

The 35 m/s course correction to refine the Earth flyby was made in two parts on 9 April and 11 May. This was the largest propulsive maneuver of the mission ahead of the one that would be made after the probe was released, 6 months from Jupiter. Galileo approached Earth from a direction almost precisely opposite to the Sun,

Galileo took these pictures of Venus through a violet filter between 4 and 6 days after its flyby. (JPL/NASA/Caltech)

presenting an opportunity to make valuable scientific observations as it traveled 'up' the terrestrial magnetotail. The particles and fields instruments were switched on 30 days prior to the encounter, and detected entering the magnetotail 560,000 km from the planet. Galileo reached a minimum altitude of 960 km over Africa at 20:35 UTC on 8 December 1990. Its heliocentric velocity was increased by 5.2 km/s, and it left in a 0.90 × 2.27-AU orbit which would return it to the vicinity of Earth on 8 December 1992. The flyby geometry also increased the orbital inclination relative to the ecliptic to set up the flyby of asteroid Gaspra. On the outbound leg, Galileo was finally able to view the day-side of Earth, and a number of images were taken featuring Australia and Antarctica for calibration and public relations purposes. The near-infrared spectrometer observed very-high-altitude clouds, and put constraints on models of the 'ozone hole'. Starting 2.5 days after closest approach, the camera took images of Earth through a sequence of six filters once every minute over an interval of 24 hours. These were assembled into a stunning movie covering a full planetary rotation. In addition, scientifically useful images and other measurements of the Moon were taken, both as a thin crescent when inbound and of the illuminated

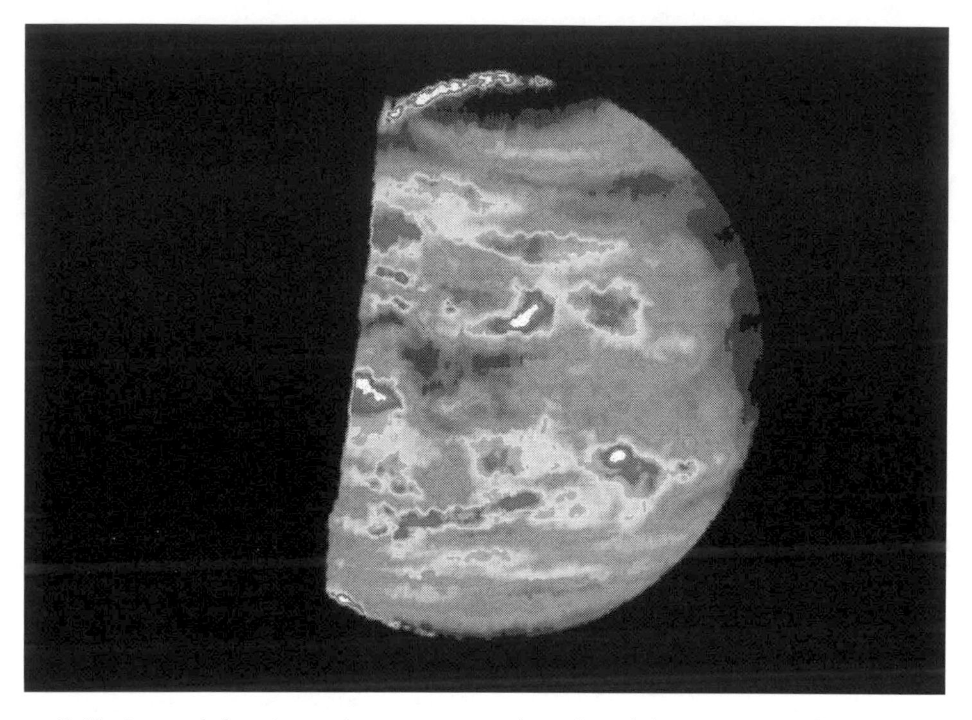

Galileo's near-infrared mapping spectrometer investigated the temperature field of the middle atmosphere of Venus. (JPL/NASA/Caltech)

leading hemisphere when outbound. Excellent images were assembled into multi-spectral mosaics to show differences in the composition of the surface, with the 'bull's eye' of the Orientale basin particularly well positioned at the center of the disk. Observations continued for a week after the encounter. In addition to a wealth of particles and fields data, almost 3,000 images were taken. Tracking established that Galileo had received an unexpected additional increase in velocity of 4.3 mm/s. Although minuscule in terms of the spacecraft's mission, the anomaly was important for the reason that it could not be explained in terms of gravitational effects unless some measurements of the Earth's physical parameters contained implausible errors. Just as with the better-known 'Pioneer Anomaly', a variety of exotic theories have been advanced to explain this effect.[171,172]

On 19 December 1990 Galileo refined its aim for Gaspra. It passed perihelion on 11 January 1991, and then, precisely 3 months later, on 11 April, was directed to unfurl its high-gain antenna. The system utilized a pair of redundant motors on the spacecraft's centerline that drove a set of ball-screws, carrier rings and pushrods to rotate each of the 18 ribs into position and simultaneously stretch out the mesh in a symmetrical manner. As the motor started, 13 ribs were released by their locking pins, but the others jammed in their pin-and-bracket mechanisms. An additional rib was released by a further ball-screw turn, as was another after two more turns. But three adjacent ribs remained in place, and as the rods increasingly pushed on their

A mosaic of Galileo images of the Ross Ice Shelf in Antarctica, taken during the spacecraft's first Earth flyby. (JPL/NASA/Caltech)

bases they bent, bowing and causing the locking pins to 'dig' into their receptacles and reinforce the grip. When the ball-screw bent too, the mechanism jammed. The ball-screw had traveled only 1.5 cm of the 8.6-cm stroke required to fully open the antenna. The deployment ought to have required 3 minutes, but after the motor had been on for 8 minutes a software timer switched it off. If the antenna had opened as planned, the conservation of angular momentum would have caused the rate at which the spacecraft was spinning to slow down. When the engineers noticed that the microswitch that was to have sensed the completion of the deployment had not been tripped, they inferred from the fact that the spin had slowed that the antenna had indeed opened and that for some reason the sensor had failed. However, closer study showed that the spun section had not slowed as much as predicted, indicating that something had gone awry with the antenna deployment. An 'anomaly team' of engineers at JPL noted from the telemetry that the deployment motor had operated for longer than intended, and that a Sun sensor was being periodically occulted by one of the partially deployed ribs. By careful analysis, they were able to deduce the configuration of the antenna. Early ideas for how the deployment might have been fouled included inadvertently exposed adhesive tape and the ribs entangling in the Sun shade at the end of the antenna tower, but tests showed that the motor should have had sufficient force to overcome such obstacles.[173]

The trivial cause of the problem was soon found. The high contact stress of the pin and socket could cause plastic deformation that would damage the coating of the pin. During the repeated trips across the United States, and its years in storage, the antenna had been kept closed and horizontal with the three stuck ribs on top of the assembly and subjected to the greatest vibrations. This caused a loss of the dry lubricant intended to ease the disconnection between the pin and socket, which in turn had enabled it to develop sufficient friction in the vacuum of space to become 'cold welded'. Unfortunately, no one had had the foresight to lubricate the pins in storage. No such anomaly had impaired the antennas of the TDRS satellites (which were transported by air rather than by truck as was the case for Galileo) and whose antennas were lubricated shortly prior to shipment. For Galileo, lubricant had been applied to the pins only once, a full decade before the eventual launch! Moreover, the antenna had not been fully tested, as there was no replacement in the event of its being damaged. Although NASA briefly considered a 'crash' program to build a relay satellite that would be launched by a Titan IV and sent on a fast trajectory to Jupiter to save the mission, it was instead decided to attempt to coax the pins of the antenna to deploy by exploiting the different thermal expansion coefficients of the central tower and the ribs.[174] The spacecraft would vary its orientation in order to alternately put the antenna tower in sunlight and in shadow for days at a time in the hope that a series of hot/cold cycles would release the pins. Other options for the future were to 'hammer' the pins by running the deployment motors or firing the thrusters in bursts; if necessary after increasing the spacecraft's spin for additional centrifugal force. The drive mechanism of the antenna had not been designed to be closed and reopened, and tests showed that to attempt to do so would degrade its mechanical characteristics. Furthermore, the mesh would probably become fouled if an attempt were made to close the antenna in order to start afresh. The first two

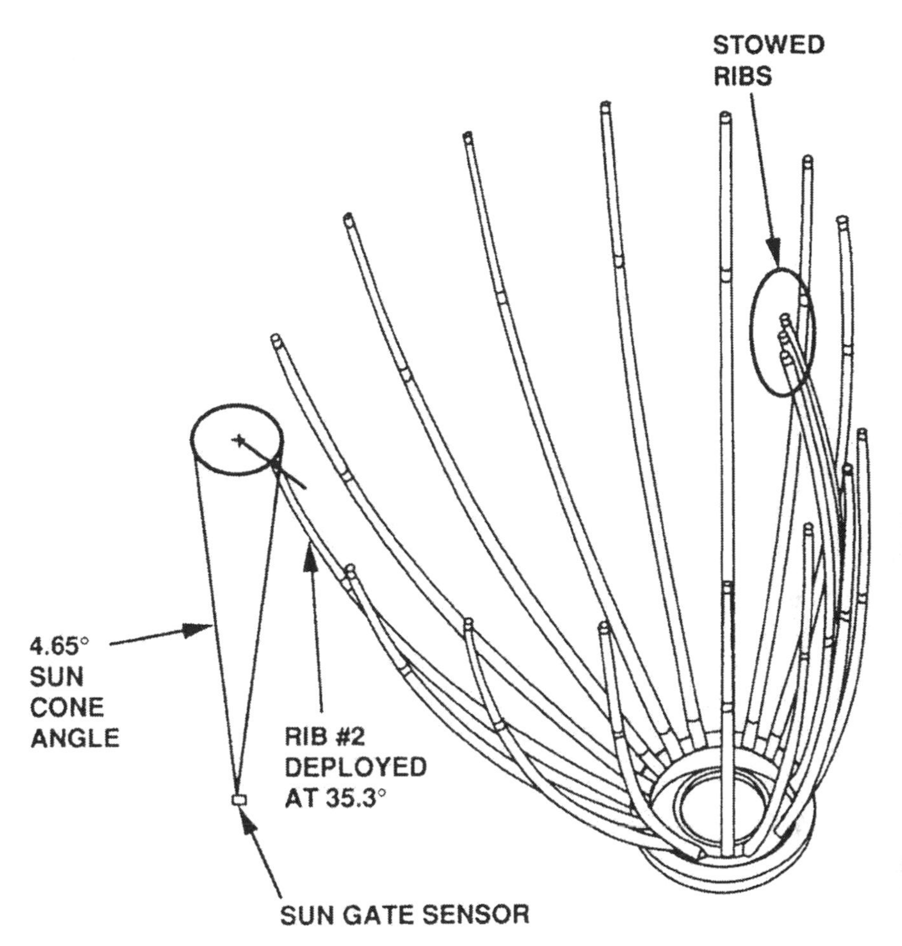

The inferred configuration of the ribs of Galileo's high-gain antenna in its partially opened state.

'cold soaks' were performed in July and August 1991, but to no avail, and further efforts were postponed until after Galileo made its flyby of Gaspra on 29 October 1991.[175,176]

After Gaspra was announced as the target for the first-ever asteroid encounter, astronomers readily compiled a list of everything that was known of it, which was not much, and then set about making further observations. As a result, it was now understood to be a 10 × 11 × 18-km object that was rotating 'end over end' with a period of 7 hours. It also appeared to belong to a large family of asteroids whose dynamical and orbital properties were similar to those of (8) Flora, suggesting that they were all fragments of a larger body which was shattered by an impact. Galileo would pass by Gaspra at a range of about 1,600 km at a relative speed of 8.0 km/s on a trajectory that would provide good illumination and contrast for imaging at high resolution. The greatest operational challenge was to reduce the uncertainty in the asteroid's ephemeris, as its position was initially known only to within several hundred kilometers, which was much larger than the field of view of the camera at the flyby range. In September, Galileo started to take navigational images of the asteroid against the background stars. Prior to the failure of the high-gain antenna to deploy, it had been planned to take about 40 such images, but the number had to be reduced to four. However, the camera was used in its multiple-exposure mode, which could mosaic up to 64 exposures onto a single frame. Based on the results of this effort, the spacecraft made two course corrections to refine its aim. Galileo did not have the autonomy or the software to locate and track a target by itself (as had, for example, the Halley missions). Instead, the location of the target was predicted by the ground team, and the appropriate attitude of the spacecraft and the angle of the scan platform calculated and uploaded.

On the day of the encounter, with the camera and infrared spectrometer already making observations, the particles and fields instruments were switched on to try to sense the asteroid's presence. Half an hour before closest approach, Galileo took a 9-frame mosaic which spanned an area of the sky sufficiently large to ensure that at least one color image was obtained of the entire object with a resolution of 160 meters. It then took a mosaic of 49 monochrome frames. As the target loomed, what would turn out to be the highest resolution mosaic was composed of a pair of images taken at a range of 5,300 km 10 minutes before closest approach. The final frame, with a 50-meter resolution, was taken a few minutes later but captured only a partial view. The scan platform's fastest rotation rate of 1 degree per second was insufficient to 'pan' to maintain the object in the field of view, so no imaging was scheduled for the 1,604-km flyby, which occurred at 22:37 UTC on 29 October. It had been estimated that there was a 95 per cent chance of the asteroid appearing in the middle of the 9-frame color mosaic. In November this was 'sampled' over the low-gain antenna at the agonizingly slow rate of 40 bps to verify that Gaspra was in the field of view, then the relevant lines for each of the three filters were sent (at that rate it would have taken 3 days to transmit the complete color image). In all, 150 camera frames were stored for transmission either when the high-gain antenna was deployed or when the spacecraft made its second Earth flyby in 1992.

As expected, Gaspra proved to be an irregular body; in fact, the most irregular yet visited by a spacecraft. Near-infrared observations of compositional differences between its two ends implied that it was actually two smaller blocks in contact. A number of small craters, the largest of which was about 1.5 km in diameter, were superimposed on a smooth surface. An analysis of the high-resolution pictures inferred from the fact that the craters were mostly small and fresh that the asteroid's surface was about 200 million years old. There were also subtle grooves and ridges associated with depressions, and two large concavities that could be due to fractures in the body from which Gaspra was separated.[177,178]

In January 1992 Galileo reached aphelion and passed through solar conjunction. Near aphelion it was again turned with the high-gain antenna facing away from the Sun for a 'deep cold soak' lasting 50 hours, but this failed to release the stuck pins. Simulations suggested that 6 to 12 warm-cold cycles would be required to free the pins, but it had been necessary to make assumptions as to the positions of the pins

Galileo was the first spacecraft to inspect an asteroid. This mosaic of the highest resolution images of (951) Gaspra has a resolution of about 54 meters per pixel.

and their frictional loads, and these might have been wrong. Six times the low-gain antenna at the end of the 2-meter mast was retracted in the hope that the jolt when it reached its hard-stop might nudge the pins, but to no avail. By September 1992, after 7 thermal cycles had failed to release even one pin, concern was expressed that each turn to reposition the antenna was consuming 4 kg of propellant.[179]

As an alternative, engineers suggested moving the antenna motor in bursts, each of which should rotate the ball-screw a fraction of a degree and increase the force designed to release the pins. The thermal expansion of the antenna's tower would be maximum for 3 months around perihelion, and it was thought that simultaneous 'hammering' of the ball-screw might release one rib, thereby increasing the loads on the others enough to snap them free too. The technique was tested in October 1992. The natural proper frequencies of the ribs and antenna were also investigated in an effort to fine-tune the forces applied. It was also decided that if the problem had not been resolved by March 1993, the high-gain antenna would be written off and the remainder of the mission conducted using only the low-gain antenna. The communications strategy developed for this eventuality relied on using the tape recorder to store data from the spacecraft's encounters with the Jovian satellites near the time of each periapsis of its orbital tour, and transmitting it at 10 bps during the leisurely cruise to apoapsis – a rate only slightly faster than that of Mariner 4 during its Mars flyby in 1965. As it would take weeks to return an image at 10 bps (and the data collected during a single satellite encounter would take years) algorithms for onboard compression of the data (in particular imagery) would be implemented by reprogramming the main data computer. In combination with upgrading the Deep Space Network, including arraying its antennas, the onboard improvements would boost the effective rate at which data would be returned 100-fold, sufficient for good coverage of the Jupiter approach phase, nearly continuous coverage of routine engineering and particles and fields data, a full tape dump for six periapses of the main tour, and partial data for the four periapses when Jupiter was on the opposite side of the Sun from Earth. It was estimated that even though only 4,000 of the originally envisaged 50,000 images would be able to be returned, it should still be possible to achieve 70 per cent of the scientific objectives set for the mission. However, as comprehensive mapping would be impracticable, the analyses would have to be undertaken on the basis of combining low-resolution overviews and high-resolution imagery of selected features.

Meanwhile, in mid-1992, as Galileo was approaching Earth for its final gravity-assist, NASA formalized the 2-year 'mini-tour' of the Jovian system which would provide encounters with each of the large satellites except Io, whose orbit was too deep in the planet's radiation belt for the spacecraft to routinely visit. It would last 23 months, and include four close encounters each for Ganymede and Callisto and three for Europa, in addition to several distant encounters, a 150-Rj flight down the Jovian magnetotail, and a survivable total radiation dose. As a bonus, because the spacecraft had to make its orbit-insertion burn within the orbit of Io, the approach trajectory would provide a close flyby of this intriguing volcanically active moon. This set the date of arrival as 7 December 1995, and allowed the propellant margin to be accurately estimated. Armed with this estimate, and based on the experience gained during the Gaspra flyby, NASA sanctioned an encounter with Ida in 1993. In fact, the agency announced that all future outer planet missions would include at least one minor planet flyby during their interplanetary cruise in order to build up a survey of such bodies.[180]

The downloading of the Gaspra data resumed in June 1992 with recovery of a pair of the highest (50-meter) resolution images, with 20 per cent of the asteroid in one frame and the rest of it in the adjacent one. The remainder of the science data was downloaded over a 2-day period in November. The near-infrared data showed the thermal inertia of Gaspra to be neither that of 'bare rock' nor of a body covered by a mature regolith. Instead, it might be covered by a mixture of rocks, regolith and coarsely and finely grained material. Overall, the imaging verified ideas about the 'look' of an asteroid. The great surprise was the magnetometer data, which showed two disturbances in the solar wind. The first disturbance occurred 1 minute prior to closest encounter, and the second was 4 minutes later. In between these events, the magnetic field vector rotated towards Gaspra. While the possibility of a fortuitous coincidence could not be ruled out, it was generally presumed that the asteroid was weakly magnetized. This could be explained if the progenitor from which Gaspra derived had been sufficiently large to undergo thermal differentiation and become magnetized as the iron was concentrated in its core. This conclusion was supported by the infrared spectrometer finding metal-bearing compounds on the surface, as these could 'fossilize' a remanent magnetic field. In a sense, the Gaspra encounter marked the transformation of such objects from 'asteroids' (literally, 'looking like stars') to 'minor planets'.[181,182]

Soon after completing the Gaspra download, Galileo began the approach phase of its second Earth flyby. Because the data rate of the low-gain antenna increased as the range decreased, it was possible to calibrate some of the bandwidth-hungry instruments and give the entry probe a health check. As previously, Galileo flew up the Earth's magnetotail. This time astronomers who studied near-Earth asteroids managed to spot the spacecraft telescopically at a record range of 8.06 million km. On the way in, it passed 110,300 km above the northern hemisphere of the Moon, enabling it to make valuable observations of areas that had received scant attention by the lunar missions of the 1960s and for which, therefore, mapping coverage was poor. The ultraviolet spectrometer detected weak emissions from the hydrogen that some scientists believed was enriched in permanently shadowed craters. It also briefly sought cometary emissions from the near-Earth asteroid (4179) Toutatis, which was making a particularly close pass by Earth at the time. At 15:09 UTC on 8 December 1992 Galileo passed by Earth, with the 3.7-km/s increment giving it the heliocentric velocity of 39 km/s required to reach Jupiter. However, at an altitude of 304 km the atmosphere was able to impart sufficient drag on the vehicle to rule out an attempt to verify the 'anomaly' detected during the first flyby. The point of closest approach was over the Atlantic between South Africa and Argentina. As it departed, the spacecraft took high-resolution images of the Andes and once again studied clouds over the south pole and the ozone layer in the stratosphere. With its new orbit ranging between 0.98 and 5.30 AU, in effect Galileo set off on a 3-year transfer to Jupiter similar to that which would have been produced by the Shuttle–Centaur.

On both Earth flybys, Galileo was used to conduct some unusual experiments. On the first encounter, an experiment by Carl Sagan attempted to demonstrate that a probe could detect life, and in particular intelligent life. In addition to detecting the presence of water, oxygen and methane in the atmosphere and of geometric patterns and "a widely common red-absorbing pigment", i.e. chlorophyll, on the surface, images were taken of the night-side in search of city lights. However, the results were only weakly indicative of the presence of life. Indeed, the only firm proof of intelligence were bursts of narrow-bandwidth radio waves.[183] On the way out after the second encounter, Galileo tried to detect laser beams aimed at it by telescopes, to demonstrate the feasibility of optical communication in deep space. As a final bonus, from the spacecraft's point of view 8 days after the encounter the Moon crossed in front of, and just above, Earth, and over an interval of 14 hours a unique 'family portrait' sequence of 50 color images was taken that showed the Moon gliding through the frame as Earth turned on its axis. In all, more than 6,800 pictures were taken during the flyby.[184,185]

As Galileo passed perihelion, and the tower of the high-gain antenna was at its maximum thermal expansion, controllers mounted a final campaign to release the stuck ribs. Over 15,000 'hammer' pulses were applied between 29 December 1992

Shortly after the second Earth flyby, laser beams were fired at Galileo at a range of 1.4 million km to evaluate using a laser to communicate with a deep-space mission. The camera repeatedly scanned Earth to record the pulses of light fired in darkness from the Air Force Phillips Laboratory's Starfire Optical Range near Albuquerque in New Mexico (the line of brighter dots to the right) and from the Table Mountain Observatory in California (to the left). (JPL/NASA/Caltech)

and 19 January 1993. The first pulses caused the ball-screw to make more than one full turn before it stalled again. This caused the free ribs to open a little further, but the pins on the others remained stuck. In March, the hammering was resumed and the spacecraft spun up to its maximum rate of 10.5 rpm, again to no avail. NASA then conceded that there was "no longer any significant prospect of deploying" the antenna. Nevertheless, an engineering test was carried out in which the transmitter of the high-gain antenna was activated for the first time, to determine whether the recalcitrant antenna yielded some 'gain lobe' that could be exploited. In the spring of 1993 Galileo, Mars Observer (traveling to Mars) and Ulysses (on its way to pass over the south pole of the Sun) collaborated in a search for gravitational waves passing through the solar system.

After Ida was announced as the target for the second asteroid flyby, telescopic studies had shown it to be an elongated object with a mean diameter of 28 km and a rotational period of just 4.63 hours. It belonged to the Koronis family produced by the disruption of a 100-km-sized body. As dynamical studies suggested that this event occurred only 20 million years ago, the surface of Ida was expected to be much younger-looking than that of Gaspra. The Ida encounter sequence would be similar to that at Gaspra, but because the spacecraft would not be returning to the vicinity of Earth all the data would have to be sent at a low rate in the ensuing months. To minimize the transmission of pixels representing black sky, a 'jailbar' technique was to be used whereby only a few lines of each picture would be sent to start with, then, after the position of Ida in the frame had been determined, only the relevant pixels would be returned, with this being done when the Earth-spacecraft distance reached its next minimum in the spring of 1994 in order to use a rate of 40 bps. There were several minor problems during the encounter. The first involved brief short circuits in the articulation between the spun and despun sections, and by entering 'safe mode' the computer canceled all of its programmed activities. There had been three such incidents in 1991, then a clear spell until 2 months before the encounter. Owing to the time required to restore the spacecraft to normal, if a short were to occur a few hours prior to the flyby, then it would sail past the asteroid in a passive state. Fortunately, the only effect that the shorts had on the encounter was to prevent two of the five images intended for navigation purposes. However, the three that were obtained proved sufficient. In fact, they were so good that only one course correction was required to target the encounter. An additional issue was that a week before the Ida flyby the largest antennas of the Deep Space Network had to discontinue tracking Galileo to search for Mars Observer, which had fallen silent as it approached Mars. Although 75 per cent of the final navigation image was lost when the receiving antenna was requisitioned for a higher priority task, the partial image was sufficient to confirm the trajectory. Priority reverted to Galileo 52 hours before the encounter. One final thrill occurred 4 hours 16 minutes before the flyby, when the spacecraft mysteriously adopted 'cruise mode' and stowed its scan platform! The situation was recovered by precisely timed commands sent over the next hour, with only modest degradation of the imagery.

At 16:52 UTC on 28 August 1993 Galileo flew by Ida at a range of 2,410 km and a relative speed of 12.4 km/s. A total of 150 frames were taken, half of which were devoted to the wide-area mosaics designed to allow for the uncertainty in the asteroid's position. A 75-frame sequence taken over an interval of 5 hours was to document, albeit at low resolution, just over one full rotation of Ida to characterize its entire surface. This was followed by four mosaics at increasing resolution, the second of which also included infrared spectrometry. Because the fourth and final 15-frame mosaic covered only part of the ellipse representing the uncertainty in the ephemeris, the chances of capturing even part of Ida were estimated at only 50 per cent. The 'jailbars' for the 30-frame monochrome third mosaic were transmitted a few days later, in order to locate Ida. It was straddling the edge of five frames that were taken close in, and therefore at almost the highest resolution of the sequence. As a result, when the first mosaic of Ida was unveiled in September, it comprised frames with resolutions ranging between 31 and 38 meters/pixel.[186] It showed the asteroid to be an angular object with dimensions of 54 × 24 × 21 km, dominated by craters of a wide range of sizes, and while some were fresh-looking others were distinctly degraded with impacts on their rims and floors. In fact, the crater density was five times greater than for Gaspra, implying an age for Ida of at least 1 billion years. Evidently either the dynamical studies of how the progenitor of the Koronis family broke up were flawed, or Ida appears older than it actually is because it was showered by debris during the break up of its progenitor. The eastern limb on view in the first mosaic showed a group of five craters ranging in size from 5 to 8 km (a significant fraction of the asteroid's mean diameter), inside which were clusters of boulders exceeding 100 meters in size. Grooves in the same area, some running for several kilometers, could represent fractures in the underlying 'bedrock' produced by the impacts that made the large craters.[187]

By late September, Galileo was so close to the Sun as viewed from Earth that the data playback was suspended until February. Meanwhile, a course correction in early October put the spacecraft back on track for Jupiter. The downloading of Ida data resumed on 16 February 1994 and continued until 26 June, initially at 10 bps and later, as the range reduced, at 40 bps. The most amazing discovery of the Ida flyby was made on the second day, while returning the 'jailbar' version of a frame in the second color mosaic. This was taken 14 minutes before the flyby, and from a range of 10,870 km. As camera-team member Ann Harch scrutinized the 'jailbar' rendition, she noticed a small object off to one side of Ida. The chances of Galileo being in place to witness another asteroid passing by were implausibly low, so this other object had to be an asteroidal satellite – the first ever seen. Its existence was confirmed several days later by the infrared spectrometer, whose field of view had also included the satellite. The plan for retrieving the encounter data was revised to ensure that all data on the satellite was also recovered – it was present in a total of 47 frames. It proved to be a 1.6 × 1.2-km egg-shaped object with dozens of craters ranging up to 300 meters across. The flyby geometry precluded obtaining enough good positional measurements to derive the satellite's orbit, but constraints could be imposed by arguing that the orbit could not be too low, as otherwise instability would long since have caused the satellite to hit Ida, and the orbit could not be too high, as otherwise a perturbation would have caused the satellite to escape. Further constraints were provided by the Hubble Space Telescope, which, upon observing

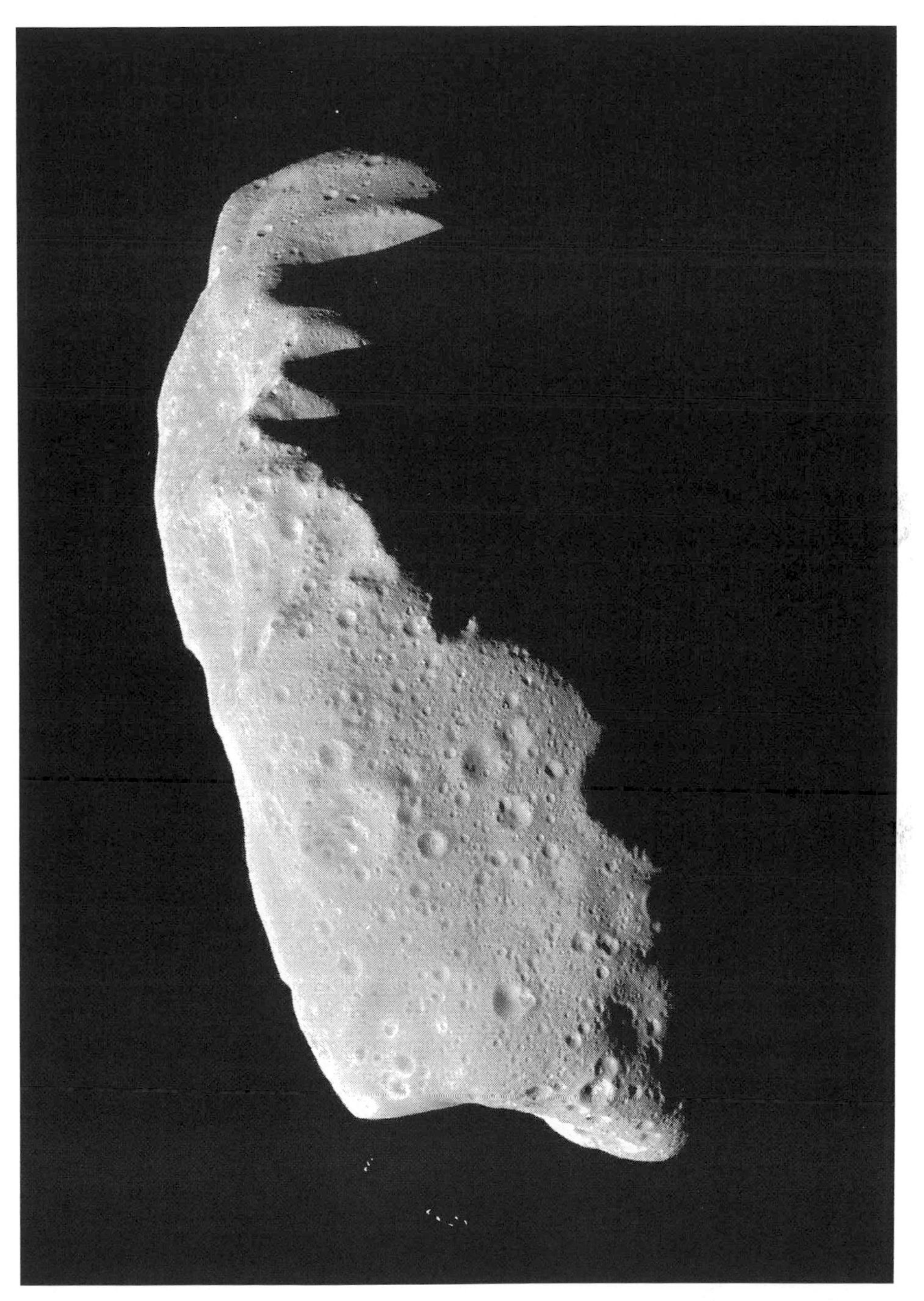

A high-resolution mosaic of the heavily cratered asteroid (243) Ida, which is about five times larger than Gaspra. Note in particular the size of the craters along the terminator. (JPL/NASA/Caltech)

Ida in April 1994 failed to detect the satellite. Taken together, these constraints also provided a well-bracketed measurement of the mass and density of Ida, which appears to be relatively porous, suggesting that it is an accumulation of fragments rather than a chip off a dense parent. As more data arrived, it helped to refine Ida's shape. One end was angular, and the opposite end was a distinctive 'bottle-shape' dominated by a large depression. Only one of the final frames had included part of the limb, showing it at a resolution of 24 meters per pixel. Thermal data indicated the presence of a thick layer of regolith, which was consistent with the smoothness of the craters and other features. The near-infrared spectra showed the surfaces of both bodies to possess an iron-rich silicate composition, with only minute differences between them. This similarity implied that Ida and its satellite were related. As at Gaspra, the magnetometer noted a deflection in the direction of the interplanetary magnetic field. The International Astronomical Union named Ida's satellite Dactyl, after the Dactyli Idaei of Greek mythology, these being creatures which lived on Mount Ida. How could the Ida–Dactyl system have formed, and remained stable? Theories ranged from their both being fragments of the Koronis progenitor which promptly became gravitationally bound, through to Dactyl having been chipped off Ida.[188,189] As late as the 1980s the idea of asteroidal satellites had been contentious, but dozens of such discoveries soon followed, including some orbiting the Trojans of Jupiter and near-Earth asteroids, and even several examples involving a pair of satellites.[190]

Shortly after Galileo had completed downloading the Ida data it was called into action again, this time to observe one of the most amazing astronomical events of the 1990s. During a photographic search for near-Earth asteroids and comets using a telescope on Mount Palomar in March 1993, Carolyn and Eugene Shoemaker, David Levy and Philippe Bendoya discovered a fuzzy bar-shaped object close to Jupiter that gave the appearance of being a small comet which had been squashed. Comet Shoemaker–Levy 9, as it became known, proved to be a string of about 20 pieces of a recently disrupted cometary nucleus. As measurements accumulated of the positions of the train of fragments, it became evident that the nucleus had been captured by Jupiter several decades earlier, and that as its orbit evolved it flew so close to the giant planet at periapsis in July 1992 that the weak internal structure of the body, which was probably some kilometers across, yielded to tidal stresses and disaggregated like a veritable rubble pile. When discovered, it was in a 2-year orbit and approaching apoapsis. Furthermore, owing to perturbations by the Sun and the satellites of Jupiter, most if not all of the fragments would fall into Jupiter over the period of a week in July 1994! As such an event had never been witnessed, no one knew what to expect. So little was known about the make-up of the comet, and the physics of impacts, that estimates of what might be seen ranged from nothing at all to fireballs which left long-lived spots in the planet's atmosphere. To the immense disappointment of astronomers, it transpired that whilst all of the fragments (which were designated alphabetically from A to W) would strike Jupiter, in each case the impact site would be just beyond the morning limb of the planet, and so not visible from Earth. However, owing to Jupiter's rapid rotation, when the Sun rose several tens of minutes later, the impact sites would come into view. Many Earth-orbiting

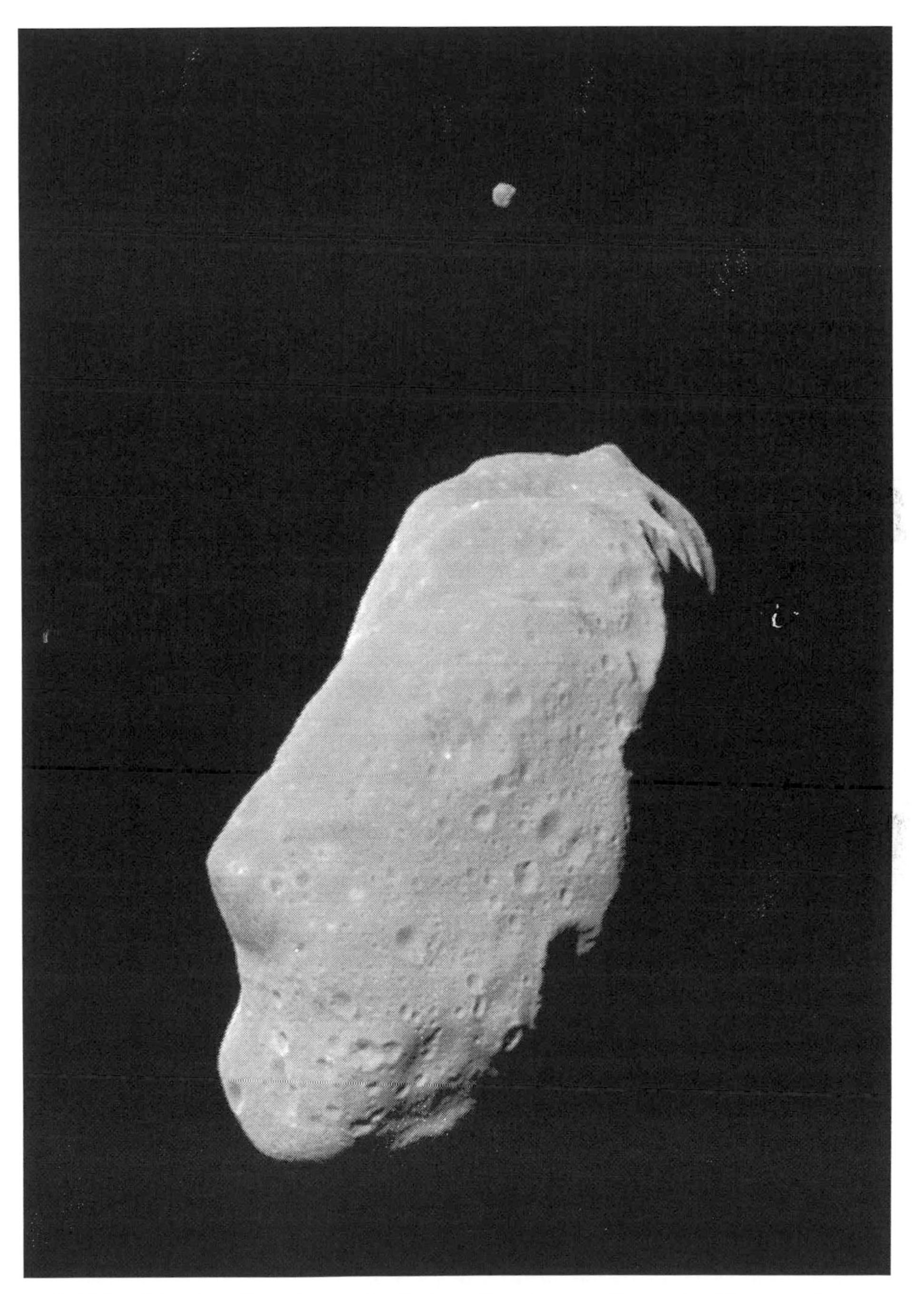

Orbiting Ida, tiny Dactyl was the first confirmed case of an asteroid possessing a satellite. (JPL/NASA/Caltech)

astronomy satellites, including the recently refurbished Hubble Space Telescope, were called upon to make observations of the impacts and their aftermaths. Of the spacecraft in interplanetary space, Voyager 2's line of sight would allow it to view the night-side of Jupiter, but unfortunately the software to operate its cameras had been deleted and the imaging team had dispersed, and in any case the range would have been so great that Jupiter would have spanned only a handful of pixels and in effect only photometric measurements of the impacts would have been possible. The Clementine spacecraft operated by the US Department of Defense was on its way to encounter an asteroid, and would have been able to observe the impacts on Jupiter, but 2 months beforehand it suffered a malfunction and was lost. As regards Galileo, it was initially believed that the impacts would occur just beyond the limb of the planet, but as better orbit data became available early in 1994 it was realized that the sites would be observable. At a range of 238 million km, the planet would span about 60 pixels of the CCD. Ironically, had Galileo been in orbit around the planet at the time, scientists may not have been so lucky, since the spacecraft could easily have been near the apoapsis of its orbit on the opposite side of Jupiter, or perhaps its tape recorder might have been full of other data.

Five instruments were to make observations of the impacts: the camera, infrared spectrometer, ultraviolet spectrometer, polarimeter and plasma-wave experiment. If enhanced dust streams were left by the cometary fragments, the spacecraft's dust counter would detect these in the months following impact. A complicating factor was that the time of each impact could be predicted only to within an interval of a few minutes, so a means of dealing with uncertainties in timing was needed. It was decided that the camera would operate in two modes: on some impacts, it would use the mosaicking mode by which up to 64 exposures taken at intervals of several seconds would be stored side by side on a single frame; on other impacts, Jupiter would be allowed to drift across the field of view, with several such 'bands' being recorded on one frame. Whereas the first technique would hopefully provide some amazing images of an impact and the resulting fireball, the second would produce an essentially continuous (but smeared) record of the impact which would enable scientists to determine the exact time at which the fireball (if any) appeared. Once the timing was known, it would be possible to save on the downlink resources by transmitting only the data which was pertinent to studying the rapid evolution of a fireball.[191] In fact, the impacts surpassed the most optimistic predictions and when the sites emerged over the limb astronomers were astonished to find that most of the fragments had produced dark 'black eyes' in the planet's atmosphere, in many cases accompanied by circular rings of ejecta which were dark visually but glowed brightly in the infrared. In some cases the Hubble Space Telescope saw the plume of expanding gases catching the Sun even before the site rotated into view. Galileo observed many of the impacts. For some (fragments K and N) it used the camera in drifting mode, timing the impact to within a fraction of a second and recording the evolution of the fireball's brightness. Fragment K (the largest identified) created a fireball which lasted for 37 seconds, and at its peak was as bright as 10 per cent of the planet's entire disk seen through a methane filter. The multiple-exposure mode managed to obtain a time sequence spanning the impact of fragment W on 22 July,

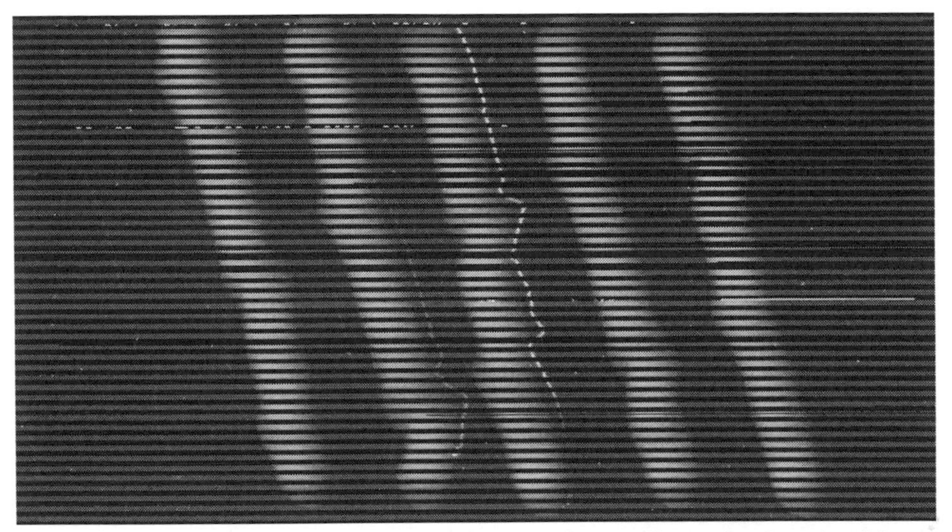

A drift-scan image taken by Galileo of fragment K of comet Shoemaker–Levy 9 striking Jupiter. In this mode, the planet was allowed to drift through the camera's field of view in order to provide a detailed reconstruction of the light-curve of the fireball.

the flash of which saturated the CCD for several seconds. Other impacts, including fragments G, H, L and Q1, were monitored by the photopolarimeter by measuring the total brightness of the planet four times every second, thereby producing 'light curves' that enabled the temperatures of the fireballs to be calculated; in the case of fragment Q1 this exceeded 10,000K. The infrared spectrometer enabled the sizes, altitudes and temperatures of the fireballs to be inferred: for example, seconds after the impact of fragment G, the fireball was expanding at over 1 km/s and glowing at about 2,000°C.[192,193,194]

GALILEO BECOMES A SATELLITE OF JUPITER

In late January 1995 Galileo finished replaying the Shoemaker–Levy 9 data selected by the scientists. In February new software was loaded into the main computers, marking a significant 'first' for a mission in deep space. In March the atmospheric probe was tested in preparation for its imminent release. A course correction made after the Ida encounter had set the spacecraft on a collision course with Jupiter, but for the probe to survive its penetration of the Jovian atmosphere it was crucial that it approach along a very narrow corridor. On 12 April a series of 64 pulses of the thrusters produced an 8-cm/sec refinement which ensured that the probe would fly down the center of its corridor. After it was confirmed that the probe's coast timer and g-switches were functional, on 11 July the umbilical from the main spacecraft was cut. The next day Galileo orientated itself so that when the probe was released it would be in the attitude at which it was to enter the atmosphere, adopted its one-spin mode and increased the rate of spin to 10.5 rpm to provide the probe with the

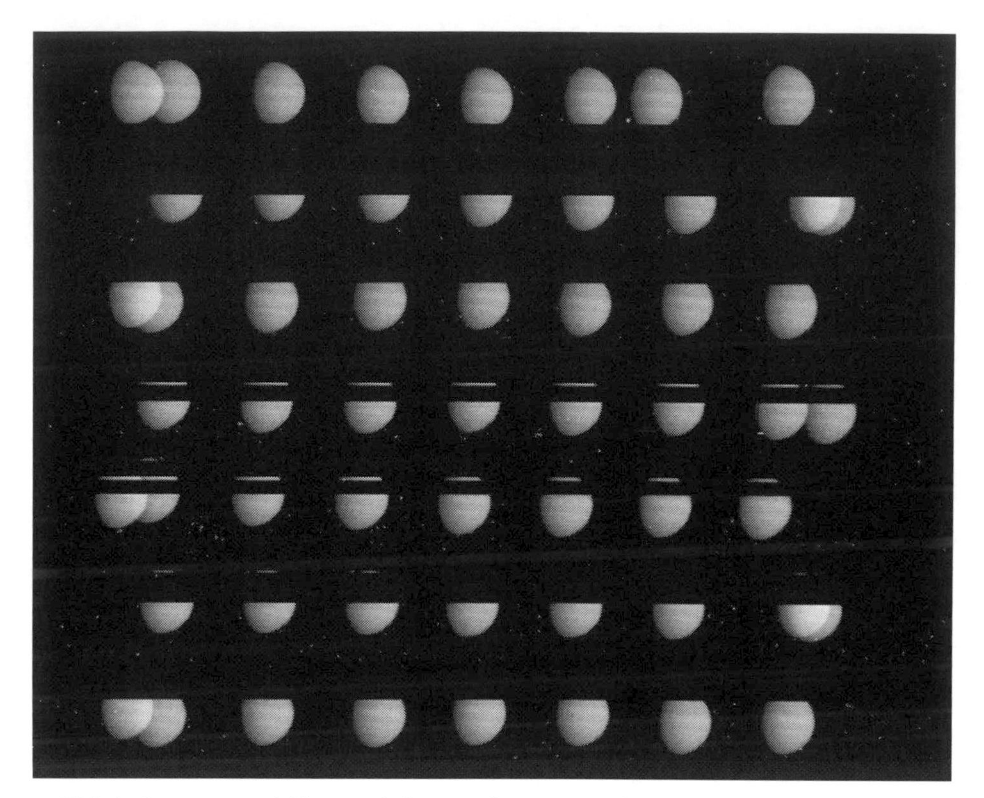

This is the raw mosaicking mode image taken to record the impact of fragment W of comet Shoemaker–Levy 9. The saturated fireball is visible beside the night-side limb in the last three frames of the top line.

greatest possible stability once free. At 05:30 UTC on 13 July three explosive bolts were detonated to release the probe and it was pushed away by springs at a relative speed of 0.3 m/s. The velocity change of the orbiter was the first indication that the operation had succeeded. On the probe, only the coast timer was active. If all went to plan, this would start an activation sequence 5 months hence, a few hours before the probe reached Jupiter. The navigational data showed that both the latitude of the entry site and the angle of entry were within a fraction of a degree of the values required, and that the probe would arrive just 2 seconds later than planned. Having adopted a trajectory to enable the probe to make an 82-million-km ballistic fall to Jupiter, Galileo now had to maneuver to avoid the same fate. The release of the probe had finally cleared the nozzle of the main engine. In order to test the engine and calibrate its performance in advance of the orbit-insertion maneuver, it was to perform this orbiter deflection burn. After a 2-second firing on 24 July to 'clear its throat' it increased the speed by 62.2 m/s on 27 July, in the process consuming some 40 kg of propellant. This was the largest maneuver that the spacecraft would perform during the entire cruise, and the 23rd course correction of some 26 needed to reach Jupiter. When a pressurant check-valve failed to close after the burn, this raised the prospect of oxidizer fumes migrating through the plumbing, reaching the fuel tank, and causing an explosion such as was believed to have caused the loss of Mars Observer. However, engineers believed that they could control the migration by monitoring the temperature and evaporation rate of the oxidizer.

On 28 August Galileo refined its trajectory to pass within 1,000 km of Io on its way in to Jupiter. In September it flew through the most intense dust storm ever encountered in space, recording up to 20,000 impacts per day. Dust had also been noted by Ulysses, which, although launched after Galileo, had flown a more direct route and reached the planet in 1992. The dust was clearly associated with Jupiter, but it would be several years before the precise source was found. In the meantime, speculation focused on Io's volcanoes or impact ejecta, Jupiter's main ring or the 'gossamer' ring, and dust left by Shoemaker–Levy 9.[195,196]

On 11 October, Galileo took its only approach image. It exposed several frames using different filters for a color picture showing the 'half phase' Jovian disk, with the probe entry site in view, plus the satellites Ganymede and Io.[197] The next task was to rewind the tape recorder in order to play back the frames, which would take several weeks to transmit over the low-gain antenna. It was at this point that a new hardware failure occurred. Although the motor of the recorder ran, the tape did not move. If the tape had snapped, this would be disastrous since the strategy devised to overcome the loss of the high-gain antenna relied upon the recorder. Without the recorder, Galileo would be limited to taking data that could be transmitted in real-time. By ruling out imaging, this would turn the mission into a particles and fields investigation. A test 9 days later showed that the tape was able to advance but was sticking on a 'dummy' eraser head, and if run in reverse it lost tension and slipped. In case the tape had been worn whilst stuck, it was decided to advance it a little to redefine the start-of-tape point. It was also decided that the recorder should not be used until it had stored and replayed the data transmitted by the atmospheric probe. Denying the use of the tape recorder during the approach meant losing not only the recently taken picture of Jupiter, but also a study of the probe entry site planned for the last 2 days of the approach, all of the imagery of Io, imagery of the south pole of Europa which would not be able to be viewed during the orbital tour (in fact, the arrival date had been chosen specifically to allow these observations, which would have greatly expanded upon the Voyager coverage of these fascinating bodies), and distant views of the moonlets Amalthea and Adrastea. Although the particles and fields data from the plasma torus occupying Io's orbit would still be recorded, the final three course corrections were canceled, as there was no longer a requirement to precisely point the scan platform at Io. Losing the imagery of the probe's entry site was significant because the fact that Galileo would arrive just 12 days before the planet was at solar conjunction meant not only that the Sun would impair radio communications but also that telescopes on Earth would have difficulty monitoring conditions at the probe's entry site.

In November Galileo switched on, tested and calibrated its various instruments one by one, and made ultraviolet observations of the Io torus. Unlike the Voyagers, for which the approach to Jupiter had been a frenzy of activity, Galileo was almost silent, taking only dust and magnetic field data. After the first crossing of the bow shock on 16 November, there followed the usual series of events as the Jovian magnetosphere oscillated in response to the varying solar wind pressure, with the final crossing being on 26 November at a distance of 9 million km from the planet.

It had been decided more than 3 years previously that 7 December 1995 would be the day on which Galileo's probe would enter the Jovian atmosphere, and the main spacecraft, after making its closest approach to the planet and recording the data from the probe, would become the first artificial satellite of the planet. Nine hours from Jupiter, Galileo flew by the southern hemisphere of Europa at a range of 32,958 km. A little more than 4.5 hours later, at 17:46 UTC, it skimmed 898 km over the equatorial region of Io (about one-fourth of the moon's diameter, with the latitude of closest approach at 8.5°S). The close flyby of Io had been built into the mission plan not only to provide an opportunity to image this fascinating body, but also, and primarily, because it would provide a gravity-assist which by slowing the spacecraft a little would save the equivalent of 95 kg of propellant in the imminent orbit-insertion

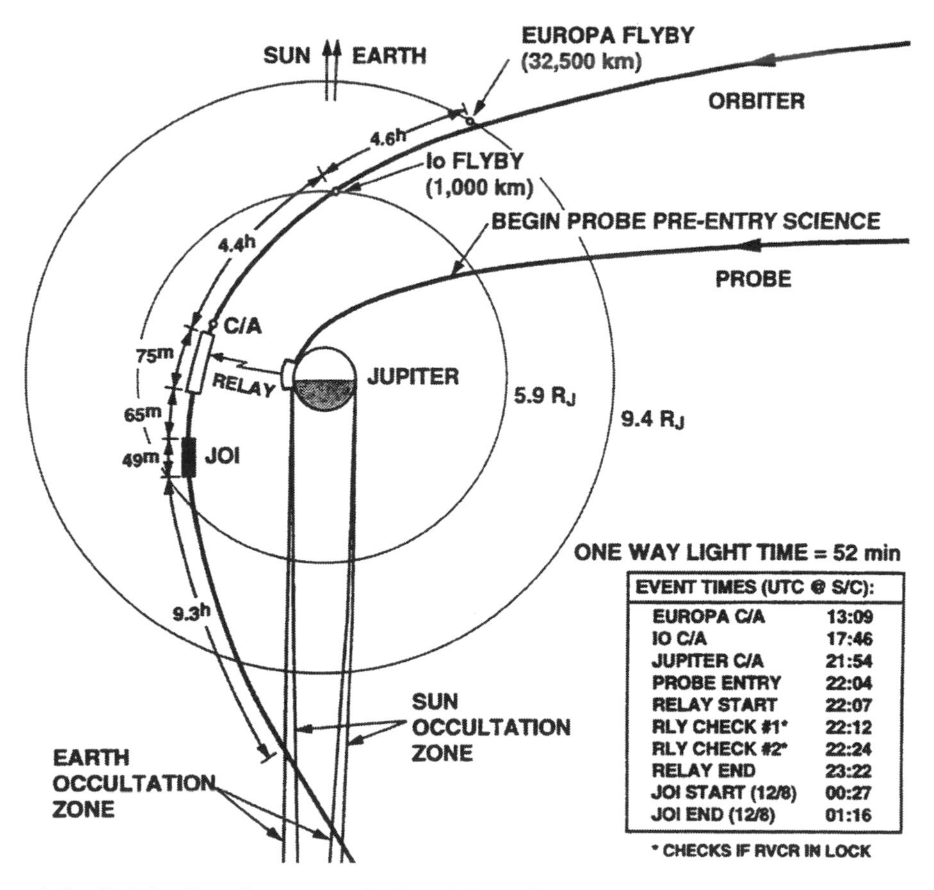

A detailed timeline of events on the day of the Galileo spacecraft's arrival in the Jovian system.

maneuver. As Galileo crossed Io's orbit, it flew through the plasma torus and deep into the radiation belts. At 21:53 it reached its closest point to the planet, flying 215,000 km above the cloud tops at a planetocentric distance of 4 Rj. Fourteen minutes later, it began to listen for the transmission from its probe.

At 16:01 UTC the timer that had been set running prior to the probe's release in July, timed out just before it crossed the orbit of Io, with 6 hours remaining before the predicted moment of atmospheric contact. As planned, the timer started a chain of events that included powering up the radio systems and calibrating the various instruments. Three hours later, the energetic-particle instrument started to measure the fluxes of electrons, protons, helium nuclei and heavy ions down to an altitude of 8,000 km, much closer to Jupiter than the main spacecraft's particles and fields instruments would ever sample directly. Sixteen minutes prior to entry, the probe started to monitor the accelerometers and resistance detectors which measured the ablation of the heat shield material. To save the battery for sampling inside the atmosphere, the probe was not yet transmitting, and this early data was stored in solid-state memory. Atmospheric contact was arbitrarily defined as the moment when the probe reached a point 450 km above the 1,000 hPa pressure level. Later analysis showed that this occurred at 22:04:44 UTC. It entered the atmosphere at an angle 8.6 degrees below the local horizontal, traveling at a relative speed of 48 km/s. When the deceleration reached a peak of 228 g some 58 seconds later, the pressure of the atmosphere was at 'Martian' levels and the temperature of the heat shield peaked at 16,000°C. After another 122 seconds the g-switches commanded the mortar to deploy the drogue chute, and once this had slowed the rate of descent to 120 m/s the aft cover was released and within an interval of less than 2 seconds the 2.5-meter-diameter main chute was deployed and opened. Several seconds later the heat shield was jettisoned. This was noted by the downward-looking sensor of the net-flux radiometer. The 152-kg heat shield had been built with a safety margin of 40 per cent, and sensors in the material indicated that some 82 kg of it had been eroded – although this was essentially as predicted, the ablation had been less than expected at the nose and greater at the sides, perhaps due to some unpredicted heat transfer mechanism.

Once the arm with the mirrors for the nephelometer had swung open, the probe was fully configured for science, and 192 seconds after entry, almost a minute late, it began to transmit to the main spacecraft passing overhead. In fact, as would later be determined, the parachute phase events had occurred 53 seconds later than planned because a wiring fault involving the deceleration switches delayed the deployment of the parachute. It had been intended to start sampling at an altitude 50 km above the 1,000 hPa level, but as a result of the delay it started 25 km lower than planned. Until that time, JPL had no way of knowing how the probe was performing, and it would be several days before scientists could inspect 'snapshot' preliminary data. But to verify that the descent was going to plan, the orbiter sent snatches of data to Earth twice during the relay to confirm that the probe was transmitting on both of its telemetry channels and that the orbiter's receivers had locked on properly. This showed that the probe was alive and well! The scientific objective of reaching the 10,000 hPa pressure level was achieved 36 minutes after entry. Data collection and transmission continued without apparent problems until the 51-minute mark, when one of the telemetry channels was lost and the output of the other started to vary erratically as the transmitter case deformed under the rising atmospheric pressure. Finally, at 61.4 minutes, with the probe some 180 km deeper than at the start of the transmission and falling steadily at about 27 m/s, with the ambient pressure now 23,000 hPa and the temperature in excess of 150°C, the transmitter overheated and failed, concluding the mission. Deep as this penetration may seem, the probe had barely 'scratched the surface' of the planet, because it had traveled a mere 0.22 per cent of its radius. By the time that contact was lost, Jupiter's rotation was carrying the probe towards the dusk terminator, which it reached 15 minutes later. Hours later, as the ambient temperature increased further, the components and structure of the probe would have first melted and then evaporated to enrich the atmosphere with aluminum and titanium. Some 20 minutes after the probe fell silent, Galileo ended its relay program, turned off the receivers and stowed the relay antenna. The antenna had been repositioned four times during the relay to maintain contact with the probe. It had been almost perfect, with only 1 second's worth of data from one of the channels being lost to noise. A total of 57.6 minutes of data were recorded in real-time, in addition to the replay of the data which the probe had stored in solid-state memory during the approach and entry phases.

The radio carrier of the probe was also tracked by all 27 radio-telescopes of the Very Large Array, as well as by the six smaller dishes of the Australian Telescope Compact Array, to try to conduct the Doppler wind experiment independently of Galileo. Although the signal received on Earth was a billion times weaker than that received by Galileo, the fact that the relative positions and geometry of the probe, Jupiter and Earth were better known meant that the wind speeds measured in this way would have smaller uncertainties. Moreover, the fact that Galileo was almost directly overhead in relation to the probe (in order to maximize the strength of the received signal) meant that the horizontal motion of the probe would be difficult to see in the Doppler data – a given Doppler shift might be due to a small vertical motion, a zonal motion 12 times as great, or a meridional motion about 25 times as great. In contrast, the line of sight to the probe from Earth was almost parallel to the zonal winds. The two sets of Doppler data were therefore complementary, with that measured by Galileo monitoring the probe's rate of descent and that measured on Earth monitoring how it was carried by the zonal winds.[198,199]

Having adopted one-spin mode and increased its rate to 10.5 rpm for maximum stability, at 00:27 UTC on 8 December Galileo fired its main engine for the orbit-insertion maneuver, which lasted 1 second short of 49 minutes and was terminated when the accelerometers sensed that the desired deceleration had been achieved. Sadly, the pins jamming the high-gain antenna were not jarred loose by the jolts of igniting and shutting down the 400-N engine. The 645-m/s burn put Galileo into a 215,000 × 19,000,000-km orbit of Jupiter, as its first artificial satellite. The plane of the orbit was inclined at slightly over 5 degrees to the planet's equator, in which the four large moons travel. The first two encounters of the tour (with Ganymede) were to both make the plane of the spacecraft's orbit almost equatorial and reduce its period from the initial 7 months to about 70 days. Although Galileo's hardened systems meant that it was affected less by its deep penetration of the radiation belts than were the Voyagers, its star scanner was sufficiently affected to lose its lock on Canopus. Nine hours after orbit insertion, the spacecraft passed behind Jupiter's trailing limb, as viewed from Earth, but the results of sounding the atmosphere by this radio-occultation do not appear to have been published – most likely owing to uncertainties in the level of the atmosphere through which the radio beam passed, since the position of the vehicle would not have been precisely known immediately after the orbit-insertion maneuver.

Owing to concern about the reliability of its tape recorder, Galileo was directed to store as much as possible of the probe's data in solid-state memory as well. This partial data was transmitted to Earth in the week following orbit insertion, then the radio system was used to sound the Sun's corona while Jupiter passed through solar conjunction. The solid-state memory was read out twice more in early January to ensure that transmission flaws were eliminated. The scientists exploited the hiatus of solar conjunction to convene and preview the preliminary data from the probe. This showed that the parachute had deployed 53 seconds late, causing the atmospheric sampling to start deeper in the atmosphere than expected. This would complicate correlating the probe's findings with telescopic studies of the atmosphere at cloud-top level. Prior to returning the full set of data, the tape recorder was exercised to assess its operating parameters. It became stuck during the first high-speed test, but could still be operated at data rates ranging from about 8 kbps to 800 kbps. It then became stuck within 2 seconds of being started at 100 kbps. It later failed at the slowest speed! Nevertheless, engineers devised ways to use the recorder routinely. Starting in late January 1996, the taped probe data was transmitted to Earth. In the end, 100 per cent of the information was successfully retrieved, thereby achieving one of the mission's primary objectives.[200,201]

The results of the probe were extensive and in many cases unexpected. But the mission had taken so long to launch, and the interplanetary cruise was so long, that by the time the spacecraft reached its destination many of the instrument teams had lamented the deaths of some of their original members.

The energetic-particle instrument had collected data in three bursts as the probe passed through the inner part of the Jovian magnetosphere, and then continuously almost to entry. It discovered a radiation belt between the ring and the planet that extended down to within 30,000 km of the cloud tops. The particle densities were very high, and included energetic helium ions of unknown origin. This radiation belt could be the source of some high-frequency Jovian radio emissions.[202]

The entry site was on the boundary between the bright Equatorial Zone and the dark Northern Equatorial Belt, at 6.57°N, 4.94°W in terms of System III, which is a longitude system tied to the planet's rotation as indicated by its radio emissions, and hence slightly out of synchronization with the rotation of the atmosphere. As the probe, within its heat shield and thus unable to sample directly, penetrated the fringe of the atmosphere the deceleration enabled the composition and profiles of density, pressure and temperature to be inferred. The exosphere data was mostly consistent with the solar and stellar occultation results from the Voyager missions, and below an isothermal layer which extended down over 200 km there were some oscillations

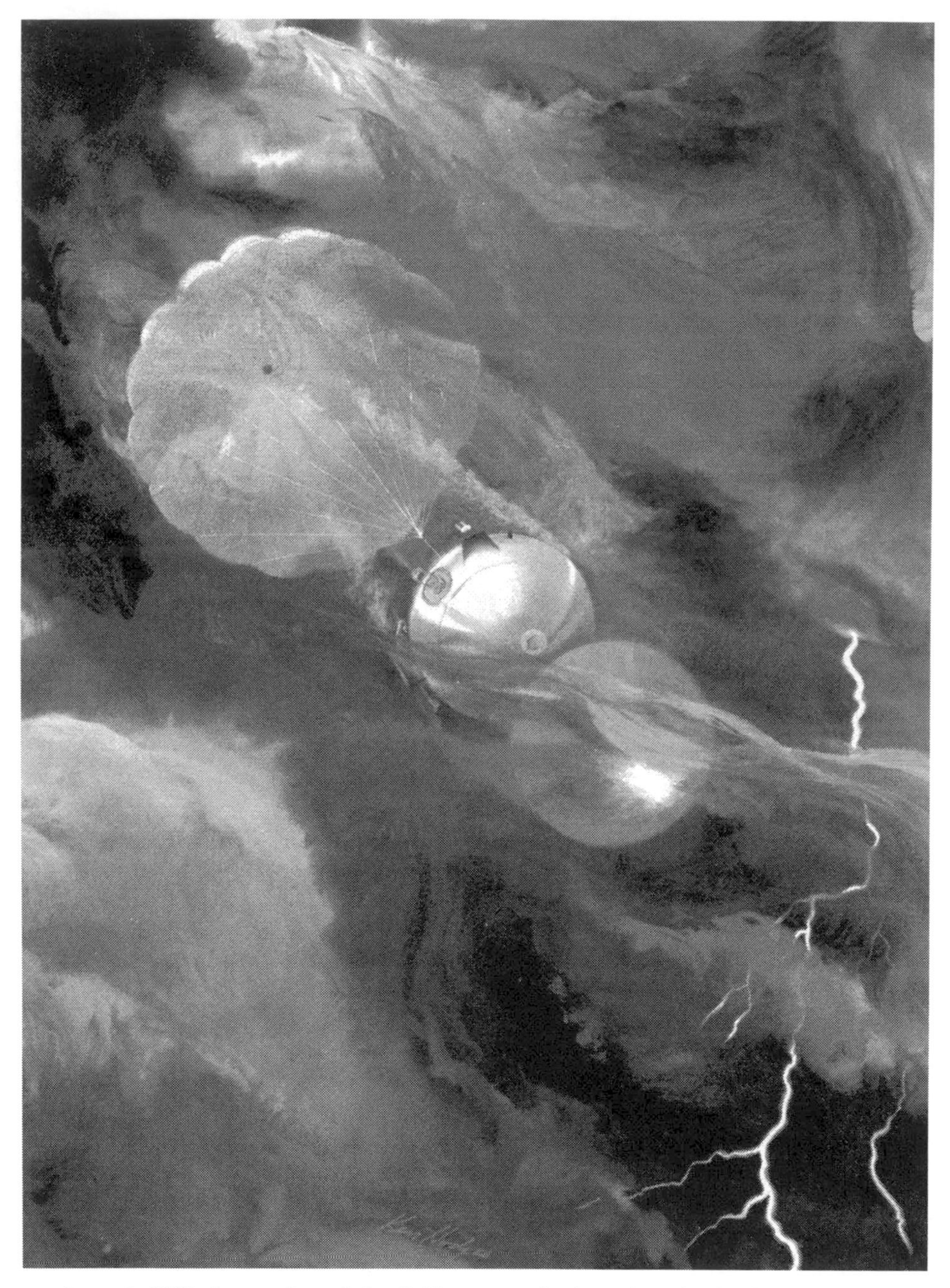

An early-1980s impression of the Galileo atmospheric probe releasing its heat shield.

in the temperature. This data assisted in reconstructing the path in the atmosphere. Since the search-coil antenna of the lightning detector swept through the planet's magnetic field as the capsule spun on its axis, it could measure the rate of spin by the modulation of the magnetic field. This confirmed that prior to entry the spin rate was the same as when the probe was released in July – i.e. 10.5 rpm. On entering the atmosphere it spun up to about 33 rpm, probably after longitudinal grooves formed on the ablative heat shield. (The influence of such 'flutes' on the aerodynamics of an entry vehicle was first seen while testing warheads for ballistic missiles.) After the spent heat shield was jettisoned, the spin rate rapidly slowed to less than 25 rpm and then more gradually to 14 rpm (the last measured value). The probe proper had spin vanes to damp rotation, and thereby reduce the Doppler shift on the transmission, and their efficacy is indicated by the fact that receivers on the orbiter were able to lock on within 35 seconds.[203,204,205]

As the telemetry from the probe was studied, it was realized that almost all the instruments had been colder at the start of the parachute descent and warmer at its end than predicted, and so required extensive recalibration before their data could be reliably processed. The limited knowledge of the atmospheric composition was not the cause of the anomalous temperatures; they probably arose from the fact that the probe was not hermetically sealed, and the air currents which flowed through it had not been properly simulated in wind tunnels.[206] A further complication was the delayed opening of the parachute, which meant the instruments began to operate at an atmospheric pressure of 420 hPa, some 300 hPa deeper than had been intended. The general expectation was that the probe would initially descend through a frigid brown hydrocarbon aerosol haze just above the white ammonia ice cloud. At some distance beneath this there would be a thin layer of ammonium hydrosulfide cloud, then a layer of water-ice crystals leading to a billowy water-rich cloud, after which the atmosphere would be clear. As a result of the delayed opening of the parachute, instead of initiating its measurements prior to penetrating the ammonia cloud the probe had already reached this level when it started. Intriguingly, the nephelometer saw no evidence of the ammonia cloud. An ammonia fog may have been detected between 460 and 550 hPa, but it was at best very thin near the probe. There were slightly denser layers between 690 and 1,550 hPa, but the visibility still exceeded 1 km and this may well have been the ammonium hydrosulfide layer – the atmospheric analyses found concentrations of ammonia and sulfur at the same level which were sufficient to form the cloud. A very thin cloud was crossed at the 1,900 hPa level, and then a series of light hazes and other weak features. Even weaker signals were produced at deeper levels, until about 40 minutes into the descent the instrument's output began to be corrupted by the high temperature. There was no evidence of a water-vapor layer. In fact, as seen by this instrument the 'weather' at the entry site was essentially clear. Whereas the nephelometer sampled the probe's immediate environment, the net-flux radiometer analyzed cloud structures in terms of optical thickness and the vertical separation of layers. The variations in brightness which abruptly ceased when the probe reached 600 hPa suggested that it had entered the ammonia cloud layer. The fact that the nephelometer did not detect ammonia in the immediate vicinity was explicable if this layer of cloud were patchy and the probe had happened to fall through a gap – i.e. even if the probe was not within a cloud, once it fell below the level of the cloud tops the fact that the Sun was very low on the horizon would make the illumination uniform. This instrument did not observe water clouds either. In fact, other instruments confirmed that the atmosphere was unexpectedly arid.

Undoubtedly the greatest surprise from the probe was that water vapor was at a concentration just 10 per cent of that which had been inferred from the evolution of the plumes produced by the Shoemaker–Levy 9 cometary impacts. Oxygen, which forms water molecules, also appeared to be present in lesser amounts on Jupiter than solar abundance. (It may have been present at greater concentrations deeper down, but this result was questionable because by that time the instruments were severely overheating.) In fact, if these results were typical of the planet, then the dryness of the atmosphere would have a great impact on theories of how the solar system was formed, in particular implying that collisions with water-bearing primitive bodies might not have played a major role in the chemical history of the planets. In the case of Earth, comets are believed by some scientists to have contributed much of the water in the oceans. The infrared and visible-light images of Jupiter by terrestrial and Earth-orbiting telescopes were the only source of insight. Unfortunately, Jupiter was approaching conjunction and the Hubble Space Telescope could not be pointed so close to the Sun in the sky, it had broken off imaging 2 months before the probe's arrival. Ground-based telescopes took pictures of the planet during November and December, and then again in January as it emerged from conjunction. Although of low resolution, these showed that the entry site was within (or at the southern edge of) a dark-blue spot which appeared to be warmer than its surroundings, just as if a gap in the ammonia cloud gave a view of the deeper, warmer – and arid – interior. By sheer fluke, the probe had entered a 'hot spot' which represented less than 1 per cent of the visible surface of the planet, and could reasonably be thought of as the equivalent of a terrestrial desert. The fact that the lightning detector did not record any flashes confirmed the absence of water clouds in the probe's neighborhood, as water precipitation is required to create thunderstorms. It did, however, detect the 'radio noise' of thousands of lightning bolts hundreds (or more likely thousands) of kilometers away. Although overall each square meter of Jupiter has less lightning than on Earth in a given time interval, the Jovian storms are much more energetic than the typical terrestrial storm.

According to theories of how the solar system formed, and to the measurements by previous missions, Jupiter was thought to have a helium-to-hydrogen molecular number ratio similar to that of the Sun's outer layers, namely 13 ± 2 per cent, but the preliminary data from the helium abundance detector gave a ratio of half this value, which raised the possibility that through the ages the helium had rained out of the upper atmosphere toward the massive core – as has occurred on Saturn – although the models of how Jupiter evolved suggested that this ought not to have happened. But once the data was recalibrated to allow for the environment being warmer than expected, the ratio was 13.6 per cent. But argon, krypton and xenon were all present in abundances greater than the solar values, implying that Jupiter must have become enriched. As it would have been too warm at Jupiter's distance from the Sun for these gases to condense out of the solar nebula when the planet formed, this prompted speculation that it may have formed farther from the Sun and spiraled inward to its current orbit. In addition to noble gases, the neutral mass spectrometer found methane in concentrations consistent with theories of how the planet developed its atmosphere, and an anomalous and very high concentration of ammonia. However, some data may have been falsified by droplets clinging to the wall of the instrument. Other than methane, ethane was clearly detected and there were traces of phosphine. (Red phosphine has long been held to be one of the main coloring agents, or 'chromophores', of the planet's atmosphere.) The paucity of organic molecules possessing at most two or three carbon atoms argued against the speculation prompted after the Voyager flybys that lightning could drive a natural Miller–Urey laboratory that synthesized complex biological molecules – there was simply not the raw material. Certainly the floating life-forms that some people had imagined could be present were now deemed to be rather implausible. The fact that carbon and sulfur exceeded solar abundances was proof that Jupiter has received, and continues to receive, a contribution from small asteroids and comets crashing into it.

Doppler tracking of the probe by both the orbiter and terrestrial radio-telescopes showed that it was carried by a jetstream of between 180 and 200 m/s, which was consistent with the wind speeds computed by measuring how atmospheric features at that latitude moved in ground-based and Voyager views. One of the objectives was to determine whether the jetstreams were confined to the upper atmosphere, or were deeply rooted. The fact that the wind remained constant far below the level of the cloud tops was a clear indication that they were powered by heat released from the core of the planet, rather than by solar energy. In fact, below the 10,000 hPa level, the ambient light was only 0.01 per cent of that at the start and the solar flux was almost zero. A down-flow of several meters per second was encountered early in the descent, but other drifts seen at greater depths were probably manifestations of the oscillator of the transmitter overheating. In fact, the Doppler tracking was so sensitive that it even noted the rocking and swinging of the probe on its parachute.

The fact that the probe fell into a gap in the clouds where there was essentially no water vapor present enabled the attenuation of the radio-link to the orbiter to be used to profile the concentration of ammonia with depth, and this showed there to be large amounts of ammonia at depths below the 6,000 hPa level; a measurement that was consistent with the net-flux radiometer's results. The tracking from Earth was able to monitor the probe to 7,000 hPa, which confirmed that no water clouds were present at the site, for otherwise this would have attenuated the signal to the detection limit much earlier.[207,208,209,210,211,212,213,214,215,216,217,218,219,220]

As scientists were completing their preliminary analyses of the probe's data, on 14 March 1996 Galileo, now near the 20-million-km apoapsis of its initial orbit, fired its main engine for the final time to speed up by 378 m/s, in effect doubling its velocity in order to raise its periapsis from 4 to 11 Rj, a little outside the orbit of Europa and clear of the most intense part of the planet's radiation belts. This left the spacecraft with about 90 kg of propellant (10 per cent of the initial supply) for 'trim' burns by the thrusters during the tour of the Jovian system. A few days later, a final attempt was made to 'hammer' the motor of the high-gain antenna, but the pins holding the three stuck ribs remained in position.

In May, the flight software of Galileo's main computer was replaced with the programs required to perform the tour. In the following weeks, as the spacecraft returned from its initial apoapsis, this software was used to transmit the particles and fields data which had been taken during the Io flyby immediately prior to orbit insertion in December. A 40 per cent decrease in Jupiter's magnetic field measured when the spacecraft was at its closest point to the volcanic moon suggested that the planetary field was being masked by a dipole field centered on the moon itself. On the other hand, unexpectedly large plasma densities were recorded near Io, including plasma that was stationary with respect to the moon, and this plasma provided an alternative explanation for the disturbance in the planetary field. The fact that the plasma extended 900 km above the surface of the moon – much further than as measured by the radio-occultation of Pioneer 10 – and included ionized oxygen, sulfur and sulfur dioxide, suggested that it derived from the moon's volcanism, and so was variable in both scale and intensity. An apparently "inescapable" answer to the magnetic field issue was provided by the Doppler tracking which had begun on 4 December and continued uninterrupted until 2 hours before the spacecraft started its 'listening' period for the penetration of its probe into the atmosphere. This gave improved estimates of the masses of Io, Europa and Jupiter, refined the ephemeris of Io and (of particular significance in this context) yielded insight into the internal structure of Io, revealing it to be differentiated and to possess a dense core of a size that occupied between 36 and 52 per cent of the radius depending on whether it was assumed to be pure iron or a mixture of iron and iron sulfide. This core could easily generate a dipole magnetic field. Io therefore became only the second solar system body after Earth to be proved to have an iron core – the Moon and Mercury were suspected of having iron cores, but the results were ambiguous. The plasma-wave experiment had measured the electron densities in the Io torus and in the vicinity of the moon itself, showing values which were almost double those measured by Voyager 1; in turn suggesting that the ultimate source of the electrons – the volcanoes – were more active than they were in 1979. The fact that small grains of dust were concentrated near the moon raised the prospect that Io was the principal source of the dust in the Jovian system. (Because the interaction of the planet's magnetosphere and the electrically charged dust ejected into space by Io's volcanoes creates a wealth of periodicities in the flux of dust particles, the data which the spacecraft would collect over the next 2 years would prove that Io was the source of the dust. Furthermore, this dust was being accelerated by the planet's magnetic field and escaping from the system to form the dust streams that the spacecraft had encountered during its interplanetary cruise.) Galileo had also made the first in-situ measurements of the 'flux tube' that links Io to Jupiter by way of the lines of the planet's magnetic field, along which currents of millions of ampères flow. The flux tube had been narrowly missed by Voyager 1, but Galileo's trajectory had been selected to intersect it. At the points where the flux tube intersects the planet's atmosphere at high latitudes, it induces intense auroral displays.[221,222,223,224,225,226]

On 23 June 1996, nearing periapsis, Galileo began to take pictures and perform ultraviolet observations of Jupiter and the Io torus; and polarimetric and infrared measurements of the planet's atmosphere, in particular the Great Red Spot, and of the Galilean moons. On 26 June it crossed the orbit of Callisto (although this was not in the neighborhood) and a little more than 24 hours later it made its first flyby of Ganymede, passing by at an altitude of 835 km. The geometry of the encounter placed the anti-Jovian hemisphere (i.e. the 'far-side', as the rotation of the moon is synchronized with its orbital period) on view, and the sub-Jovian hemisphere in darkness. The imaging objectives were to determine whether the brightest parts of Ganymede had been resurfaced by icy volcanism, since their smooth appearance in the Voyager views suggested they were plains of ice; to characterize the tectonism which shaped the surface; to characterize the impact craters, ranging from young bowl-shaped ones to old flattened 'palimpsests'; and to investigate the composition of the surface. The flyby also had the effect of shortening the spacecraft's orbital period from 210 to just 72 days. A veritable marathon started at this point, as all the data from one periapsis passage had to be downloaded in time to start the next, whilst continuing to provide engineering and particles and fields data in real-time. Accordingly, all remote-sensing sequences ceased on 30 June, and playback began the next day. The first data returned, which was presented at a press conference on 10 July, provided the first close-up images of the Jovian system since the Voyager flybys 17 years earlier. The first pictures were of Uruk Sulcus, the bright area that lies between the dark Galileo and Marius regions (respectively named in honor of the

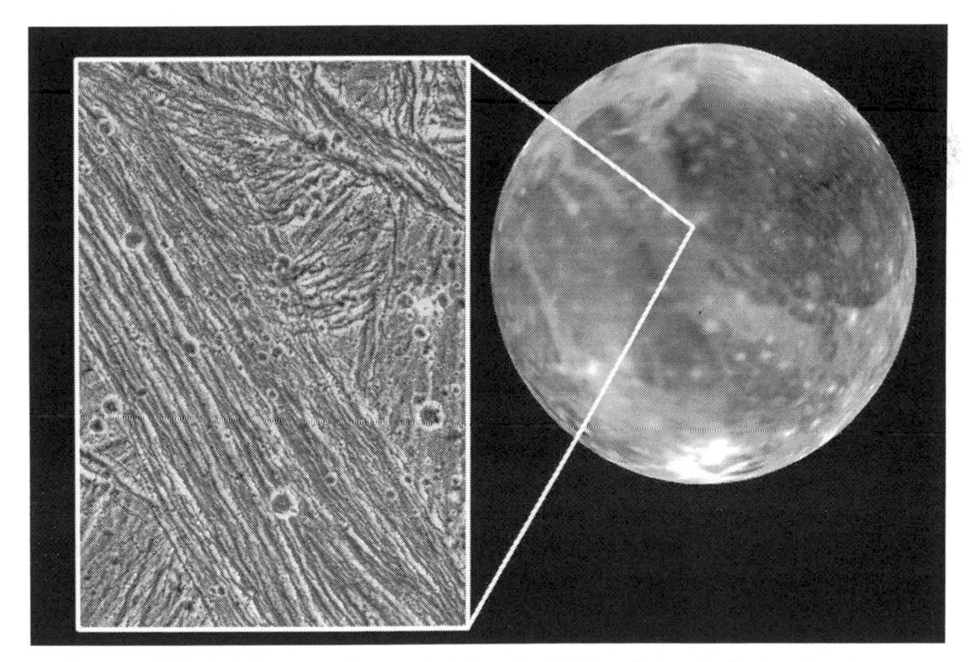

A view of Uruk Sulcus on Ganymede taken by Galileo during its first periapsis pass. (JPL/NASA/Caltech)

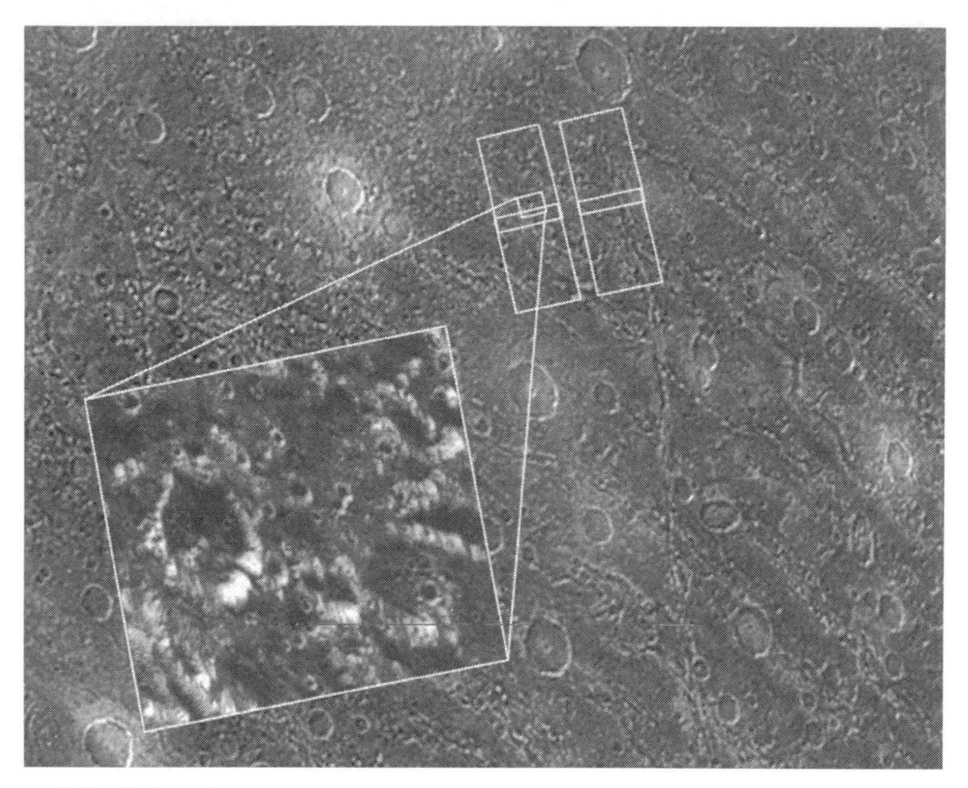

A detail of a 4-frame mosaic of Galileo Regio on Ganymede taken by Galileo during its first periapsis pass. (JPL/NASA/Caltech)

discoverer and possible independent co-discoverer of the fact that Jupiter has four large satellites). The 75-meter resolution was sufficient to reveal that Uruk Sulcus was not as smooth as believed, and was actually extensively ridged and grooved in a manner which suggested that it formed as the icy crust was extended, broken and sheared.

In the ensuing 2 months, at least portions of 127 (out of a total of 129) pictures were returned to Earth, in addition to a number of other observations that included the remaining data taken on 7 December 1995 while passing through the Io torus, but the data from the infrared spectrometer was not returned since in the interim it had become corrupted, making it impossible to compress.[227] The resolution of the images of Ganymede varied between 13 km at long range and 11 meters in close. Galileo Regio displayed an ancient cratered surface together with hummocky hills and other structures which hinted at icy volcanism. The fact that some craters were reshaped by faults and fractures indicated that tectonic processes had been at work. The floors of the many valleys were dark. The floors of the craters were also quite dark, but their rims displayed many bright spots (so bright that they saturated the CCD). The infrared spectrometer saw abundant water on the surface, in particular on the sulci. The fact that the temperatures measured by the radiometer on the day-side were in the range 90–160K meant that the surface ice must be as hard as rock, but it was probably very brittle. The dark regions were also icy, but displayed more hydrated minerals. The close flyby enabled Galileo to detect an ionosphere. Later, a spectroscopic study by the Hubble Space Telescope found a tenuous atmosphere of oxygen. However, the greatest surprise concerning Ganymede was made by the plasma-wave experiment and magnetometer, which detected bursts of radio noise and magnetic effects that were consistent with the moon having a magnetic field that was several times stronger than the planetary field at that distance from the planet – sufficiently strong, in fact, for it to form a well-defined magnetosphere nested inside the planetary magnetosphere.[228,229]

On Jupiter, sequences captured the Great Red Spot with a resolution as fine as 30 km, either at the center of Jupiter's disk or on the limb over an interval greater than one planetary rotation. The displacements of the individual features within the anticyclonic system were used to measure wind speeds and directions. Pictures at differing wavelengths penetrated to different depths in the atmosphere. The Great Red Spot appeared to stand 20 km above its base in the ammonia cloud deck, and its base was surrounded by a warmer ring. The bright spots that formed nearby and (in Voyager imagery) were often sheared apart, were found to be the tops of anvil-shaped thunderstorms projecting many kilometers above the ammonia cloud tops. The winds inferred from wavelike features seemed to rule out the Great Red Spot being deeply rooted.

Galileo also made observations of a number of infrared 'hot spots' which had been identified by infrared telescopes on Earth, and as such resembled the site at which the probe had entered the atmosphere. The spacecraft's data confirmed them to be arid at least to a depth corresponding to the 8,000-hPa pressure level. Similar observations of colder areas of the atmosphere were shown to contain thousands of times as much water. Although various theories had been advanced to explain why the probe had found so little water, these observations convinced most scientists that

A reprojected six-frame near-infrared mosaic of Jupiter's Great Red Spot. (JPL/NASA/Caltech)

the probe had sampled an atypical area and that there was therefore no flaw in our understanding of the Jovian atmosphere. Unfortunately, shortly after making these observations the infrared and polarimetric instruments both fell temporary victim to radiation-induced problems.

Soon after periapsis Galileo had turned its attention to Europa. At 155,000 km the range was only slightly less than for Voyager 2, but the camera was better. It took 12 pictures at a fairly good resolution of areas with dark bands of a type seen in the Voyager views. These were revealed to have subdued rather than sharp edges – as if geysers had sprayed out dark material. Indeed, all other evidence confirmed that the narrow dark lines and broader bands were formed as water and dirt oozed out of cracks in the icy crust. A previously unobserved 30-km-diameter crater was also spotted. The paucity of such features indicated the surface to be very young. There was a remarkable landscape to the south of the crater where long curved fissures in the crust had opened to allow dark material to rise to fill the gaps. This area also included several flexi of the type seen in the Voyager pictures. The fact that these curved and spiked fractures resembled 'cycloid curves' was undoubtedly a clue to how they were formed (a cycloid is the curve traced by a point on a circle which revolves and translates without slippage), but the precise mechanism had yet to be discovered.

The first quality images of Io since the Voyager flybys had been obtained on the inbound leg and during the periapsis pass. Three sets of full-disk color images, two sets of crescent images, and two clear images of Io in Jupiter's shadow were taken. Even at the relatively low resolution of the full-disk images it was clear that there had been substantial changes, including the appearance of new bright-yellow flows and ballistic deposits, although most of the source vents seemed not to be currently active. In particular, there were new flows and deposits around Ra Patera, a shield volcano near where the plume named Marduk was seen by the Voyagers. Changes in this area had been suspected from observations by the Hubble Space Telescope, whose resolution was poorer owing to the range. If Galileo had been able to take pictures of Io during its close flyby in December 1995, Ra Patera would have been one of the main targets. New deposits were seen near Euboea Fluctus, and dozens of smaller features. But there were few changes near other active high-temperature sites, such as Loki. The different morphologies of these deposits hinted at different types of volcanism at work. While Ra Patera was believed to erupt sulfur-rich lava at a moderate temperature, Loki could well be erupting a silicate-rich lava which was very similar to that issued by terrestrial volcanoes. The deposits surrounding Pele seemed to have changed little, although their intense red color suggested that the volcano may have been active recently. There were now Pele-type deposits around Surt and Aten. Only two active plumes were clearly seen in the limb imagery: the 100-km-tall plume of Ra Patera (dormant during the Voyager flybys) and a plume near Volund. The plume over Pele was only tentatively detected in pictures taken using a green filter. Loki appeared to be inactive. The images of Io taken in eclipse showed at least five 'hot spots' whose locations matched Pele, Marduk and three smaller spots in Colchis Regio. The fact that Pele was the 'brightest' of these sites, and therefore probably the hottest, suggested that it, like Loki, was a silicate vent.

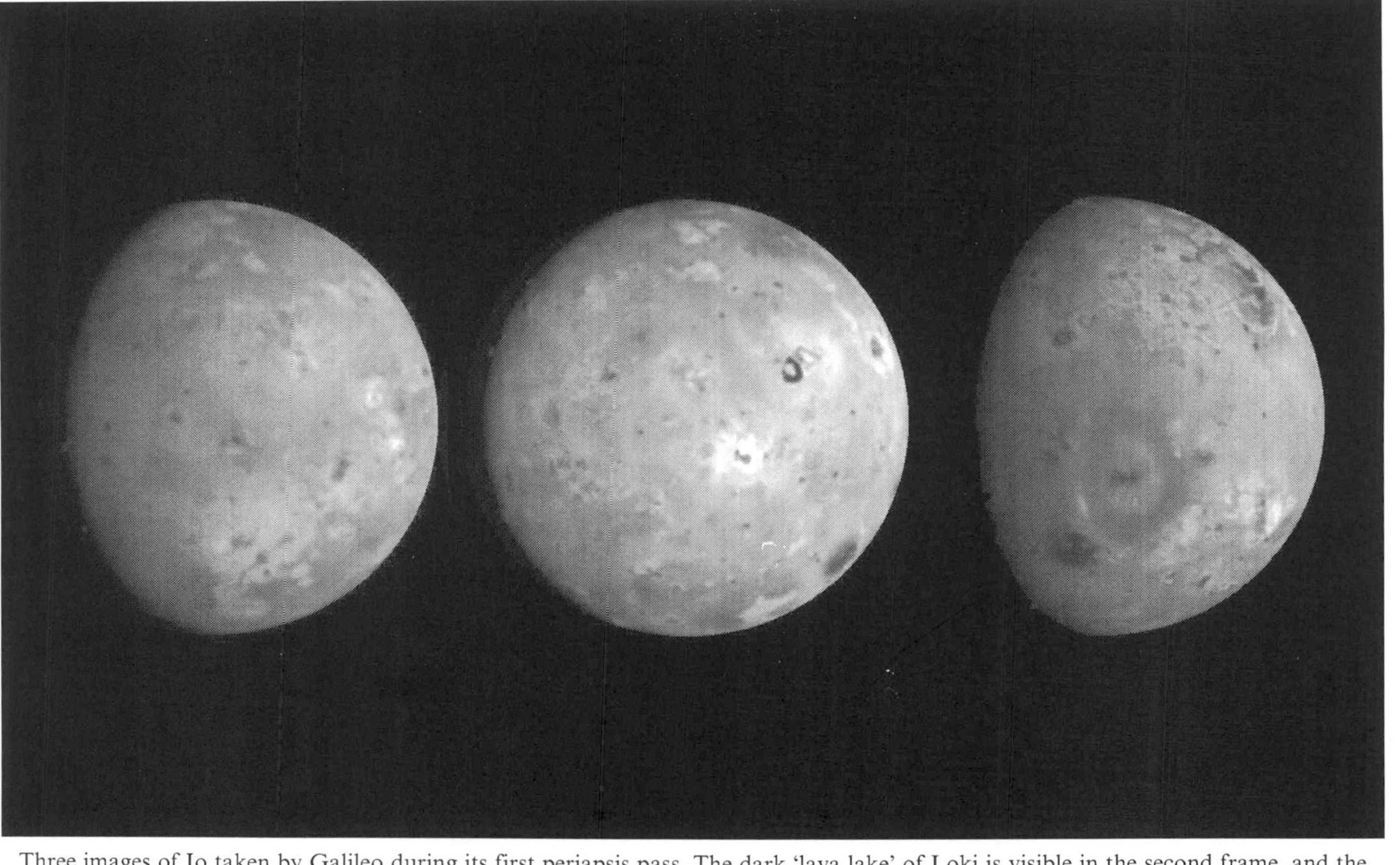

Three images of Io taken by Galileo during its first periapsis pass. The dark 'lava lake' of Loki is visible in the second frame, and the characteristic 'heart-shaped' deposit of plume material of Pele is in the third frame. (JPL/NASA/Caltech)

There was also a faint glow on the limb in the eclipse pictures produced by auroral and airglow emissions of oxygen and sulfur. The eclipse observations of Io proved so effective in revealing fresh lava deposits glowing in the dark that it was decided that the spacecraft should take such pictures on subsequent periapses.[230,231,232]

The downloading of data was interrupted on 24 August by the command and data subsystem entering 'safe mode'. Since a course correction to refine the next encounter with Ganymede was imminent, the maneuver was commanded using the backup system and successfully executed on 27 August. Although full control of the spacecraft was regained on 29 August and normal operations were resumed on 1 September, the data download was suspended to prepare for the second periapsis observations.

On 6 September 1996 Galileo flew by Ganymede on a trajectory that yielded an excellent view of the bright 'polar spot' situated to the north of Galileo Regio. At an altitude of 260 km, this would be the closest encounter with any satellite during the spacecraft's 2-year primary mission. In addition to slightly reducing the orbital period to about 2 months, the flyby aligned the orbital plane with that in which the four large satellites travel, and so facilitate flybys of moons other than Ganymede. The bright floor of a crater at high latitude that had appeared in the Voyager views to have been resurfaced by icy volcanism was revealed only to be masked by frost. In fact, the frost was particularly prevalent in the shade of crater rims and in pits. High-resolution images of a 350-km-diameter palimpsest showed subtle brightness variations across the structure, but no relief around its periphery – it was flat. This prompted the suggestion that instead of being the isostatically relaxed rims of large craters, palimpsests might be thick deposits of ejecta. Images were taken of Uruk Sulcus at a different viewing geometry than that of June to provide a 3-dimensional reconstruction of its topography. On this flyby, pictures were also taken of Nippur Sulcus, adjacent to Marius Regio. This network of intersecting grooves and ridges showed signs of shearing and rotation. In addition, the presence of a magnetic field around this moon was confirmed.

The fact that the Doppler data from the first two Ganymede encounters spanned a broad range of latitudes enabled the gravitational characteristics of the satellite to be calculated. This suggested a 3-layer model of its interior that included a metallic core some 400–1,300 km in diameter, a silicate mantle and an outer shell of water ice – with the latter perhaps incorporating layers of liquid water and a 'slush' made of a mixture of ice and liquid water. If the core was sufficiently fluid for a dynamo to operate then this would explain the observed magnetic field. However, the core of a body as small as Ganymede should long since have solidified. If the field were indeed produced by a dynamo, then a process would have to be providing the heat required to maintain the core in a fluid state. One speculation was that Ganymede was passing through a temporary period of orbital resonance with Europa in which the gravitational tides had melted its core. If the core were solid, then the magnetic field must be the result of another process. The 'patchy' nature of the gravitational field indicated the presence of relatively large mass concentrations (mascons) such as had been identified on other rocky planets and moons whose gravitational fields had been studied in detail. Of the two most prominent such features on Ganymede, one

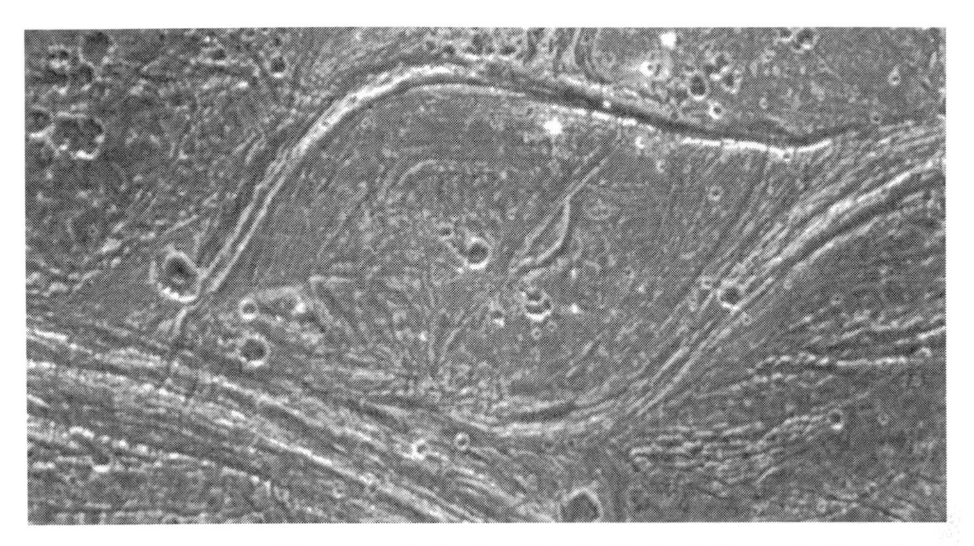

An 80-km lens-shaped feature on the border of Marius Regio of Ganymede viewed by Galileo on its second orbit. Tectonic stress is indicated by shearing and rotation. (JPL/NASA/Caltech)

(a positive anomaly) was at high northern latitude and another (negative) was at low latitude, but there was no obvious correlation with geological features on the surface.[233,234,235] (Interestingly, as the mission unfolded, no mascons were found on any of the other Galilean satellites, making Ganymede unique in this respect.)

The second periapsis of the tour also included fairly close encounters with the other large satellites, and for the first time the spacecraft was able to take pictures of Amalthea. Although the infrared spectrometer again had problems, it was able to recover and managed to locate carbon dioxide frost on the surface of Callisto. It also saw a 'hot spot' on Io near Malik Patera, and recorded several other eruptions. The optical imaging of the limb captured a plume over Prometheus and possibly another over Culann. Although a stuck filter wheel prevented the photopolarimeter from being used on this periapsis, the wheel was later recovered by subjecting the instrument to an ingenious series of heating and cooling cycles – but unfortunately the wheel jammed again later on.[236]

On 4 November, on its third periapsis, Galileo flew by Callisto at an altitude of 1,136 km. Of all the Galilean satellites, this had been the least favorably surveyed by the Voyagers. Galileo approached from the anti-Jovian side of the moon, which was in darkness, then crossed the terminator to the day-side. It imaged the Asgard and Valhalla multiple-ringed basins. A surprising discovery was that although the enormous Valhalla structure appeared to be several billion years old, its center was remarkably flat and free of impacts – at least down to the limiting resolution of 60 meters – and the few craters present had 'relaxed' to such a degree that their rims were essentially level. In fact, it would become evident that some kind of fluid had reached the surface and flattened the floor of the basin. This impression contrasted with the conclusion drawn from the Voyagers that Callisto was the most heavily cratered body in the solar system. When observations revealed there to be a dearth of craters smaller than a few kilometers in size, this prompted speculation that on this scale some process was resurfacing the moon. Galileo also took pictures of a chain of craters in northern Valhalla. Before the discovery of Shoemaker–Levy 9, the process by which such a 'catena' might have formed would have been a matter for debate, but having observed the fragments of a disrupted cometary nucleus rain down on Jupiter it was evident that this chain of craters was produced in a similar manner. The fact that the walls of some craters showed landslides hinted that there was much rock and dust mixed in with the ice. In contrast to Ganymede and Io, the Doppler tracking indicated Callisto to be a fairly homogeneous mix of water and rock. Evidently there had not been sufficient thermal differentiation to form a core of rock. Despite the encounter trajectory being ideal for its detection, there was no evidence of an intrinsic magnetic field. However, a concentration of plasma could have indicated the presence of a tenuous atmosphere. Unlike Ganymede, which is subjected to tidal heating from Jupiter and the nearby moons, Callisto orbits much further out and is much less stressed.[237,238,239]

On its way to periapsis Galileo flew by Io at a range of 244,000 km, which was the closest that it would approach this moon during the primary mission. The infrared spectrometer detected many of the known volcanoes, and some smaller 'hot spots'. (Unfortunately, shortly thereafter one of the instrument's 17 detectors failed.) The medium-resolution imagery by the main camera revealed a number of mountains on the anti-Jovian hemisphere. A map of the distribution of sulfur dioxide on the surface enabled the locations of vents and their various depositions to be mapped.

The outbound leg provided a relatively distant encounter with Europa at a range of 34,800 km. In fact, this was the first of the 'non-targeted' encounters – in the sense

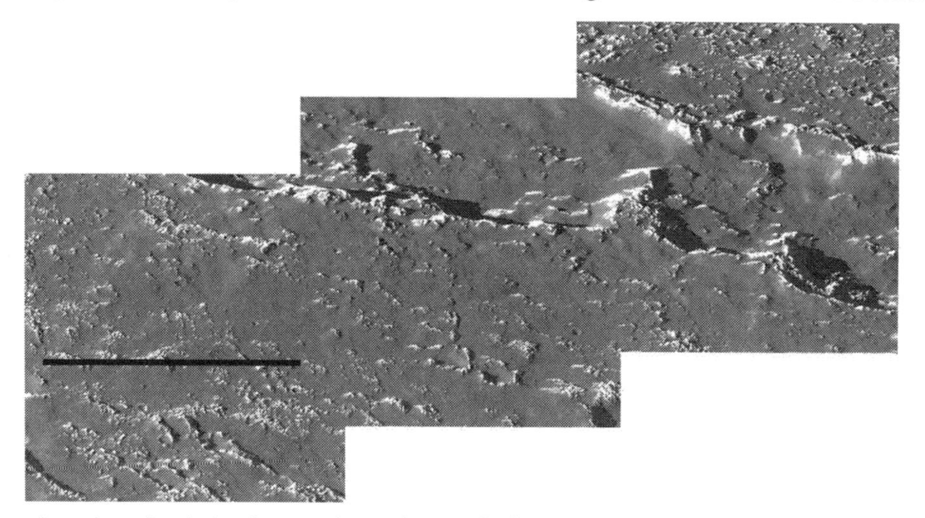

A section of a chain of craters in northern Valhalla on Callisto imaged by Galileo on its third orbit at a resolution of 160 meters per pixel. Note the paucity of small craters, the blanket of dark material and the brightness of the peaks and slopes. (JPL/NASA/Caltech)

A high-resolution view of a section of the fault scarp that forms the outermost ring of Valhalla on Callisto. (JPL/NASA/Caltech)

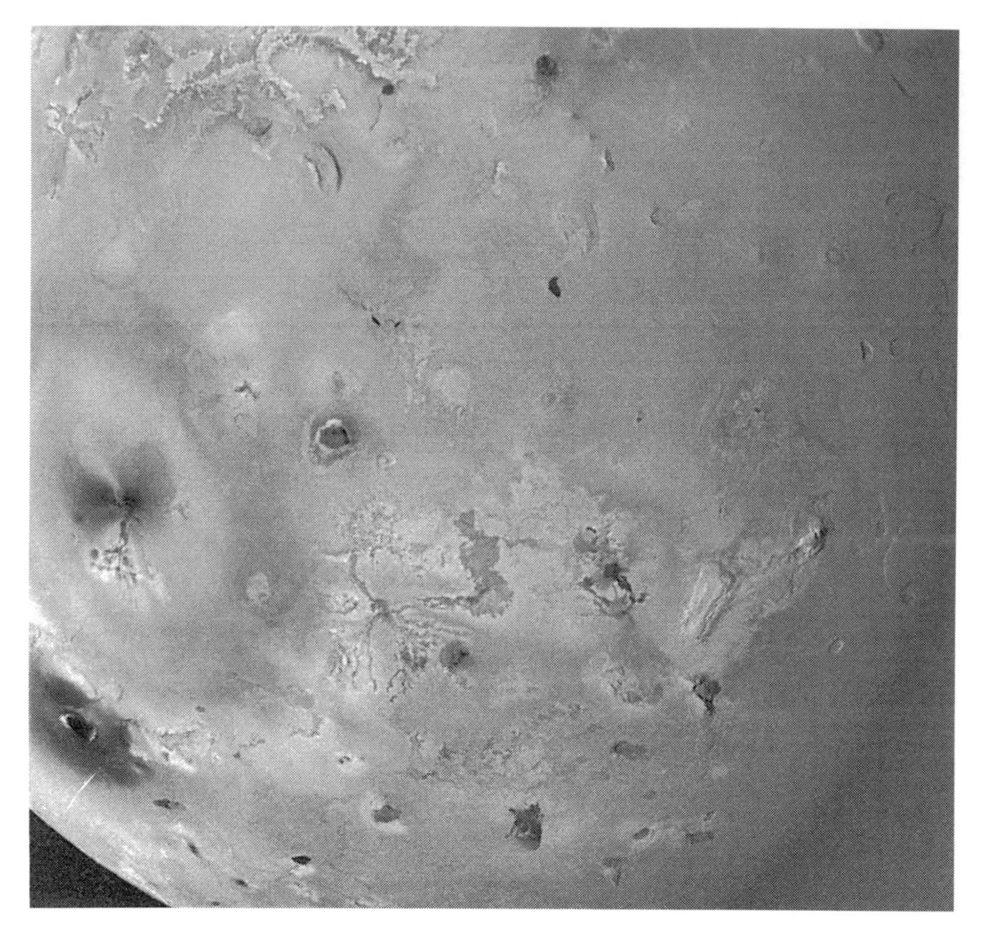

On its third orbit, Galileo approached within 244,000 km of Io, and took some of the best images of the primary mission of this volcanic moon. The ring of plume material around Pele is on the left. (JPL/NASA/Caltech)

that the trajectory was not chosen specifically to set up that encounter, as for a 'targeted' encounter. The objective was to assemble global coverage of the large satellites at the moderate resolution provided by imaging within 100,000 km. The non-targeted encounters would not require control of the encounter geometry, and would not be used to shape the evolving orbit. In this case, views of 'dark wedges' at a resolution of 400 meters indicated that these were tectonic spreading features where the icy crust had been pulled apart to allow aqueous fluid to rise to the surfacc and freeze.

As Galileo receded from Jupiter, it entered first the Earth-occultation zone and then the solar-occultation zone. Because the latter occurred when the spacecraft was just outside the orbit of Callisto and lasted 4 hours, it was the best solar occultation of the primary mission, very similar to that of Voyager 2 many years before. In addition to

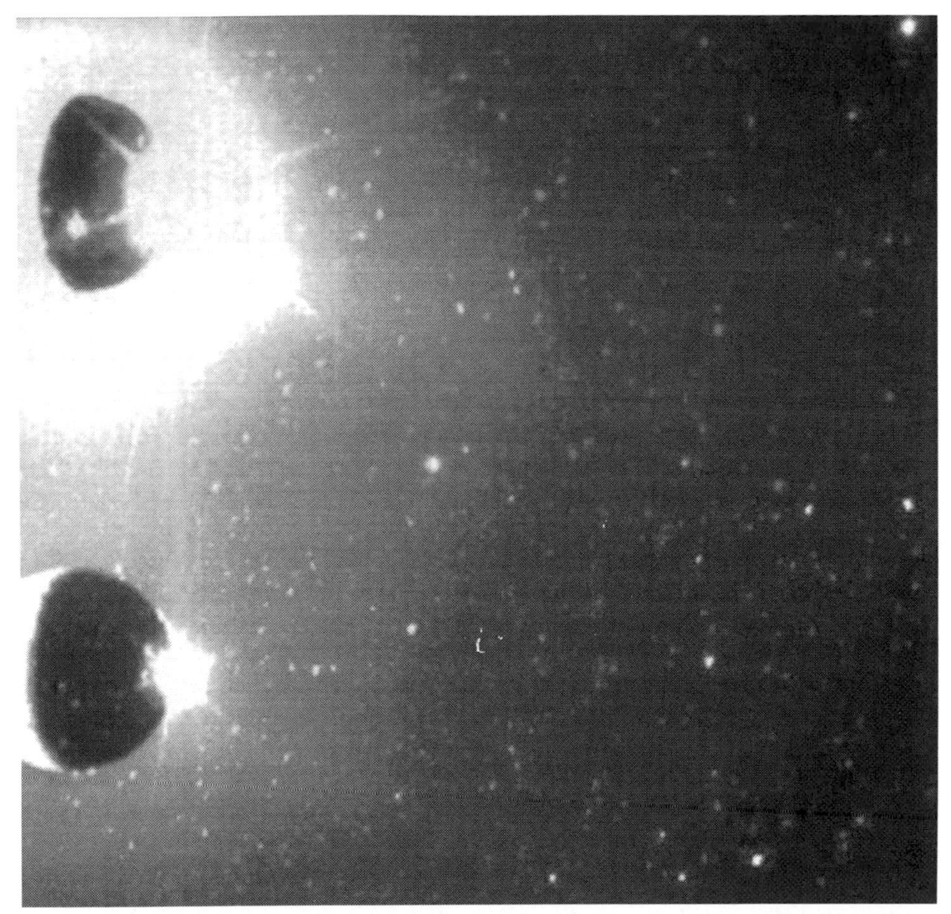

A double exposure of Io taken through a clear filter (top) and a yellow-green filter. The bright spot in darkness is thermal emission from Pele. The glow on the limb is Prometheus's plume. Much of the diffuse sky emission is the glow of sulfur in the space around the moon. (JPL/NASA/Caltech)

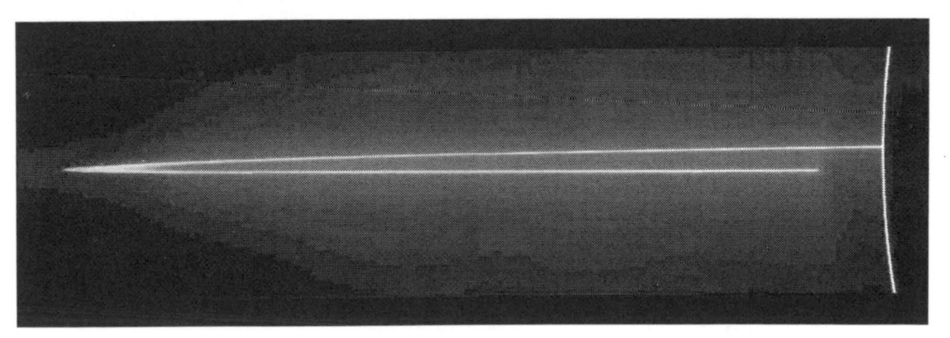

A mosaic of four images taken when Galileo was in Jupiter's shadow showing the ring in forward-scattered sunlight. Note the fine detail at the ansa. (JPL/NASA/Caltech)

viewing the Jovian ring in forward-scattered sunlight, the spacecraft made observations of the night-side of the planet, detecting lightning flashes at intermediate latitudes close to a westward jet – just as were those seen by Voyager 1. Auroras were also detected at higher latitude, at the foot of Io's flux tube. Long exposures were taken of the crescent of Io to study the glow of sulfur in the space around the moon, and although the camera was blinded by the bright plume of Prometheus the glow of sulfur also stood out clearly.

The fourth periapsis of Galileo's tour was one of the most eagerly awaited, for the reason that it provided the first opportunity to obtain close-up views of Europa, the mysterious ice-enshrouded moon which had replaced Mars as the solar system body (apart from Earth) believed most likely to harbor life. A major objective was to characterize how the moon interacted with Jupiter's magnetosphere, in the hope of determining whether there was an ocean of liquid water beneath the icy crust. Another mystery was whether Europa's rotation period was synchronized with the orbital period. Several dynamical characteristics suggested that its spin rate might be slightly faster than synchronous, in which case a deep ocean might decouple the outer shell from the core. In particular, the orientation of lineae with ages and the absence of an asymmetry in impacts between the leading and trailing hemispheres strongly hinted that the rotation was not synchronous.[240] The trajectory would also permit routine studies of Io (the closest point of approach being a relatively remote 320,000 km); observations of the smaller satellites orbiting nearer Jupiter; an Earth occultation and a solar occultation by Jupiter; several solar occultations that would provide opportunities to seek an ionosphere or atmosphere around Europa; the best occasion of the primary tour to study the wake made by Europa in the Jovian magnetosphere; and distant infrared data to characterize the surface compositions of Ganymede and Callisto. However, because Jupiter was on the far side of the Sun from Earth, with conjunction in January 1997, the data rate from the spacecraft for this and the next two orbits would be greatly reduced, with the result that the data cache of 69 Mbits from this fourth periapsis sequence would be by far the least of any orbit of the primary mission.

Galileo encountered Europa on 19 December 1996. Owing to the illumination it observed the trailing hemisphere, and the closest point of approach at an altitude of 692 km was over the night-side. The imaged area was one poorly resolved in the Voyager imagery – the closest encounter of which had been at a distance 300 times as great. On the way in, medium-resolution images of one of the few impacts were obtained. Named Pwyll, this was a fresh crater surrounded by brilliant rays of ice. Low-resolution images of the same hemisphere taken during the second orbit had shown a dark spot 100 km across. Seen at high resolution this spot, Callanish, was revealed to be a multiple-ringed basin similar to those on Callisto and Ganymede. The plasticity of the ice had allowed isostasy to modify the crater into a central patch of jumbled terrain and a succession of circular structures. Testifying to its old age, Callanish was later crossed by ridges and lines; but these were then modified by its erosion. Images taken at 100-meter resolution resolved ridges and lines broken and interrupted by smooth, craterless plains and chaotic terrain. Only a handful of small craters were visible on the oldest ridged terrain. A 26-meter-resolution image taken

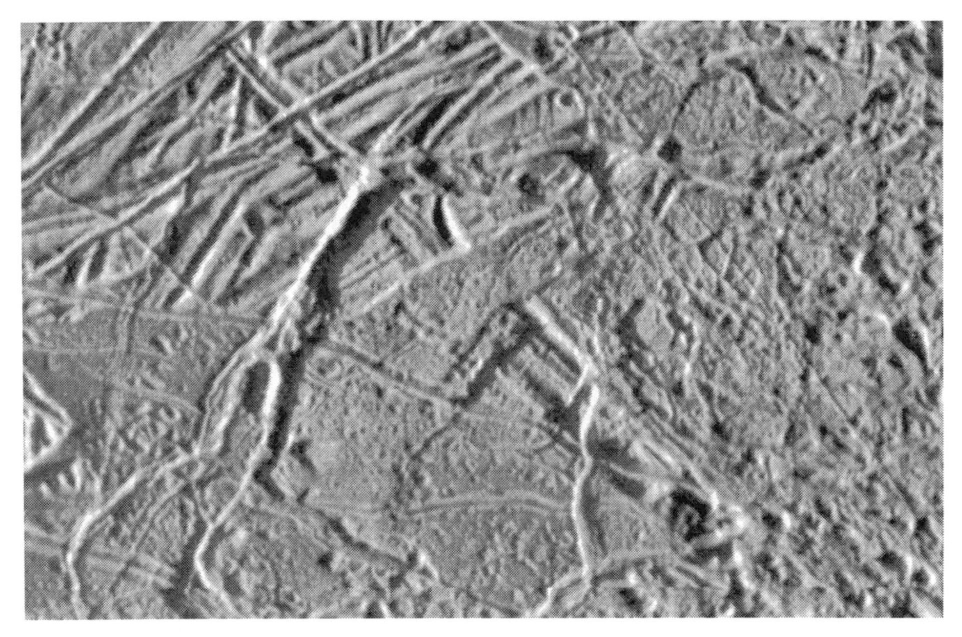

A 100-meter-resolution view of Europa taken on Galileo's fourth orbit, showing cross-cutting ridges, grooves, knobs, rare craters and a wedge-shaped ice flow. (JPL/NASA/Caltech)

closer in showed an area that included crisscrossing lines, a hilly knob and a circular patch where a ridged terrain seemed to have been transformed into a small pond, and right at the center of this 'ice rink' there was a 250-meter-diameter crater. The magnetometer noted an abrupt field rotation as the spacecraft approached Europa, a magnitude change near closest approach, and another as it passed in front of the moon. Although these magnetic perturbations could have been caused by electrical currents connecting Europa to the Jovian magnetosphere, it was also possible that the moon had an intrinsic magnetic field, in which case the fact that the field had a large quadrupole component in addition to a dipole meant that (as for the magnetic fields of Uranus and Neptune) it could be represented as a dipole whose axis was offset from the center of the body – which was a particularly interesting possibility because a subsurface ocean would create such a signature. Hopefully, observations on future encounters would settle the issue.[241,242] During the whole of the periapsis passage the spacecraft was occulted by Europa no fewer than three times, on 19, 20 and 25 December, and the data from most of the ingresses and egresses suggested the presence of a tenuous ionosphere.[243]

After apoapsis, Galileo headed back in and passed through solar conjunction on 19 January 1997, the day before the fifth periapsis of the tour. As communication with the spacecraft was impractical during the 10 days centered on conjunction, no satellite encounters were planned this time. The fact that the Sun suffered some coronal mass ejections at this time which happened to cross the line of sight between Earth and Galileo yielded a bonus for solar science.[244] Once contact was re-

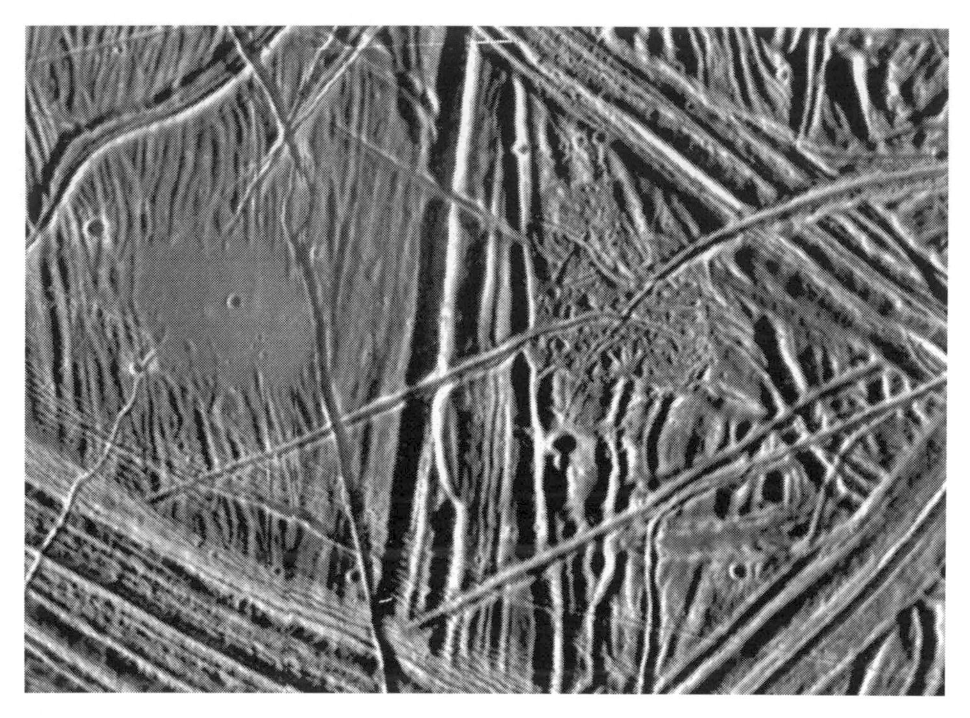

Details of Europa at a resolution of 26 meters per pixel. Note the remarkable 'ice rink' on the left. (JPL/NASA/Caltech)

established, the remainder of the 1-month orbit was dedicated to returning data from the previous Europa flyby. The data rate was boosted by arraying the antennas of the Canberra and Goldstone stations of the Deep Space Network together with the Parkes radio-telescope, which was a regular contributor to solar system exploration. Prior to the arraying, reception of real-time particles and fields was guaranteed at a low data rate when Galileo was within 50 Rj, with possible extended coverage either recorded or in real-time while further from Jupiter. Arraying increased the downlink capacity sufficiently for particles and fields data to be continuously received in real-time for most of the orbit.

Two months after its first encounter with Europa, Galileo made its second, just prior to the sixth periapsis of its tour. Unfortunately, whilst the spacecraft had been out of contact during solar conjunction the magnetometer's processor had stopped, and the fact that this remained off during the whole of the next orbit meant that the instrument provided no data to further investigate the possibility of Europa having a magnetic field.

The encounter sequence started on 17 February 1997, and ran for a full week. The scientific objectives for this pass included the collection of particles and fields data; continued monitoring of Io (the closest point of approach being at a range of 400,000 km) and of the Io torus; and using all of the remote-sensing instruments to inspect the 'white ovals' in the South Temperate Belt of Jupiter, which appeared to be anticyclonic storm systems like the Great Red Spot. The Europa flyby would be geometrically similar to the first for magnetospheric observations and a Doppler two-way radio-tracking experiment to determine the gravitational field and internal structure of the moon. Moreover, four radio-occultation experiments were planned for this orbit: two by Europa and one each by Io and Jupiter.

On 20 February Galileo flew by Europa at an altitude of 586 km, and a mere 16 seconds after closest approach the spacecraft was occulted by the moon, remaining behind it for 12 minutes. Because the geometry of the encounter was similar to that of the first flyby, the same areas were presented for viewing. The best-resolution imaging was dedicated to the dark region of Conamara, near the 'X' which marked the intersection of two prominent lineae. This was resolved into polygonal slabs of ice that had been broken apart, shifted and tilted to form a jumble reminiscent of terrestrial pack ice. The motion of the 'icebergs' seemed to be inconsistent with their having floated on an icy medium, instead indicating a liquid medium. If this were the case, then the icy crust in this region might be just a few kilometers thick. Small mounds several kilometers across could mark sites where the crust had been uplifted by some process, possibly involving diapirs of warmer ice rising with the crust, perhaps extruding liquid water when they broached the surface.[245] The few small craters in view were primarily located on top of the oldest displaced blocks. The fact that the matrix between the blocks was craterless might indicate that the process that displaced the blocks occurred within the last million years.[246] But the age of the Europan surface was disputed, with some scientists arguing that it could be billions of years old. The fact that an older basin like Callanish seemed to have punched through the crust and into the ocean, whereas a fresh crater like Pwyll had not, suggested that the thickness of the crust had increased in the intervening time. Asterius Linea, one of the branches of the 'X' adjacent to the Conamara region, was revealed to consist of a series of parallel ridges, each no more than 180 meters tall, and together about 6 km wide. There was a double-ridged linea to the north of Conamara superimposed on a complex terrain of crisscrossing ridges in various states of degradation.

The Doppler data from the two Europan flybys provided insight into the moon's gravity field and internal structure. It appeared to have a deep interior consisting of either a metal core and a rocky mantle, or undifferentiated rock and metal. In any case, the relatively low bulk density implied that there was an outer shell of water 100–200 km in thickness. Although the gravity data could not distinguish between the solid and liquid phases of water, because they have similar densities, it seemed possible that there might be more liquid water in Europa's subsurface ocean than on the Earth's surface.[247]

It was a particularly interesting time to make an observation of the 'white ovals' on Jupiter, because as they independently drifted in longitude two had trapped and deformed a lower-lying cyclonic system. High white clouds had formed where the cyclone was disturbed by one of the anticyclonic ovals.[248] Unfortunately, another detector on the infrared spectrometer started to operate anomalously during this periapsis.

Although Galileo passed through Jupiter's shadow, no observations had been

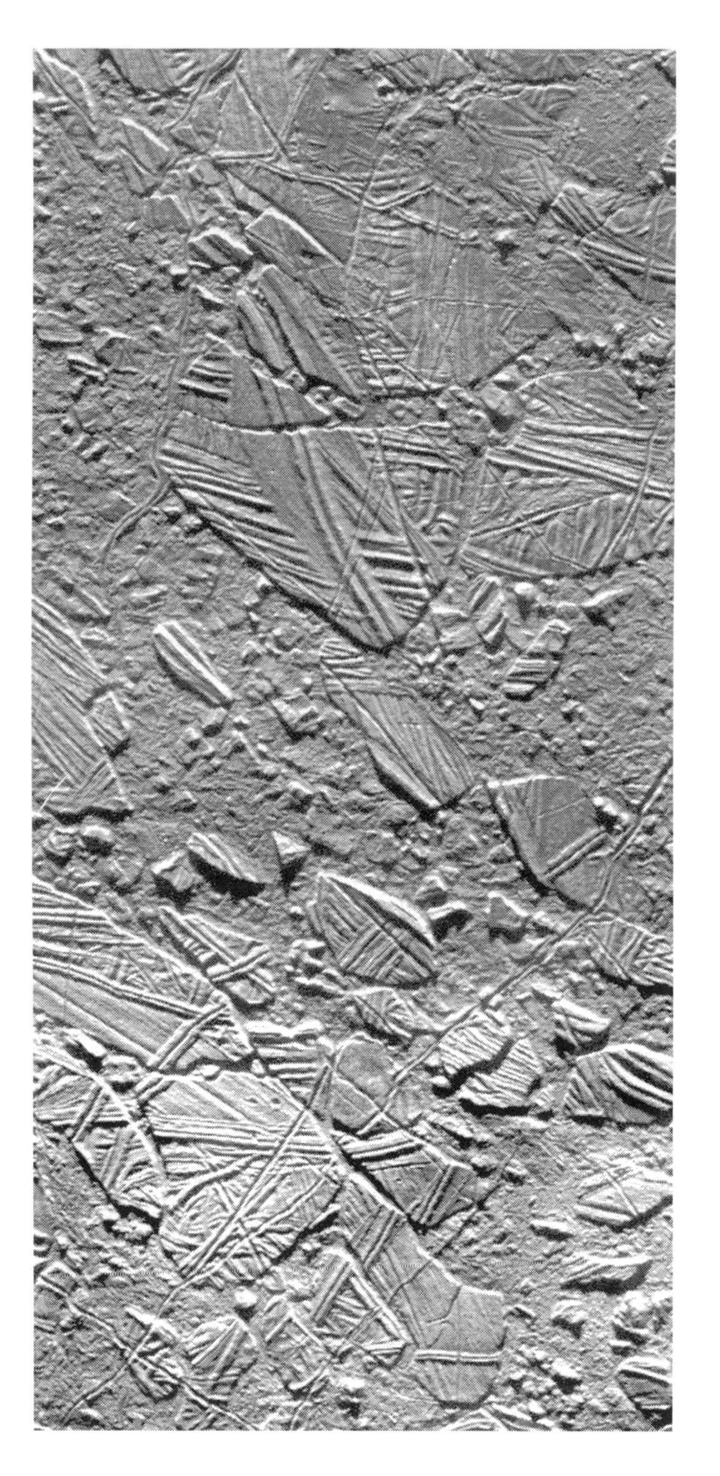

On its sixth orbit Galileo made a close pass by Europa and imaged Conamara at a resolution of 54 meters per pixel, revealing it to be comprised of broken and displaced ice rafts which once floated in a fluid matrix. (JPL/NASA/Caltech)

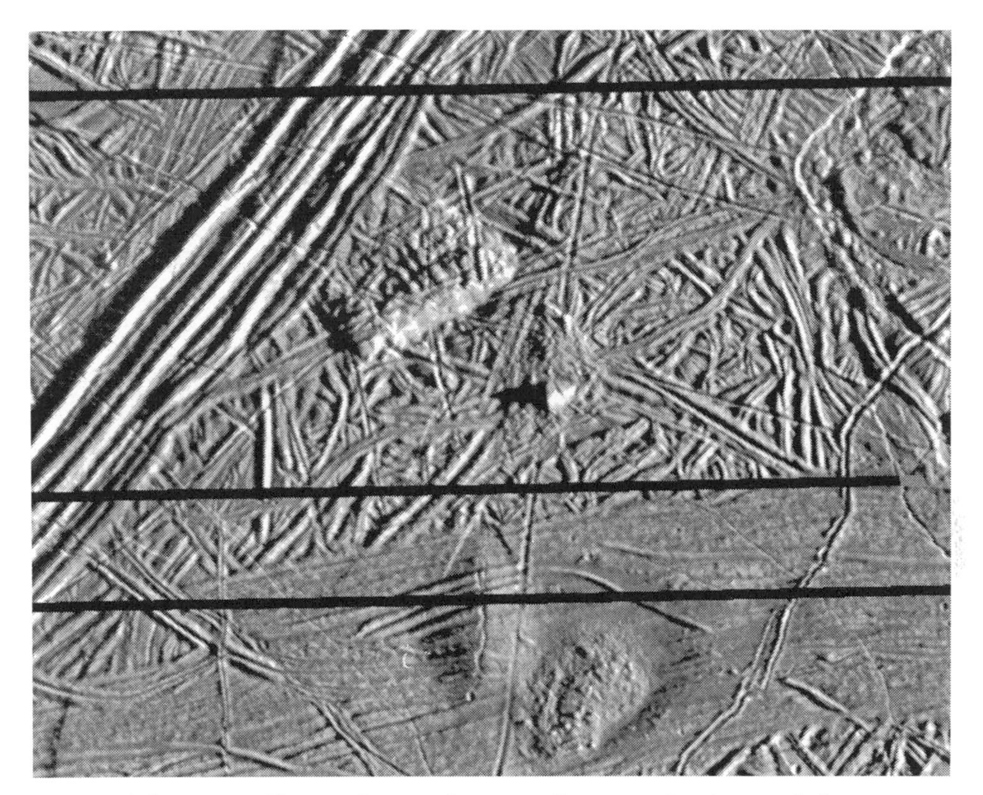

A partially corrupted image of an area in western Conamara showing a peak that casts a shadow, a rectilinear block with ridges on its top that seems to have been uplifted, a hummocky mound in a depressed area and the 'triple band' of Asterius Linea. (JPL/NASA/Caltech)

assigned this time because the spacecraft would have been obliged to turn to target the planet, and the planners had decided to save fuel for later in the mission. On 26 February Galileo was occulted by Io at a range of 3 million km, and this provided new data on that moon's ionosphere and tenuous atmosphere – in fact, this was the first such opportunity since Pioneer 10 was occulted by Io in 1973, which was long before it was discovered that the moon was volcanically active.

The flyby of Europa increased the period of the spacecraft's orbit to a little over 6 weeks. The next periapsis was to obtain imagery to extend the global mapping of the large satellites, as well as views of three of the smaller satellites, a non-targeted inspection of Europa at a range of 23,500 km and a flyby of Ganymede on 5 April at a range of 3,102 km. In fact, because the moon's rotation was synchronized with its orbital motion and this would be the only Ganymede encounter of the primary tour to occur after periapsis, this was the only opportunity to inspect some features at high resolution. Nicholson Regio appeared to differ from similarly dark terrains seen previously, suggesting an ancient landscape which had been heavily fractured by tectonic processes. In some cases, old craters were split by grooves. Individual craters

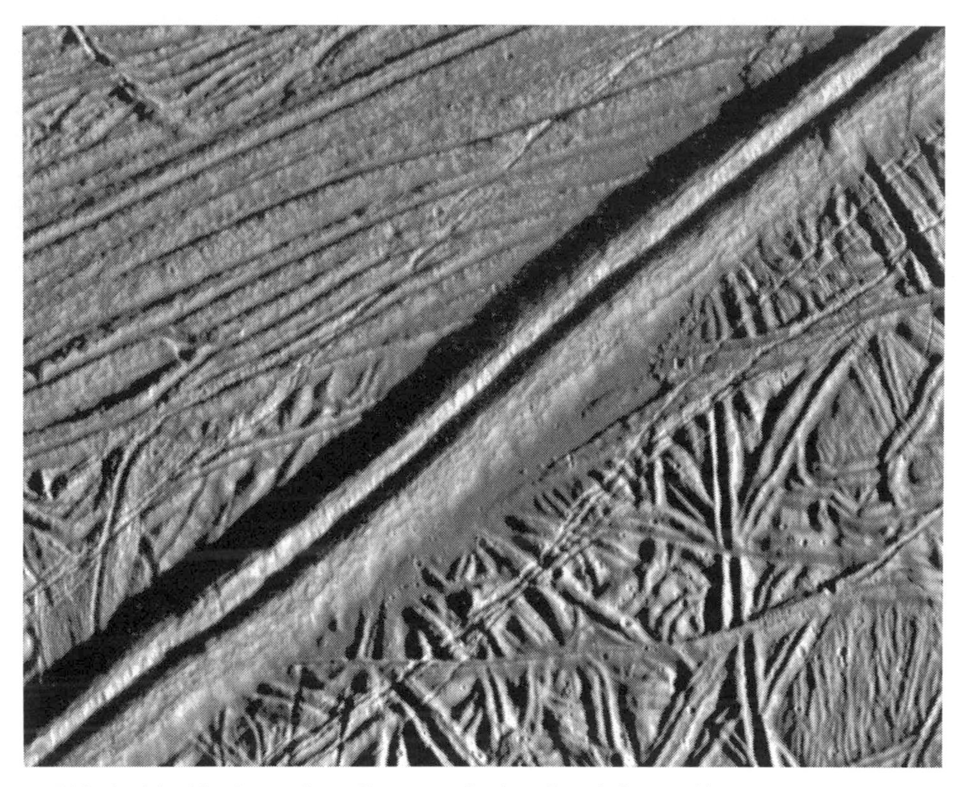

This double ridge in northern Conamara is virtually pristine, and hence young. It is 2.6 km wide and rises 300 meters above the adjacent terrain. (JPL/NASA/Caltech)

were targeted, including some in the polar region which were on 'pedestals' of icy crust that seemed to have been melted by the impact and refrozen around the crater's periphery, and Enki Catena, which comprises no fewer than 13 craterlets strung out in a straight line. The medium-resolution pictures of Europa revealed Tyre Macula to be a 140-km-diameter multiple-ringed structure. (In the Voyager imagery it had appeared as a dark spot; now that it was established to be the site of an impact the classification of macula would be deleted.) The infrared spectrometer also targeted Tyre, and found not only the signature of water but also of a chemical that could be a type of salt.[249]

On Jupiter, the infrared spectrometer took swaths of the northern hemisphere showing the alternating belt and zone structure in amazing detail, and temperature maps of the Great Red Spot at the highest resolution yet achieved. In addition, this instrument made a low-resolution mineralogical map of Callisto.[250]

Io was observed both in eclipse and in full sunlight. To the surprise of scientists who had telescopically monitored Loki in eruption in recent months, there were no significant changes visible.

To date Galileo had imaged one or two of the moons when there ought to be at least one bright star in the background. These pictures were processed onboard to

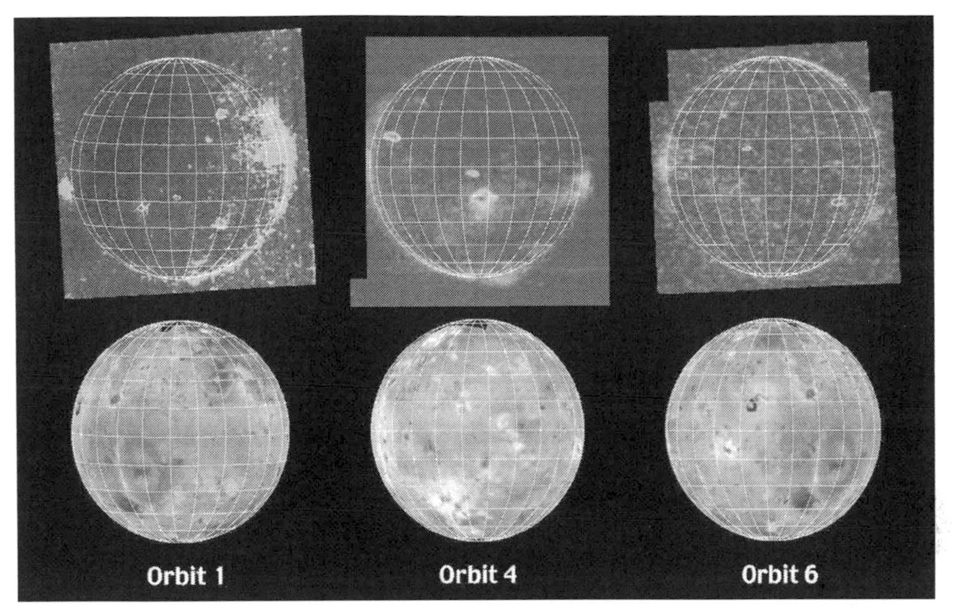

Images of Io in eclipse captured during Galileo's early orbits. The globes on the bottom row provide context for these observations. Note the correlation between known volcanoes and the 'hot spots' seen in eclipse. (JPL/NASA/Caltech)

On its sixth orbit Galileo took this mosaic showing a balloon-shaped vortex being squashed and stretched between a pair of 'white ovals' in Jupiter's atmosphere. (JPL/NASA/Caltech)

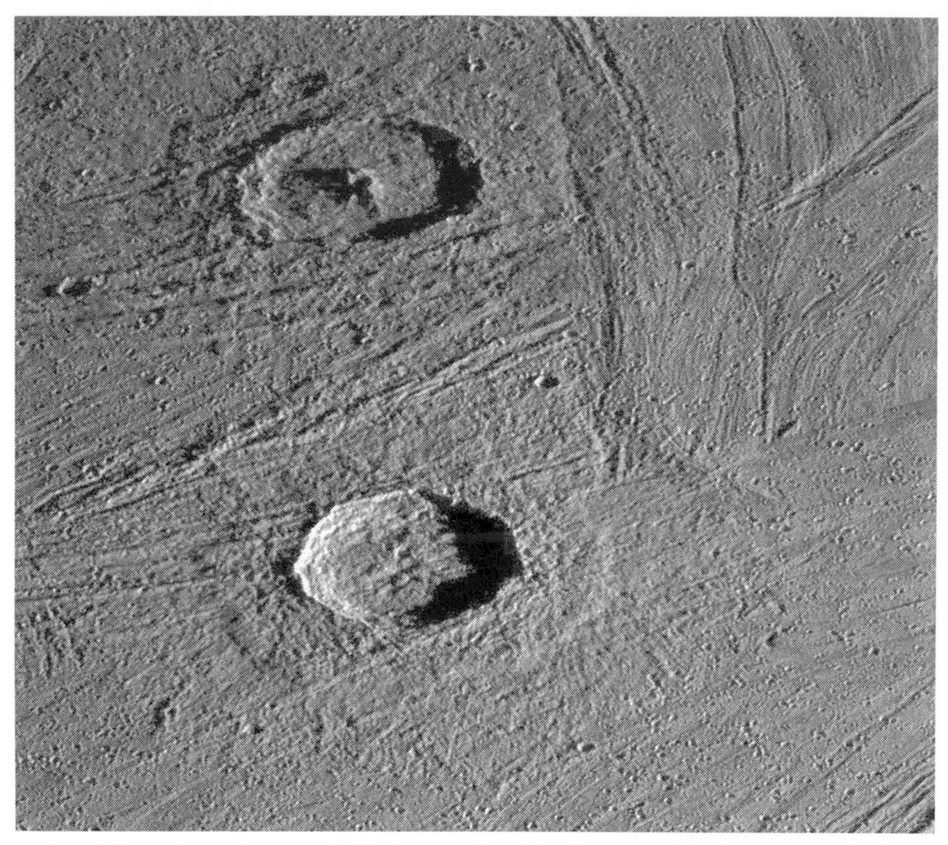

An oblique view taken on Galileo's seventh orbit of two fresh-looking craters in the grooved terrain near the north pole of Ganymede. The lobate ejecta deposits imply that these impacts melted substantial amounts of ice, which froze as it splashed out. (JPL/NASA/Caltech)

select only slices of the satellite's limb and around the star's predicted position, and once this data had been analyzed on Earth three trim maneuvers were defined for each orbit: one near apoapsis, one immediately before the targeted encounter of the next periapsis, and the third just afterwards. However, after this Io encounter the ephemerides of the large satellites were accurate enough for optical navigation to be discontinued, which in turn released space on the tape recorder and downlink for scientific data.

On 7 May Galileo flew by Ganymede at a range of 1,603 km, for its fourth and final targeted encounter with this moon of the primary tour. In addition to the sulci Tiamat, Lagash, Erech and Sippar in Marius Regio, it imaged the bright palimpsest Buto Facula. The sulci displayed interesting differences. For example, Erech was clearly a tectonic feature similar to a terrestrial rift valley, but the smooth Sippar, which transected Erech almost perpendicularly, appeared to have been flooded by ice. In fact, a 3-dimensional elevation model showed that in Sippar icy volcanism had

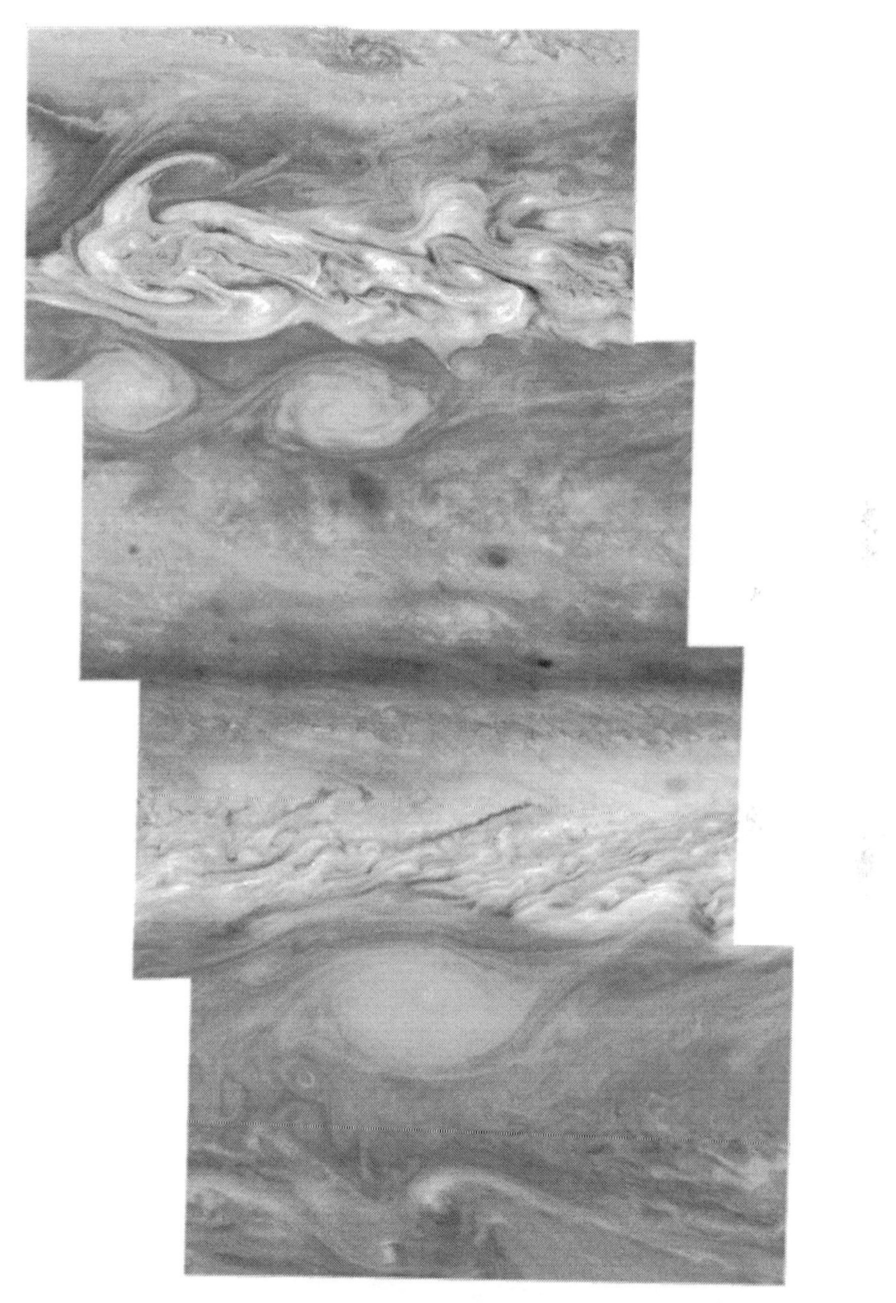

A mosaic of Jupiter's atmosphere taken by Galileo on its seventh orbit, featuring the alternating jetstreams of the zonal circulation system in the northern latitude range 10 to 50 degrees. (JPL/NASA/Caltech)

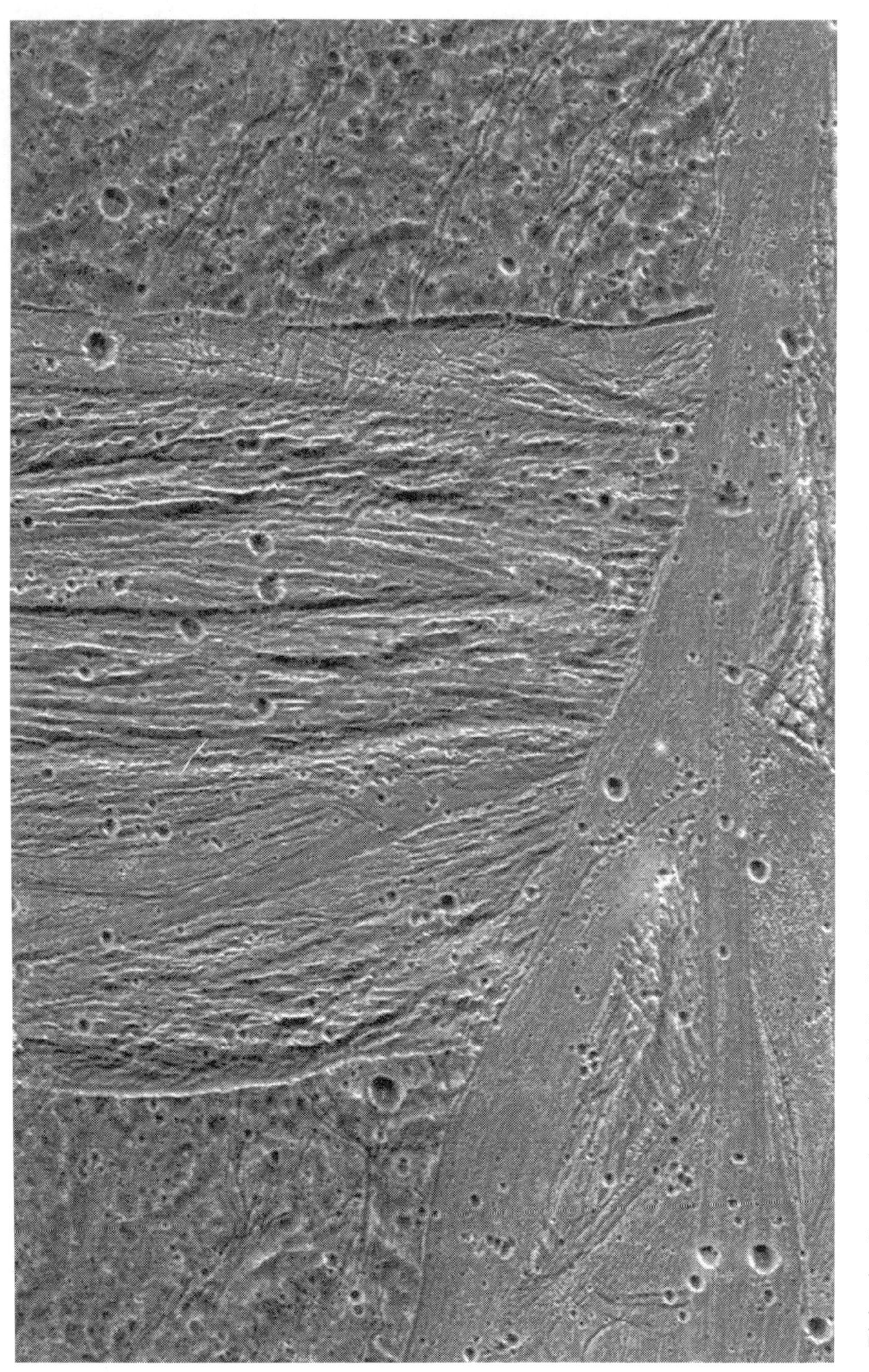

Flying by Ganymede on its eighth orbit, Galileo imaged the intersection between the southern end of the extensively grooved Erech Sulcus and the much smoother (and lower lying) Sippar Sulcus. (JPL/NASA/Caltech)

formed bright smooth pools in low-lying areas between the darker and rougher terrain. Targeted imagery of a horseshoe-shaped feature in Sippar some 55 km in length proved to be the best evidence of past icy volcanism on Ganymede.[251,252]

The four close encounters with Ganymede suggested that the moon's magnetic field lines were 'open' over the poles, merging with those of the Jovian field. This configuration enabled charged particles circulating in the planet's magnetosphere to impinge on the surface in the polar regions. Remarkably, the boundary between open and closed field lines approximated the boundary between the bright polar caps and the darker regions at lower latitudes, suggesting that a radiation-liberated frost might coat the polar regions.[253]

Although the closest approach to Io was at a range of almost 1 million km, this time the geometry was particularly favorable and provided day-side, night-side and eclipse observations of its 'hot spots'. Europa received minimal attention because the spacecraft approached no closer than 1 million km, and no pictures were taken. A non-targeted flyby of Callisto at a range of 33,100 km viewed a portion of this moon that had previously not been seen by either the Voyagers or by Galileo itself, thus enabling the south polar region to be mapped. A number of Jovian remote-sensing and particles and fields observations were also undertaken. Other highlights of this periapsis included a mosaic of the southern hemisphere of Jupiter to match that of northern latitudes taken previously, and ultraviolet observations of Elara, one of a number of small satellites orbiting far from the planet, to seek evidence that would support the hypothesis that these are captured asteroids or comets.[254]

During its ninth periapsis, Galileo had a non-targeted encounter with Ganymede at a range of 79,700 km, and a targeted flyby of Callisto on 25 June at an altitude of

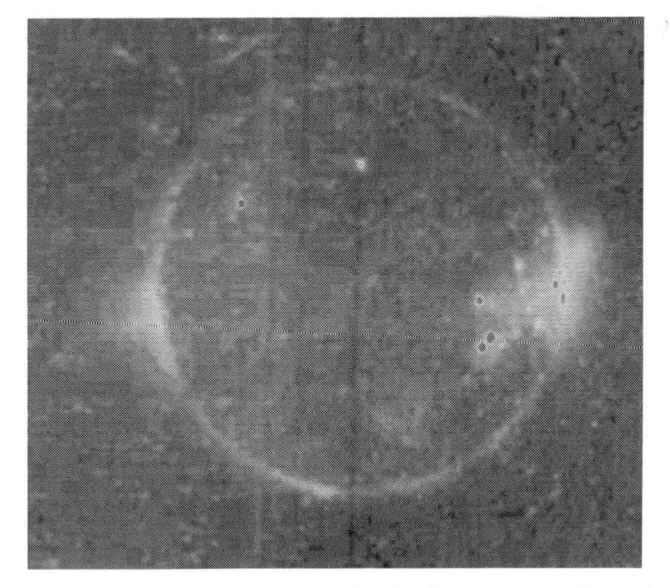

Another image of Io in eclipse, showing volcanic 'hot spots' on the surface and the associated sodium airglows. (JPL/NASA/Caltech)

418 km which featured the heavily cratered sub-Jovian point, enabling this area to be mapped in detail. Moreover, as the spacecraft approached Callisto it not only passed through the moon's shadow, but also offered the only radio-occultation by this moon of the primary tour. In fact, this orbit would provide an unusually large number of occultations: one by Jupiter, three by Ganymede and no fewer than five (including a grazing one) by Io. The camera made observations of Io both in and out of eclipse. Plumes were seen over Pillan and Prometheus, and since the latter was viewed from almost directly overhead the long shadow cast by the tall plume was evident.[255] Europa was also observed. On Jupiter, the turbulence to the west of the Great Red Spot was viewed near the limb, with the spot itself just out of sight. The highlight of this orbit, however, was not the moons, nor the periapsis passage, but the apoapsis at a range of 143 Rj directly down-Sun to study the tail of the planet's magnetosphere. The Callisto flyby had been used to set up this apoapsis, and most of the 3-month orbit was spent in the magnetotail. Although Voyager 2 had probed this region fairly deeply, and Pioneer 10 had swept by its extremity when almost as far from the Sun as the orbit of Saturn, it was a relatively unexplored region of the Jovian system. This orbit had been timed to occur when Jupiter was at opposition, and at its closest to Earth to maximize the data rate and permit almost continuous real-time transmission of particles and fields data.

On 17 September, having returned from its trip down the magnetotail, Galileo made its third targeted flyby of Callisto, the final one of the primary mission. The moon was at the same position in its orbit as for the encounter in June, and hence identically lit, but this time the 539-km point of closest approach was over the day-side rather than the night-side. As there would be no radio-occultations around this periapsis, it was possible to undertake continuous Doppler tracking to measure the gravity field and internal structure of the moon. In fact, once the data from this and the previous flyby had been analyzed, it became evident that although the interior of the moon had initially appeared to be undifferentiated it had in fact undergone partial thermal differentiation, prompting the remark that it was "half baked". The main imaging target for this flyby was a 700-km swath along a line radial to the Asgard basin, running from the central plain out to the periphery of the structure. Interestingly, the 50-km-wide crater Doh situated on the central plain had a dome at its center which spanned fully 25 km. Along the radial, there were landslides on the rims of some craters, bright icy bumps, ridges marking the inner ring structure of the basin, mysterious pits and smaller craters which did not seem to be impacts and perhaps were the result of endogenic activity, and troughs marking the outer edge of the basin.[256] Just after the flyby, the infrared spectrometer made a 6-minute scan of the day-side limb in search of airglow emissions. It detected an extremely tenuous atmosphere of carbon dioxide with a surface pressure one-billionth of that at the Earth's surface – so thin, in fact, that it would be lost within decades without replenishment, either by outgassing or by the Jovian particle radiations (which also create tenuous atmospheres on Europa and Ganymede).[257] A remarkable discovery was made by the magnetometer, which finally found the signature of a very weak field. This seemed to be similar to that of Europa, in that it was caused by electric currents in the Jovian magnetosphere interacting with a subsurface layer of liquid

On its ninth orbit Galileo imaged the 50-km-wide impact crater Har near Callisto's sub-Jovian point, revealing its floor to be domed. (JPL/NASA/Caltech)

water. Because Callisto's interior had undergone only partial differentiation, it was unlikely to possess a deep ocean in the manner of Europa, but it was possible there was a transitional layer of slushy ice. However, the fact that there were no surface features resembling those of Europa meant that this fluid must be situated at much greater depth – modelling suggested that if the slushy layer were 10 km thick then it must be 100–200 km below the surface. Also, since Callisto was not subjected to the same gravitational stresses as Europa, the heat in its interior would be primarily due to decaying radioactive isotopes, and because these would not be concentrated into a rocky core the sources of heat would be more widely distributed.[258]

This periapsis included remote-sensing of the north pole of Jupiter in search of high-altitude hazes and auroras, ultraviolet observations of Himalia (another small outer satellite) and further monitoring of Io's volcanoes. In July the Hubble Space Telescope had seen Pillan in eruption, and scientists were eager to inspect the site for changes. The new imagery showed there to be a dark circular patch of erupted material some 400 km in diameter. The fact that this deposit displayed none of the colors (red, white, yellow) characteristic of a sulfur-rich lava indicated a different composition. Because the Voyager data had suggested the vent temperatures were 700K, this had ruled out silicate lava and prompted the belief that the vents erupted sulfur-rich lavas. But in the 1980s terrestrial telescopes had measured temperatures exceeding 900K, and since liquid sulfur cannot exist in vacuum at this temperature it was evident that at least some volcanoes must be erupting silicate-rich lava. (As infrared scans by Galileo and pictures of Io in eclipse would later establish, some of the lavas on Io were even hotter than lavas erupted onto the surface of Earth!) It identified no fewer than 30 such high-temperature spots, Pillan Patera being one of them.[259,260] However, the repeated passes through the intense radiation were taking their toll, and on this periapsis one of the functions of the plasma-wave instrument was severely degraded, and later lost.

On 5 October, Galileo had a 19-hour solar occultation. Because this occurred at a considerable distance from Jupiter, there was ample time to make observations of the ring by forward-scattered sunlight, of the night-side of the planet, and of Io and Europa. In addition, a long radio-occultation provided an opportunity to sound the atmosphere.[261] Highly processed Voyager pictures had hinted at the presence of a 'gossamer' ring exterior to the main ring. Imagery by Galileo in solar occultations not only proved that this existed but also resolved it in two nested rings, one denser than the other. A detailed study revealed that the rings comprised dust ejected by impacts on the inner moonlets. The main ring appeared to originate from Adrastea and Metis, which occupy similar orbits lying just beyond the ring. The gossamer ring appeared to be formed by a similar process, but involving the larger Amalthea and Thebe.[262,263] As it entered the planet's shadow cone, Galileo also took a series of images of the night-side of Europa in search of evidence of icy volcanism such as water geysers. The results were negative in this respect, but the long-exposure images happened to capture the ring system almost in front of Europa, providing some of the most aesthetically pleasing images from the Galileo mission.[264]

Returning after a short foray into the magnetotail, on 6 November 1997 Galileo made the final targeted encounter of its primary mission with a 2,042-km flyby of

Europa. One feature surveyed at high resolution was a ridged plain, part of which had been broken to form a jumble of ridges, grooves, mounds and peaks, possibly by the onset of melting similar to that which appeared to have made the 'icebergs' at Conamara. A new patch of chaotic terrain was also observed, and nearby was a long wavy gray band. Like the 'dark wedges', this band was evidently a spreading feature which had been sealed by material which oozed up through the fault as the crust was pulled apart, over time forming a series of dark and bright lines running parallel with its axis. The presence of 'tributaries' indicated that the crustal blocks had been subjected to rotational as well as extensional forces.[265]

Spectra of Europa taken on the first 11 orbits showed peculiar broad absorption bands which implied the presence of molecules in addition to those of water ice. In fact, this 'non-ice material' was concentrated in the darker regions, including some of the lineae. A comparison with laboratory data suggested that the minerals were hydrated salts, with a mixture of sulfates and carbonates providing the best match. Such salts could have been formed by hydrothermal or volcanic vents on the floor of the ocean, and then made their way up through fractures in the icy crust onto the surface. Of particular interest to exobiologists was the possibility that this process could lace the ocean with the ions required to support life.[266] On the surface, the radiation bombardment would transform the sulfur salts into sulfuric acid – which was also evident in the spacecraft's spectra. In fact, the darkest regions appeared to match the areas of highest acid concentration, and since sulfuric acid is colorless this correlation suggested that these areas had been coated by sulfur polymers created by interactions with the Jovian radiation. Furthermore, there was evidently a sulfur cycle at work on the surface, with a radiation-induced equilibrium between various forms of sulfur-bearing molecules.[267] However, it was not possible to say for certain that the sulfur was endogenic in origin, it could be material that was blasted into space by the volcanoes on Io and later 'painted' Europa; although if the latter were true then Ganymede ought also to be coated to some degree with sulfur. The endogenic origin was appealing, because the presence of salts would make a good electrolyte to conduct the electric currents induced in the subsurface ocean by Jupiter's magnetosphere, and so create the observed perturbations of the magnetic field. The infrared spectrometer also found rather large concentrations of what appeared to be hydrogen peroxide on the surface of Europa. The fact that this molecule is extremely reactive implied that it must be being replenished on a continuous basis, probably by radiation bombardment (or 'radiolysis') of water ice.[268]

The significance of these observations was that discoveries made in the Earth's oceans in the 1980s and 1990s meant that the prospect of life having developed in an ocean beneath Europa's icy crust was plausible. It had been believed that all life on Earth derived from a food chain based on the process of photosynthesis, which, as its name implies, requires the presence of light. For this reason, the ocean floors, beyond the reach of sunlight, were presumed to be lifeless, or at best populated by creatures that fed on organic material raining down from the ocean's surface. But then research submersibles discovered complex ecosystems thriving on the ocean floor near volcanic vents which issued superheated water that was black because it was enriched in minerals. The food chains of these isolated ecosystems were based on

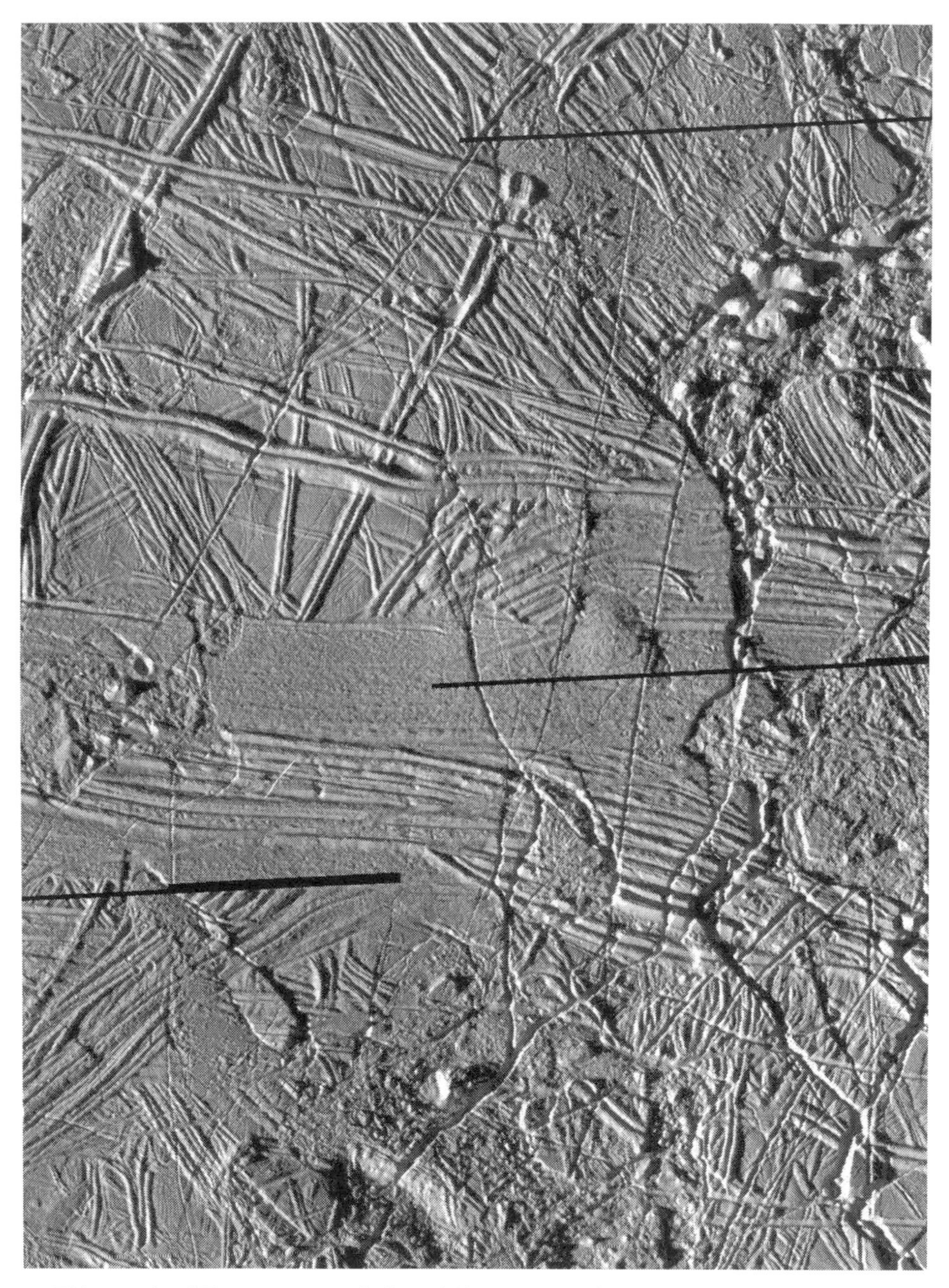

This mosaic of Europa at a resolution of 68 meters per pixel was taken on Galileo's 11th orbit. It shows ridges, grooves, chaotic terrain, mounds, plains and a number of icy peaks rising about 500 meters. (JPL/NASA/Caltech)

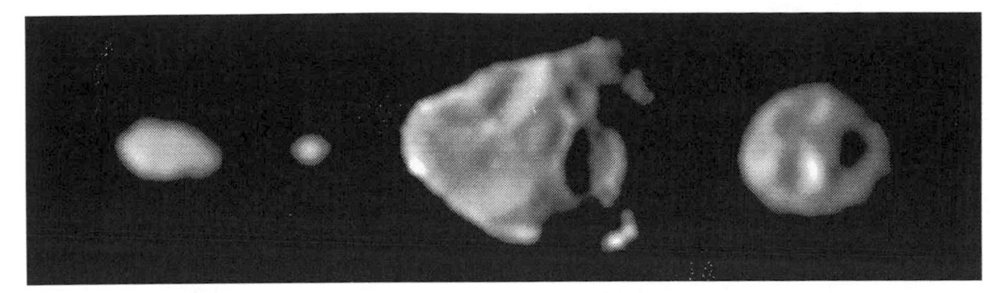

Images of Jupiter's small inner moons taken by Galileo: left to right Metis, Adrastea, Amalthea and Thebe. (JPL/NASA/Caltech)

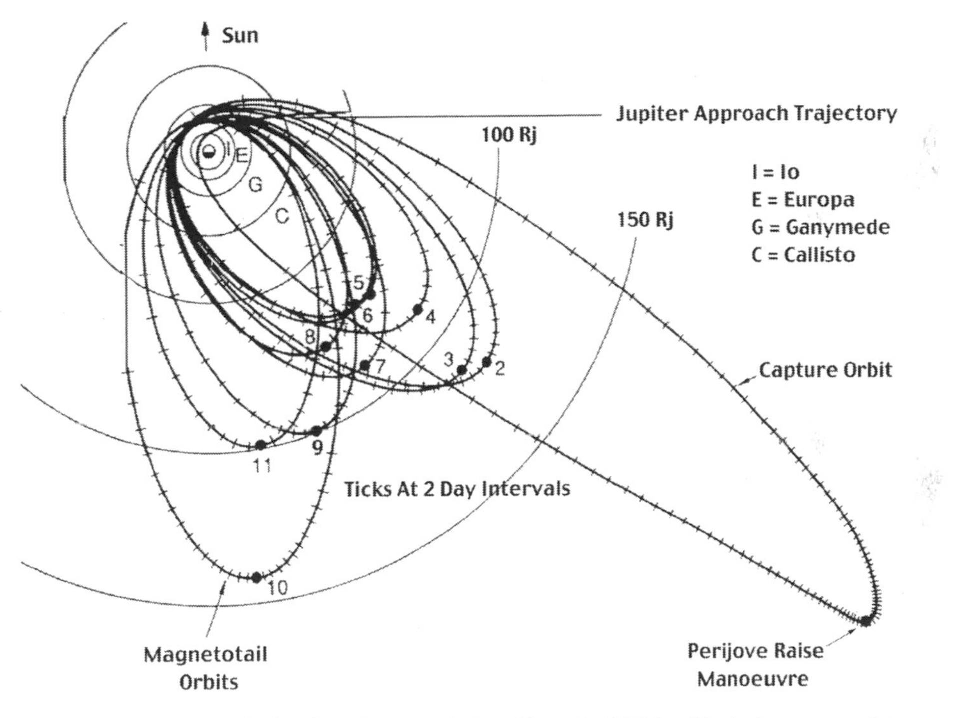

Galileo's trajectory during its primary mission. Note the highly elliptical capture orbit and the 10th orbit which flew down the magnetotail, opposite to the Sun.

thermophylic bacteria which could survive such high temperatures and exploit the nutrients provided by the vents. If there are hydrothermal vents on the floor of the Europan ocean powered by tidal heating in a process similar to that of Io (albeit on a reduced scale) there could be similar ecosystems living entirely independently of sunlight.[269]

During the 11th and final periapsis of the primary mission, Galileo also imaged all the small inner satellites (Amalthea, Adrastea, Metis and Thebe) at resolutions ranging between 5 and 9 km/pixel; took long-range pictures of Io and Callisto; and of the Jovian atmosphere and ring. As the spacecraft withdrew, a radio-occultation 'sounded' the mid-northern latitudes of the planet's atmosphere.[270]

RETURNING TO EUROPA AND IO

On 7 December 1997, Galileo completed its primary mission, although the replay of the data from the 11th orbit would take a few more days to complete. During its 2 years in orbit of Jupiter, the spacecraft had returned a total of 2.1 Gbits of data which included a total of 1,645 images of the planet, its satellites and ring system. The extremely accurate navigation had limited positional uncertainties to less than 10 km, which in turn had minimized the propulsive maneuvers needed to shape the ever-changing orbit. As a result, there was almost 60 kg of propellant remaining. Although the intense radiation had caused a number of glitches, the spacecraft was faring better than expected – for example, only one memory cell had failed, instead of the expected twenty. While Galileo was still on its interplanetary cruise towards Jupiter, JPL had drawn up a plan to extend the tour of that system beyond the 2-year point, but this did not attract much support and NASA was unimpressed by the relatively high cost. But once Galileo was safely in orbit around Jupiter, a small team evaluated a number of options for extending the mission – including focusing on comprehensive studies of Europa and Io; a high-inclination particles and fields survey; flybys of the small satellites that orbit far from the planet and about which almost nothing was known; using Galileo as a radio relay for a Europa lander; and even escaping from Jupiter in order to investigate the Trojan asteroids at one of the Lagrangian points located 60 degrees ahead of or in trail of the planet in its orbit of the Sun. It was decided that the most scientifically interesting mission would be to make follow-up observations of Europa and conclude with close-up studies of Io. This Galileo Europa Mission (GEM) would run for 2 years, to 31 December 1999, and involve eight flybys of Europa to determine the age and thickness of the crust, to seek evidence of ongoing icy volcanism, and to investigate whether there was a subsurface ocean; then four of Callisto to lower the periapsis for six passes through the Io torus as a prelude to two flybys of Io to characterize its volcanism, internal structure, atmosphere and magnetic field. Despite an eagerness to make up for the close inspection of Io that was lost owing to the problem with the tape recorder as Galileo was first penetrating the system, this moon was left to last because it was in the most intense part of the Jovian radiation belt. The periapsis sequences would be shorter than during the primary mission, lasting two days instead of seven, but would be more focused. Although the risk of Galileo being disabled by radiation grew with accumulating exposure, the spacecraft was thought to have a fighting chance of surviving its encounters with Io.

The initial phase of the extended mission would be so focused that most of the observations at periapsis would be devoted to Europa, on average accounting for at least 80 per cent of the returned data. Continuous particles and fields data coverage would no longer be assured throughout an orbit, in part due of the revised aims of the mission but also because the Deep Space Network would sometimes be called upon to support higher priority missions. The $30 million budget for the extended mission would be achieved by a range of cost-saving measures, including reducing the workforce – although it would be necessary to retain people having knowledge of the spacecraft's idiosyncrasies. Most of the planning for the extended mission was done before the science team was reduced at the end of the primary mission. Since only the final details remained to be filled in, this would simplify the usually time-consuming process of generating the observation sequences. But the planning for the Io flybys was left open, so as to be able to react to last-minute discoveries.

On approaching periapsis on the first orbit of its extended mission Galileo had a non-targeted flyby of Ganymede at a range of 14,389 km, during which it obtained four images of Gilgamesh, the multiple-ringed structure that appeared to be the youngest impact basin on that moon, and then, on 16 December 1997, it skimmed by Europa at an altitude of 205 km on a trajectory just south of the equator which provided an excellent opportunity to study the moon's magnetic field. One target for the camera on Europa was a 'dark wedge' spreading feature on the anti-Jovian hemisphere. Several pictures were obtained of where the wedge was transected by double ridges which had bright material on their crests which was probably clean ice. There was darker (dirtier) material on the wedge and in the trough between the ridge crests. When considered in the light of the paucity of impacts, the plethora of wedges, ridges and grooves clearly indicated that the surface had been subjected to considerable tectonic activity in the 'recent' past. In addition, very-high-resolution (at best 6 meters per pixel) pictures were taken for an east-to-west swath across Conamara to study the matrix between the fractured blocks. The results suggested that it was turbulent fluid which rapidly froze. The edges of the 'icebergs' formed steep cliffs, at the base of which were accumulations of debris. The westernmost part of Conamara was peppered by secondary impacts from the creation of Pwyll, located some distance to the south. Statistics of secondary cratering indicated that the Pwyll impact occurred 10–100 million years ago. Stereoscopic pairs of images were taken to facilitate 3-dimensional studies of the 'dark wedges' and Pwyll. One intriguing issue was how Pwyll had managed to retain a central peak against the force of isostasy. Computer simulations showed that in order to support the peak the crust must be at least 3 or 4 km thick.[271] The Doppler data accumulated over the various encounters refined knowledge of the moon's gravity field, oblateness and interior. Whilst the data was consistent with an undifferentiated mixture of metal and silicate below the ice crust, a differentiated interior was also plausible and this was the consensus view. If there had been any drag from a gaseous envelope, this should have been detectable in the data at closest approach, but no such effect was found.[272]

In addition, infrared and ultraviolet scans were made of Jupiter, and long-range observations of Io at every available wavelength. Meanwhile, engineers discovered that the electronics of one of the attitude control gyroscopes had been giving false reading of how the spacecraft was turning. When comprehensive tests proved this to be a casualty of the radiation, it was decided to alter the encounter sequences to use the star sensor instead of the gyroscopes for stabilization, even although in this mode the spacecraft would be a less stable platform for remote-sensing work and high-

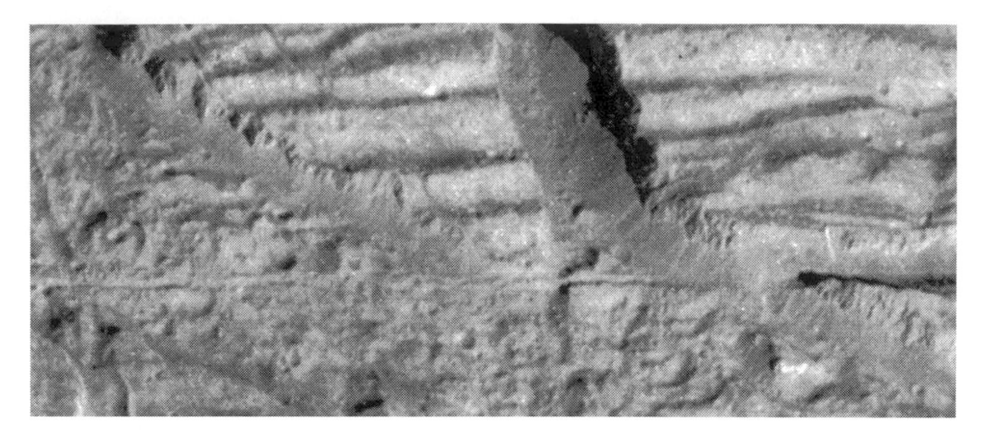

A view of the edge of an icy raft in the Conamara region of Europa at the very high resolution of 9 meters per pixel taken during Galileo's 12th orbit. (JPL/NASA/Caltech)

energy particles hitting the sensor would prompt false alarms. Unfortunately, the time spent on the engineering effort to diagnose and correct this problem meant that 30 per cent of the data taken during the 12th periapsis could not be returned. Other hardware was suffering from the frequent passes through the radiation belts. In particular, the memory of the energetic-particle detector was corrupted, halting its operation on this periapsis passage. The plasma-wave instrument also stopped, but fortunately not until after it had obtained the important magnetospheric data in close to Europa. Both instruments were able to be restored once the spacecraft had receded from the planet.

Galileo returned to Europa on 10 February 1998, but because Jupiter was near solar conjunction and communication was limited no remote-sensing observations were made. Doppler tracking was performed, but at a flyby range of 3,562 km this was of limited value. The next flyby was on 29 March, just after periapsis on the 14th orbit, at a range of 1,645 km. The illumination provided excellent images of Tyre, revealing this to be a 40-km-diameter impact crater surrounded by a succession of ridges forming rings up to 140 km wide, with the entire structure having relaxed isostatically. Rhomboidal dark features were also imaged, showing them to mark where the crust had split to allow dark slushy ice to reach the surface and freeze. There were several such features, linked by faults. Higher resolution imagery of a dark spot near the equator revealed it to be a low-lying patch of smooth ice which probably formed in a similar manner to the other 'ice rinks'. Interestingly, the pool had submerged a number of bright ridges and small secondary craters, establishing that it postdated these other features.[273] In addition, Galileo took long-range color imagery of Ganymede to provide fresh data on its radius, shape and photometric properties. The routine monitoring of volcanic activity on Io was supplemented by a relatively close 250,000-km flyby which produced the best images to date of the polar regions at a resolution of 3 km per pixel. This encounter sequence concluded with a unique infrared scan of the trailing hemisphere of Callisto taken from a range of 1,200,000 km.

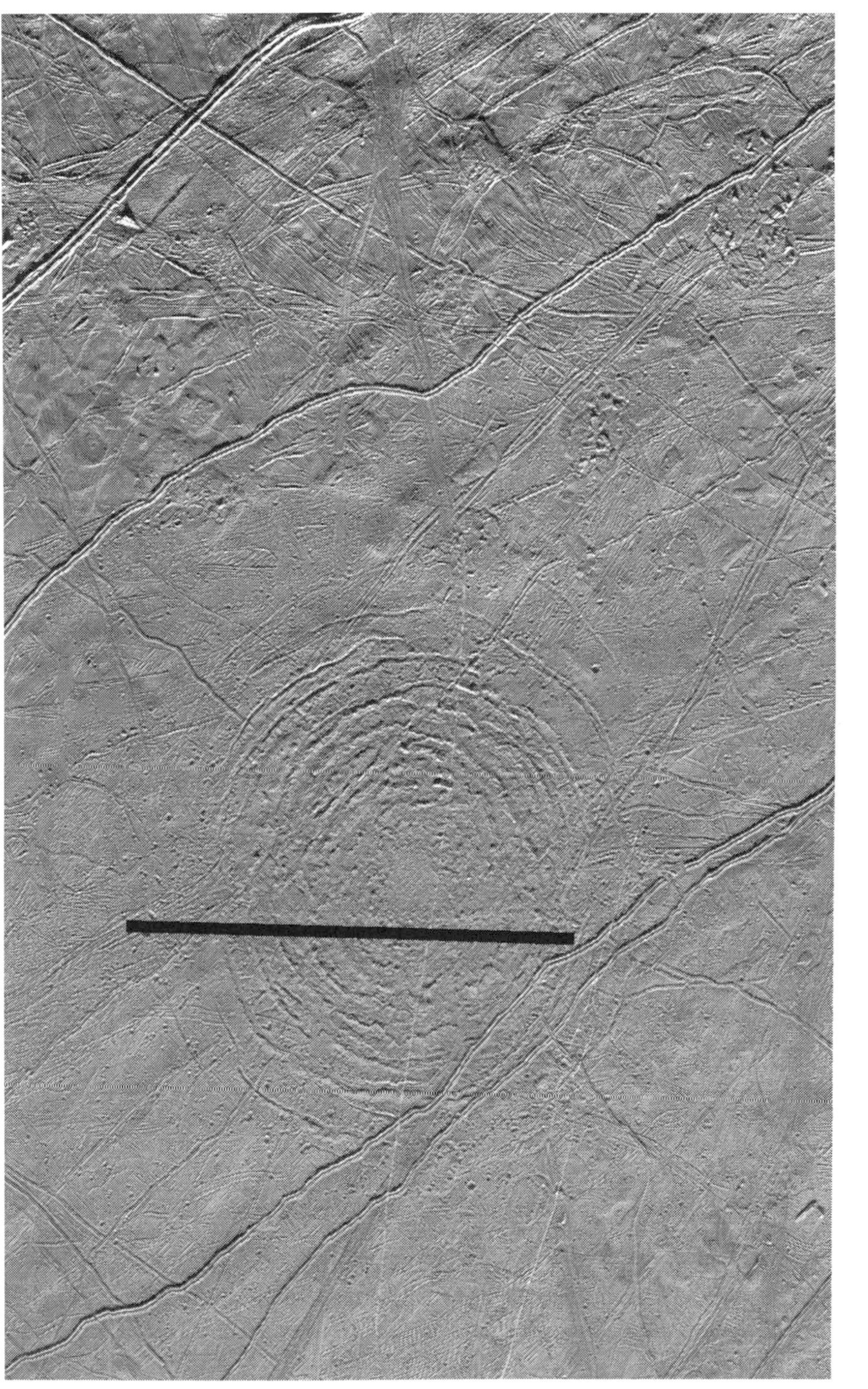

A mosaic of moderate-resolution images of the Tyre multi-ringed basin on Europa taken by Galileo on its 14th orbit. (JPL/NASA/Caltech)

Meanwhile, an analysis of the gyroscope problem had found that an electronic component was responding disproportionately to inputs from the gyroscope, and it was decided to develop a software patch that would apply a corrective scale-factor, the value of which would have to be redetermined for each periapsis passage. Two days before the data playback was due to end, the attitude control system suffered an anomaly during a trim burn and this put the spacecraft into its 'safe' mode.

Galileo made a 2,515-km flyby of Europa on 31 May, one day before the 15th periapsis. Imagery was taken to facilitate a 3-dimensional study of Cilix. This dark circular spot had appeared to be a crater in the Voyager pictures, but a mound in a long-range view taken by Galileo during a non-targeted encounter. It was now revealed to be a 23-km crater possessing a central dome. Because astronomers had decided after the Voyagers that the 182nd meridian passed through its center, this crater served as a point of reference in drawing the Europan grid system. Owing to the limitation on its downlink, the spacecraft could return only a few color images. A mosaic of color images of an area near Minos Linea illustrated that whereas the smooth plain was pure water ice, the double ridges that crossed it had extruded a mineral-rich fluid. This observation served as a point of calibration for interpreting such terrain elsewhere on the moon. An area to the southeast of Tyre was pitted by secondaries, was crossed by ridges trending in various directions, and hosted both a patch of rough matrix which contained 'rafts' up to several kilometers in size and another patch that seemed to have been more thoroughly melted and had produced a finely textured matrix containing only a few small rafts.[274] Distant infrared scans were made for compositional mapping. Doppler tracking was performed to gather more data on the moon's gravity field and magnetic field.

This periapsis passage also provided two opportunities to observe Io in eclipse. An analysis of 16 eclipse images taken during this and the previous passes enabled a number of phenomena to be characterized: a blue glow of sodium oxide marked the sites of volcanic plumes; a red glow of atomic oxygen was particularly bright over one or other of the poles, depending on Io's position in relation to its plasma torus; and a green glow of atomic sodium was over the night-side. Airglows were seen to diminish over time as Io penetrated deeper into Jupiter's shadow, probably as the tenuous atmosphere collapsed and froze, but of course the volcanic glow did not.[275] A few days before Galileo's observations of Io, astronomers noted the onset of a major eruption by Loki. (Lasting 8 months, this would prove to be one of the longest such eruptions ever observed.)

The periapses of the 16th to 19th orbits were all devoted to Europa, with mixed success.

Several hours prior to the 16th periapsis on 20 July, the primary command and data computer had a series of 'transient bus resets' similar to that which occurred shortly before the Ida encounter in 1993. The backup computer was automatically activated, but it suffered resets which put the spacecraft into 'safe' mode, thereby terminating all science operations. This was the first instance of both the primary and secondary computers suffering resets. By the time the engineers at JPL were able to restore Galileo to normal operations two days later, it was well beyond Europa. Apart from a small amount of data obtained prior to the resets, all the observations had been missed. The targets for the 1,837-km flyby had included Agenor Linea, Thynia Linea, the 'dark spot' Thrace and the crater Taliesin. However, some early infrared and photometric scans of Jupiter were on tape, and these were returned to Earth. The loss of the Europa data was frustrating, because the fact that Jupiter was near opposition would have maximized the data rate. But this opportunity was not wasted, because particles and fields observations were interleaved with replaying some of the lower priority data from the previous encounter – which had obviously not been overwritten.[276,277]

The 17th periapsis was almost lost as well. Less than a day before the encounter sequence was to start, the gyroscope electronics malfunctioned and was isolated by the self-diagnosis system, which left no other option but to rely on the star sensor for attitude determination. Twenty hours before the spacecraft reached Europa, the infrared spectrometer obtained long-range images and spectra of the moon's rarely seen Jupiter-facing hemisphere. The flyby on 26 September was 3,582 km over the southern temperate latitudes, and to some extent made up for the observations lost on the previous orbit. One of the high-resolution imaging targets was Astypalaea Linea, which ran for many hundreds of kilometers in the south polar region. It was first seen in the Voyager 2 pictures, and had been interpreted as a strike-slip fault marking where the icy crust had split and the two blocks displaced horizontally in a manner similar to the famous San Andreas fault in California. Galileo imaged the northernmost 300 km of the 800-km feature at a resolution of about 40 meters. By matching up features on each side of the linea it was not only possible to confirm that it was a strike-slip fault but also determine that it had undergone about 50 km of displacement. Along much of its length the actual fracture is marked by a very narrow ridge. As the fault moved, its zig-zagging line made openings that enabled fluid to rise and create angular features similar to the 'dark rhomboids' inspected previously. The fact that ridges transected the linea indicated that it was not a fresh feature. Although Astypalaea is a strike-slip fault like the San Andreas, the latter is a special case of a such a fault associated with the process of plate tectonics known as a transform fault, whereas Astypalaea is a fracture whose motion was the result of diurnal tidal stresses imposed by Jupiter. As Europa approached periapsis, the fault would have opened, with the migration of the tidal bulge (due to the eccentricity of the moon's orbit) causing the fault to move, then close again on nearing apoapsis, thereby building up an appreciable relative motion of the two sides over the ages. A similar, but more rapid process produced the cycloidal 'flexi'. As the surface deformed under tidal stresses, cracks formed perpendicular to the direction of the tidal bulge, then propagated along a curved course with the changing stress field, halting only when the diminishing tensile stress decreased below that which the ice could resist. Over a succession of orbits, the process produced a series of cycloids of ever smaller radius until the stress ceased to reopen the crack. For this process to operate, there had to have been a substantial ocean beneath the crust, and a tidal amplitude sufficient to impart the stress needed to open the crack.[278,279] Although it is not known how recently these features formed, the fact that each arc in the series would have opened during a single orbit around Jupiter (85 hours), and only a few dozen orbits would have been required to complete the structure, made this one of the fastest non-instantaneous types of geological feature known.

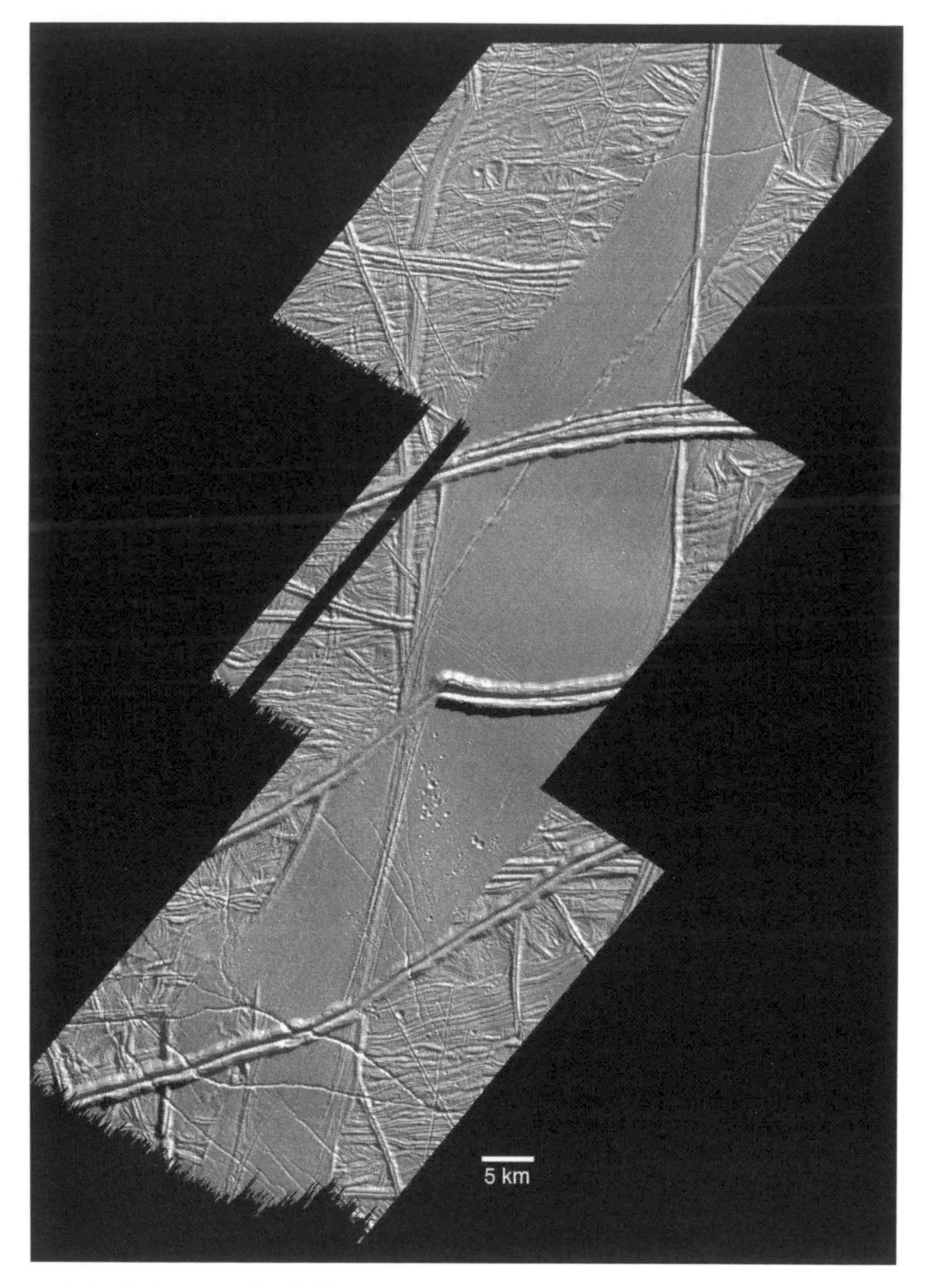

A detail of compression folding along Astypalaea Linea in the southern hemisphere of Europa taken by Galileo on its 17th orbit. (JPL/NASA/Caltech)

Imagery was also taken to produce 3-dimensional models of the two large dark spots Thera and Thrace. The results proved Thera to be somewhat lower lying than the adjacent terrain, while Thrace was higher. Whereas Thera had large slabs in its interior which suggested it formed by the collapse of the surface, Thrace had only hummocky knobs suggestive of a process of uplift. The fresh flow of brownish ice where Libya Linea transected Thrace attested to the thinness of the icy crust in this area. (For this reason, Thrace may be an ideal landing site for a mission designed to attempt to penetrate the ice to sample the ocean beneath.) A section of Agenor Linea, an unusually bright 'triple band' to the west of Thera and Thrace, was found not to be a ridge, but to be fairly flat. In fact, it comprised several parallel bands, only one of which was very bright, and fine striations suggested crustal spreading. On the empiricism that bright meant fresh ice, Agenor had been predicted to be one of the youngest features on the moon, but the fact that it was transected by a variety of features and even marred by very small craters established it to be rather older.

The infrared spectrometer took spectra to continue mapping the composition of the non-ice material on Europa. The photopolarimeter and radiometer instrument measured the temperatures on the night-side, with the surprising discovery that the lowest value was near the equator and increased away from it. Several suggestions were advanced to explain this. The most intriguing idea was that endogenic activity was warming the crust and it was calculated that for the ice to have warmed to the temperature observed, the crust must be less than 1 km thick. However there was no evidence of the proper 'hot spots', which should be detectable for decades after such activity ceased.[280] Doppler data was taken continuously for 20 hours in order to gain further insight into the moon's gravity field.

Observations of Jupiter were added to the 17th periapsis. In 1939 a trio of dark brown streaks had formed in the South Temperate Belt, which was itself unusually bright, and over the years astronomers had watched as these streaks stretched out longitudinally, compressing what remained of the bright belt into three ovals. The ovals were confined in latitude, but free to migrate independently in longitude. In February 1997 Galileo had observed as two ovals trapped a lower-lying storm, and when this 'lost power' in February 1998 it allowed the two ovals to merge. At that time, the spacecraft had been inactive while Jupiter was in solar conjunction. The result of the merger was now observed. A 'dark spot' that had recently formed was also investigated. These were rare, and this was one of the darkest ever seen, being even darker than the 'black eyes' created by the impacts of the Shoemaker–Levy 9 fragments. The photopolarimeter revealed it to be a break in the clouds, similar to that which was penetrated by Galileo's atmospheric probe. Finally, the brightness of the Jovian ring was measured at differing phase angles.

Unfortunately, about 2 hours before the 18th periapsis on 22 November, Galileo suffered another spurious reset which put it into 'safe' mode and canceled most of the observations planned for the 2,273-km flyby of Europa scheduled 4 hours after periapsis. Lost were observations of pits and plateaus at the south pole of Europa, as well as massifs, dark and bright plains, polarimetry and composition data, etc. Also lost were observations of Ganymede and Io, and a magnetospheric survey. The Doppler tracking was achieved, though, and long-distance ultraviolet, infrared and

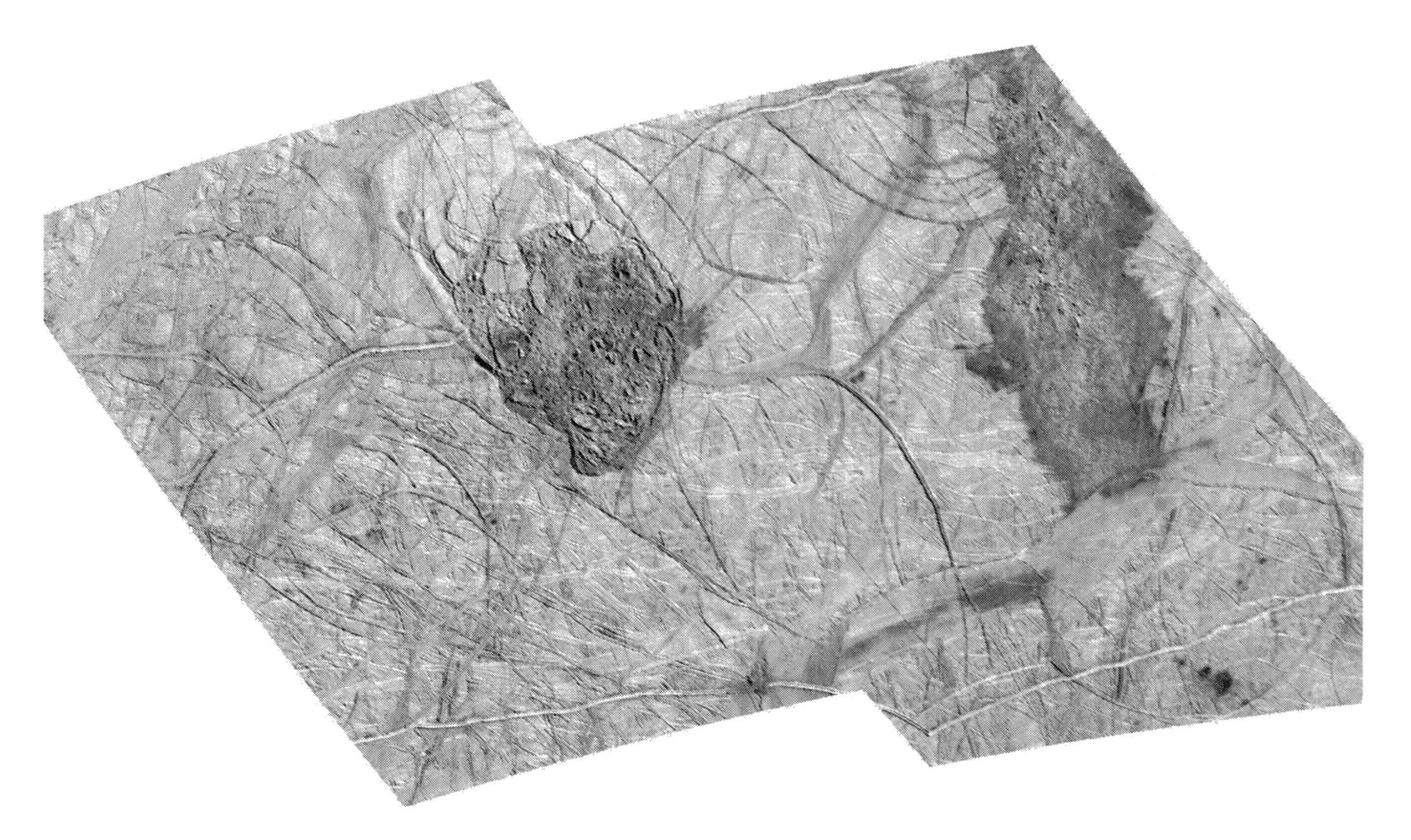

The two dark maculae Thera (left) and Thrace (right) at a resolution of 220 meters per pixel taken by Galileo on its 17th orbit. Libya Linea transects the southern tip of Thrace. (JPL/NASA/Caltech)

Two unusual pictures taken by Galileo. Left: the night-side of Europa glowing in 'Jupiter shine' and the planet's ring in the foreground, taken on the 10th orbit. On the right is Saturn, pictured on the 18th orbit.

radiometric scans of Jupiter's merged 'white oval' taped on the inbound leg were salvaged. On the plus side, a safing event before the principal observations of a sequence released most of the playback time during the long climb to apoapsis to other observations, and on this occasion one such task was a remarkable camera-calibration exercise involving targeting a conjunction of Saturn and its moon Titan with Uranus and Neptune.

The main objective of the flyby of Europa on 1 February 1999, during the 19th periapsis, was a thermal scan of the sub-Jovian hemisphere in darkness, but it had also been decided to attempt to use the infrared spectrometer to measure how the surface refracted sunlight with a grazing angle of incidence, a measurement which would indicate whether the ice was crystalline and hence young, or amorphous as a consequence of prolonged exposure to the charged particles that circulate in the Jovan magnetosphere. The sequence for this pass was unusual in that it required an unprecedented three attitude turns in order to provide the instruments carried by the scan platform with a view unobstructed by the appendages of the spun section. The imaging targets were to include Radamanthys Linea. The close (1,439 km) and relatively low-latitude (31°N) flyby also gave a new geometry to study the moon's magnetic signature. Although Galileo safed itself 4 hours after periapsis (this time because it could not re-acquire the Sun during a slew) by then most of the closest-approach observations of Europa had been made. Most of the observations of Jupiter, Ganymede and Io were lost, however. Limb scans of Europa at a high phase angle were undertaken to look for geyser plumes, while the ultraviolet instrument sought atmospheric emissions. It was later determined that particulate debris was accumulating in the slip-ring connector between the spun and despun sections, and that this was producing momentary short circuits that could cause the spacecraft to enter safe mode. Once the cause was understood, software patches were written to recognize and ignore such transient events.

As a result of this flyby of Europa, the apoapsis of Galileo's orbit was increased to 11 million km. Meanwhile, Jupiter passed solar conjunction. On 5 May 1999, as it headed back in, the spacecraft made a 1,315-km flyby over the equatorial zone of Callisto. This 20th orbit began the Perijove Reduction Campaign in which a series of four such flybys, one before periapsis and three afterwards, would progressively lower the periapsis. This time not only was the periapsis lowered by 593,000 km to place it between the orbits of Europa and Io, but the apoapsis was also cut to about 8 million km. At Callisto, Galileo imaged the hemisphere opposite to Asgard. The imagery verified the paucity of craters less than about 1 km in size. The three main remote-sensing instruments made coordinated studies of the 100-km crater Bran to determine the composition of the 'deep crust' by the material that this impact had excavated. A reset again occurred, but by this time Galileo recognized it for what it was, ignored it, and proceeded to make infrared and ultraviolet scans of Europa in the planet's shadow; but unfortunately the ultraviolet data was partly corrupted by the radiation. The most significant outcome from this periapsis was insight into Jovian meteorology. Atmospheric scientists were still debating the source of the energy that powered the rapid winds and created the belt and zone structure seen at the surface. On this pass, Galileo tracked a storm crossing from the night-side onto the day-side. The storm was situated just west of the Great Red Spot in the South Equatorial Belt, spanned about 4,000 km and rose to a height of 50 km. The results showed that the vast thunderstorms and lightning seen at night corresponded to the towering anvil clouds observed in daylight. It was calculated that these convective systems were drawing a thousand trillion watts of energy from beneath the visible cloud deck, and this drove the high-speed winds. A similar process converts heat into kinetic energy in the Earth's atmosphere.[281]

As the Perijove Reduction Campaign continued, Galileo would fly ever deeper into the inner magnetosphere, providing the opportunity for in-situ observations of the Io torus. The 21st periapsis featured a 1,047-km flyby of Callisto on 30 June, on which imagery and other data were obtained of the floor of the Valhalla basin with an almost vertical illumination in search of clues to the paucity of small craters. In addition, models of the moon's gravity field and internal structure were refined by Doppler tracking. This periapsis passage also provided a relatively close encounter with Io, with the closest point of approach occurring at a range of 124,000 km. The spacecraft spent nearly a day inside Io's torus, which was sampled by the particles and fields instruments. Two sequences of Io images were taken, one at long range providing global coverage and the other closer in providing regional coverage at a resolution of 1–2 km to monitor changes on the surface and plume activity; with a plume being detected over Masubi. In addition an infrared map was made showing the surface composition at the unprecedented resolution of 60 km. Infrared scans were also obtained of Jupiter in order to monitor thermal and chemical variations at the level of the cloud tops over time. Meanwhile, technical issues, minor and major, continued to mark periapsis passes. For example, attitude control problems were caused by the star sensor mistaking a radiation spike for a star, but this was only to be expected in that environment. Another issue was that the ability of the star sensor to detect faint stars diminished as the radiation progressively fogged its optics.

For the 22nd periapsis it was decided not only to take Doppler data during the 2,296-km flyby of Callisto on 14 August but also to exploit occultations of both the Sun and Earth near closest approach in order to 'sound' the moon's ionosphere and measure the vertical distribution of free electrons. Although this time Galileo only made it to within 730,000 km of Io, it made a particularly deep foray into the torus. In addition to pictures of Io and Jupiter, pictures were obtained of Amalthea at a resolution of 9 km. In the infrared, global scans of the trailing hemisphere of Europa were obtained, particularly of concentrations of dark surface material, and on Io the 'hot spots' and plumes were monitored to assist in planning the flyby that was due two orbits later. Unfortunately, some of the observations were lost owing to radiation-induced glitches. In fact, this deep penetration of the inner magnetosphere prompted an unprecedented three bus resets, including one which stopped the tape recorder. In addition, a failure of the encoders of the grating drive of the ultraviolet spectrometer prevented the instrument from taking spectra. During this periapsis pass Galileo again took calibration images of distant Saturn and of its moon Titan, although this time the task was performed one day before the Callisto flyby rather than during the cruise to apoapsis.

The fourth and final Callisto flyby of the Perijove Reduction Campaign was at a range of 1,057 km and occurred on 16 September, during the 23rd periapsis. There was a Callisto occultation and extreme-ultraviolet observations of the Io torus and of the Jovian auroras, but no observations were assigned to the camera, the infrared spectrometer or the photopolarimeter. In fact, this pass was primarily dedicated to magnetospheric science, with 6 hours of particles and fields data being obtained of the Io torus – which had recently received a substantial injection of material from a surge of volcanic activity. As the spacecraft drew away from Jupiter, its periapsis had been reduced to 393,000 km, which meant that upon its return its trajectory would narrowly intersect Io's orbit.[282]

The planning of the Io encounter on the 24th orbit started early in the extended mission, but the deterioration of the gyroscopes and the star sensor's susceptibility to radiation-induced signals imposed some constraints. In particular, owing to the problem with the gyroscopes the Io encounter had to be flown using the stars for attitude determination, backed up by the Sun sensor. Also, in view of the problems at orbit insertion when the star sensor was so 'blinded' by the radiation in close to Jupiter that it lost track of Canopus, one of the brightest stars in the southern sky, stars were chosen as points of reference that were sufficiently bright for the sensor to stand a good chance of remaining locked on even when flooded by radiation; and a software patch was written to enable the spacecraft to perform its operations using a single star – although there would have to be a backup plan in case it were lost. The software was tested on the 22nd periapsis, and used with remarkable success on all subsequent orbits with no star misidentification or loss. As regards imaging while deep inside the Jovian radiation belt, it was decided to use a 'fast mode' by which within 2.6 seconds of an image being taken, the camera's software would perform averaging, pixel summation and compression, and store the data on tape in order to minimize the time that the data spent on the CCD and susceptible to corruption by radiation. Another method was to take fast, full-resolution strips using only the upper part of the CCD array. Tests to regain control of the ultraviolet spectrometer confirmed it to be unusable, so it was decided to leave it switched off on this pass. (Unfortunately, it remained unusable.) And by now the energetic-particle detector was also exhibiting unpredictable behavior. Because the extended mission offered only two flybys of Io, Galileo would have to work very hard to address the many aspects of this remarkable moon: geology, gravity, structure, atmosphere, surface composition, magnetic field, and torus. In fact, the encounter plan had also to take into account the need to make observations which would complement those which were made inbound to orbit insertion, so as to better characterize the moon's upper atmosphere and ionosphere.

The eagerly awaited flyby of Io occurred on 11 October 1999 just after Galileo had passed periapsis at 5.5 Rj, as it crossed the moon's orbit heading out. It made its approach from upstream with respect to the wake which Io makes in the rapidly rotating inner magnetosphere, viewing the night-side of the moon, passing almost directly over Loki and then over the dawn terminator at the equator. The closest point of approach was at an altitude of 612 km. The departure provided a view of the day-side. The hope was that it would chance to pass through a plume and make unprecedented in-situ observations.

The early particles and fields observations were lost to a safing event 19 hours before closest approach. This also threatened the remote-sensing observations of the inbound leg. Fortunately, a software patch to avoid a memory bit that had been damaged by radiation was able to be developed and uploaded to the spacecraft just in time for the encounter. This bit would have been used by the photopolarimeter and infrared spectrometer, both of which made thermal scans while approaching the night hemisphere to locate 'hot spots' and measure their temperatures in order to impose constraints on the composition of their lavas. Although the particles and fields observations at periapsis had been lost, these instruments were able to start measuring 50 minutes from Io to investigate whether the moon did indeed have its own magnetic field.

When the imagery was replayed, it was found that radiation had corrupted the 'fast mode' of the camera, and it took extensive analysis and processing to restore the badly scrambled data – although even then portions of some frames were not able to be retrieved at all, and the inability to rely on the photometry meant that it was impracticable to make color pictures. However, because taking fast, full-resolution strips using only the upper part of the CCD array worked properly, it was decided to use only this method on the next encounter. The infrared spectrometer was also affected by radiation, when a stuck grating restricted it to just 13 wavelengths. On the one hand the constrained spectral range ruled out both determining whether the dark areas were indeed silicate lavas and the prospect of discovering previously undetected molecules on the surface; on the other hand the fact that the instrument made many more observations per wavelength in the available range increased the signal-to-noise ratio, which improved the temperature measurements.

On the way in, the infrared spectrometer scanned the remarkable 'lava lake' of Loki in darkness. Its field of view first captured the island within the lake and its warmer 3.5-km-wide dark crack, then the scan proceeded to the southwestern corner of the lake, across the wall of the caldera and out onto the plains beyond while transferring to the next target. Data showed that the temperature of the caldera was uniform, and more than 100 degrees warmer than the surrounding terrain. A high-temperature spot probably corresponded to the site of the eruption that had started in late August. Other areas appeared to be warm, but cooling, and could have been active in the second half of 1998.

The next target was Pele, which was located on the terminator. Because this had been persistently hot without producing a lava flow, it too was believed to possess a lava lake, even though this had never actually been resolved. Numerous pictures were taken as Galileo passed overhead. One image through a near-infrared filter showed a dark surface crossed by a bright sinuous line that ran for over 10 km, yet was only several tens of meters in width. The brightness of this line indicated that its temperature exceeded 1,000K, and it was inferred to be hot lava exposed where the lava lake was in contact with the rim of the caldera. Lower-resolution pictures taken over the next few hours showed Pele was in eruption, suggesting that Galileo had narrowly missed viewing the eruption center. The Hubble Space Telescope was monitoring Pele at about this time, and detected the presence of sulfur and sulfur dioxide in the plume. Molecular sulfur would be polymerized by solar ultraviolet to produce the characteristic red and pinkish aureole around Pele which so amazed the Voyager team when it was first seen.[283] Next was Pillan Patera, situated on the outer rim of the Pele aureole and, at that time, just over the terminator and hence in daylight. It had issued plumes in recent years, and given rise to the most dramatic surface changes. Galileo took pictures at resolutions ranging from 10 to 30 meters of a recent lava flow which had pooled in a low-lying area, and a patch nearby that combined smooth areas and rough areas dotted by pits, domes, channels and cliffs which could have originated when hot lava overran a volatile-rich surface such as a snow of sulfur dioxide. Galileo's data established that Pillan had indeed extruded a silicate lava.

Zamama had not been on view to the Voyagers, but it had been seen in eruption by Galileo in long-range 'volcano watch' observations. It had produced a long lava flow, and a sequence of 13 images were taken along its length. Although corrupted by radiation upsetting the 'fast mode' of the camera, when reconstructed on Earth the images showed a feature that resembled smooth basalt known by the Hawaiian name of 'pahoehoe' (meaning 'lava you can walk on'). At the center of the daylit hemisphere was Prometheus. Galileo's 120-meter-resolution imagery of it showed in

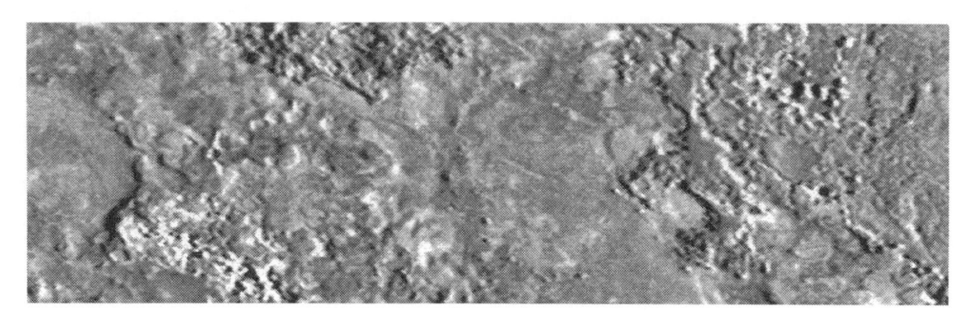

A very-high-resolution image of a recent lava flow from Pillan Patera on Io taken by Galileo on its 24th orbit, showing details 9 meters across. (JPL/NASA/Caltech)

Another very-high-resolution imaged by Galileo on its 24th orbit of a smooth patch near Pillan Patera showing pits and domes. (JPL/NASA/Caltech)

particular the semicircular 28-km-diameter caldera and its associated network of sinuous flows. The easternmost part of the volcano was quiescent, but the western section was active. This site had been erupting continuously since its discovery in 1979. Galileo viewed straight down into the omnipresent plume, the base of which was displaced some 80 km from where it had been for the Voyagers, and there was a dark lava flow linking the two locations. Evidently the plume was created by the vaporization of sulfur snow as it was overrun by the lava flow, which was pooling in a low-lying area. This was supported by the fact that the infrared spectrometer detected gaseous sulfur dioxide in the plume. (A similar phenomenon occurs when molten lava interacts with water either on or near the Earth's surface.) In the paper that reported discovering the "wandering plume" of Prometheus, the authors joked "the Prometheus of legend was bound to a stake, whereas Prometheus on Io has not been bound by any physical or mythological constraints".[284]

As Galileo withdrew, it viewed Amirani and Maui. These had been believed to be distinct, but Maui proved to be the active front of a 250-km-long lava flow that originated from Amirani. Nearby were some of Io's most impressive mountains. In the Voyager coverage it seemed that the mountains, which can stand 16 km above their surroundings, were unrelated to volcanism, since they had no summit vents or flanking lava flows. It appeared that the rapid resurfacing rate had compressed the crust, fracturing it into large blocks that were uplifted in such a manner as to form ramps that terminated in cliffs. Imagery at 500-meter-resolution of Skythia Mons and Gish Bar showed both young angular and rounded older peaks, with evidence of slumping on the ramps. Surprisingly, in many cases the mountains seemed to be located near calderas, hinting at a relationship. There were no volcanic domes or viscous lava flows to suggest the presence of silica-rich lavas sufficiently buoyant to form a permanent crust. Instead, the entire surface would seem to be a very thin crust that is rapidly recycled back into the mantle. There were no impact craters, even in the highest resolution views. Although impacts must occur, the craters must soon be buried by volcanic deposits.[285,286,287,288]

The imaging sequence concluded with a global color image and an observation as Io entered Jupiter's shadow. Owing to the safing event at periapsis, almost none of the planned remote-sensing observations of either the planet or the other moons were able to be made. Galileo took a total of 156 frames of Io, 122 of which were corrupted to some degree by radiation; but many of these were rectified and a total of 135 images were usable. Fortunately, the damaged memory bit was not in the section used for the preliminary image processing and compression! In addition, the spacecraft obtained good particles and fields data. In particular, plasma waves corresponding to sulfur monoxide ions were clearly detected. These observations, together with those made inbound to orbit insertion and from Voyager 1, which flew by Io at a range of ten times the moon's radius, helped to clarify the dynamics of its exosphere and ionosphere and the mechanism by which material infuses the torus.[289]

The spacecraft made a maneuver at apoapsis to turn the second flyby of Io from a pass over the equatorial zone (as was the case on the recent flyby) to a pass over the south pole. Remarkably, volcanoes on Io seemed to be divided into two types, with those located at or near the equator such as Loki, Prometheus and Pele being primarily long-lived, while those at high latitudes had brief eruptions which issued larger volumes of lava. Although there were hints that plumes must occur at high latitudes, as yet none had been seen. Both close encounters would target the anti-Jovian hemisphere, for which global color context images had been taken on the 21st orbit. Inbound to periapsis and to Io on the 25th orbit, Galileo flew by Europa at an altitude of 8,642 km, observing the northern polar region for the first time and also the rarely seen sub-Jovian hemisphere at a resolution of about 1 km. The sub-Jovian point, which was both imaged and scanned by the polarimeter, was dominated by lineae and other types of crack that probably derived from the stress on the icy crust imposed by gravitational tides. By now there were only small gaps in the coverage of Europa remaining to be filled by future missions.[290]

Continuing towards periapsis on 26 November, Galileo started a 6-hour session recording particles and fields data at high time resolution as it entered the Io torus, and the photopolarimeter joined in by taking a thermal scan of the night-side of Io. But halfway through the torus the spacecraft safed itself and ceased observations. Time was of the essence, because just 2 hours remained to periapsis and twice that to Io. To make matters worse, at JPL it was 4 p.m. on a national holiday. When the engineers reported that they could not identify the fault, but had also determined that no hardware would be put at risk if the unknown cause were to strike again, it was decided to allow the spacecraft to proceed anyway with a modified sequence. It resumed making observations a mere 4 minutes after skimming by the moon at an altitude of 300 km! It was later determined that the safing had been caused by an interaction between the software patch installed to avoid the damaged memory bit during the previous Io flyby, and an already present but undetected software bug. Although the inbound observations of the moon's night-side were lost, those of its day-side were gained. Unfortunately, the lost sequences included an investigation of whether Io possessed its own magnetic field, which was the primary reason for selecting a polar pass, together with high-resolution imaging of the south pole and of several high-latitude calderas. However, Emakong Patera was able to be imaged at medium-resolution. It was one of the largest examples of a caldera with recent lava flows that showed no 'hot spot' in the infrared, meaning that it was dormant at the time. The bright lava flow of low-viscosity fluid was probably rich in sulfur. The most amazing observations of this encounter were of a caldera in Tvashtar Catena. Because this is located at high northern latitude, Galileo had only an oblique line of sight, but the result was a dramatic view of an eruption projecting a line of 'fire fountains' to a height of 1.5 km; 100 times higher than a typical fountain on Earth. This 'fire curtain' was so hot and bright that the CCD was saturated, and electrons spilled over from a pixel to the adjacent ones to produce a 'bloom' of white pixels. It also saturated the infrared spectrometer, so the temperature could not be directly measured, but the hottest unsaturated pixel indicated at least 1,000K. Within hours NASA's Infrared Telescope Facility and the 10-meter Keck Telescope in Hawaii were observing this eruption and their observations, in combination with Galileo's data, suggested a lava temperature as hot as 1,800K. The hottest terrestrial lava is only 1,500K. Lava as hot as that on Io has not been erupted on Earth for billions of years. As a planetary interior melts, the amount of iron and magnesium in the melt increases with the extent of melting. Depending on the pressure, a lava at 1,800K would indicate up to 30 per cent melting, which suggested that the interior of Io is a partially molten 'mush' of crystals in magma.[291] The images of colorful Culann Patera came as an anticlimax! It was a 'hot spot' in all of the spacecraft's previous observations, and at times a faint plume had been seen. The infrared spectrometer resolved it into two separate centers of activity.[292,293,294]

During this periapsis passage, Galileo was also able to image Amalthea at the unprecedented resolution of 3.6 km. These images resolved not only a large crater,

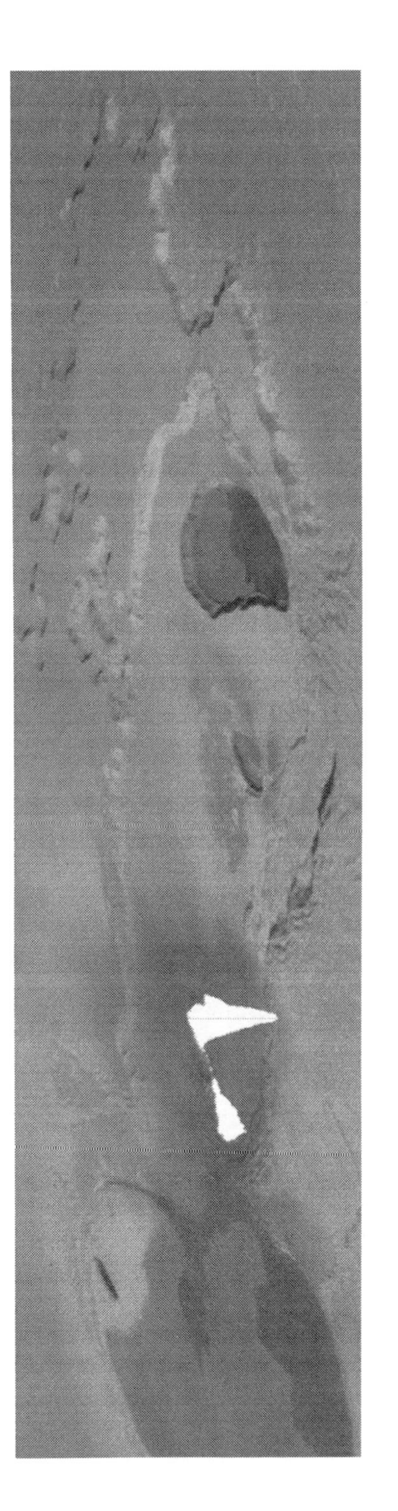

As Galileo made its second close pass by Io, it caught a 'fire curtain' in one of the calderas of the Tvashtar catena. The lava was so hot that it saturated the CCD, producing the 'bleeding' artifact.

but also features such as ridges. Although there were bright spots and streaks that may have been deposits of debris from nearby craters, these areas seemed to be the furthest from the center of mass, where the weaker gravity could be expected to have enabled the ejecta to escape most readily.

Although this orbit concluded the Galileo Europa Mission, it led directly into a second extension. The highlight of this Galileo Millennium Mission would be a joint study with Cassini, as this flew by heading for Saturn. Between October 2000 and March 2001, they would monitor how the gusty solar wind interacted with the planet's magnetosphere. Because when Cassini made its fly by on 30 December it would be 9.7 million km from Jupiter, it would remain outside of the bow shock. At that time Galileo would be at periapsis. A number of observatories, both on the ground and in space, would also study Jupiter at wavelengths ranging from radio to ultraviolet. In particular, the Hubble Space Telescope would study the Jovian auroras. Cassini would also image Jupiter and its satellites from long range on a routine basis, which was something that Galileo had been prevented from doing by the failure of its high-gain antenna to deploy. When not participating in this joint activity, Galileo would continue to investigate Io and Europa to assist in planning future missions, in particular collecting new data on the magnetic field of Europa which would hopefully resolve the question of whether there was an ocean beneath the icy crust. The spacecraft had survived the radiation remarkably well. Whilst it had always been expected that there would be an extended mission, the fact that a second extension was practicable was due primarily to the extreme accuracy of the navigation, which had minimized propulsive maneuvers to such an extent that the spacecraft had 37 kg of propellant remaining. The other constraint was power. Due to the decay of the plutonium fuel and the degradation of the thermocouples which transformed heat in electricity, the output of the RTGs declined by 7 W per year, but they were still delivering 450 W in 2000. Nevertheless, in order to save power it was decided to permanently switch off all systems that were no longer required, including the catalyst bead heaters of the main engine and the heaters of the failed ultraviolet spectrometer.

The 26th periapsis, the first of the Millennium Mission, was extremely limited in scope, in part due to the fact that the flight control team had been reduced, but also because there was little time available in which to download the data before the encounter in February. In fact, part of this time was to be spent replaying some of the data from the second Io encounter, including the images of Tvashtar.

On 3 January 2000 Galileo had its final encounter with Europa, passing 351 km above a position just to the southwest of the prominent impact crater Pwyll. This time the software designed to ignore resets worked perfectly, and no observations were lost; yielding high-resolution imaging with grazing illumination of possible icy volcanic flows, of the intersection between two ridges and of the ejecta around Callanish. But the primary purpose of the encounter was to further investigate the moon's magnetic field. The axis of the Jovian magnetic field is inclined relative to the plane in which Europa orbits, with the result that sometimes the moon is in the northern part of the magnetosphere and sometimes in the southern part. Since the field lines point away from the planet on the northern side of the magnetic equator and towards the planet in the south, the fact that this time the phase and orientation of the Jovian field would be opposite to that of the previous flybys would facilitate observations to determine whether the Europan field was intrinsic or induced. The magnetometer found that the dipole field had made an almost complete half turn to point away from Jupiter, as would be the case for an induced field. Although there are many conductive materials that could cause this response (including gold, iron, etc) the only plausible one in this case was the presence of a near-surface ocean of salty water.[295,296] In addition to providing further tracking data to probe the interior of the moon, this flyby also included a radio-occultation that placed constraints on the extent of its atmosphere.

Just before periapsis, Galileo took the highest resolution images yet obtained of Amalthea, Thebe and Metis. Pictures of Thebe taken on this and other passes, at a maximum resolution of 2 km per pixel, showed there to be a 40-km crater (Zethus) on the otherwise relatively smooth surface of the 60-km-wide moonlet. Metis was revealed to be a small elongated object, but even the best resolution was too poor to see any surface detail. Adrastea remained a barely resolved speck of light. As the spacecraft withdrew from Jupiter it flew by Io at a range of 212,000 km, taking new long-range images and infrared scans of Loki.

Galileo made its third close approach to Io on 22 February, passing just 198 km above its volcanic surface. It suffered two reset signals, but ignored them and was able to obtain its observations. While viewing the night-side, the photopolarimeter scanned Loki and the vicinity of Daedalus Patera, and the infrared spectrometer mapped the Daedalus, Loki, Pele and Mulungu lava flows at high resolution. By design, most of the targets were features and phenomena seen on the previous two encounters, to enable changes to be identified. After ground-based telescopes had shown Loki to be heating up, Galileo's photopolarimeter had been directed to scan its lava lake. In comparison to the previous October, it had dramatically changed. The southwestern part of the lake, which had been hot in October, had cooled, but the remainder of the lake was 30–40K warmer, as if the heat from the earlier 'hot spot' had been redistributed across the caldera by the spread of a lava flow.[297] The most astonishing imagery of this pass was of the floor and wall of Chaac Patera at a resolution of 10 meters, whose green deposits had been dubbed the 'golf course'. A comparison of the new images of Prometheus with those of the first encounter (134 days earlier) showed that an area of about 60 square kilometers was covered by fresh lava, with a similar area being covered by bright sulfur-rich deposits. This rate of coverage was about 10 times faster than that of Kilauea, which is one of the most active volcanoes on Earth. In fact, pictures and infrared measurements taken during this and previous encounters enabled scientists to develop a detailed model of the magma chamber, fissures, lava tubes, flows and deposits of Prometheus. Additional pictures were taken of Zal and Shamshu Paterae, Tohil Mons and the Amirani-Maui complex. A color image of Tvashtar was also taken. Although the lava curtain seen on the previous encounter was no longer active, and the site had cooled to 500–600K, there was now a high-temperature flow present at another position in the catena.[298,299,300]

As Jupiter was near solar conjunction, the reduced data rate limited the amount of data that could be returned. Only a small amount of particles and fields data was obtained inbound, but a near-occultation of Jupiter 'sounded' the uppermost levels of the atmosphere in the north polar region, and Doppler data was taken during the Io encounter to improve models of its gravity field and internal structure. The only camera activity apart from at Io, was a color image of some rarely seen longitudes of Europa. After this a bus reset interrupted the data collection, but since most of the planned observations had been made, it was decided to forego scans of Jupiter and the Io torus using the extreme-ultraviolet spectrometer on the spinning section (the ultraviolet instrument on the scan platform was now inoperable).

Galileo's next encounter, on 20 May, was the first of two Ganymede flybys that would set up the trajectory for the joint observations by Galileo and Cassini later in the year. The imaging objectives included: investigating the nature of dark terrains at high resolution; stereoscopic studies under grazing illumination of the smoothest bright terrains; studies of the transitional regions and the relationships between the various types of terrain; high-resolution studies of the smooth bright lanes and their relation to grooved and dark terrain; and moderate and high-resolution studies of a scalloped depression to determine whether it was icy volcanism – in the latter case the results showed a smooth interior and mottled texture which was consistent with icy volcanism, and the depression appeared to have collapsed over a melt reservoir and then undergone tectonism but there was no evidence of a flow front or similar structure.[301,302] On the 809-km flyby, the spacecraft penetrated the region in which the strength of the moon's intrinsic magnetic field matched that of the Jovian field, permitting a study of how the two magnetospheres interacted. The results revealed that the moon's field comprised two signatures, one being the known intrinsic field and the other apparently an induced field caused by a conductive layer only several kilometers thick located at a depth of 200–300 km. As in the cases of Callisto and Europa, this layer was most likely to be a subsurface ocean of salty water. One line of evidence to support this hypothesis was the fact that infrared spectra had shown the presence of a non-ice material on the surface of Ganymede that appeared to be some type of hydrated salt.[303]

The periapsis sequence included two occultations, one by Jupiter and the other by Ganymede. In addition to Ganymede, the photopolarimeter targeted Io, Europa and the upper atmosphere of the planet, but the infrared spectrometer investigated Ganymede and Jupiter only. In particular, the trajectory allowed a scan of the limb of Ganymede in order to characterize its tenuous atmosphere. The primary imaging targets on Jupiter were the Great Red Spot, which had been neglected for months, and the merged white ovals. In addition, the ring was imaged at an unprecedented oblique angle. No new observations of Io were made because the tape recorder still held a lot of high-resolution data from the previous encounter. However, as Galileo drew away from Jupiter it made the observations of the Io torus that had been lost on the previous orbit, with the extreme-ultraviolet spectrometer observing both the day-side and night-side ansae of the torus for approximately 17 days.

The Ganymede flyby raised Galileo's periapsis to between the orbits of Io and Europa in order to reduce its exposure to radiation. It also marked the start of the joint study of the Jovian magnetosphere with Cassini, during which Galileo was to

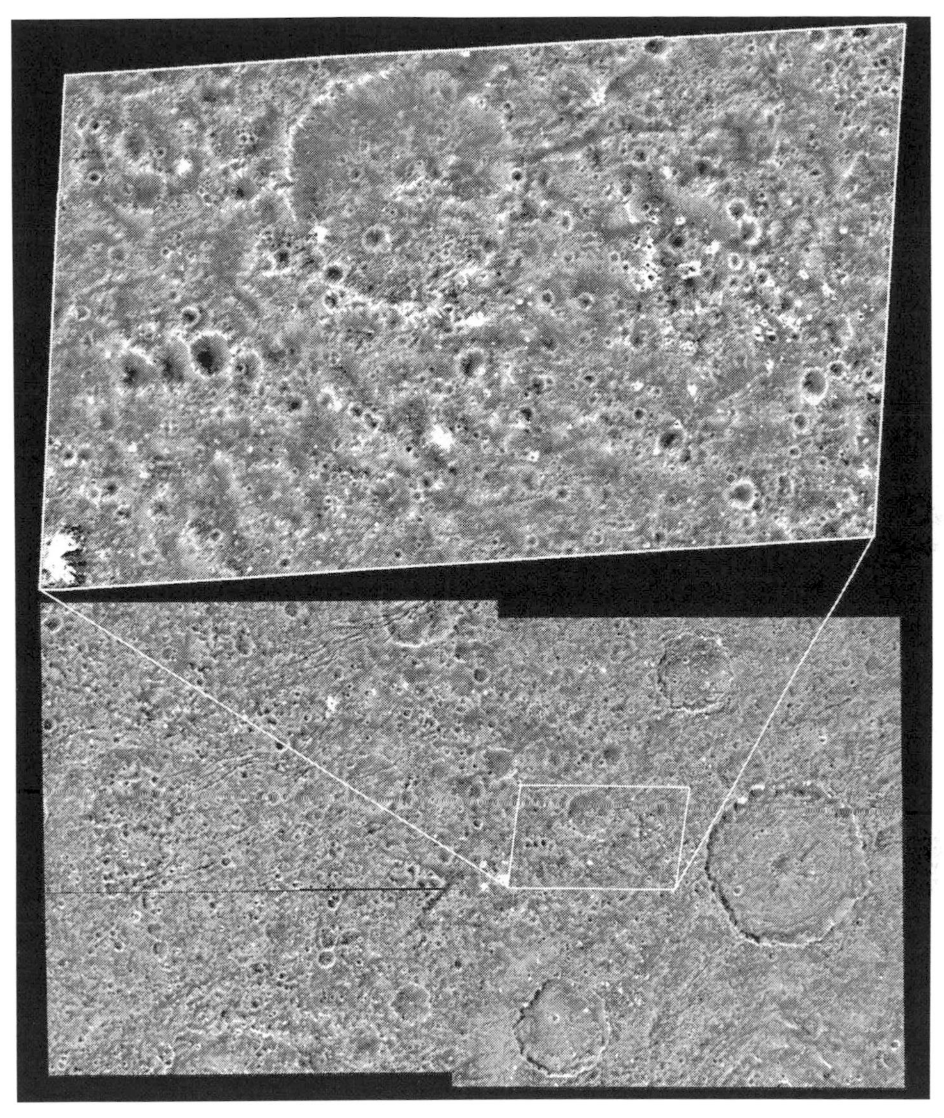

A high-resolution view of the dark terrain of Nicholson Regio of Ganymede taken by Galileo on its 28th orbit. (JPL/NASA/Caltech)

make just two orbits. The first orbit would last 7 months, with its apoapsis on the relatively unexplored 'dusk side' of the magnetosphere some 20.7 million km from the planet (290 Rj). In addition to collecting particles and fields data, the backlog of taped data, in particular the Io imagery, would be downloaded. However, owing to the recent renaissance of deep-space missions it was no longer possible to assign Galileo routine continuous coverage for real-time transmission of high-resolution particles and fields data.[304] In June the spacecraft made an unexpected contribution

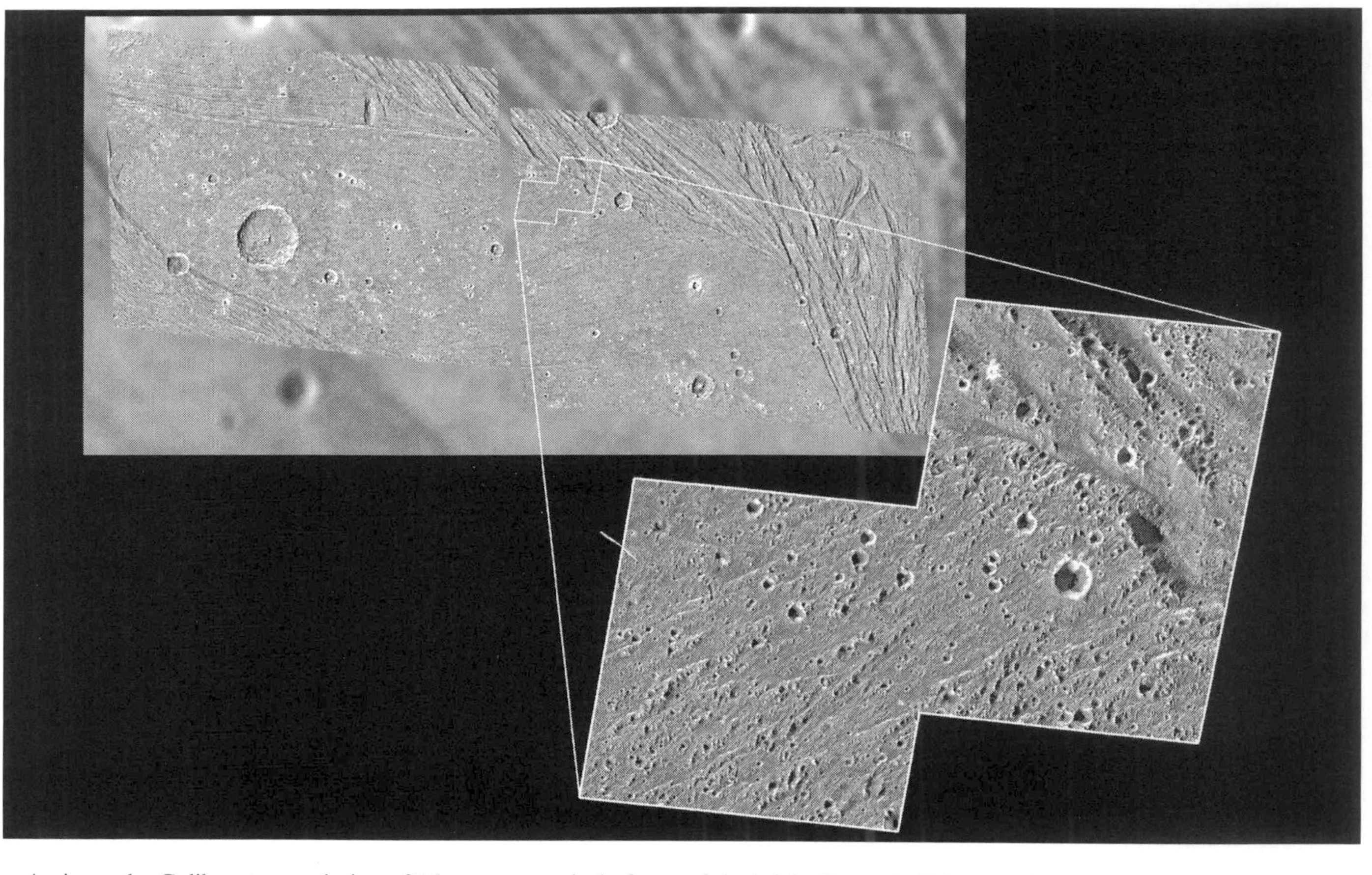

An image by Galileo at a resolution of 16 meters per pixel of part of the bright Harpagia Sulcus on Ganymede, which had appeared smooth in the lower resolution Voyager view. (JPL/NASA/Caltech)

to variable-star astronomy. The star sensor temporarily lost delta Velorum, which was one of the 50 brightest stars in the sky, then recovered it. A similar event had occurred shortly after Galileo began its cruise in 1989. Intrigued, a member of the control team queried an association of amateur astronomers that specialized in variable stars, who pointed out that delta Velorum is a double star system in which one star occults the other every 45 days. This behavior was confirmed by Galileo, which purposefully aimed its star sensor at the star just before a predicted eclipse in February 2001.[305] On 26 October 2000, when Galileo was some 200 Rj from the planet, it began a 100-day particles and fields survey of the magnetosphere. In fact, at that time the spacecraft was outside the magnetosphere. It took data as it passed through the bow shock and magnetopause, made its periapsis passage, and climbed back out into the solar wind. Meanwhile, Cassini monitored conditions in the solar wind upstream of Jupiter.

Galileo began the encounter sequence for its 29th periapsis on 28 December, with a radio-occultation that grazed the north pole of Jupiter and provided profiles of the density of electrons in this part of the ionosphere. Shortly after emerging from the occultation it passed by Ganymede at a range of 2,337 km, with the time of closest approach 15 minutes after the moon entered the planet's shadow for a 109-minute eclipse. The photopolarimeter collected thermal data of how the surface cooled on being plunged into darkness and warmed up as sunlight was restored, to provide insight into the thermal inertia and texture of the surface. Meanwhile, the camera attempted to image the aurora at the pole of the moon in eclipse. The schedule also included monitoring 'hot spots' and aurorae on Io and tracking features on Jupiter in concert with Cassini, but most of these observations were lost when radiation-induced anomalies resulted in the images being completely saturated – although a series of workarounds involving memory reloads and cycling the power on and off were partially successful. The Io observations were particularly affected, but some global images were salvaged for the 'volcano watch', and these complemented the continuous imaging by Cassini at lower resolution owing to the greater distance. A long-exposure of Prometheus was obtained which, although it suffered from pixel bleeding, clearly captured the plume. The great surprise occurred in global color imagery which revealed that Tvashtar had produced a Pele-style ring some 720 km in diameter. It was calculated that the plume which created it must have risen to an altitude of almost 400 km. This was the first evidence of a volcano at high latitude having produced a plume of any size.[306] The imagery of Jupiter was also lost, but that to investigate the out-of-plane structure and other characteristics of the ring system was unaffected. Meanwhile, the infrared spectrometer made observations of all the Galilean satellites, in addition to observations of the Jovian atmosphere in concert with Cassini.

The particles and fields observations by Cassini and Galileo clarified the effect of gusts in the solar wind on Jupiter's magnetosphere and auroras, as well as on the flux tube that links Io to the polar regions of the Jovian ionosphere. In particular, on 10 January 2001 the two spacecraft encountered the magnetospheric bow shock within a half hour of each other when a shock front between low and high pressure portions of the solar wind was passing Jupiter. This provided insight into how the planet's

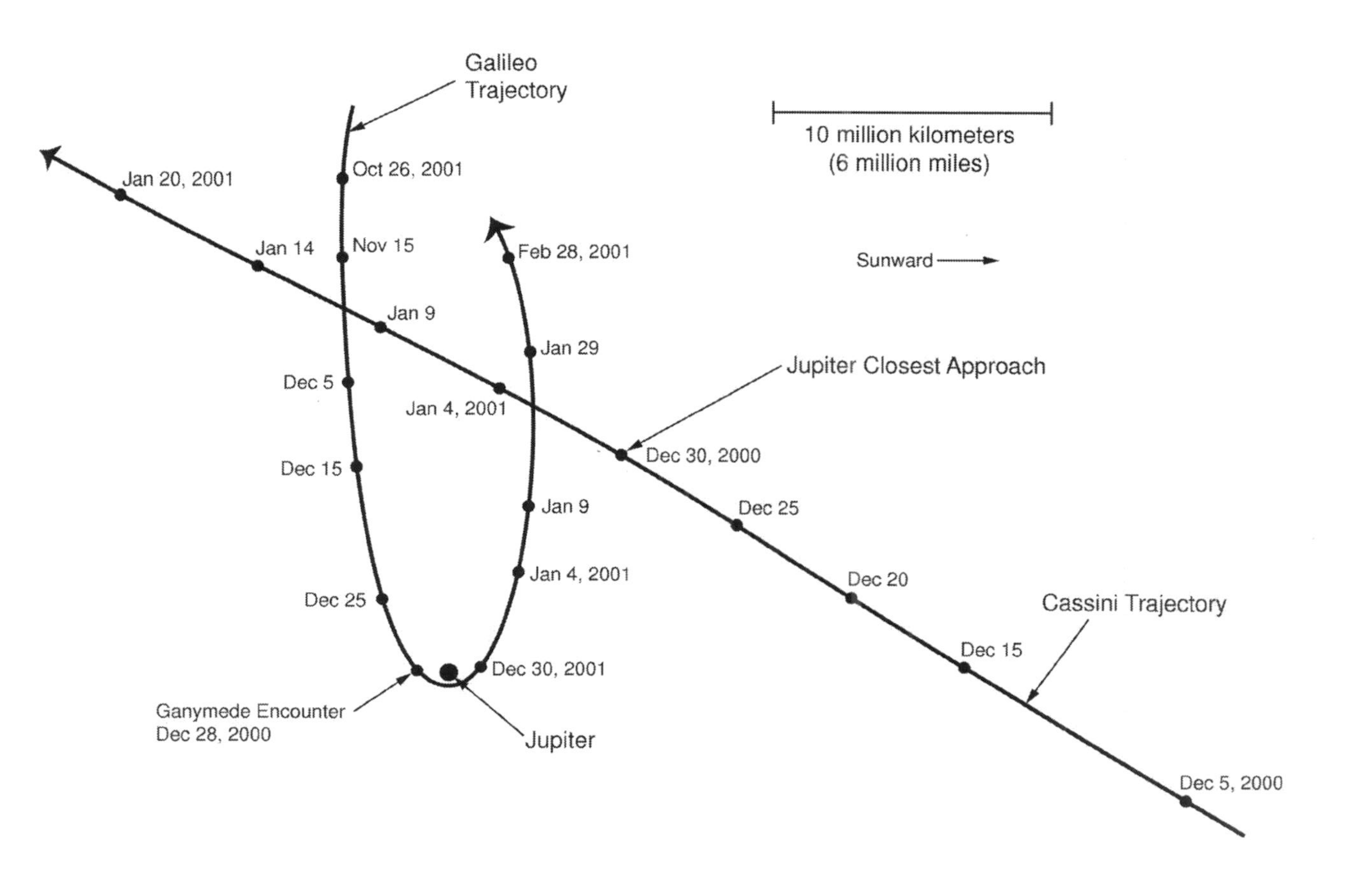

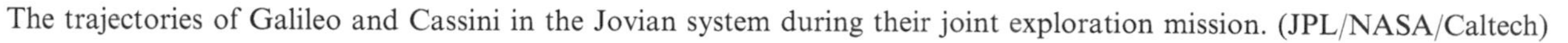
The trajectories of Galileo and Cassini in the Jovian system during their joint exploration mission. (JPL/NASA/Caltech)

magnetosphere reacts to such an event. On that occasion, the bow shock in the vicinity of Cassini was still facing the low-pressure wind and was inflated, but at Galileo's location the bow shock had been squashed by the high-pressure wind. There must have been a 'kink' somewhere in between, marking the passage of the shock front. Three days later, the Hubble Space Telescope saw a small but bright auroral oval on the planet, and there were bursts of radio noise, possibly caused by the passage of this shock wave in the solar wind.[307,308,309]

Meanwhile, plans were being drawn up to bring Galileo's mission to an end. In its final few orbits it would follow up some key questions concerning Io, in particular whether its magnetic field was intrinsic or induced in origin; a measurement of its heat budget to clarify the process by which tidal stress provides the heat that drives its volcanism; details of the eruption mechanism; and the origin of its mountains. However, the first encounter of this final phase of the mission was to be a flyby of Callisto to return the periapsis close the orbit of Io in preparation for a final three flybys of that moon: the first over the north polar region, the second over the south polar region and finally over the equatorial zone.

The sequence for the 30th periapsis started with a 2.5-hour radio-occultation by Jupiter that 'sounded' previously unexplored latitudes of the northern hemisphere. Although three software reloads and on/off cycles had been built into the camera's program in an effort to enable it to recover from glitches, most of the images, and in particular all the images of Io, Jupiter and Amalthea, proved to be saturated. The Io imagery had included 'hot spots' on the sub-Jovian hemisphere in eclipse, color images of the leading hemisphere, and ultraviolet-filtered frames of the plume over Tvashtar. Only a reload commanded from Earth was successful in salvaging one of two Ganymede observations and all of the Callisto images. The Callisto encounter on 25 May was unusual in that the approach was made by flying along its shadow cone, with the spacecraft spending nearly 1.5 hours in eclipse. In addition, a radio-occultation lasting about an hour 'sounded' its ionosphere. The flyby occurred at an altitude of 138 km. The infrared spectrometer mapped the surface composition, in particular targeting regions hitherto not well observed, as well as the crater Bran and the Asgard basin. As on the previous encounter, the photopolarimeter scanned along the dawn terminator to determine the thermal properties of the material that had just started to be warmed by the Sun. The camera took a number of very-high-resolution images of locations at the terminator, and a color sequence of the entire leading hemisphere – in fact, the first such sequence of the orbital tour. The terrain antipodal to the Valhalla basin was imaged at a resolution of 100 meters in order to search for hummocky features such as those that exist on Mercury antipodal to the Caloris basin.[310] Between 4 and 24 June, Galileo was once again out of contact as Jupiter passed through solar conjunction. In truth, this final flyby of Callisto had yielded comparatively little data, and most of it was downloaded in July.

With its propellant almost exhausted, Galileo would soon have to terminate its mission. Although scientists believed that any stowaway micro-organisms would have been killed by the Jovian radiation, it had been decided to ditch the spacecraft in the planet's atmosphere in order to rule out the possibility of the expired vehicle striking Europa and contaminating any ecosystem that this moon may possess. The

31st periapsis would provide the first of three close Io flybys which would not only provide opportunities to get new data on this moon, but also set up the trajectory to dive into Jupiter. It was calculated that by that time the spacecraft would have only 2.4 kg of hydrazine remaining, representing a mere 0.25 per cent of the original supply. The first Io encounter of this final phase of the mission was unusual from the operational point of view, in that Galileo was not able to be tracked for several hours near periapsis because the only Deep Space Network antenna with a line of sight, Madrid, was offline for upgrade work. Meanwhile, engineers had figured out why the camera had been creating saturated images. The radiation was causing an operational amplifier to saturate in certain conditions, such as when the flood lamp which evenly illuminated the CCD chip to erase the previous image was activated. In an effort to salvage as much of the Io science as possible in the event of continuing problems, it was decided to add long-range imaging to the sequence to characterize large-scale surface changes. (In fact, this precaution would pay off, as these were the only usable Io images from this encounter!) Observations by all three surviving remote-sensing instruments targeted Jupiter, in particular the 'white oval' and both polar regions, for these were of particular interest in the study of the dynamics of the atmosphere.

Approaching Io's night-side on 6 August, the infrared spectrometer inspected several known 'hot spots'. The closest approach occurred at an altitude of 194 km at 78°N, just as Galileo crossed the terminator. As a result, the departure view was of the fully illuminated disk. The camera's first task was to take long-range images to determine whether Tvashtar had a plume. It saw nothing, but perhaps if the plume existed it was not rich in gas but a 'stealthy' all-vapor plume. However, the camera did detect a plume over a previously unrecognized high-latitude volcano that was named Thor. Rising to an altitude of 500 km, this was the tallest plume yet seen on Io. The spacecraft flew directly over Tvashtar, and would have passed through the fringe of its plume if this were present. Although the plasma experiments detected a tenuous gas, this probably represented the Thor plume. All of the images made at closest approach were lost, including close ups of Masubi and the Amirani-Maui complex. Also targeted by the infrared spectrometer during the approach was a 'hot spot' in the southern hemisphere which had been discovered on the previous orbit. Next, low-resolution global images were taken of the illuminated disk. These were unaffected by the camera problem, and revealed the presence of new large deposits of plume fall-out around Dazhbog and Surt. This increased to four the number of giant plumes, or their deposits, known at high latitude. This periapsis passage also yielded images of the sub-Jovian hemisphere of Callisto and of Amalthea – in the latter case to refine the ephemeris of the irregular moonlet in preparation for the 34th and penultimate periapsis, when it was to be closely inspected.[311,312]

In view of the risk of Galileo succumbing to the radiation and precluding the plan to prevent the spacecraft ever striking Europa, the two remaining Io flybys were carefully orchestrated.

On the 32nd periapsis, the flyby of Io occurred on 16 October, with the closest point of approach at about 79°S at an altitude of 184 km. Prior to the encounter, the photopolarimeter observed Jupiter's northern polar region and the 'white oval' in

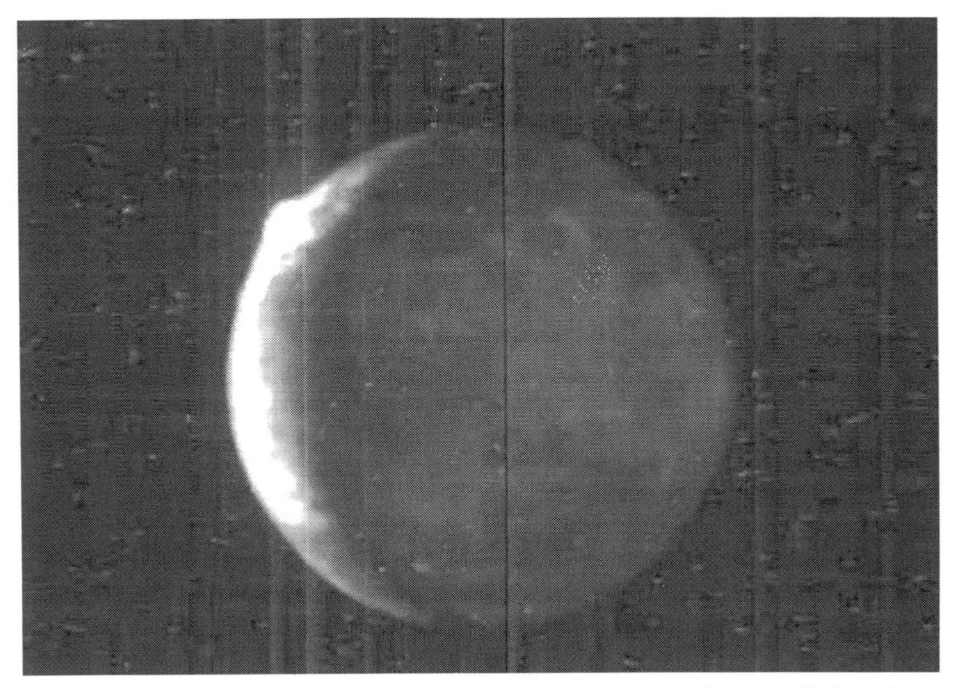

An image of Io taken on 4 August 2001 revealing the 500-km-tall plume of Thor, the highest plume yet observed.

the southern hemisphere. The Io sequence began at long range, with the moon in eclipse. The photopolarimeter took several hours of high-resolution radiometric temperature data of northern-latitude 'hot spots'. A radio-occultation 'sounded' the planet's ionosphere and atmosphere from an altitude of 25,000 km above the 1,000 hPa level. Meanwhile, the infrared spectrometer scanned Pele and Loki and a number of southern-latitude 'hot spots', and mapped the distribution of sulfur dioxide in the south polar regions.

When Io emerged from Jupiter's shadow, the camera started work. Its software had improved self-diagnosis and repair functions, and the commanding cycles had been revised to ensure that as many as possible of the images were taken. Its main objective was to recover observations which had been lost on earlier passes. High-resolution images of Loki near the terminator showed recent lava deposits which looked flat and glassy, and due to reflected light appeared bright rather than dark. The patera's rim was measured to be no more than 100 meters high. There were peaks nearby that rose 1,000 meters. A high-resolution temperature map of Loki obtained by the infrared instrument showed areas of the lava lake that were in the process of cooling. This lava must have been extruded within the last 80 days, and spread at a rate of about 1 km/day. Models of how this lava cooled suggested it was only about 1 meter thick.[313] Next, a sequence of 5 night-time images of Pele showed an area of lava at a temperature of about 1,400K, which indicated it had a silicate composition. A cliff in Telegonus Mensae showed down-slope movement. High-resolution views of a lava channel near the rim of Emakong Patera showed islands of bright material, encrusted areas, etc. Simultaneous infrared data showed that the floor of the patera was warm, indicating recent activity, but imaging on previous orbits had not shown any changes. Tohil Mons was imaged near the terminator and measured at over 9 km tall. A linear ridge near its peak might have been a caldera, but there were no lava flows emanating from it. Nearby Radegast Patera might be recycling debris which fell into it from Tohil. It seemed sufficiently warm to have erupted recently. Color mosaics of Tupan Patera showed red sulfur-rich areas and dark areas that were probably silicate. Greenish areas probably marked the sites of chemical reactions between the two types of eruption. Seen in close-up, Tvashtar appeared to be quiet. It was marked by radial streaks from the recent eruption. The dark flows issued by the eruption seen during the 27th orbit appeared to have faded as they cooled. Mosaics of the patera and peaks of Gish Bar were obtained with the illumination almost overhead. The new dark lava flows that ran for 30 km across part of the patera might have been erupted in October 1999, during an outburst of activity detected from Earth. The final observations included two swaths across the terminator that revealed topographic details in the Mycenae and Colchis regions and at the sites of new plumes. These included Tsûi Goab Tholus which, although it is an unspectacular mound, appears to be a rare case of an Ionian shield volcano (lava flows on Io rarely build edifices). Other frames captured fractures and deposits in the Thor area. Two global low-resolution color mosaics ended the sequence.[314,315]

The magnetometer data obtained during this and the previous encounter showed that, contrary to indications gained at orbit-insertion, Io does not have an internally generated intrinsic magnetic field. But tracking data confirmed it to have a metallic core. Remarkably, a new model of Io's magma composition and interior had just been proposed in which convection in the metallic core was suppressed, resulting in no magnetic field.[316] This left the stationary plasma near Io as the cause of the observed field. Finally, new data was obtained on interactions between the moon, its torus and its flux tube.

This periapsis also provided a single image of Amalthea and a final look at the gossamer ring, and the particles and fields instruments collected a large amount of data spanning, depending on the experiment, between 12 and 15 consecutive days. As the spacecraft withdrew from Jupiter, a test revealed that the photopolarimeter, which had suffered several problems throughout the tour, had mysteriously healed itself! As a result, on the very last encounter, the instrument would be able to apply its full radiometric capabilities. On the other hand, during a conditioning sequence on 13 November 2001 the tape recorder showed the first indication of a renewal of an old problem – although it was possible that because the point at which the tape became stuck was a 'stop' position defined after this problem developed in 1995 it had simply become worn as a result of the heads routinely parking there.

The final flyby of Io occurred on 17 January 2002. The sequence began with a radio-occultation lasting 2 hours 42 minutes which 'sounded' the northern latitudes of the Jovian atmosphere. The spacecraft approached the night-side of Io, crossed the terminator and departed over the day-side. The particles and fields instruments collected data throughout, in particular making a detailed survey of the structure of

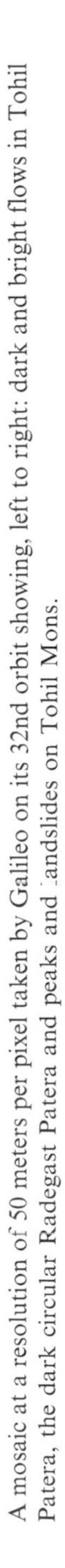

A mosaic at a resolution of 50 meters per pixel taken by Galileo on its 32nd orbit showing, left to right: dark and bright flows in Tohil Patera, the dark circular Radegast Patera and peaks and landslides on Tohil Mons.

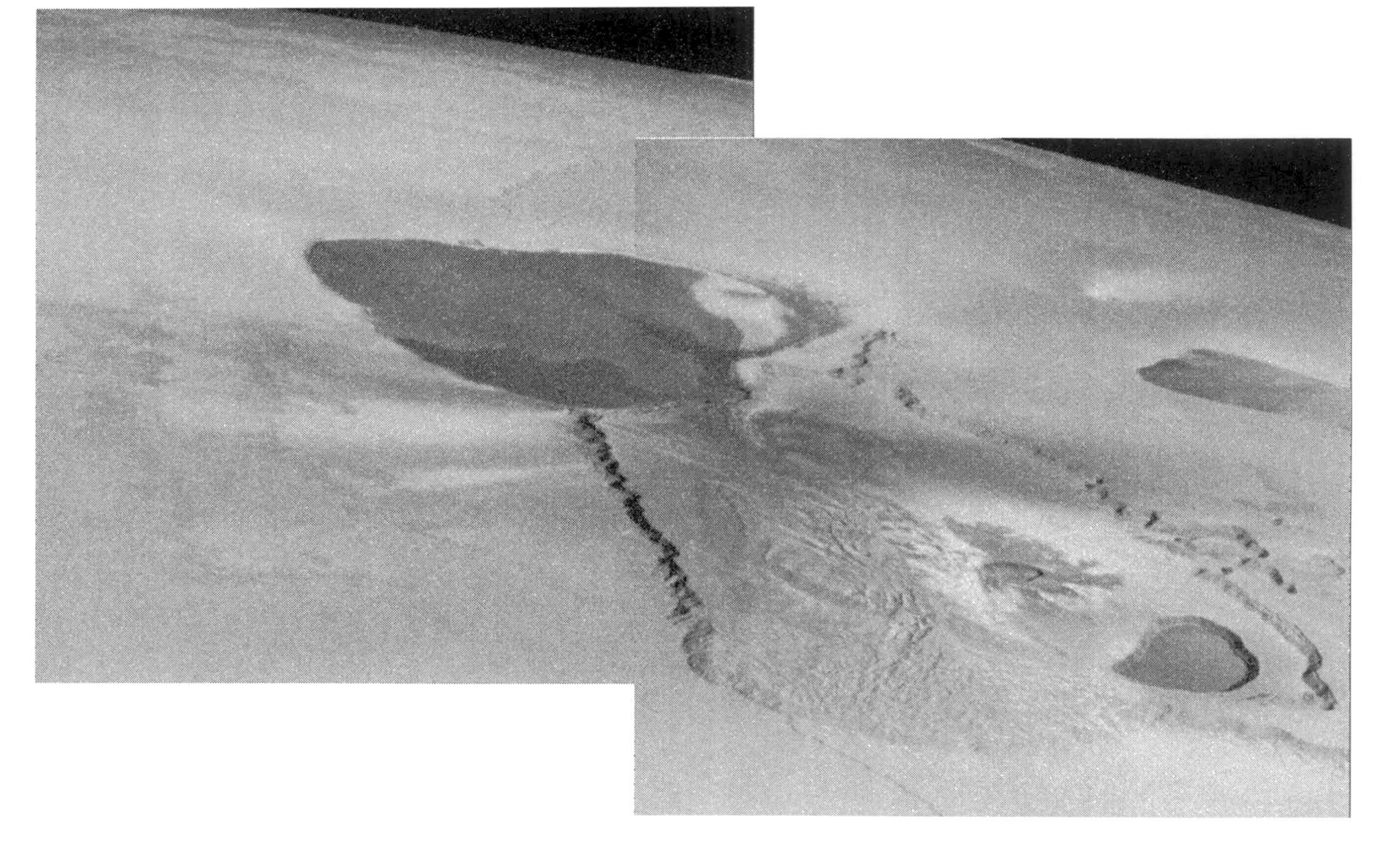

The volcanic Tvashtar catena imaged at a resolution of 200 meters per pixel by Galileo on its 32nd orbit, showing how it had changed since the earlier flybys.

the Io torus. The remote-sensing observations started with a thermal scan of the 'hot spots' of Prometheus and Marduk. About 28 minutes before closest approach, a bus reset occurred of a type which the modified software was not programmed to work around, and Galileo safed itself. This was unfortunate, as this encounter had offered the first and only opportunity to image the sub-Jovian hemisphere of Io at high resolution, and color imagery of an area at 1-km-resolution that had not been viewed at better than 10 km per pixel by any spacecraft.[317] As Galileo passed over the intermediate southern latitudes at an altitude of 101.5 km, it was inert. In fact, this was the closest approach that it had made to any of the Jovian satellites on its entire tour. The reason for this extremely close approach was to gain the slingshot required to raise the apoapsis to 350 Rj, which was sufficiently far from Jupiter for the spacecraft's trajectory to be perturbed by the Sun in a manner that would result in it falling into Jupiter at a chosen future data. Doppler tracking was unaffected by the safing, and the data contributed significantly to knowledge of Io's gravity field. The lost remote-sensing included infrared scans and visual imaging of mountainous and volcanic features of Io, images of Thebe, and polarization studies of both Jupiter and Europa. However, engineers were able to restore Galileo in time to obtain two new pictures of Amalthea in order to further refine its ephemeris. The camera then took a color sequence of the Jupiter-facing hemisphere of Europa. In fact, this was the very last science data that the camera would provide. The infrared spectrometer made its final global mosaic of Jupiter, which was a multispectral map to characterize the morphology of the clouds.

As Galileo withdrew from Jupiter, the extreme-ultraviolet experiment initiated a Lyman-alpha survey of the sky that ran through almost the whole of the next orbit. A bus reset occurred in February at 180 Rj, which was the farthest from the planet that this had happened, and since the software to deal with such an event was not used during the data playback this safed the vehicle. Meanwhile, the cause of the 'unexpected' reset at the recent periapsis had been identified, and a new software patch designed for the final orbits. In March, the camera and infrared spectrometer were calibrated to wrap up Galileo's remote-sensing operations. By now the tape in the data recorder had become stuck on the 'dummy' erase head at the start of the tape and on one or both of the two playback heads. After 2 months of diagnosis, it was decided to order the tape to advance at its fastest speed, and this successfully released it. Over solar conjunction in July the seventh coronal sounding in eight conjunctions was made. Only that of 1998 had been missed, for technical reasons. Three of the conjunctions were made at solar maximum, three at minimum, and the remaining one in the ascending phase.[318] Meanwhile, the engineers devised measures to keep the recorder operating for the Amalthea flyby – the first and only close encounter with a non-Galilean satellite. It was fortunate that the renewal of the tape issue had not occurred earlier. Despite media calls for NASA to provide the $1.15 million to exploit the unique opportunity to observe Amalthea up close, no remote-sensing was scheduled for the 34th periapsis because it was unlikely to yield useful results.[319] (In fact, the trajectory also offered the best geometry for imaging Io's sub-Jovian hemisphere, including Pillan, Pele and Loki – the latter at the unprecedented resolution of 500 meters per pixel.) However, the particles and fields instruments would make a survey of the inner radiation belts in a region which, owing to the presence of ring dust, the mixture of plasma and dust approximated conditions in a protoplanetary disk around a young star. The dust detector would make in-situ observations of the gossamer ring. Because Amalthea's shape had been determined from Voyager and Galileo images its volume was known, and once the tracking of Galileo during its flyby provided an estimate of the moon's mass it would be possible to calculate its density and gain insight into its origin. Attitude control during the encounter was a particularly difficult issue. Despite the fact that no remote-sensing was to be done, precise information about the axis and period of the spacecraft's spin was required to interpret other data. However, the gyroscopes could not be reliably used, and the fact that the vehicle would spend an hour in Jupiter's shadow near periapsis meant that a Sun sensor could not be used. And, of course, in so close to Jupiter the star sensor would suffer so much interference from radiation as to lose its reference. It was decided to have the sensor monitor Vega, which is one of the brightest stars in the sky, schedule software restarts and repeating data sequences, and accept that some small drift in the vehicle's attitude was inevitable.[320]

In late October 2002 Galileo began to report real-time particles and fields data. It was in the solar wind, approaching the bow shock. Once in the magnetosphere, it made six crossings of the current sheet that marked the boundary between the two polarities of the planetary magnetic field. So extensive was the survey that three of the tape recorder's four tracks were used to store data. The flyby of Amalthea on 5 November occurred at a distance of 244 km and a relative speed of 18.4 km/s. The radiation was so intense that the two-way radio link was unable to be maintained, and only limited Doppler data could be gained from the backup one-way carrier. Nevertheless, after a lengthy analysis Amalthea's density was estimated to be less than 1.0, which suggested that it was a porous mix of rock and water ice, probably formed in the outer solar system and later captured by Jupiter.[321] Sixteen minutes after Amalthea, Galileo (which had by this time endured in excess of four times its designed total radiation dosage) started to operate erratically and, recognizing this, safed itself, terminating its data gathering at about 2.3 Rj inbound to periapsis. The problem was a bus reset, but instead of originating in the despun section, as on all the previous cases, this time it was in the spun section, even though this was better protected from radiation. The Jovian radio-occultation started 6 minutes later, but no usable scientific data was obtained. Some 65 minutes after the Amalthea flyby, Galileo reached periapsis, 71,500 km (2.0 Rj) above the Jovian cloud tops. Only Pioneer 11 and Galileo's atmospheric probe had ventured closer to the planet. By the nature of the programming, the star sensor operated even after the safing event had prompted the deactivation of the scientific instruments, and intriguingly it saw a number of flashes at around the times of closest approach to both Amalthea and Jupiter. Assuming that these flashes were not simply radiation interfering with the sensor, then the data suggests that there were a number of moonlets about 5 km in size within 5,000 km of the spacecraft at those times.[322] The sensor readings also provided an indirect measurement of particle fluxes. After spending about 9 hours inside the orbit of Io, Galileo set off for its final apoapsis. As the recorded data was being checked 3 days later, the tape again showed signs of sticking. But this time there were also indications that the motor had stalled. This could have been caused by radiation damaging the hardware or fogging the optical sensors that controlled the motion of the tape. Extensive tests failed to identify the actual cause, but it proved possible to use the recorder reliably in sessions lasting no longer than 20 minutes, and the Amalthea data was able to be replayed.

On 14 January 2003, its mission almost finished, Galileo adopted the attitude in which it was to enter the Jovian atmosphere. In early March the science sequences that were to transmit real-time data during this terminal dive were installed. For the next 6.5 months the spacecraft would be contacted only once per week, and only to verify its status. On 14 April it reached the apoapsis of its extremely elliptical orbit at 26.4 million km (370 Rj). In August it passed through its final solar conjunction. Meanwhile, it fell towards Jupiter in an orbit which had a notional periapsis inside the atmosphere, 9,700 km below the 1,000 hPa level. The science activities began at 14 Rj, some 19 hours prior to impact, with an emergency sequence ready to take over in the event of trouble. With about 4 hours to go, Galileo crossed the orbit of Io. About an hour later, the increasing radiation caused the star sensor to lose its target. An hour later still, at 3 Rj, the magnetometer gave its final reading before it became saturated. Galileo then crossed the orbits of the inner moonlets, where the star sensor had been specifically programmed to detect flashes in order to further study the possible presence of small bodies. At 18:50:54 UTC on 21 September, the spacecraft, now just 9,283 km (0.13 Rj) above the cloud tops, crossed the limb and 7 minutes later, out of sight from Earth and in darkness, emulated its probe by diving into the atmosphere at 48.2 km/s. In this case the entry site was 0.2 degrees south of the equator at a longitude of 191.6 degrees.[323,324] Galileo had been in orbit of Jupiter for 3 months short of 8 years, and had accompanied the planet for three-quarters of a circuit of the Sun.

BEYOND THE PILLARS OF HERCULES

The plane of the Earth's orbit around the Sun defines the plane of the ecliptic, and the fact that the rotational axis of the Sun is inclined about 7 degrees to the ecliptic meant that the Pioneer and Helios missions that were put into heliocentric orbits in the inner solar system coincident with the ecliptic were restricted to sampling the solar wind streaming from the near-equatorial solar latitudes. Scientists wondered whether this region was representative of the star as a whole. In fact, observations of the corona as revealed at the time of a solar eclipse, and of the tails of comets immersed in the solar wind, had shown that the character of the solar wind varied with solar latitude. At a latitude of 40 degrees, the solar wind appeared to be much calmer, and probably faster. Also, sunspots seemed to be restricted to a relatively narrow belt centered on the equator. And the solar magnetic field seemed to be divided in hemispheres of opposite polarity, with the field lines directed towards the Sun in one hemisphere and directed away from it in the other. In the equatorial plane they were isolated by an essentially planar 'current sheet'. Spacecraft which orbited in the ecliptic passed through this current sheet many times, moving from one polarity region to the other. At the Earth's distance from the Sun, this occurred typically every 7 days. The region of constant polarity at higher latitudes was only marginally sampled by Pioneer 11, whose trajectory between Jupiter and Saturn was inclined at almost 20 degrees to the ecliptic. Knowledge of the structure of the bubble formed in the interstellar medium by the streaming solar wind – that is, the heliosphere – would also be useful for particle astrophysics, because the solar wind moderates the extent to which galactic cosmic rays can reach Earth.[325] The idea for an Out-Of-Ecliptic (OOE) mission to gain a new vantage point was first suggested in 1959 by a round-table discussion by American scientists. This would have been a difficult and costly mission for the technology of that time. Indeed, it remains so today!

For example, to insert a spacecraft into a circular 1-AU orbit perpendicular to the ecliptic it would be necessary to cancel the velocity in the ecliptic plane that it inherited from the Earth, which is about 30 km/s, and then it would be necessary to reinstate this velocity in a direction perpendicular to the ecliptic. Summing the two effects, this is a total velocity change of 42 km/s directed against the motion of the Earth and at 45 degrees to the ecliptic. This maneuver is far beyond the capability of even the most powerful of launch vehicles. Even if the mighty Saturn V were to have been fitted with an additional Centaur stage, the steepest out-of-ecliptic angle for a mass of 580 kg at 1 AU that it could have achieved is 35 degrees, which is a velocity change of about 18 km/s. In the 1960s American and European experts in celestial mechanics discovered a cheaper alternative, but it has the disadvantage of producing an orbit far larger than 1 AU. It uses the gravitational slingshot of a Jovian flyby to alter the plane of the spacecraft's orbit. By carefully arranging the approach to the giant planet, it is possible to choose from a wide variety of out-of-ecliptic inclinations and perihelion distances, although the aphelion will remain in the vicinity of Jupiter's orbit.[326,327]

In the late 1960s and early 1970s NASA studied out-of-ecliptic missions, in all cases envisaging using Jupiter to shape the orbit. The first case would have tested the TOPS (Thermoelectric Outer Planet Spacecraft) in preparation for the Grand Tour by such vehicles. Another case was as a possible target for the spare Pioneer Jupiter spacecraft. Meanwhile, in his novel *2001: A Space Odyssey*, Arthur C. Clarke referred to "High Inclination Probe 21, climbing slowly above the plane of the ecliptic". In the 1960s European scientists proposed putting a spacecraft into a 13-degree orbit that would venture as much as 35 million km above and below of the ecliptic. In 1968 a joint US/UK mission was proposed by the British National Committee on Space Research that would start off in an orbit near the Earth and then use an ion engine to slowly increase its inclination relative to the ecliptic. When this attracted no support in either nation, in January 1971 it was submitted to the European Space Research Organisation, which studied the possible launch vehicle, orbit design and payload. The preliminary mission design was for an ion-drive spacecraft with a 'dry' mass of 336 kg, including a scientific payload of 20 kg. It would be launched by a Europa III (which was not built, but went on to form the basis of the Ariane 1), Atlas–Centaur or Titan III. Once in heliocentric orbit, it would slowly increase its inclination with the objective of reaching a solar latitude of almost 50 degrees by the conclusion of its 3-year mission.[328] By the end of 1973, an out-of-ecliptic mission was listed as one of the priorities for the 1980s. It became one of several possible cooperative ventures with NASA, with ESRO preferring the ion-drive option and NASA preferring a Jovian slingshot. At the end of 1974 NASA unilaterally decided the issue by ordering JPL to cease work on ion propulsion for this mission. W.I. Axford of ESRO proposed that if two spacecraft were built, one by each agency, then it would be possible to perform a stereoscopic study of the Sun. Meanwhile, ESRO was incorporated into ESA, and in 1977 this agency formally approved the two-spacecraft mission. When submitted to the US Congress in 1978 it was as the International Solar Polar Mission (ISPM); the name change reputedly having been made by NASA Administrator Robert A. Frosch to avoid having to explain the meaning of the term ecliptic to the legislators.[329]

The ISPM would consist of two similar spacecraft, one built in the US by JPL and the other in Europe. Each would weigh 330–450 kg, be spin stabilized, hold its antenna axis pointing at Earth, and be powered by RTGs. They would carry similar payloads, but the American spacecraft would also have a despun platform carrying a visible-light coronagraph and a telescope for the X-ray/ultraviolet range. From a polar vantage point, the optical instruments could monitor the evolution of features from the moment that they developed until they expired. This was not possible from an ecliptic orbit because after becoming visible on one limb a feature remained observable for at most 13 days before it crossed the far limb. The spacecraft were to be stacked in tandem on an IUS, and carried into Earth orbit by a Shuttle during a 14-day launch window in February 1983. Once Jupiter-bound, the spacecraft were to separate and adopt trajectories such that one would fly over the planet's north pole and the other over its south pole. On the way, they would make simultaneous measurements of the interplanetary environment. The Jovian slingshots in May 1984 would deflect one vehicle north of the ecliptic and the other south of it, on trajectories that would pass over the polar regions of the Sun at a distance of about 2 AU. After reaching perihelion on crossing the ecliptic on the far side of the solar system, each would make a similar pass over the opposite pole and head back out towards the orbit of Jupiter, although by the time they arrived the planet would not be in the vicinity. The mission would nominally end in September 1987.[330,331]

But in 1980 the mission suffered the first of a string of setbacks, when NASA announced the delay of some science programs in an effort to relieve the financial pressure of developing the Shuttle. The launch of the ISPM was slipped from 1983 to 1985. However, ESA was already well advanced in building its spacecraft and, after reviewing its options, decided to proceed as planned and place the spacecraft in storage. In retrospect, this decision by ESA was a blessing in disguise, because as events transpired the spacecraft spent two long periods in storage – and the first provided valuable experience for the (unscheduled and longer) second. In 1981, in response to the dramatic reduction in funding for space science demanded by the incoming administration of Ronald Reagan, NASA decided to cancel its part of the ISPM mission. Although the European spacecraft would be able to visit both poles a year apart, the fact that simultaneous observations would no longer be possible greatly reduced scientific interest in the mission. Moreover, this decision was a public relations disaster for the mission, since there would be no spectacular imagery for press releases. In fact, years later, the mission managers would regret that the European spacecraft did not have a camera.[332] Although a study was made into ESA building both spacecraft using subsystems and instruments provided by NASA, this was not seriously considered. Fortunately, by transferring some of the instruments earmarked for the canceled spacecraft to the surviving one, it would be possible to achieve most of the scientific objectives.[333] In the new plan ESA would provide the spacecraft and half the payload, and oversee operations, while NASA would provide the remainder of the payload, an RTG power unit, a Shuttle launch, the Deep Space Network and flight operations by JPL. Shortly before canceling its ISPM spacecraft, NASA had terminated the development of the IUS and switched the mission to a cryogenic Centaur G-prime with an additional solid-fuel escape stage, but the latter would not be needed to dispatch a single spacecraft. As a result of the switch to the Centaur, the launch was slipped another year, to 1986. Unfortunately, a Jovian slingshot in late 1987 was not particularly favorable, as the planet would be just past perihelion and the spacecraft's relative velocity would be at maximum, which would slightly reduce the maximum attainable solar latitude. The spacecraft was built by a consortium of European industries led by Germany's Dornier, tested and accepted by ESA in 1983, then partially disassembled and stored in a nitrogen-filled transport container. The instruments were returned to the science teams for safekeeping and periodic calibration checks.[334,335,336,337,338]

The scientific objectives covered disciplines from planetary to solar science, to astrophysics. The properties of the interplanetary medium were expected to change at high solar latitudes. The character of the solar wind appeared to be related to the large X-ray-dark areas called 'coronal holes' that are centered on the poles at the minimum of the 11-year cycle. And whereas the rotation of the Sun causes both the solar wind and the associated magnetic field to develop a complex spiral in the ecliptic, it could be expected to become smoother (and probably oriented radially) as the latitude increased. The mission was expected to build upon the stereoscopic radio-burst observations begun with the French STEREO experiment on the Soviet Mars 3 spacecraft. It was also to contribute to stellar astrophysics by measuring the cosmic rays which penetrate the solar system in the magnetically simple regions in the polar regions of the heliosphere, and by carrying gamma-ray burst and gravity wave detectors to augment the interplanetary network.[339]

The spacecraft comprised a simple aluminum honeycomb box, 3.2 meters long and 3.3 meters wide, which held a platform for some of the electronics and a tank for 33 kg of hydrazine. The RTG power unit was installed on a truss mounted on one side of the box. This truss was designed not only to withstand the loads of the launch environment, but also an emergency landing after an aborted launch. On the opposite side (as far as possible from the RTG) were the electronic systems, a pair of redundant tape recorders and most of the scientific instruments, together with a 5.56-meter-long radial boom (on a double hinge, initially in a folded configuration alongside) to carry some of the instruments. On each of the two remaining sides was the deployment mechanism for a radial wire boom that would be deployed by centrifugal force to give the spacecraft a total span of 72 meters. On top, beams

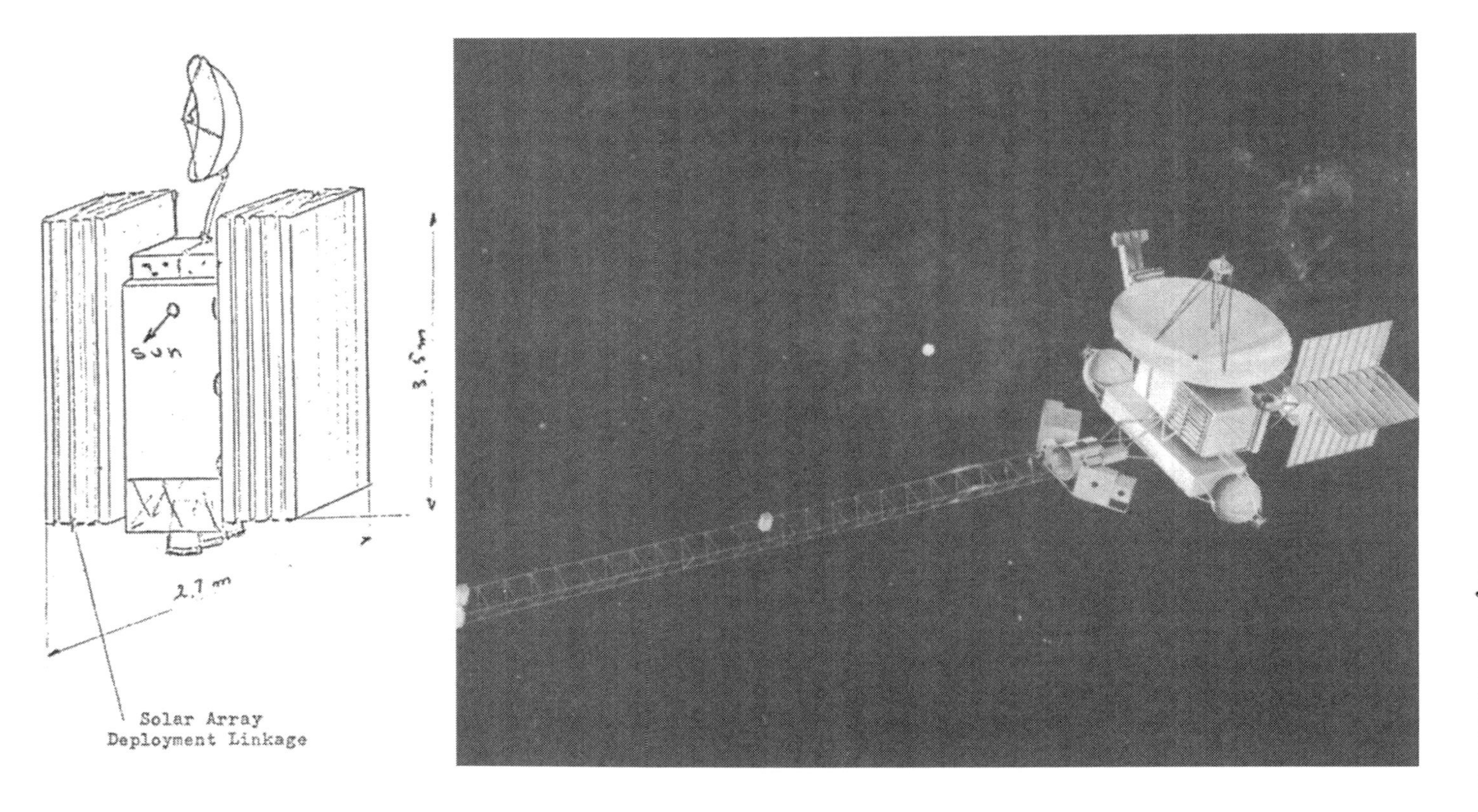

A sketch of the original ESRO ion-propelled spacecraft for the international out-of-ecliptic mission that was to reach high solar latitude starting from a near-Earth heliocentric orbit (left), and the NASA contribution (right) that was canceled by the drastic space science cuts of the Reagan administration in 1981. (ESA and JPL/NASA/Caltech)

The definitive ESA Ulysses out-of-ecliptic spacecraft, atop the controversial Centaur G-prime stage developed for the Shuttle.

mounted on the main instrument platform carried a 1.65-meter-diameter high-gain antenna capable of operating in both the X-Band and the S-Band, and, mounted on the latter's feed, a low-gain antenna for use when the spacecraft was in the vicinity of the Earth. On the underside of the box there was a second low-gain antenna on a short boom, a 7.5-meter-long monopole antenna and thermal regulation radiators. Because the conditions in the cargo bay of the Shuttle at launch were ill-defined when the structural design of the spacecraft started in 1979, it was decided to build in sufficient margin to be readily adapted to the Shuttle–Centaur, the Shuttle/IUS, and even the Titan family. Like most missions dedicated to particles and fields studies, it was to spin at 5 rpm for stability. Attitude determination would use Sun sensors, and the spin axis would be maintained by slightly offsetting the axis of the high-gain antenna from the spin axis and utilizing a system akin to the CONSCAN of the outer solar system Pioneers. Eight 2-N thrusters mounted on a pair of pods on the sides of the spacecraft would control the attitude and make small course corrections. (As there was no 'main engine', there would be little scope for making a major correction to the trajectory inherited from the Earth-escape stage, so every effort would have to be made to ensure that this was accurate.) The American-supplied RTG was of the same type as on the Galileo mission. It contained 10.75 kg of plutonium oxide, and produced about 285 W at the start of the mission, reducing to 250 W at its nominal end. The scientific instruments accounted for 55 kg of the 370-kg launch mass. To ESA, the total cost of the hardware was $57 million.[340,341]

Of the nine scientific instruments, four were US-built, three were from Europe and two were international projects. Two packages measured the density, speed and direction of travel of the solar wind particles; one for electrons and the other for ions. Another experiment could identify the composition of ions with masses from hydrogen to iron. There were two magnetometers mounted on the radial boom; one of the fluxgate type and the other of the helium type. Three experiments were to monitor energetic particles of solar and cosmic origin at a wide range of energies. One of these experiments would also observe interstellar helium entering the solar system. A radio experiment using the radial wire booms and the shorter axial boom would detect solar radio-noise bursts and locally generated plasma waves. Gamma-ray bursts and solar X-rays were the subject of a separate two-detector instrument. A hemispherical sensor would measure the mass, speed, electric charge and flight direction of dust particles. The communication hardware would be used to perform additional experiments, including 'sounding' the corona during solar conjunctions and attempts to detect gravity waves from violent events elsewhere in the universe passing through the solar system.[342]

While the spacecraft was in storage, it was renamed Ulysses in reference to the Greek hero who (as related in Dante's *Commedia*) decided to follow the Sun and venture beyond the Pillars of Hercules at the edge of the known world.

After almost 2 years in storage, Ulysses was re-integrated, tested and shipped to NASA for launch in May 1986. It was undergoing its final tests, including checks for compatibility with its Centaur, when STS-51L exploded, killing its crew. When it became apparent that the remaining Shuttle fleet would have to be grounded for a

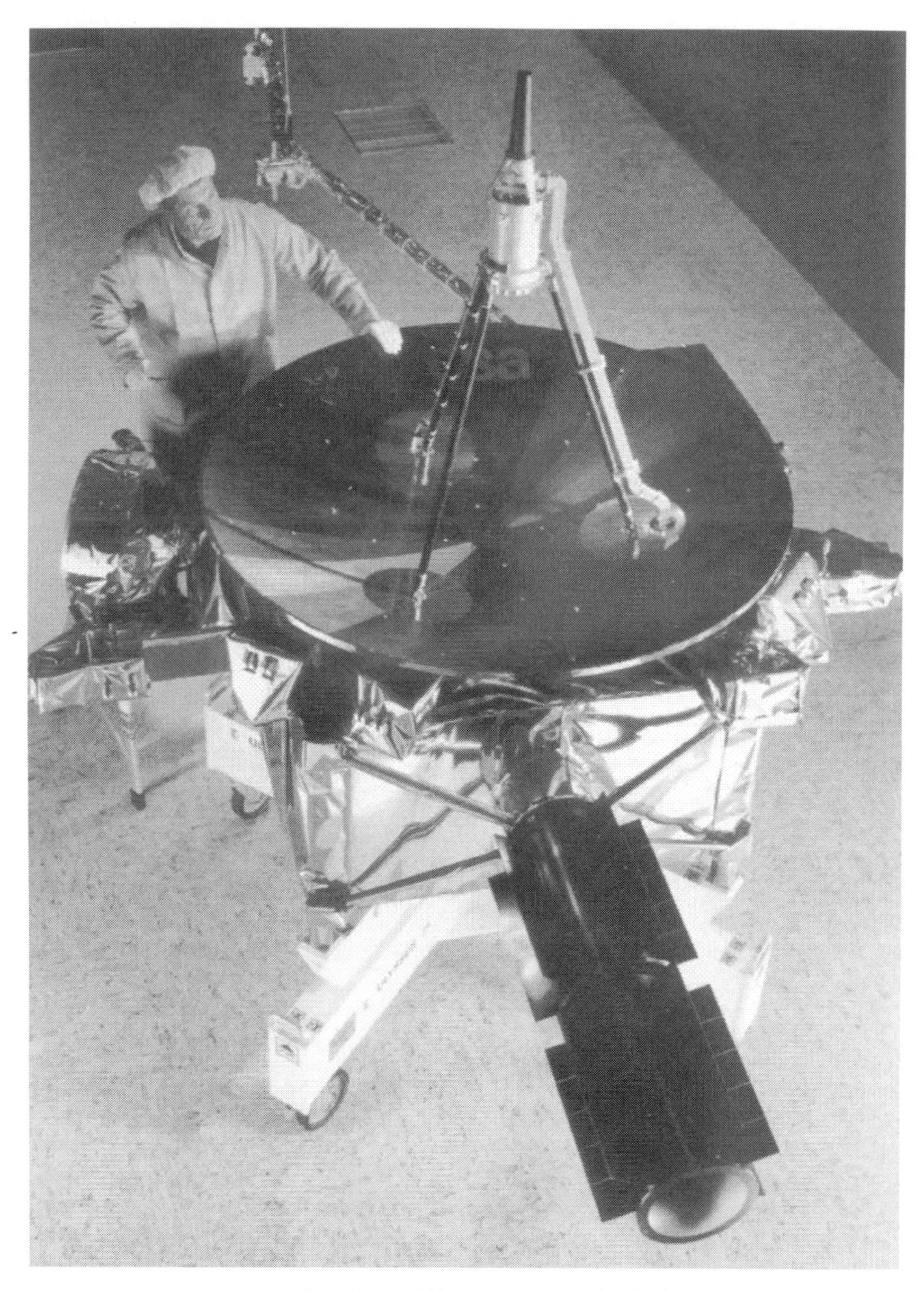

The Ulysses spacecraft, with the RTG on the boom in the foreground. (ESA)

The circular cover of the cosmic dust analyzer on Ulysses. The arcuate object in the background is the solar wind ion composition spectrometer. (ESA)

considerable time, Ulysses was returned to Europe and placed back in storage, this time for an indeterminate period. NASA decided to cancel the Centaur and use the IUS as the escape stage for Ulysses, Galileo and Magellan. Although the IUS was less powerful than the Centaur, with the addition of an extra PAM-S stage it would be able to send Ulysses on a fast trajectory to Jupiter. As a result of the new safety measures imposed on the Shuttle, it was unlikely to fly two planetary missions in a single launch window for Jupiter. NASA scheduled Galileo for 1989, and Ulysses for a window between 5 and 23 October 1990. On being retrieved from storage in mid-1989, Ulysses was subjected to a new series of qualification tests to certify it fit to fly. It was delivered to NASA in May 1990, mated with its IUS on 31 July, and inserted into the payload bay of Discovery. The final operation prior to closing the payload bay doors was to install the RTG.[343,344,345,346] Even after Discovery was rolled out to the pad, there were doubts about whether it would be able to meet its launch window. This was because a series of hydrogen leaks on Columbia and to a lesser extent on Atlantis, initially from the lines between the External Tank and the Orbiter and later from the main engines, raised the unwelcome prospect that there might be a generic problem that would ground the other Shuttles. But one by one the problems were resolved, and on 6 October Discovery lifted off flawlessly for mission STS-41. Interestingly, although Ulysses was essentially the first ESA deep-space probe, it had spent so long on the ground that it flew much later than the agency's Giotto mission to Halley's comet.

After Ulysses and its IUS had been checked out and the tilt table elevated to its deployment angle, the stack was released. The two stages of the IUS and then the PAM fired in succession, and 7 hours 25 minutes after lift off, Ulysses was bound for Jupiter. Discovery landed on 10 October after a faultless flight.[347,348,349,350] This was the third deep-space mission to be dispatched by Shuttle and, to the surprise of the skeptics, each had achieved its window. However, given the shortcomings of the Shuttle, including its high operating cost, it had been decided that future missions would be launched by 'expendable' rockets.

On being given the greatest departure speed of any deep-space mission, Ulysses was in a 0.996 $\times$ 17.038 AU orbit, although it would never reach this aphelion in the vicinity of the orbit of Uranus. It refined its course twice in the first month, and again on 3 July 1991, to place its flyby of Jupiter within 250 km of the aim point. A problem arose with the axial boom, which became slightly warped in the heat of the Sun and induced the spinning spacecraft to wobble, which made it difficult to hold the high-gain antenna pointing at Earth, but over time the boom increasingly penetrated the spacecraft's shadow, which eased the problem.[351] Meanwhile, the instruments were calibrated and the first observations of the interplanetary medium in the familiar environment of the ecliptic were made. During December 1990, the spacecraft was in opposition to Earth, and tested a method which would later be used to try to detect gravity waves. If they existed, such waves would leave a subtle signature in the tracking data. The fact that the vehicle was wobbling complicated the analysis of the data.[352,353] On 21 August 1991 Ulysses was in conjunction, passing behind and slightly above the Sun as viewed from Earth, and dual-frequency radio soundings of the corona were made.[354,355] Three months after it emerged from conjunction the spacecraft began to prepare for its encounter with Jupiter, with the final calibration of its instruments and the preliminary scientific observations.

Ulysses crossed the bow shock on 2 February, when still 113 Rj from Jupiter, then penetrated the magnetopause. Based on the state of the upstream solar wind, the models developed from the previous flyby missions had led scientists to expect the magnetosphere to be squashed. Inside it, the solar wind spectrometer and other instruments found sulfur and oxygen ions which had originated from the volcanoes of Io. These 'Iogenic' ions proved to permeate the magnetosphere at every latitude. Ulysses was approaching from ahead of the planet, and slightly above the plane of its solar orbit, and hence reached relatively high magnetic latitudes. On achieving a latitude of just over 40 degrees at about 9 Rj, the cosmic-ray detectors sensed the possible penetration of a polar cap region in which the fluxes of charged particles suddenly reduced to those typical of the interplanetary medium. As in the case of

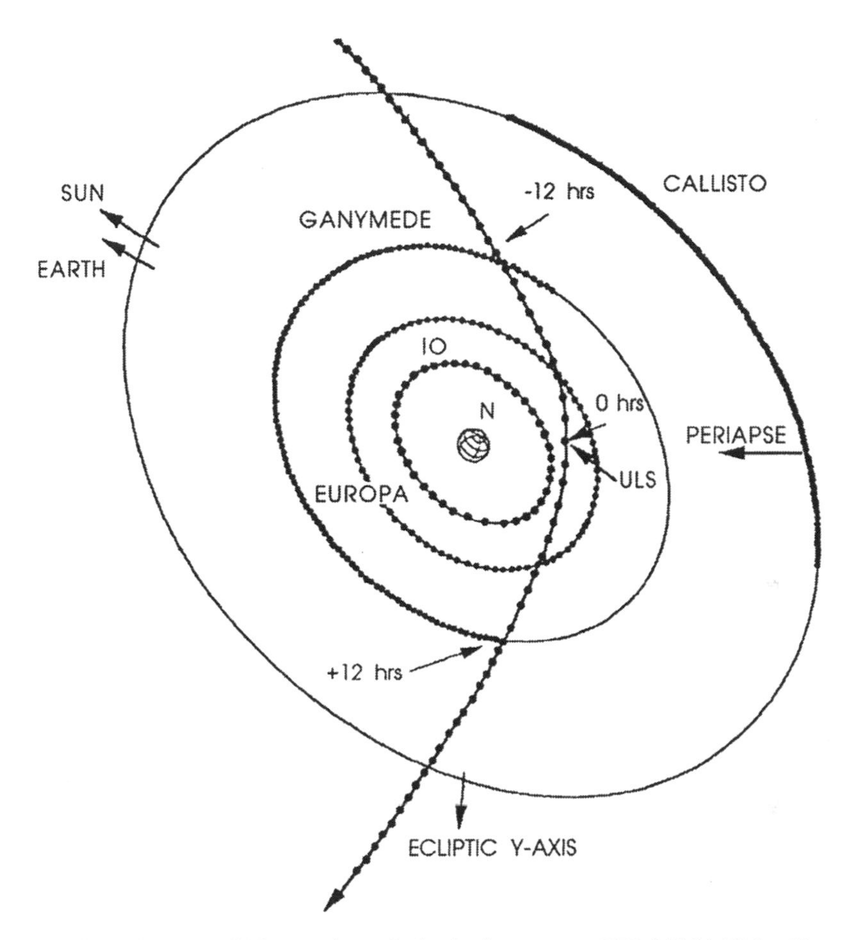

The trajectory of Ulysses through the Jovian system. (JPL/NASA/Caltech)

Earth, the Jovian polar cap might be a region where the magnetic field lines 'open' into interplanetary space. Meanwhile, ground-based telescopes and satellites were providing contextual data. In particular, the Hubble Space Telescope was used for the first time in support of a planetary encounter, and although it still suffered from spherical aberration it provided ultraviolet images of the north polar aurora at the time of the encounter which displayed a morphology dissimilar to any previously observed. At 12:02 UTC on 8 February 1992 Ulysses became the fifth spacecraft to reach Jupiter, flying by at an altitude of 379,000 km (5.31 Rj) above the cloud tops. Less than 2 hours later, it spent 5 hours passing from north to south through of the plasma torus which occupies Io's orbit. The spacecraft's radio 'sounded' the torus to derive the density of electrons along the line of sight to Earth. The results revealed unexpected inhomogeneities in the form of denser and hotter sectors with such large variations as to challenge models of the torus. Meanwhile, telescopes on Earth were monitoring Io – the most prominent volcano, Loki, was quiescent, and the moon's volcanic emissions appeared to be near minimum. The flyby deflected the spacecraft's trajectory to the south and to the dusk side of the planet. On the outbound leg it passed through a scantly explored region of space, and surprisingly found that the magnetic field was already swept back towards the magnetotail by the solar wind. The fact that the dust detectors noted only nine impacts during the entire traverse confirmed that the planet's magnetic field and intense fluxes of charged particles efficiently swept dust out of the inner magnetosphere, in particular the dust originating from Io. Ulysses recrossed the bow shock at 109 and 149 Rj before finally re-entering the interplanetary medium on 16 February. This concluded the Jovian encounter. Some of the instruments had been switched off at times to protect them from the intense radiation in close to the planet, and the spacecraft emerged unscathed.[356,357,358,359,360,361,362,363] It left Jupiter in a 1.341 × 5.408-AU orbit inclined to the ecliptic at 79 degrees. When the angle of the Sun's spin axis to the ecliptic was factored into the equation, this trajectory meant that Ulysses would make its polar passage at a solar latitude in excess of 80 degrees, which was excellent.

Although the dust in the Jovian environment had seemed benign, in the months during which Ulysses withdrew from the planet it noted collimated bursts of dust coming from the direction of the planet with a periodicity of about 28 days, which was similar to the rotational period of the outer regions of the Sun and its corona. As related earlier, Galileo would find that the dust in these streams originated from the volcanoes of Io. On becoming electrically charged, the particles of dust were 'picked up' by the planet's magnetic field, accelerated and sent into interplanetary space. An even more surprising finding was the presence of particles whose speeds and directions were consistent with dust from interstellar space which was passing through the solar system and not bound by it. This was confirmed by the fact that most of this material came from the direction in which the Sun travels around the galaxy – known as the solar apex. It had been believed that the solar wind and its magnetic field would prevent interstellar dust from reaching the inner solar system, but beyond 3 AU the flux of micron-sized interstellar dust grains exceeded that of similarly sized material belonging to the solar system.[364,365,366]

The gravity-wave experiment was resumed by 28 days of continuous tracking, concluding on 18 March. A similar session was held in March 1993, this time with Galileo and Mars Observer joining in.[367] On 26 June 1994, some 28 months after leaving Jupiter, Ulysses began its first south polar pass of the Sun. In fact, this was arbitrarily defined as exceeding a solar latitude of 70 degrees. The pass ended with recrossing this latitude 132 days later, on 5 November. Because the spacecraft was also heading for perihelion, its heliocentric distance decreased from 2.8 to 1.9 AU, and was 2.3 AU at the most southerly latitude of 80.2 degrees on 13 September. During the pass, the radio-wave detector listened for emissions from Jupiter as the Shoemaker–Levy 9 fragments struck it. As predicted, between early October 1994 and February 1995 the Sun–Ulysses–Earth geometry again exposed the axial boom to sunlight, reintroducing the wobble, but an active damping procedure devised for the CONSCAN system meant that very little data was lost. At essentially the same time as Ulysses reached perihelion on 13 March, it crossed the ecliptic between the

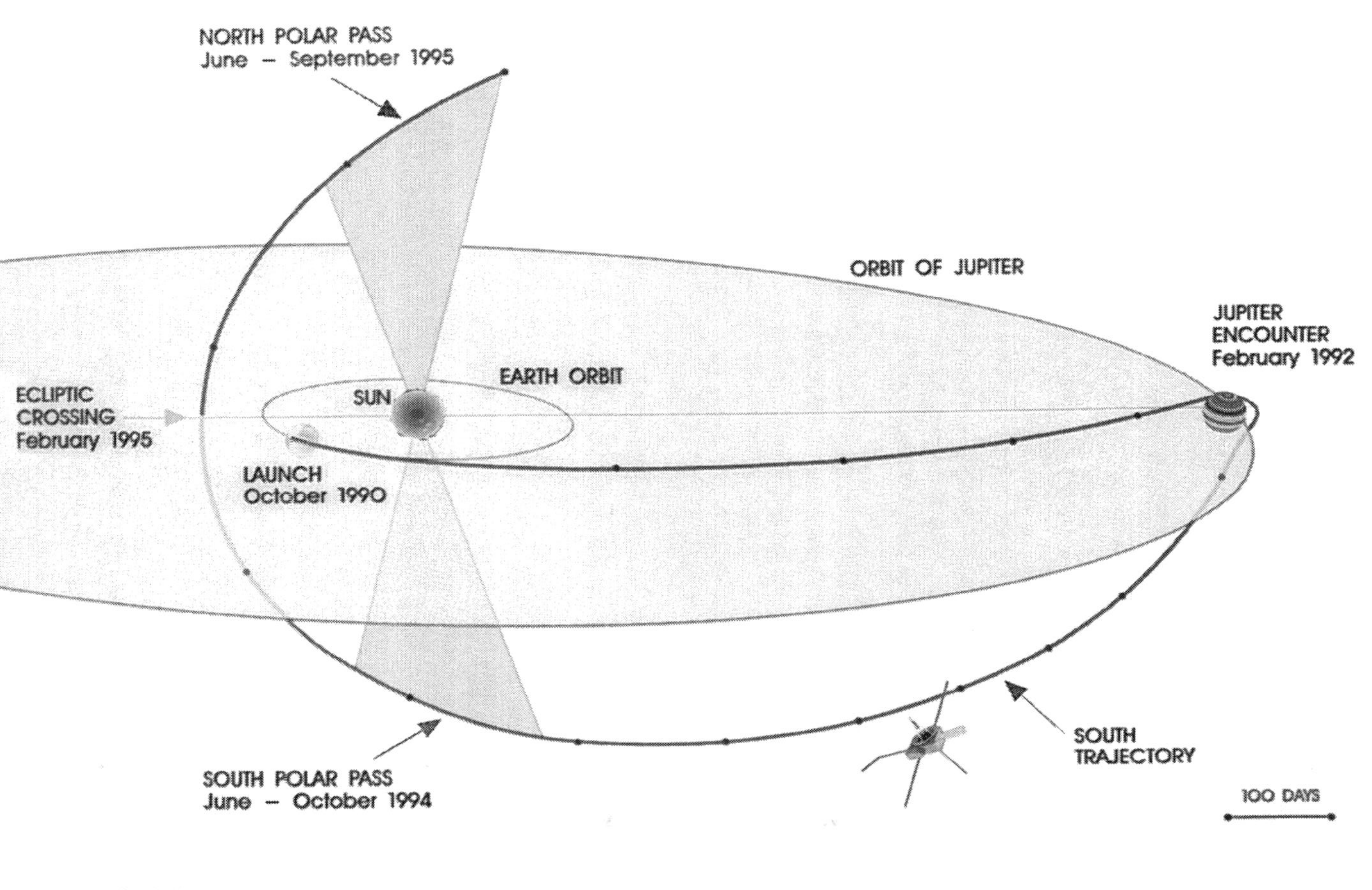

The heliocentric trajectory of the Ulysses spacecraft viewed from 15 degrees above the plane of the ecliptic. (ESA)

orbits of the Earth and Mars, heading north. While it was in conjunction the radio system provided a fast latitude-scan of the corona. The north polar pass began with crossing latitude 70 on 19 June and ended on 29 September 1995, with the limit of 80.2 degrees being attained on 31 July. Both solar passes occurred near the time of minimum activity in the 11-year sunspot cycle, when the corona was dominated by large relatively cool 'holes' at medium to high latitudes which, based on what had been observed in the plane of the ecliptic when these reached low latitudes, were expected to emit a fast solar wind. This was exactly what Ulysses saw as its solar latitude increased. First, a fast solar wind appeared once every solar rotation, with the familiar slow equatorial wind in between. The speed of this 'interstream' wind gradually increased until, at a latitude of just over 40 degrees, the spacecraft was totally immersed in a fast 750 km/s solar wind – more than twice as fast as close to the ecliptic. This persisted during the whole of the south polar pass, and continued until February 1995, when the equatorial wind abruptly reappeared at a latitude of 22 degrees. (In fact, Ulysses had been expected to exit the fast wind earlier because satellite observations had shown that the coronal holes approached no closer than 60 degrees to the equator.) These results showed that, at least close to the Sun, the characteristics of the solar wind were not defined in a purely radial manner. After 2 months of respite the spacecraft re-entered the fast solar wind as it approached the northern pass, where the pattern was repeated.

The solar wind composition instrument showed that the fast solar wind actually originates in regions of the atmosphere 100,000 degrees cooler than at the equator, and the polar wind is significantly richer than the equatorial zone in hard-to-ionize elements such as oxygen. Although the results mainly confirmed predictions, the magnetic fields over the poles, which were expected to display a dipolar structure (like a bar magnet) proved to be much more complex. This obliged scientists to rethink how magnetism is carried and redistributed equatorward by the solar wind. The dipole field was expected to allow cosmic rays easy access over the poles, but only a relatively small increase in the flux of cosmic rays was observed – probably because these 'cosmic-ray funnels' were closed by irregularities in the magnetic field, in particular by its abrupt variability in direction. Another discovery was the possible presence of 'corotating shocks' (or at least their effects) at high latitudes. Where jets of fast wind impacted previously released slower wind, regions having stronger magnetic fields, higher densities and temperatures were seen rotating at the same speed as the Sun. Corotating shocks had been expected to form only at or close to the ecliptic, but their effects were observed up to 70 degrees of latitude. Ulysses also found interstellar particles which had been ionized upon entering the solar system and then been 'picked up' by the Sun's magnetic field. It was the first detection of carbon, oxygen, nitrogen and neon ions from the interstellar medium. The spacecraft also measured the helium density in interstellar space in the vicinity of the solar system. The flow of interstellar dust was persistent, even over the poles where interplanetary dust, which is most concentrated along the ecliptic plane, was almost non-existent.

After Ulysses had completed its north pass over the Sun, ESA decided to extend the mission to the end of 2001 in order to make one more pass over each pole, this time near solar maximum. Furthermore, when Ulysses was at aphelion out near the orbit of Jupiter in 1997–1998, it would be able to make coordinated observations with the agency's SOHO solar observatory, which was to be placed in heliocentric orbit close to Earth.[368,369,370,371] Ulysses made an astonishing observation on 1 May 1996 at a heliocentric distance of 3.7 AU, when it noted a sudden decrease in the density of the solar wind coincident with a magnetic field disturbance and the appearance of heavy ions. Two years later, it was realized that the spacecraft had encountered the distant tail of one of the two 'great comets' of the decade. On that date, C/1996B2 Hyakutake was making its perihelion passage at 0.23 AU, and the spacecraft happened to penetrate the outer fringe of its magnificent ion tail. In fact, this would prove to be just the first of several such incidents. Enhancements in the interplanetary magnetic field in 1990, 1995 and 2001 appeared to be due to comet 122P/De Vico, suggesting that a dense stream of dust may occupy the orbit of this Halley-type comet. (However, one of these enhancements might be associated with the periodic comet 12P/Pons–Brooks.) Then in 2000 the spacecraft crossed the tail of C/1999T1 McNaught–Hartley, and perhaps also that of C/2000S5 SOHO. For a period of 4 days in February 2007 a decrease in the speed of the solar wind and a large flux of heavy ions of cometary origin were noted as Ulysses passed through the tail of C/2006P1 McNaught, marking the longest ever encounter with a comet. It revealed that even though the distance from the nucleus was a large 1.6 AU, the cometary ions had not yet reached equilibrium with the solar wind.[372,373,374]

As Ulysses approached the Sun for the second time, it became evident that there were differences. The simple and repetitive solar wind structure had been replaced by a more complex configuration, and instead of a uniform fast solar wind above a certain latitude, this occurred only in sudden bursts which originated from isolated coronal holes. During the south polar pass and most of the subsequent flight across the ecliptic there was no clear relationship between solar latitude and the speed of the wind. The conclusion was that the wind was slower on average at all latitudes, and much more variable. Only when the spacecraft reached high northern latitudes did the formation of a new polar coronal hole produce a high-speed wind similar to that observed at solar minimum. Departing the pole, conditions reverted to those of solar maximum. A characteristic of solar maximum is the development of 'coronal mass ejections' in which enormous amounts of plasma are blasted out of the Sun at very high speed. At perihelion in May 2001, Ulysses encountered the shock wave of a mass ejection and measured the most intense interplanetary magnetic field and the densest solar wind ever observed by a spacecraft during the entire space age. It was also able to observe directly the inversion of magnetic field polarity that was known to occur during a solar maximum. In fact, when the spacecraft reached the north pole, it found the same polarity as had been present the previous year over the south pole. Moreover, increases in energetic particles once every solar rotation were no longer seen.[375,376]

As Ulysses started its second series of polar passes, ESA and NASA approved a second extension of the mission until September 2004. On approaching aphelion, it was one of many spacecraft to report on the series of powerful solar flares which occurred in October and November 2003, returning first-class data on what became known as the 'Halloween' flares. Only Ulysses was sufficiently far from the Sun to be able to measure the energy output without its instruments saturating. In the time that it took Ulysses to make two orbits of the Sun, Jupiter made slightly more than one orbit, so when the spacecraft reached aphelion the planet was on the same side of the Sun. The nearest that Ulysses approached Jupiter was on 5 February 2004 at a range of 120 million km (1,684 Rj; 0.8 AU) some 7 degrees north of the planet's orbital plane. In contrast to 1992, the RTG was giving only just sufficient power to keep the spacecraft running, and so ESA and NASA decided to provide continuous tracking for 40 days in order that scientific data could be returned in real-time; thus not wasting power operating the tape recorder. This commitment is remarkable in view of the fact that the Deep Space Network was at the time tracking at least eight other deep-space missions, most of which had higher priorities. To accommodate the dwindling power output, a time-sharing strategy for hardware and instruments had to be devised. Although only a few of the results have been published, at least 17 different streams of dust were found emanating from Jupiter (more than twice as many as on the first encounter); and radio bursts were detected from the polar regions – which were also monitored for auroral emissions by satellites such as the Hubble Space Telescope and the Chandra X-ray Observatory.

In 2005 an unusual phenomenon was noted. The flow of interstellar dust usually appeared to arrive from the solar apex, but now its source had mysteriously shifted 30 degrees in a southward direction.[377,378]

During the second Jupiter encounter, ESA had authorized a third extension of the mission to achieve another pair of polar passes over the Sun prior to ceasing operations in March 2008. However, securing a commitment from NASA to provide the tracking was difficult, as that agency, seeking to save money, was myopically considering ending some of its oldest missions despite the fact that they were still providing useful data. In fact, NASA's support was obtained only at the end of 2005.[379]

On approaching the Sun for the third time, Ulysses continued to send interesting scientific data. In particular, even although the Sun was closer to the minimum of its 11-year cycle than in 1993, the spacecraft had to reach a much higher latitude before it left the variable equatorial solar wind behind. It flew over the south pole in February 2007, and then over the north pole in January 2008. In the interim it had gained another (and probably final) extension, this time of 1 year to March 2009. But a potentially mission-ending mishap threatened this. Owing to the spacecraft's low power margins, care had to be taken to prevent the hydrazine for the attitude control thrusters from freezing (which it does at temperatures only slightly higher than for water), and in order to allocate power to the heaters of the thermal control system only minimal scientific activity could be undertaken whilst at large heliocentric distances.[380,381] At the time of the north pole pass, it was decided to switch off the back-up X-Band transmitter temporarily (the primary had failed in February 2003) to divert 60 W of power to the hydrazine heaters and scientific instruments. When it was powered on again on 15 January 2008, the transmitter did not start. The cause of the failure was traced to the power supply, and it meant that the excess power could not even be rerouted to the heaters. After many fruitless attempts to reactivate the X-

Band, it was reluctantly concluded that the remainder of the mission would have to be done using only the slower S-Band. Nevertheless, a subset of instruments were able to provide some scientific data for a few hours per day at 128 bps completing the north polar pass on 15 March 2008. In the absence of high-rate telemetry on the temperature of the spacecraft, it was difficult to estimate solely from thermal models when the hydrazine would freeze. The fuel was allowed to "bleed" slowly through the pipes in an effort to stave off freezing. By the time it freezes, the orbital motion of Ulysses and Earth will bring our planet out of alignment with the antenna, severing contact within a week. The engineering efforts focused on obtaining as much data as possible before the nominal termination of the mission on 1 July 2008. However, consideration was given to changing the orientation of the spacecraft in order that at the next perihelion in 2013 its high-gain antenna would be pointing at Earth, in the hope that communications would be able to be restored after a 5-year hibernation, because by that time the heat of the Sun may well have thawed out the hydrazine to enable control over the spacecraft to be regained for at least a few months. Even if it is unable to be revived, the mission has fully justified its development, launch and operating cost of $1.2 billion. Designed to last only half a solar cycle, it actually returned data for a full cycle and a half, and thus effectively studied the Sun in *four* dimensions.[382]

Long after it has fallen inert, Ulysses will probably make a close encounter with Jupiter which will greatly perturb the spacecraft's orbit, possibly even ejecting it from the solar system.

THE DARKEST HOUR

To follow up on the successful Viking missions of the 1970s JPL studied several 'Purple Pigeon' missions, but these failed to attract funding. In the early 1980s JPL proposed the Mars Geoscience/Climatology Orbiter (MGCO) mission. This was to be launched by Shuttle in August 1990, and adopt a near-circular polar orbit of the Red Planet at an altitude of 350 km to perform an in-depth reconnaissance over a period of at least one Martian year (687 terrestrial days) to map the topographical, elemental and mineralogical characteristics of the surface; study the behaviour of volatiles and dust in the atmosphere; study the structure of the atmosphere; and chart the gravity and magnetic fields. At the same time, NASA was working on the 'low cost' Planetary Observer and Mariner Mark II concepts in which standardized buses would perform a variety of interplanetary missions. The Planetary Observer was to be a commercial satellite bus with the "minimal modifications necessary for the planetary mission", fitted with an instrument pallet which was to be mounted in such a manner that no complex mission-unique integration activities would have to be performed. One proposal was to fit the spin-stabilized Hughes HS-376 bus with a de-spun platform that would carry about 60 kg of scientific instruments. But the winning proposal was by General Electric (which in 1993 became Martin Marietta Space and later became Lockheed Martin Astronautics). This combined the bus of the 3-axis-stabilized Satcom K communications satellite with subsystems and components from the DMSP (Defense Meteorological Satellite Program) and Tiros meteorological satellites.[383] In 1984 the MGCO was approved as the first mission of the Planetary Observer series, and in 1986 it was renamed Mars Observer. In an innovative financial arrangement, the manufacturer's profits would depend on the amount of scientific data returned over the life of the spacecraft. Unfortunately, the cost of the Mars Observer development soon began to spiral out of control; almost doubling to $500 million by 1987. Meanwhile, NASA had accepted a redesign of the payload to accommodate more instruments – thereby confounding the rationale for the Planetary Observer architecture. Furthermore, part of the cost inflation was due to the fact that some of these instruments were not 'off the shelf', but designs which required extensive development.

After the Challenger tragedy in January 1986, scientists urged NASA to offload Mars Observer from the Shuttle to an 'expendable' launch vehicle in order to meet the launch window in 1990, but by the time that the agency decided to transfer it to a Commercial Titan III it was evident that the launch would have to be postponed. In fact, it suited the budget to slip the mission to the next window, in the autumn of 1992.[384] In a remarkable case of short-term thinking, $70 million was saved over a 2-year period at the expense of a $125 million increase in the overall cost! In the meantime, Mars Observer's budget was cut by $50 million in order to contribute to the cost of returning the Shuttle to flight and the ballooning cost of designing the Space Station. In fact, JPL exploited the delay to improve the capabilities of the spacecraft, thereby further increasing its cost. As a result, by August 1993 the cost of the mission – $831 million ($959 million according to another source) including the launcher – was thrice the original sum. An independent study sponsored by JPL later found that the laboratory had abandoned the low-cost Planetary Observer concept within 6 months of Mars Observer being approved.[385,386]

The body of the Mars Observer spacecraft was a rectangular bus of 1.1 × 2.2 × 1.6 meters. To this was affixed a 5.3-meter-long boom for the 1.5-meter-diameter high-gain antenna; a single 3.7 × 7-meter solar array made of six small panels that would not be fully deployed until the spacecraft was in Mars orbit, and would then yield 1,130 W; and long narrow trusses for a magnetometer and a gamma-ray spectrometer. Attitude determination would be by gyroscopes, star sensors and Sun sensors. Attitude control was to be achieved by four reaction wheels and a system of monopropellant thrusters. A bipropellant propulsion system would make major maneuvers. This incorporated two redundant pairs of 490-N main engines and four 22-N thrusters. It burned monomethyl hydrazine and nitrogen tetroxide. The fuel was stored in a tank housed within the main thrust tube of the bus, and the oxidizer was in two separate tanks. The attitude control thrusters were completely independent of the bipropellant propulsion system.[387] The launch mass of 2,565 kg included 166 kg of scientific instruments and 1,440 kg of propellant.

Although the Apollo missions (and to a lesser extent some Soviet orbiters) had used gamma-ray spectrometers to map elements such as uranium, thorium, silicon, iron and potassium in the near-equatorial zone of the Moon, and the Soviet Mars 5 and Fobos 2 orbiters collected a small amount of such data, Mars Observer was the first NASA planetary mission to be equipped with a gamma-ray spectrometer. In addition, a thermal emission spectrometer was to analyze the composition of rocks and ices; a laser altimeter was to measure topography to a high vertical resolution; an infrared radiometer was to measure temperature profiles of the atmosphere and monitor their variations by location and time; an ultrastable oscillator was to be used for radio-occultations; a magnetometer and a French-built electron reflectometer were to investigate the magnetosphere (whose existence was disputed) and the manner in which the solar wind interacted with the planet. However, this being a JPL craft, the most important instrument was the imaging system. This comprised a pair of 11-mm-focal-length cameras for horizon-to-horizon low-resolution meteorological maps of the planet's disk using blue and red filters, and a narrow-angle telescopic camera to take pictures of the surface at a resolution of about 1.5 meters per pixel. However, the fact that this was the highest-resolution imaging system yet flown on a planetary mission meant that a 96-megabit solid-state memory had to be included to store the imagery, and the downlink capacity would limit the mission to imaging an area equivalent to at most 1 per cent of the planet at high resolution. The cameras used linear 'push broom' CCD detectors. All of the imaging and remote-sensing instruments were mutually boresighted, and in fixed positions on the Mars-facing side of the bus in order to eliminate the need to maneuver between observations. Understandably, scientists seized upon the opportunity to send instruments to Mars for the first time in over 15 years, and provided a suite which was more extensive, complex and expensive to develop than was initially intended. For example, high-resolution imaging capabilities were not required for the climatological objectives which were the original motivation for the mission. In fact, some of the 'add on' instruments were deleted in order to regain control of the budget. These included a visible and infrared spectrometer and a radar altimeter – the latter being replaced by a laser one. But the Mars Balloon Relay Experiment was retained. This was a French package to enable the spacecraft to act as a communications relay for the balloons and surface stations that the Russian Mars 94 mission was scheduled to deliver. Although none of Mars Observer's instruments were built by the Eastern bloc, ten Russian scientists had held positions on many of the science teams since at least 1989, and the collapse of the Soviet Union served only to enhance the spirit of cooperation.[388,389,390]

Mars Observer was delivered to Florida in mid-1992, and in early August was mated with the first 12-tonne TOS (Transfer Orbit Stage), which was essentially a single-stage IUS, that had been named *USS Thomas Paine* in honor of the recently deceased former NASA administrator. At the end of the month the spacecraft and its propulsive stage were mated with the Commercial Titan III. But then Hurricane Andrew swept in, and even although the spacecraft was enclosed in the launcher's aerodynamic fairing it was contaminated with dust, requiring it to be removed and thoroughly cleaned. There was a great sense of urgency, since the launch window opened on 16 September and would last only until 13 October.[391] It lifted off on 25 September. Although there was some concern when the telemetry system of the TOS failed, the spacecraft communicated at the scheduled time and reported that it was on course. It had a few minor anomalies: the antenna took longer than expected to lock into place, an attitude control engine developed a slight leak, and a failed Sun

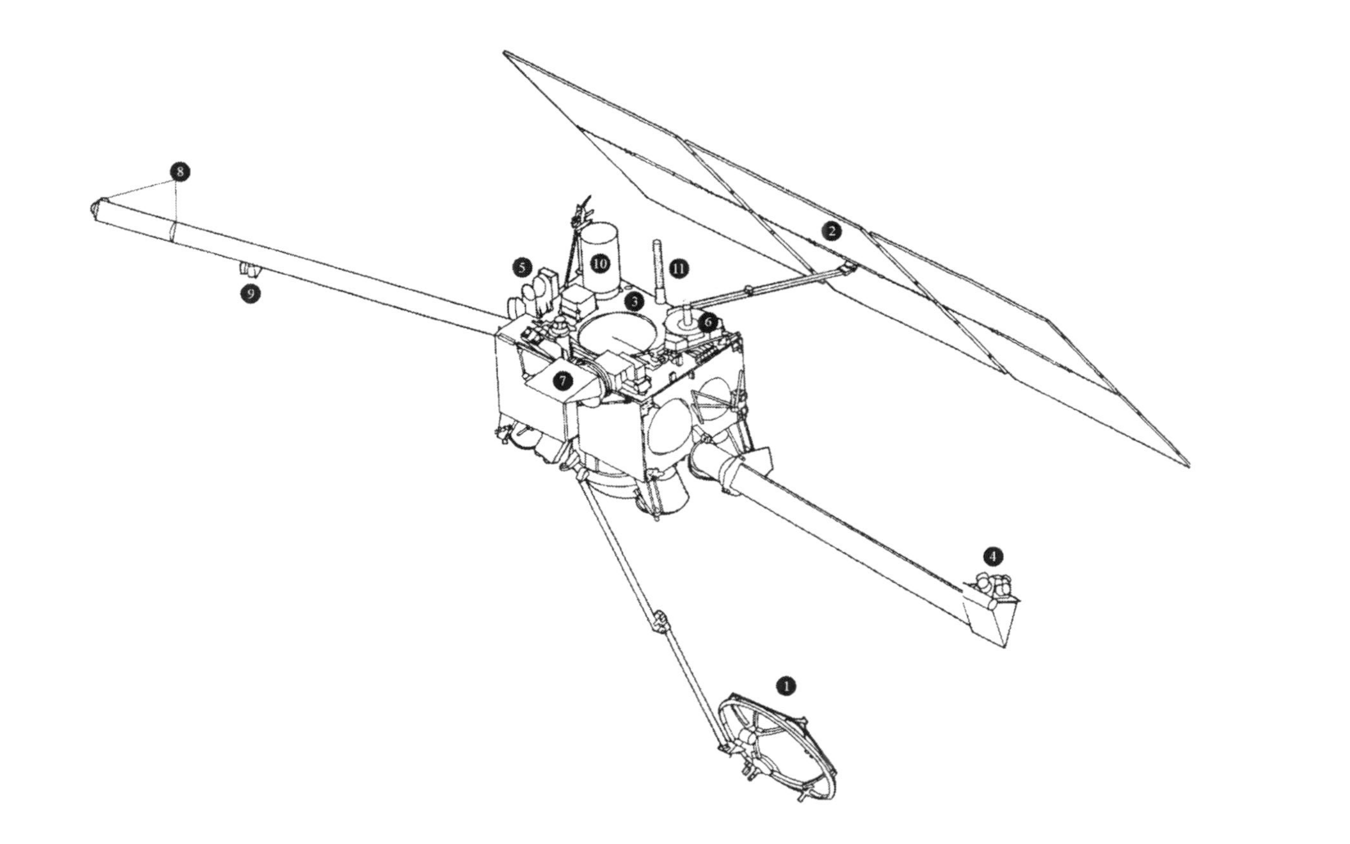

The Mars Observer spacecraft: (1) high-gain antenna; (2) solar panels; (3) nadir panel; (4) gamma-ray spectrometer; (5) thermal emission spectrometer; (6) laser altimeter; (7) pressure modulator infrared radiometer; (8) magnetometer; (9) electron reflectometer; (10) camera; (11) balloon relay antenna. (JPL/NASA/Caltech)

The ill-fated Mars Observer mounted on the TOS stage. The large cylindrical object on the top is the narrow-angle camera.

sensor was replaced by its backup.[392,393] Seven months before launch, it had been decided to postpone the pressurization of the propellant tanks of the bipropellant propulsion system from the intended 5 days after launch until a few days prior to arriving at Mars; the rationale being to preclude the problems which Viking 1 had suffered as a result of having had its regulators and valves pressurized throughout the interplanetary cruise.[394] The course corrections on 8 February and 18 March 1993 were therefore made using the propulsion system in 'blow down' mode. The flight plan included the option of a correction 20 days prior to Mars arrival, but this was canceled as unnecessary.[395,396] The spacecraft also briefly lost contact four times, possibly owing to a flaw in the stellar attitude determination system, but engineers were confident of solving the problem before the spacecraft reached its destination. During the interplanetary cruise, it joined other spacecraft in collecting data on two completely different astrophysical phenomena. In March and April it cooperated with Galileo and Ulysses in an effort to detect the passage of gravitational waves through the solar system. In traveling through space a radio signal traces a straight line, but if the fabric of space is disturbed by a gravitational wave this will have a

Mars Observer took this narrow-angle camera image on 27 July 1993, when 5.8 million km and 28 days away from the planet. Although the spacecraft was making its approach from the south, the south pole was in darkness. The well known dark area of Syrtis Major can be seen at the center of the illuminated portion of the disk. (JPL/NASA/Caltech)

subtle Doppler effect on the transmission from the spacecraft.[397] Then the gamma-ray spectrometer performed joint observations with the gamma-ray burst detectors on Ulysses and NASA's Earth-orbiting Compton Gamma-Ray Observatory. Up to 11 events were detected by all three spacecraft, including a possible 'repeater' and a burst whose position on the sky could be constrained to within a 'box' spanning 1 × 4 arc-minutes.[398] Starting in June the camera was tested by imaging star fields and Jupiter, and the first images of Mars on 26 July from a range of 5.8 million km showed the atmosphere to be extremely clean of dust, which held out the promise of good results from the mapping activity.

Mars Observer was to fire its bipropellant propulsion system at 20:42 UTC on 24 August for 1,730 seconds in order to slow down by 761.7 m/s and thereby enter a highly eccentric orbit in a plane essentially perpendicular to the planet's equator, and with a periapsis of 498 km and a period of 75 hours. Then 11 days later, it was to reduce the period to 23 hours and await the desired orientation and illumination conditions for the mapping orbit. During this 32-day hiatus, it was to power up all of its instruments (apart from the laser altimeter) and take its first scientific data on Mars – and possibly also on Phobos. In October, the orbit was to be circularized in two steps to attain the final orbit at an altitude of 378 km and with a period of 118 minutes. After completing the checkout of the instruments, the main mapping was to start in December.[399] Circular orbits are common for Earth-orbiting satellites but planetary missions had hitherto used elliptical orbits, largely owing to the fact that circularization is expensive in terms of energy. Indeed, to maneuver from its initial orbit into its mapping orbit Mars Observer would need a total change in velocity of 1,367 m/s, which was almost double that of the arrival burn.

By 22 August Mars Observer was 400,000 km from its destination and, with orbit insertion 68 hours away, it was time to pressurize the bipropellant propulsion system by detonating pyrotechnics to open the valves. As sensitive components in the radio electronics had not been certified to survive the mechanical shock of such pyrotechnics whilst in a powered-on state, the spacecraft was told to switch off its transmitter during the operation. The telemetry ceased as expected at 00:40 UTC, but did not reappear 14 minutes later. But all was not necessarily lost, because the spacecraft was programmed to attempt the insertion on its own to attain an orbit of some description prior to working through a series of options designed to regain communications; but nothing was heard from it. At the end of September a signal was sent to tell the spacecraft to switch on the balloon relay, in the hope that this weak (1-W) signal would reveal something about the condition of the spacecraft. The 70-meter antennas of the Deep Space Network were augmented by the Jodrell Bank radio-telescope, but nothing was detected.[400] In the hope that communication would be regained and it was discovered that the orbit-insertion maneuver had not been attempted, a plan was drawn up which called for a major course correction 30 days after the Mars flyby to arrange a second opportunity 10 months late.[401] But all these efforts were to no avail. Evidently the spacecraft had been crippled by some kind of catastrophic event as it pressurized its bipropellant propulsion system. The planetary flyby would then have deflected the trajectory of the spacecraft, or perhaps its debris, into a heliocentric orbit ranging between 169 and 241 million km with a period of 585 days.

The loss of Mars Observer came soon after two other embarrassing and costly failures for NASA, namely the spherical aberration of the Hubble Space Telescope and the failure to deploy of the high-gain antenna on the Galileo spacecraft. It was only the fourth deep-space mission to be lost in flight by JPL; the last being a lunar mission in 1967. Two investigating committees were formed, one internal to JPL and the other appointed by NASA administrator Daniel S. Goldin.

Early speculation accused the transistors in the master and backup clocks, which proved to be from a manufacturing batch which caused other failures; most notably on the NOAA 1 satellite barely an hour after the Mars Observer 'anomaly'. The backup clock could have failed at any time, as it was never powered on during the interplanetary cruise. The pressurization shock could have caused the master clock to fail.[402] The results of the investigations were published in January 1994 and highlighted four areas where the fatal problem could have developed: (1) an unlikely double failure of both the primary and backup transmitters; (2) the remote possibility that a high return-current as a result of firing the pyrotechnic valves had disabled both of the main computers; (3) a massive short in the electrical system that caused a loss of power; and (4) a problem involving the bipropellant propulsion system. In the latter case, three possible scenarios were identified. The first scenario was an oxidizer leak. It was calculated that during the 6 weeks that the spacecraft spent on the ground after being loaded with propellant, and the 11 months it spent in space, at most 2 grams of nitrogen tetroxide could have leaked from a valve. If this reached the hydrazine plumbing during the pressurization process, it could have reacted with the fuel and caused the tube to crack, which would have allowed the hydrazine to escape along with the helium pressurant. Such venting would have significantly destabilized the spacecraft. It must be noted that this scenario could never be replicated in ground tests, and so was rated as being credible but unlikely. The second scenario was that the valve of the nitrogen tetroxide tank's pressure regulator had become corroded and blocked, causing the tank to exceed its maximum design pressure, and then burst after an interval of between 30 and 200 seconds. It must be noted that the regulator was carried over from the propulsion system of the satellite bus on which Mars Observer was based, and as such had been designed in the expectation that it would be opened within a few days of launch, as opposed to having to endure the very different thermal environment of deep space and corrosive chemicals for almost a year prior to being fired. Subsequent ground tests confirmed this to be the most credible cause of the loss. In the third scenario, one of the 15-gram squibs of the pyrotechnic charges designed to open the valves of the pressurization system could have malfunctioned and been expelled at anything up to 200 m/s and thus punctured the adjacent hydrazine tank. In either the second or third scenario the spacecraft would have experienced "immediate critical physical damage". The lack of telemetry at the time of the tank pressurization precluded the determination of the actual dynamics of the accident. The decision to power off the transmitter during the tank pressurization followed from the decision by the program managers to save $375,000 by not doing tests to determine whether the delicate electronics could be safely operated.[403,404,405,406,407]

Martin Marietta, "in light of the unsuccessful mission", returned the $17 million that it had been paid, and renounced its claim to $21.3 million for the operational performance expected of the spacecraft while orbiting Mars.[408] There was a rumor that Mars Observer 2 would be built from spare components, but this was not to be. NASA also rejected a proposal to re-fly some of the instruments on a low-cost spacecraft based on the Department of Defense's Clementine mission (of which more later) because the agency was against military "contamination" of its civilian program. But this proposal led to the Mars Surveyor program, thereby breathing new life into Mars exploration, which had appeared to have been compromised by the loss of Mars Observer.[409]

Remarkably, whilst the anomalies suffered by Galileo and Mars Observer had highlighted the risk of mounting single-spacecraft missions – in comparison to the earlier strategy of dispatching pairs of spacecraft – JPL did not modify this policy, which had been imposed by the use of the Shuttle, a launch vehicle that was not only unlikely to be able to fly two missions during a single launch window but was also, as Carl Sagan put it, "almost absurdly expensive".[410]

OVERDUE AND TOO EXPENSIVE

In view of the successful Viking orbiter/lander combinations, JPL began to plan two similar missions, one of which would deliver a lander with a rover capable of traveling across the surface for hundreds of kilometers, and the other with a lander that would collect a sample and then lift off to deliver this to Earth. But these were soon recognized to be prohibitively expensive: the rover mission alone, which was the simpler of the two, would have cost at least $1.5 billion.[411] Nevertheless, these objectives continued to be ranked highly, and with the revival in the mid-1980s of studies for planetary missions, JPL began to plan a program to exploit the expected success of Mars Observer. It soon settled on a sample-return mission as the most logical next step, short of a human mission. This would land with a mobile robot which would deliver a representative collection of rocks, dust and dirt to the ascent vehicle. This reasoning reflected the lesson learned from the efforts of the Viking landers to test for life on the planet – namely that whatever might be learned by in-situ instruments, more would be learned by returning samples to Earth in order to apply the full range of analysis techniques.

After much discussion, NASA decided to combine various aspects of several of these studies into the long-term Mars Rover and Sample Return (MRSR) program. Of course, the mission would have to be launched by the Shuttle. The plan was for a Centaur G-prime propulsive stage to make the Earth-escape maneuver. However, to launch the spacecraft and the stage together on the Shuttle would limit the mass of the spacecraft to 7,800 kg, which was far too little to realistically undertake such a task. It was therefore decided to have one Shuttle deliver the integrated orbiter, lander, rover, ascent vehicle and return vehicle into Earth orbit, and have a second Shuttle deliver the Centaur. This strategy would enable the mass of the spacecraft to be increased to 12,750 kg. Because it was expected to take a decade to develop the various elements required for this mission, launch windows between 1996 and 2005 were identified. The 2001 window was preferred because the return would occur at the time of a Martian 'great opposition', and minimizing the energy of the Mars-escape maneuver would enable the mass of the return vehicle to be maximized. The round trip would last 3 years. After the Challenger tragedy of 1986, the plan was revised. Now the Shuttle would launch only the orbiter and the lander/rover. One Titan IV would supply the Mars ascent vehicle and return vehicle, and another Titan would deliver the propulsive stage. In order not to have to carry propellant for Mars orbit-insertion, the revised plan was to use a risky aerocapture maneuver in which penetrating the atmosphere at a 15-degree angle below the horizontal would create sufficient drag to cause the spacecraft to enter a highly elliptical capture orbit. The rover and orbiter would be encapsulated together in a shell that would lead the way during this maneuver. In fact, because the capture orbit of 6 × 2,000 km would not be stable, the spacecraft would have to perform a small burn at its first apoapsis in order to lift the periapsis out of the atmosphere prior to circularizing the orbit at 500 km.

NASA planned to use a reduced-scale model of the Martian heat shield to test aerobraking and aerocapture maneuvers in Earth orbit. In 1990 the agency issued a $45-million contract to McDonnell Douglas for the Aeroassist Flight Experiment on STS-82 in 1994. Once deployed, the vehicle would use its propulsion system to perform aerobraking tests, and then be recovered and returned to Earth.[412]

Once the spacecraft was in its operating orbit of Mars, the orbiter would detach,

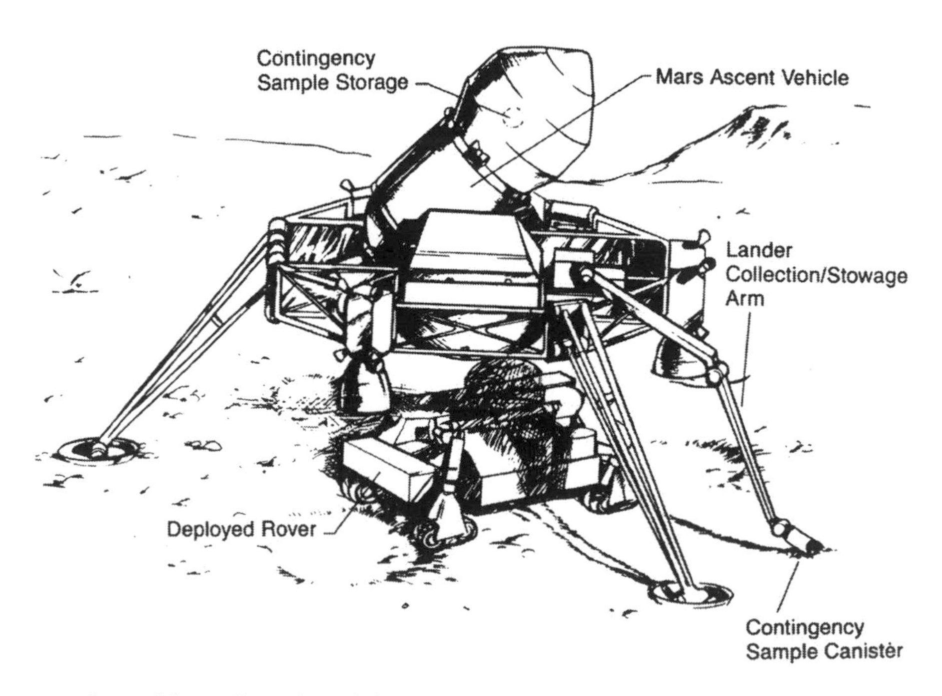

A possible configuration of the Mars Rover and Sample Return spacecraft.

together with the return vehicle. The integrated lander, rover and ascent vehicle would use the shell and aerodynamic lift to precisely target the landing site, which was to be flat and within 30 km of at least two areas of geological interest. The final phase of the descent would use retrorockets for braking. On Mars, the rover would roam for about 300 days and then transfer to the ascent vehicle the samples it had collected. Rover designs studied by JPL included an articulated train with a trio of 2-wheeled cabs, provided with a 280-W RTG, stereoscopic cameras, robotic manipulators, drills and a sample storage system. The 1-meter-diameter wheel and articulation design would enable it to roll over 1.5-meter-tall obstacles and operate on slopes of up to 30 degrees. Using a semiautonomous guidance system, it would travel several kilometers per day. One option was to release several rovers, each of which would explore as wide an area as possible while the primary rover collected samples. The ascent vehicle was expected to have two or three solid-fuel stages. It had to be able to lift 5 kg of samples and attain orbital velocity of 4.5 km/s. But a more advanced concept envisaged an In-Situ Propellant Production (ISPP) module in which an RTG would dissociate oxygen from carbon dioxide in the atmosphere, and this would burn methane carried from Earth for greater energy. About 1 tonne of oxygen would be required. After liftoff, and once in the vicinity of the orbiter, the ascent vehicle would release the sample canister, which the orbiter would autonomously track in order to rendezvous and dock. The sample would then be transferred to the return vehicle, which would use a solid-rocket to head for Earth. On approaching Earth, the sample would be inserted into orbit by either a solid-rocket or an aerocapture maneuver, to await collection by astronauts from the Space Station. Of course, the lander and rover(s) would have cameras, meteorological instruments, experiments to detect water, etc, to supplement the analyses of the material that would be conducted on Earth.[413,414,415,416]

As envisaged by NASA, the mission would be the occasion for testing a variety of planetary mobility and locomotion techniques. Unlike terrestrial or lunar robots that would be simply remotely controlled from Earth, or 'teleoperated', rovers on Mars would have to be as autonomous as possible owing to the time (up to 45 minutes) that it would take for a command to be sent from Earth, received by the rover, executed and verified back on Earth. A Mars rover would therefore have to be able to navigate on its own, recognize obstacles and 'understand' which tasks it could perform in any given situation, and those which it should not attempt. In order to investigate the issues of mobility, JPL used a prototype of a rover General Motors developed in the mid-1960s for the Surveyor Block II lunar landers which were unfortunately never funded. It was a 6-wheeled vehicle with three compartments linked by spring joints. Refitted with modern electronics, stereoscopic cameras and rudimentary navigational software, the vehicle became known as the Blue Rover. The trials, which were sponsored in part by the US Army, began in the mid-1980s in Arroyo Seco, near the JPL campus.[417] On conceiving the Mars Rover and Sample Return mission, NASA created the Pathfinder Planetary Rover Navigation Testbed program to demonstrate the feasibility of autonomous navigation. Under the aegis of this program, and exploiting its experience with the Blue Rover, JPL designed a large demonstrator that it named Robby after the robot of the 1950s classic science

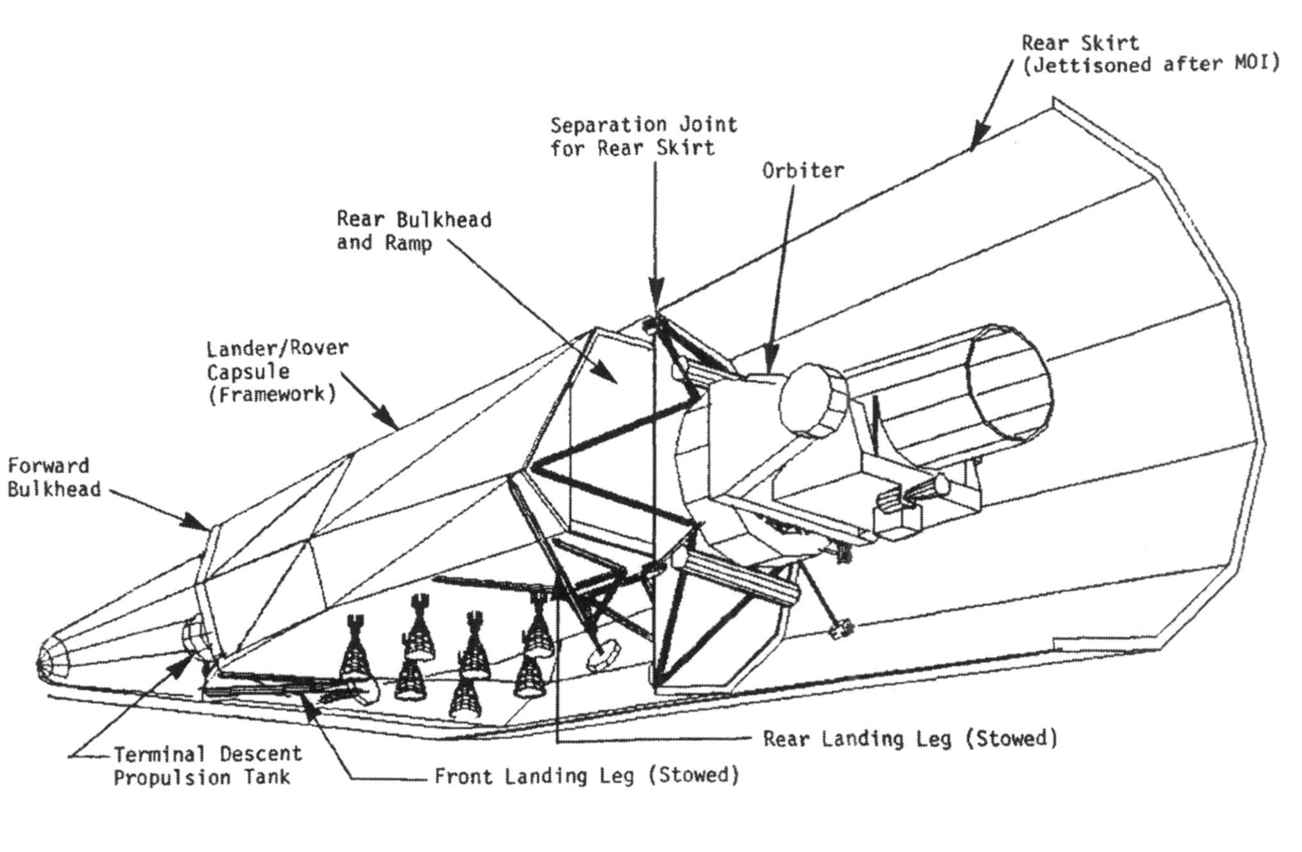

The Mars Rover and Sample Return orbiter and lander in their aerocapture shell.

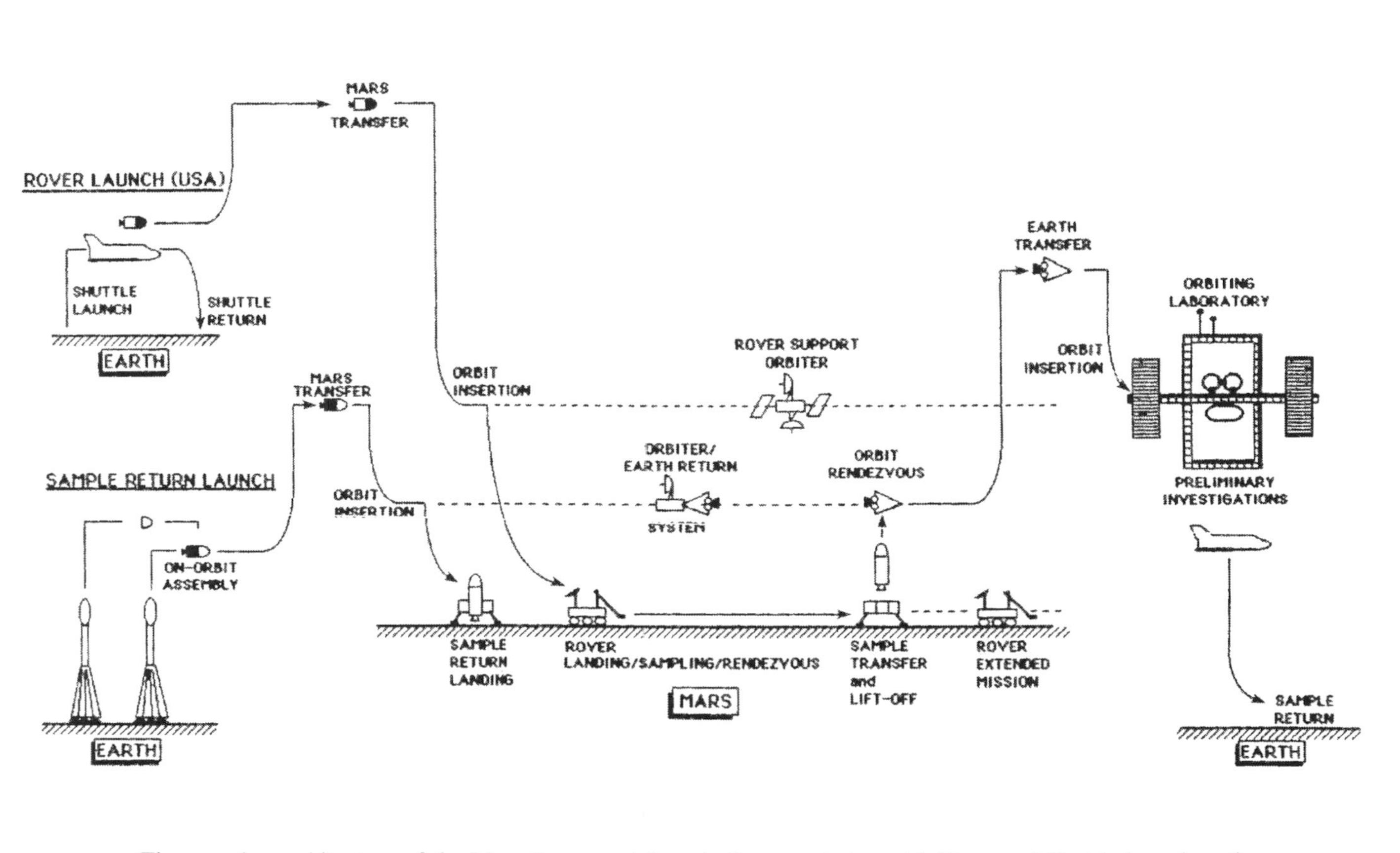

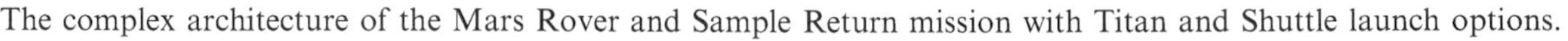
The complex architecture of the Mars Rover and Sample Return mission with Titan and Shuttle launch options.

fiction film *The Forbidden Planet*. Specifically, it was to evaluate technologies for navigation and hazard avoidance in rugged terrain. Like the Blue Rover, the new vehicle had three articulated cabs, but for enhanced mobility each cab was able to steer and roll with respect to the others and encoders relayed the relative positions, making control easier and more precise. The leading cab had a commercial robotic arm in order to demonstrate sample collection; the middle cab had a tall electronics enclosure housing the hardware and two pairs of stereoscopic cameras on top for navigation; and the rear cab had batteries. Overall, Robby was about 4 meters long, 2 meters wide, 2.5 meters tall and weighed 2,000 kg.

But developing mechanical systems to operate in rough terrain represented only a small part of the challenge. The main issue was navigation, and in particular the software. JPL studied three control strategies involving differing degrees of human intervention. Computer-Aided Remote Driving (CARD), developed originally for the Blue Rover, relied on analysis on Earth of the stereoscopic imagery to identify safe paths and create control sequences for the rover to execute. Semi-Autonomous Navigation (SAN) would interpret a projection of the terrain ahead with range and obstacle information determined by the rover, perhaps augmented by knowledge of the landscape gained from imagery provided by an orbiter, to autonomously plan a route. Behavior Control (BC) would provide the rover with a desired end point, an approximate route and heading, and leave the vehicle to attempt to make its own way there. Equipped with stereoscopic vision and optimized SAN software, Robby proved able to traverse 100 meters of unknown terrain in about 4 hours.

The only serious alternative to a wheeled rover for collecting samples on Mars was the spidery walking robot devised by the Carnegie Mellon University. Known as Ambler, the prototype had six legs stacked at different heights along two central axes; each leg having two 'ankle' and 'knee' articulations in the horizontal plane and an extensible foot with which to hold the whole robot level. The central body, to which the shafts were connected, also supported four electronics and computing racks. A laser rangefinder was used to identify the 3-dimensional structure of the terrain in front of Ambler more rapidly and reliably than could have been attained using stereoscopic cameras. Progress was monitored by force and torque sensors, inclinometers and various other sensors. One advantage of a legged robot such as Ambler was that it was required to find only a limited number of points of contact with the ground, whereas a wheeled rover had to find a fully traversable path. Due to its configuration, a Mars robot based on Ambler would be up to 5 meters tall and 7 meters across, and it would take 600 W to maintain the 3,000-kg vehicle at a pace of 7.5 cm/s. Ironically, this design was reminiscent of the walking machines in the invasion of Earth by Martians depicted by H.G. Wells in his novel *The War of the Worlds*.[418,419]

Other technologies studied included a hazard recognition and avoidance system which would avoid boulders, rocks, scarps and other obstacles during the landing on Mars; and an autonomous system to enable the orbiter to rendezvous and dock with the sample container once this had been delivered to orbit around the planet.

The MRSR program received a short-lived boost in July 1989 when, speaking at the 20th anniversary of Apollo 11's landing on the Moon, President George H. Bush

The prototype for the Mars Rover and Sample Return was the huge, three-bodied wheeled rover Robby. (JPL/NASA/Caltech)

set NASA the objective of returning humans to the Moon and then venturing to Mars. However, there was concern that this Space Exploration Initiative (SEI), which was focused on human missions and would cost at least hundreds of billions of dollars, might derail the scientific robotic exploration in much the same manner as the development of the Shuttle did in the 1970s; or that at the very least it might deflect the robotic program from the scientific objectives that had been specified in 1983 by the Solar System Exploration Committee, thereby limiting it to the issues pertinent to preparing for a human mission. Indeed, the fact that a human mission to Mars was in development would in itself undermine the case for sending a robot to retrieve a sample of the planet – even though such a sample would characterize the biological hazard posed to humans by that environment. After several months of study, NASA proposed a human-oriented Mars exploration architecture which included a Mars Observer 2 in 1996; a Mars Global Network in 1998 comprising 24 penetrators which would be delivered planet-wide to measure the distribution of subsurface water, characterize the seismicity and provide long-term meteorological data; two Mars Sample Returns with smaller rovers in 2001; a high-resolution site reconnaissance and a communication orbiter in 2003; Mars Sample Returns with enhanced capabilities every two years between 2005 and 2009; two Mars Sample Returns with smaller rovers in 2011 to well-selected sites; and finally a mission in 2015 with a 5-person crew. It was certainly an ambitious plan. But Bush's speech had been merely inspirational, for there was no prospect of commensurate funding, and within a year of it being drawn up the SEI had been consigned to NASA's archive of unrealistic dreams.[420,421] With it went the Mars

Rover and Sample Return mission, largely because the estimated cost of returning a 5-kg cache of Martian samples to Earth was at least $10 billion. Nevertheless, the desire to examine pieces of Mars in terrestrial laboratories remained high.

REFERENCES

1 For US Post-Viking Mars exploration projects see Part 1, pages 256–261
2 Sagdeev-1994e
3 Cunningham-1988i
4 Lenorovitz-1986b
5 AWST-1986b
6 Butrica-1996b
7 Lardier-1992b
8 Clark-2000
9 Furniss-1987c
10 AWST-1987a
11 Surkov-1997b
12 VnIITransmash-1999
13 VnIITransmash-2000
14 Mudgway-2001b
15 Surkov-1997b
16 Lenorovitz-1989
17 Sobel'man-1990
18 Garcia-1991
19 For gamma-ray bursts see Part 1, pages 201–202 and 277–278
20 D'Uston-1989
21 Riedler-1989
22 Grard-1989a
23 Grard-1989b
24 Lundin-1989
25 Rosenbauer-1989
26 Shutte-1989
27 Rosenbauer-1989
28 Avanesov-1989a
29 Ksanfomality-1989
30 Bibing-1989
31 Selivanov-1989
32 Korablev-2002
33 Blamont-1989
34 Krasnopolsky-1989
35 Surkov-1997d
36 For Mars 5 see Part 1, pages 164-165
37 Surkov-1989
38 Surkov-1997c
39 CIA-1988
40 Surkov-1997e
41 Garvin-1988
42 Zaitsev-1989
43 Rocard-1989
44 Lenorovitz-1988
45 Grard-1989b
46 Sobel'man-1990
47 Waldrop-1989
48 Sagdeev-1994f
49 Harvey-2007b
50 Waldrop-1989
51 Krupp-2006
52 Bruns-1990
53 Flight-1989a
54 Grard-1989b
55 Dubinin-1998
56 Baumgärtel-1998
57 Selivanov-1989
58 Surkov-1989
59 Surkov-1997c
60 Mudgway, 282-283
61 Riedler-1989
62 Grard-1989a
63 Aran-2007
64 Kolyuka-1991
65 Nasirov-1989
66 Avanesov-1989a
67 Bibing-1989
68 Selivanov-1989
69 Selivanov-1990
70 Mordovskaya-2002a
71 Mordovskaya-2002b
72 Lundin-1989
73 Rosenbauer-1989
74 Korablev-2002
75 Blamont-1989
76 Krasnopolsky-1989
77 Snyder-1997
78 Zaitsev-1989

79 Rocard-1989
80 Sagdeev-1989
81 Perminov-1999
82 Oberg-2000
83 Butrica-1996c
84 Smith-1982
85 AWST-1983
86 Smith-1984
87 Young-1990
88 Dornheim-1988
89 Butrica-1996d
90 AWST-1988a
91 AWST-1988b
92 Covault-1989a
93 AWST-1989b
94 Flight-1989b
95 Kerr-1990
96 Kerr-1991
97 Dornheim-1990a
98 Dornheim-1990b
99 AWST-1990a
100 AWST-1990b
101 Saunders-1991
102 Westwick-2007i
103 Steffes-1992
104 Spaceflight-1992d
105 Spaceflight-1992e
106 Rokey-1993
107 Sjogren-1992a
108 Sjogren-1992b
109 Saunders-1991
110 Head-1991
111 Kargel-1997
112 Tyler-1991
113 Phillips-1991
114 ST-1995
115 Kerr-1991
116 For DZhVS see Part 1, pages 291–293
117 Saunders-1999
118 Stofan-1993
119 Cattermole-1997
120 Kelly Beatty-1993a
121 Phillips-1991
122 Schulze-Makuch-2002
123 Armstrong-2002
124 Doody-1993
125 Sjogren-1993
126 Kerr-1993
127 Doody-1995
128 Giorgini-1995
129 Sjogren-1997
130 Konopliv-1996
131 Banerdt-1994
132 Sjogren-1994
133 Naudet-1996
134 Simpson-1994
135 Tolson-1995
136 Butrica-1996e
137 Russo-2000c
138 ESA-1975
139 Westwick-2007j
140 Meltzer-2007a
141 Murray-1989e
142 Meltzer-2007b
143 Dornheim-1989
144 Lorenz-2006
145 Meltzer-2007c
146 Young-1996
147 Meltzer-2007d
148 AWST-1989c
149 Belton-1996
150 Geenty-2005
151 Dawson-2004
152 Cunningham-1988j
153 Bonnet-1994
154 Meltzer-2007e
155 Dornheim-1988
156 Murray-1989f
157 Meltzer-2007f
158 Meltzer-2007g
159 Meltzer-2007h
160 Covault-1989b
161 Carlson-1992
162 Carlson-2007
163 Meltzer-2007i
164 Johnson-1991
165 O'Neil-1990
166 O'Neil-1991
167 Carlson-1991
168 Belton-1991
169 Gurnett-1991
170 Harland-2000a
171 Antreasian-1998
172 McCullogh-2007
173 Dornheim-1991
174 Flight-1991a
175 O'Neil-1991
176 Meltzer-2007j

177 Belton-1992
178 O'Neil-1991
179 Meltzer-2007k
180 O'Neil-1992
181 Kivelson-1993
182 Harland-2000b
183 Sagan-1993a
184 O'Neil-1993
185 Kelly Beatty-1993b
186 O'Neil-1993
187 Belton-1994
188 Kelly Beatty-1995
189 Harland-2000c
190 Cunningham-1988k
191 O'Neil-1994
192 Spencer-1995
193 Martin-1995
194 Weissman-1995
195 Meltzer-2007l
196 Graps-2000
197 O'Neil-1995
198 Atkinson-1996
199 Folkner-1997a
200 O'Neil-1996
201 Meltzer-2007m
202 Fischer-1996
203 Lanzerotti-1998
204 Lorenz-2006
205 Seiff-1997
206 Rust-2006
207 Young-1996
208 Seiff-1996
209 Ragent-1996
210 Sromovsky-1996
211 Niemann-1996
212 Orton-1996
213 Lanzerotti-1996
214 von Zahn-1996
215 Atkinson-1996
216 Folkner-1997a
217 Atkinson-1997
218 Kelly Beatty-1996a
219 Meltzer-2007n
220 Harland-2000n
221 Kivelson-1996a
222 Frank-1996
223 Anderson-1996a
224 Gurnett-1996a
225 Grün-1996
226 Graps-2000
227 O'Neil-1996
228 Gurnett-1996b
229 Kivelson-1996b
230 Belton-1996
231 Carlson-1996
232 Harland-2000e
233 Anderson-1996b
234 Schubert-1996
235 Anderson-2004
236 O'Neil-1996
237 Gurnett-1997
238 Khurana-1997
239 Anderson-1997a
240 Geissler-1998
241 Kivelson-1997
242 Khurana-1998
243 Harland-2000f
244 Efimov-2008a
245 Pappalardo-1998
246 Carr-1998
247 Anderson-1997b
248 Harland-2000g
249 Harland-2000h
250 Harland-2000i
251 Schenk-2001
252 Harland-2000j
253 ST-2000a
254 Harland-2000k
255 Harland-2000l
256 Harland-2000m
257 Carlson-1999a
258 Khurana-1998
259 McEwen-1998
260 Harland-2000n
261 O'Neill-1997
262 Burns-1999
263 Harland-2000o
264 ST-1999
265 Harland-2000p
266 McCord-1998
267 Carlson-1999b
268 Carlson-1999c
269 Carroll-1997
270 Mitchell-1998
271 Turtle-2001
272 Anderson-1998
273 Harland-2000q
274 Harland-2000r

275 Geissler-1999
276 Mitchell-1998
277 Harland-2000s
278 Hoppa-1999
279 Hoppa-2001
280 Spencer-1999a
281 Gierasch-2000
282 Erickson-1999
283 Spencer-2000a
284 Kieffer-2000
285 McEwen-2000
286 Spencer-2000b
287 Lopes-Gaultier-2000
288 Harland-2000t
289 Russell-2000
290 Hoppa-2000
291 Keszthelyi-1999
292 McEwen-2000
293 Spencer-2000
294 Lopes-Gaultier-2000
295 Khurana-1998
296 Kivelson-2000
297 Spencer-2001
298 McEwen-2000
299 Spencer-2000
300 Lopes-Gaultier-2000
301 Head-2001
302 Spaun-2001
303 McCord-2001
304 Erickson-2000
305 Otero-2000
306 Turtle-2004
307 Kurth-2002
308 Gurnett-2002
309 Hill-2002
310 For the Caloris basin see Part 1, pages 186–187 and 193–194
311 Thcilig-2001
312 Turtle-2004
313 Davies-2003
314 Turtle-2004
315 Milazzo-2002
316 McEwen-2002
317 Turtle-2004
318 Efimov-2008b
319 AWST-2002
320 Theilig-2002
321 Anderson-2005
322 IAUC-8107
323 Bindschadler-2003
324 NASA-2003
325 Page-1975
326 Rosengren-1990
327 Bromberg-1966
328 Ulivi-2006
329 Hufbauer-1992
330 Kozicharow-1979
331 JPL-1978
332 McGarry-1997
333 ESA-1976
334 Wenzel-1990a
335 Eaton-1990
336 Leertouwer-1990
337 Russo-2000d
338 Furniss-1990a
339 Wenzel-1990b
340 Hawkyard-1990
341 Mastal-1990
342 Caseley-1990
343 Wenzel-1990a
344 Eaton-1990
345 Leertouwer-1990
346 Furniss-1990a
347 Furniss-1990b
348 Furniss-1990c
349 Bcech-1990
350 Hengeveld-2005
351 Mudgway-2001c
352 Bertotti-1992
353 Bertotti-2007
354 Marsden-1991
355 Bird-1992a
356 Barbosa-1992
357 Smith-1992
358 Bird-1992b
359 Spencer-1992
360 Caldwell-1992
361 Angold-1992
362 Marsden-1992
363 McLaughlin-1992
364 Grün-1993
365 Grün-1994
366 Krüger-2007
367 Bertotti-1995
368 Marsden-1995
369 Marsden-1996
370 McGarry-1997
371 Marsden-1997

372 Jones-2002
373 Jones-2003
374 Neugebauer-2007
375 Marsden-2000
376 Marsden-2003
377 Krüger-2005
378 Krüger-2007
379 Lawler-2005
380 McGarry-2004a
381 McGarry-2004b
382 Angold-2008
383 Steffy-1983
384 Flight-1987
385 McCurdy-2005a
386 Westwick-2007k
387 Guernsey-2001
388 NASA-1993a
389 Furniss-1992a
390 Mecham-1989
391 Spaceflight-1992f
392 Furniss-1992b
393 Spaceflight-1992g
394 For the Viking 1 pressurization problem, see Part 1, pages 228–229
395 Guernsey-2001
396 Esposito-1994
397 Armstrong-1997
398 Laros-1997
399 NASA-1993a
400 Mudgway-2001d
401 Esposito-1994
402 Friedman-1993a
403 JPL-1993
404 NASA-1993b
405 Stephenson-1994
406 Guernsey-2001
407 Westwick-2007l
408 Flight-1994
409 Westwick-2007m
410 Sagan-1993b
411 Westwick-2007n
412 Furniss-1990d
413 NASA-1987
414 Wilson-1986c
415 Smith-1987b
416 AWST-1987b
417 Mishkin-2003a
418 Weisbin-1993
419 Mishkin-2003b
420 Smith-1989
421 Henderson-1989

6

Faster, cheaper, better

THE RETURN OF SAILS

Following their non-selection for the Halley rendezvous mission 'solar sails' fell by the wayside, but in the early 1990s there was a brief resurgence of interest, and the deployment of a small sail in space was tested.

In the 1980s the government of the French Midi-Pyrenées region decided to promote a solar sail regatta with the Moon as the target. It hoped to attract entrants from across the world, but failed to gain sponsorship. A similar undertaking was proposed in December 1988 by the Christopher Columbus Quincentenary Jubilee Commission of the US Congress as part of the program to celebrate the 500th anniversary in 1992 of the 'discovery' of America. This time the race would target Mars, with an entrant from North America, one from Europe and one from the Soviet Union. By the rules of the race, the propulsion should be uniquely solar, the target launch date should be 12 October (the anniversary of Columbus's discovery), the overall mass of the spacecraft should not exceed 500 kg, it should carry a commemorative plaque, and it should have a camera to show the unfolding of the sail and other highlights of the flight. The leader of the 'Solar Sail Cup' would be determined at successive stages including successful deployment of the sail, successful escape from Earth's gravity, closest approach to the Moon and ultimate arrival at Mars.

In America, proposals were made by the Massachusetts Institute of Technology in the US and by Canada. In Europe, proposals were made by Britain, France and Italy. The British proposal, Nina, was studied by Cambridge Consultants. It was for a 250-meter circular sail that could be folded into a 4-meter-diameter cylinder for launch. Although Nina was deemed the most technically advanced and imaginative submission, the Solar Sail Cup Committee chose the Italian proposal as the "most realistic, sounder, more spectacular and more complete" of the European entrants.[1] This was designed by Aeritalia of Turin, with considerable interest from industry, research centers and universities, and named 'Capitana Italica' (Renaissance Italian

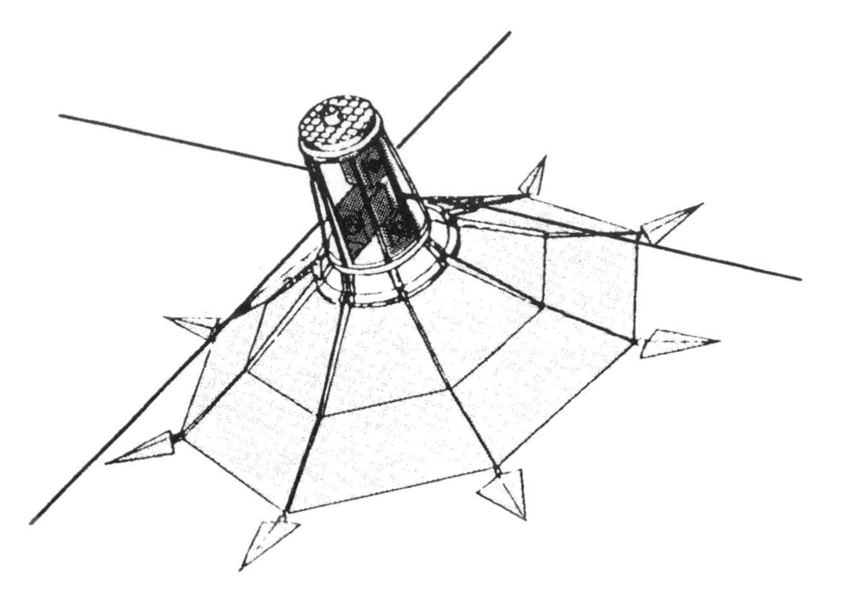

The Russian Regatta satellite, which would have employed a solar sail for attitude and orbit control.

for 'Italian Flagship'; an alternative name for Columbus's *Santa Maria*). It was a square-shaped Mylar sail possessing an area of 10,000 square meters, with a solar-powered payload capsule at the end of a 10-meter-long truss equipped with several cameras and scientific experiments for the study of Mars. It would be launched by an Ariane 4, and once in orbit around Earth it would extend four inflatable diagonal conical trusses to deploy the 5-micrometer-thick sail, and polymerize crossbeams for stability. An actuator at the base of the truss would displace the payload module with respect to the sail to generate the torque required for attitude control. The sail would slowly spiral out of Earth orbit, crossing the radius of the Moon's orbit after 288 days on a 4.8-year heliocentric cruise to Mars.[2,3,4]

However, as in the case of the 1980s solar sail regatta to the Moon, the race to Mars never received adequate funding.

The Soviet proposal was the only sail to be pursued, although in the form of a space laboratory named Regatta. Several studies were conducted into using a solar sail for attitude control, and in some cases also for orbital control. If a program of Regatta satellites operating in orbit around Earth for solar and astronomical studies proved viable, a follow-on mission might have been to make such a spacecraft orbit one of the Lagrangian points awaiting redirection to an asteroid or comet. The program had developed sufficiently far that at the time of the collapse of the Soviet Union foreign partners (including Italy) had been approached to provide scientific instruments to study the Sun. Although Regatta was not implemented, it did lead in 1993 to the successful test of the deployment of a non-propulsive solar sail from the Mir space station.[5,6,7,8]

A NEW HOPE

The Mariner Mark II and the Planetary Observer series initiated in the 1980s were intended to revitalize the US planetary exploration program, but neither ever came close to achieving this goal because only one of each type was approved and these ended up costing many times the initial estimates. As a result, despite the backlog of missions due to fly within the next few years, the program was only slightly less comatose at the end of the decade than it had been at the beginning. At a workshop in 1989 organized by NASA's Solar System Exploration Division, a special Small Mission Program Group was formed to investigate whether a new, more successful program of low-cost and narrowly focused planetary missions could be established. In view of the experience of the Mariner Mark II and the Planetary Observer proposals, which had failed to deliver their reduced-cost promises by a wide margin, this new initiative was received skeptically by scientists and engineers. At the symposium, a scientist from the Johns Hopkins University's Applied Physics Laboratory (APL), Stamatios Krimigis, drew the attention of participants to NASA's Small Explorer program of scientific satellites, which was successfully achieving scientific results at low cost. The Solar System Exploration Division therefore decided to assess the feasibility of low-cost planetary missions by initiating the Discovery program.

The Discovery program called for missions that would cost no more than $150 million to develop, $35 million to operate and $55 million to launch on a vehicle no larger than a Delta II. The management structure of the Small Explorer program was imposed, in which a single scientific Principal Investigator (PI) would take full responsibility for the mission. In fact, in contrast to the traditional way in which a mission was proposed on an institutional basis by a NASA center or laboratory, the PI would identify the goals of the mission, organise a number of lead investigators to provide the necessary scientific instruments, and obtain the support of industrial contractors to draw up the proposal. This approach greatly assisted in keeping the mission focused on its scientific objectives. It also meant that JPL, after 15 years of dominating American planetary exploration, would face competition. Some of the proposals were from Ames, which had successfully developed the various Pioneer missions and the atmospheric capsule for the Galileo mission. But for the first time academic institutions were able to play the leading role, and the first proposal to be selected came from APL.

Owing to the severe cost and time constraints, the spacecraft for the Discovery missions had to be simpler and to have fewer instruments than was previously the norm. Moreover, unlike the earlier architectures, Discovery would benefit from the military research and development effort of the 1980s; in particular that conducted for the 'Star Wars' Strategic Defense Initiative that, among other things, envisaged small or very small spacecraft for use as, for example, ballistic missile interceptors. Some of this research had been conducted at JPL itself, as a means of keeping its engineers gainfully employed during the hiatus in planetary exploration. Among its results, this effort produced propulsion systems that weighed only a few kilograms, miniaturized inertial platforms, and star trackers massing less than a kilogram. It is worth noting that, contrary to the common belief, most of the American technical advances in space during the 1980s were funded not by NASA, but by the military. What little research and development NASA conducted was mostly limited to the Space Shuttle and Space Station Freedom. As a result, instead of relying on the use of 'off the shelf' proven technology, as had been the case with the Mariner Mark II and the Planetary Observer classes, the shortcomings of which would soon become evident with the loss of Mars Observer, the Discovery program sought to reduce costs by exploiting the state-of-the-art miniaturized technologies developed by the military. In fact, the costs of the Discovery missions were artificially low because they benefited from a decade of heavy military investment – they would not have been viable if they had first to develop those technologies out of their own funding. Of course, the end of the Cold War made things easier. Another characteristic of the Discovery program was the shorter timescales of the missions. This meant that a single team of engineers and scientists could work on a project from its conception through to its completion, which in turn meant that the notoriously strict aerospace documentation requirements could be relaxed a little.

Like the defunct architectures of the 1980s, the Discovery program was thought of from the beginning as a multi-year effort that would fund a number of individual missions. The significance of this was that it would eliminate the need to propose each mission as an individual 'new start' in NASA's budget, which would not only greatly simplify the approval process but also protect the proposals from political skirmishes in Congress. This suited the desires of Daniel Goldin, who took over as NASA's administrator in 1992. As a former space industry manager and enthusiast of solar system exploration, he advocated the development of smaller scientific satellites that exploited military technology. The fact that the White House's Office of Management and Budget backed Goldin meant that, unlike Mariner Mark II and Planetary Observer, the Discovery program was given multi-year status. It moved rapidly, and in 1992 funded the first two missions. The first mission was the Near-Earth Asteroid Rendezvous (NEAR), which APL had resurrected from a Planetary Observer proposal. The second was a JPL proposal to fly a 'pathfinder mission' for the Mars Environmental Survey (MESUR) – this being a scheme to deliver a series of small scientific stations to the surface of Mars.[9,10,11] Around 100 proposals were outlined at a workshop in San Juan Capistrano in November 1992. They included Earth-orbiting planetary telescopes, lunar orbiters and landers, a flyby of the hybrid asteroid-comet (2060) Chiron which orbits beyond Saturn, and a battery powered reconnaissance mission to Pluto. There was even a Russian–US proposal for NASA to fund a Venera lander with state-of-the-art instruments.[12] Eleven concepts were picked for further study.[13,14,15]

The chosen proposals were:

- Mercury Polar Flyby: A 400-kg spacecraft similar to Mariner 10 equipped with optical imagers and radars that would make at least three polar flybys of Mercury in order to complete the preliminary mapping of its surface and characterize its polar environment. Launch windows were available in 1996 and 2000.[16]

The Venus Multiprobe proposal for the Discovery program was modeled on the Pioneer Venus mission.

- Hermes Mercury Orbiter: A JPL proposal to use existing hardware to put a spacecraft into orbit around Mercury for a reconnaissance lasting several local years of the topography and composition of the planet's surface and the nature of its atmosphere and magnetosphere. Launch was expected in 1999.[17,18]
- Venus Multiprobe Mission: This joint venture by Harvard University, JPL and Hughes was to collect data on the super-rotation of the atmosphere of Venus. Reusing Pioneer Venus components, the spacecraft would include a cylindrical bus and between 14 and 18 small atmospheric probes, each of which would carry only a transmitter. The plan envisaged launching in mid-1999 and reaching Venus in September. An initial salvo of probes would be released 37 days prior to the encounter, with the remainder being released 16 days out. The trajectories would be such that the two groups of probes and the bus itself would penetrate the atmosphere over an interval of a few hours. The results from tracking their radio signals would provide detailed data on temperature profiles and wind direction and velocity over the entire Earth-facing hemisphere. This would lead to an improved understanding of the deep

atmosphere (i.e. the region below an altitude of 20 km) which had not been accurately studied by previous spacecraft.[19,20]

- Venus Composition Probe: A proposal from the University of Colorado to exploit the experiences of a variety of missions, including Pioneer Venus, by having a probe penetrate the atmosphere of Venus, deploy a parachute in order to characterize the altitude range 75 km to 42 km, and then release a pressure vessel that would make heat balance measurements as it fell freely to the surface.
- Mars Upper Atmosphere Dynamics, Energetics and Evolution Mission (MUADEE): A University of Michigan mission calling for a small spinning spacecraft to enter Martian polar orbit to study the upper atmosphere and lower ionosphere in the altitude range 200 to 60 km.[21]
- Comet Nucleus Tour (CONTOUR): A Cornell University mission which would be launched in 2003 and encounter at least three cometary nuclei, including 2P/Encke, 9P/Tempel 1 and 6P/d'Arrest.
- Small Missions to Asteroids/Comets (SMACS): A mission proposed by the National Optical Astronomy Observatory to dispatch four spacecraft on the smallest possible booster (the air-launched all-solid Pegasus XL) in 1998–2000 to encounter a primitive object (2100 Ra-Shalom), an evolved object (asteroid 6178, listed as 1986DA), a comet (15P/Finlay) and an extinct or dormant cometary nucleus (asteroid 3200 Phaeton).
- Comet Coma Chemical Composition (C4): An Ames proposal for a simple spin-stabilized spacecraft to rendezvous with a comet at perihelion and fly alongside it for 100 days, periodically sampling the coma in order to study how this evolved. A launch date in 1999 would facilitate a rendezvous with 9P/Tempel 1.
- Near-Earth Asteroid Returned Sample (NEARS): A US Geological Survey proposal to combine hardware from the NEAR mission with a capsule used to return film from spy satellites in order to obtain and return samples from six different sites on a near-Earth asteroid. The candidates were the Apollo type (4660) Nereus or the Amor type (10302) 1989ML, which had yet to be named. Two launch windows existed in 2000 and 2002.[22]
- Earth-Orbiting Ultraviolet Jovian Observer: A Johns Hopkins University proposal to spend 9 months observing the Jovian system from a spacecraft in a distant Earth orbit.
- Solar Wind Sample Return: A JPL proposal to return samples of the solar wind during two years spent flying outside of the Earth's magnetosphere.

However, while NASA was starting its Discovery program, another US agency was working on its own low-cost planetary mission. The DSPSE (Deep Space Program Science Experiment) was sponsored by the BMDO (Ballistic Missile Defense Organization), which was the successor to the Strategic Defense Initiative. This spacecraft was built by the Naval Research Laboratory, using instruments from the Lawrence Livermore National Laboratory. DSPSE was conceived to test the optical sensors which had been developed to detect and track military targets above the Earth's atmosphere, but the shrinking budget for ballistic missile defense following the end of the Cold War denied it the planned target. Its project leader, Stewart Nozette, then approached NASA in 1992 with a proposal to conduct the tests using bodies such as the Moon and asteroids, for which tracking by NASA's Deep Space Network would be necessary. NASA assembled a team of 13 planetary scientists to consider the idea. The profile for the mission, now named Clementine, called for spending 2.5 months in polar orbit of the Moon doing a multispectral and topographical survey, prior to setting off on a trajectory to make a 100-km flyby of asteroid (1620) Geographos on 31 August 1994. Discovered in 1951, Geographos was an Apollo-type asteroid. It was about to pass within 5 million km of Earth; the closest that it would approach for at least the next two centuries. As a small fast-moving target it was ideal for testing the BMDO sensors. Moreover, there was every chance that subsequent to its encounter with Geographos the spacecraft would have sufficient propellant remaining to set course for an encounter with the Amor type of asteroid (3551) Verenia in October 1995.[23]

The payload featured four cameras for near-ultraviolet, visible and near-infrared wavelengths. Frames taken using different filters were to be combined to create multispectral images that would reveal the composition of the surface of the target body. The only other scientific instrument was a laser altimeter that would measure on an ongoing basis the spacecraft's altitude to an accuracy of 40 meters, to reveal the surface topography. Clementine had an octagonal shape with a maximum width of 1 meter. On the sides were the optics, protected by a light shield, and two solar panels. The mass at launch, including fuel for the main engine and for the attitude control engines, was 482 kg. Clementine was to be launched on a Titan II missile that had been retired in 1987 and converted into the launch vehicle configuration known as the Titan IIG. The entire mission was to cost a relatively inexpensive $75 million.

Clementine was launched on 25 January 1994 from Vandenberg Air Force Base on the Californian coast, becoming the first American deep-space mission to leave from a site other than Cape Canaveral. When it entered lunar orbit on 19 February it was the first American spacecraft to do so since 1973. The lunar mapping phase of the mission was extremely successful, producing in excess of 1,800,000 images, a global altimetry map, and data on the radar reflectivity of the polar regions to test a hypothesis that there might be water ice in permanently shaded craters.[24] Then on 3 May the main engine was fired for 4.5 minutes to leave lunar orbit and head for an Earth flyby which would put the spacecraft on course to encounter Geographos. Unfortunately, while Clementine was out of ground contact on 7 May a "software anomaly" prompted it to fire its attitude control engines for 11 minutes, thereby not only exhausting its supply of attitude control propellant but also imparting an 80-rpm spin that would preclude clear imaging of Geographos. The second part of the mission had to be canceled. Clementine was retargeted into a highly elliptical Earth orbit. Owing to its attitude control problems the batteries were soon exhausted, and when the inert craft flew by the Moon on 20 July it was deflected into heliocentric orbit. Tracking was discontinued on 8 August. At the start of 1995 the orientation of the spacecraft became favorable for recharging its batteries, and beginning on

Preparing the US Department of Defense's Clementine lunar orbiter and asteroid interceptor. (BMDO/NRL/LLNL)

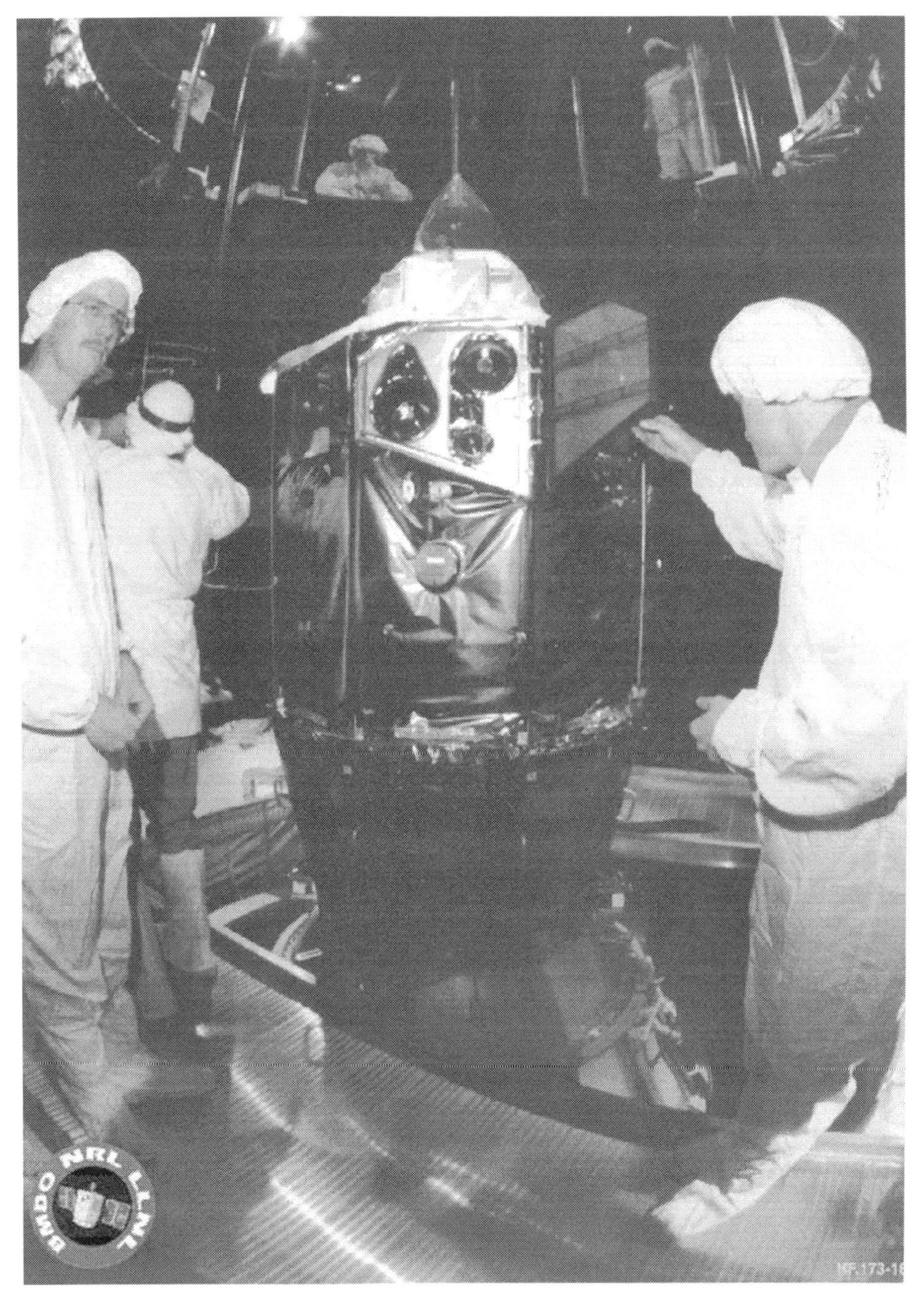

A view of the Clementine spacecraft that shows its array of cameras.

8 February attempts were made to re-establish contact, with success on 20 February. Over the following months the main engine was fired to reduce the spin rate, and engineering tests of the cameras were made. The mission was finally terminated on 10 May 1995, by which time Clementine was in a 1.023 × 1.063-AU orbit similar to that of Earth.[25] As regards Geographos, terrestrial telescopes and radars had been making observations in support of the mission. Optical observations made since the 1970s had suggested that the asteroid had an extremely elongated shape. This was confirmed by 400 radar images obtained over a 6-day period starting on 28 August 1994. From the best radar data Geographos appeared to be 5.11 km long and only 1.85 km wide.[26]

Despite the participation of NASA in the Clementine mission, some within the scientific community were concerned at the military's encroachment into the realm of solar system exploration. Critics also argued that the fact Clementine had failed fairly early in its relatively limited mission indicated that the low-cost approach to deep-space missions was flawed.[27]

In fact, the DSPSE was conceived as a two-spacecraft program. Clementine 2 would also be built by the Naval Research Laboratory, which would manage it on a mission for the Space Warfare Center of the US Air Force. It would be launched on either another Titan IIG or on a small Taurus all-solid rocket. The original proposal was to test a high-performance propulsion system using hypergolic hydrazine and

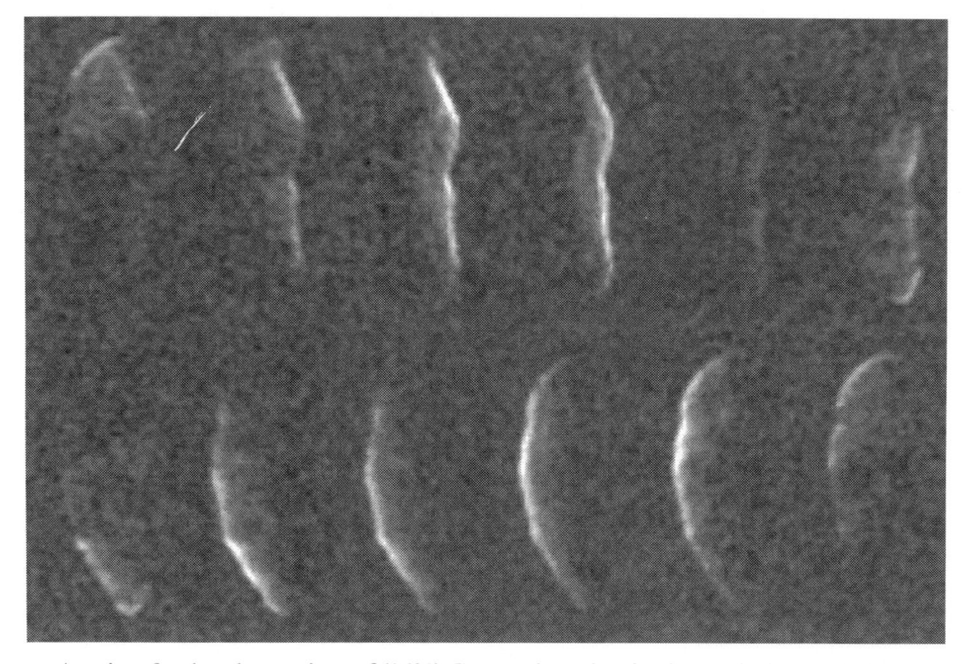

A series of radar observations of (1620) Geographos, showing its much elongated shape. Had Clementine not suffered a software fault, it would have flown by this asteroid at the time that these observations were made in August 1994 and provided 'ground truth' to assess the efficacy of radar studies. (JPL/NASA/Caltech)

chlorine pentafluoride, and for the spacecraft to carry four small kinetic 'missile-killers' known as LEAPs (Light Exo-Atmospheric Projectiles), which it would fire during encounters with several near-Earth asteroids. Other proposals, which were more acceptable politically and technically, included using the LEAPs to deliver 1-kg 'insect robots' to the Moon. But it was eventually decided to retain the profile of encountering up to three near-Earth asteroids at distances ranging between 50 and 100 km.[28,29,30] Some 3 hours prior to an encounter, Clementine 2 would release a 1-meter-long battery powered probe. Relying on an autonomous guidance system using optical and star sensors and a cold-jet thruster system, the 20-kg probe would steer to impact the target at a speed in the range 10 to 18 km/s. It would transmit close-up imagery on the way in, and a few microseconds of impact dynamics data. The spacecraft would monitor the impact site with multispectral cameras and other instruments sharing a common 30-cm telescope. Another option was to arrange for the spacecraft to fly through the cloud of ejecta from the probe's impact in order to determine its composition using a miniature mass spectrometer and a dust analyzer. Observations of the impact plume, either remote or in-situ, and of the new crater, would yield information on the composition and mechanical strength of the near-Earth asteroids; data which could be used to prepare missions to divert or destroy such an object in the event that one is found to be on a collision course with Earth. Other possible scientific objectives of the mission included monitoring gamma-ray bursts and a space-based search for small Aten asteroids which, because they spend most of their time inside the orbit of Earth, are difficult to detect from the ground.[31] The initial plan called for a 30-km flyby of (433) Eros followed by a 50-km flyby of (4179) Toutatis, which was one of the most dangerous near-Earth asteroids then known.[32] Direct observations of Toutatis could provide 'ground truth' to the radar method of imaging such bodies. In fact, high-resolution images which were obtained on 8 December 1992 when Toutatis passed by at a range of less than 10 times the radius of the Moon's orbit showed it to be comprised of two irregular components, one 4.5 km across and the other 2 km across, in close contact with each other and in a chaotic rotational state. The images were sufficiently detailed to reveal a 700-meter crater on the large component.[33] Other possible targets for this mission included (14827) Hypnos and the 530-meter-diameter asteroid (6489) Golevka.[34,35]

However, skeptical scientists and politicians argued that because the aim was to test technologies that could be used to intercept satellites and ballistic missiles, the Clementine 2 mission was really a Trojan horse for the "weaponization" of space. In response, mission engineers said that owing to the difficulty of scaling data obtained for a kilometer-sized body to a target as small as a missile or a warhead, the results would not assist in the design of missile interceptors. Another criticism was that anything likely to be learned from this exercise would be provided by the Near-Earth Asteroid Rendezvous mission of the Discovery program. In October 1997 President Clinton vetoed the $120 million Clementine 2 mission, officially owing to concern that it could be construed as a breach in the Antiballistic Missile Treaty.[36,37]

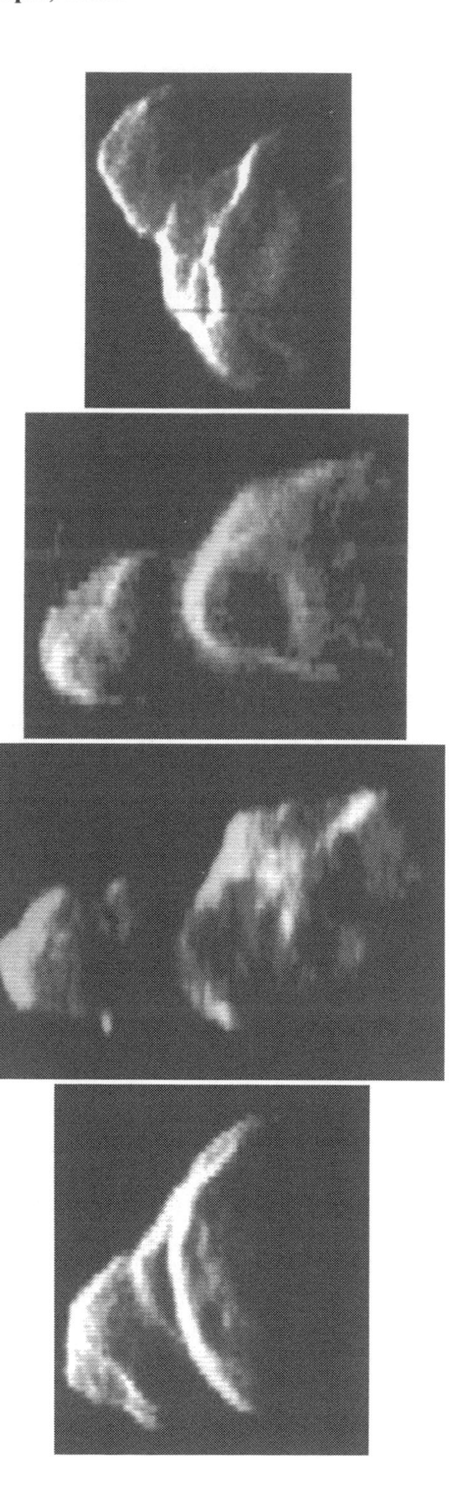

Four radar images of asteroid (4179) Toutatis obtained in December 1992. It was a possible target of Clementine 2. Better images were obtained on later approaches to the Earth. (JPL/NASA/Caltech)

IN LOVE WITH EROS

APL had never been involved in planning and managing a planetary mission, but it had developed small satellites for the US Navy since the 1960s and therefore had a tradition of producing low-cost spacecraft. Moreover, during the 1980s it had made some of the least expensive spacecraft for the Strategic Defense Initiative. APL's NEAR mission, which was the first to be approved for the Discovery program, was to determine the gross physical properties of its target asteroid, particularly its size, shape, mass and rotation; measure its composition and mineralogy; and investigate its morphology. In addition, it was to study the properties of the regolith; study the asteroid's interaction with the solar wind; search for an intrinsic magnetic field; and search its surface and immediate vicinity for evidence of recent activity. When the mission was drawn up for the Discovery program the aim was to launch in 1997 and rendezvous with (1943) Anteros, but by the time the mission was approved the launch had slipped to January 1998 and the trajectory included a flyby of the 17-km (2019) Van Albada in the main belt 9 months into the 2-year interplanetary cruise to rendezvous with 1-km (4660) Nereus; and if the mission were to be extended there were several possibilities for later encounters. But then concerns were expressed that the small size of Nereus might restrict the diversity of data returned during the 8 months that the spacecraft would spend in its vicinity, making the results "somewhat boring". An investigation was therefore started into the possibility of rendezvousing with (433) Eros, an Amor type in an orbit ranging between 1.13 and 1.75 AU.[38] Ground-based observations at particularly favorable opportunities had established it to be an elongated object with dimensions 36 × 15 × 13-km, making it the second largest of all the near-Earth asteroids. Its spectral characteristics suggested a stony S-class asteroid, which is the most common class among the main belt objects and probably related to ordinary chondrite meteorites. Optical and radar observations had measured its rotational period at 5.27 hours and determined the orientation of its spin axis.

Direct trajectories to Eros faced severe limitations owing to the relatively high inclination of its orbit, but some interesting possibilities were found. One involved launching in 2003 and utilizing an Earth gravity-assist to tilt the spacecraft's orbit, with a flyby of comet 2P/Encke on the way to Eros in 2005. Another option would allow a launch in 2000 and provide encounters with no fewer than two comets and as many asteroids, but Eros would not be reached until 2012. A trajectory was also discovered that involved inserting NEAR into a 2-year orbit, making a deep-space maneuver near aphelion, and then using an Earth flyby to tilt its orbit to reach Eros. One benefit of this trajectory was that the energy requirements at launch would be even less than for Nereus, part of the attraction for which had been its low energy requirement. In mid-1993 Eros was adopted as the baseline target, with the launch date being advanced to February 1996. If this new mission profile could be met, it would make NEAR the first Discovery mission into space.[39] In 1994 the mission designers investigated whether the initial orbit would enable NEAR to make flybys of other objects, and compiled a list of no fewer than 43 candidates. Although most of these bodies were minuscule, the trajectory passed 2.25 million km from (253)

Mathilde in the main asteroid belt between Mars and Jupiter, and it was decided to accept the penalties imposed by detouring in order to trim this range to 1,200 km.[40] Although discovered in 1885, little was known about Mathilde apart from the fact that its estimated diameter was 60 km; which would make it the largest such body as yet visited by a spacecraft. At the prospect of a flyby, Mathilde was investigated and found to be a fairly dark body with spectral characteristic similar to carbonaceous chondrite meteorites, making it a C-class, and to have an unusually long rotational period of 415 hours – in excess of 17 days. As tidal interaction with an unseen satellite was one possible reason for a very slow rotation, close-up observations of Mathilde were eagerly sought. In the event that the launch window for Eros were to be missed, two backup opportunities were identified for 1997, one to Nereus and the other to Anteros.[41]

The octagonal aluminum honeycomb body of the spacecraft was 1.47 meters in diameter and contained a propulsion system consisting of two tanks for 109 kg of nitrogen tetroxide and three tanks for 209 kg of hydrazine. The hydrazine fed four 21-N and seven 3.5-N monopropellant thrusters used for attitude control and small orbit corrections and, together with nitrogen tetroxide, the large bipropellant course correction and orbit-insertion engine which delivered a thrust of 450 N. The main engine protruded from one side of the body, which was protected by a metallic heat shield. Attitude determination was by a combination of star trackers, Sun sensors and innovative hemispherical-resonator gyroscopes that had never flown in space before. On the top of the body were clusters of attitude control thrusters, four fixed 1.83 × 1.22-meter solar panels in a 'windmill' pattern, and the 1.5-meter-diameter parabolic antenna that gave the craft a total height of 2.8 meters. The solar panels unfolded after launch, and would provide 1,880 W at 1 AU and 400 W at 2.2 AU. In fact, this would be the first solar-powered spacecraft to operate beyond the orbit of Mars. The design was simplified by the fact that the Sun–spacecraft–Earth angle would rarely exceed 40 degrees, since this enabled both the high-gain antenna and the solar panels to be installed in a fixed configuration without incurring significant performance penalties. In addition to the high-gain antenna, the spacecraft carried a fan-beam medium-gain antenna for use at times when the solar panels were facing the Sun and a high data-rate was not required; and a pair of low-gain antennas, one on the top and the other on the base.[42] Most of the body-fixed scientific instruments were on the base, along with the launch vehicle interface. The multispectral imager was based on an instrument that APL had provided for the SDI's Midcourse Space Experiment (MSX). It used 4.8-cm-diameter refracting optics, a 537 × 244 pixel CCD detector, eight colored filters, and was capable of a resolution of 10 meters at a range of 100 km. A 64-channel near-infrared spectrometer would seek evidence of minerals such as olivine and pyroxene on the surface of the target. At 27.3 kg the largest instrument was a spectrometer sensitive to X-rays and gamma rays that was to map the abundances of various elements, including magnesium, aluminum, silicon and potassium. A laser altimeter would measure distances from the surface with a resolution of 0.5 meter in order to provide a global topographical map and to assist with navigation during the orbital phases of the mission. Finally, a 3-axis fluxgate magnetometer was mounted on the high-gain antenna feed to study the solar wind

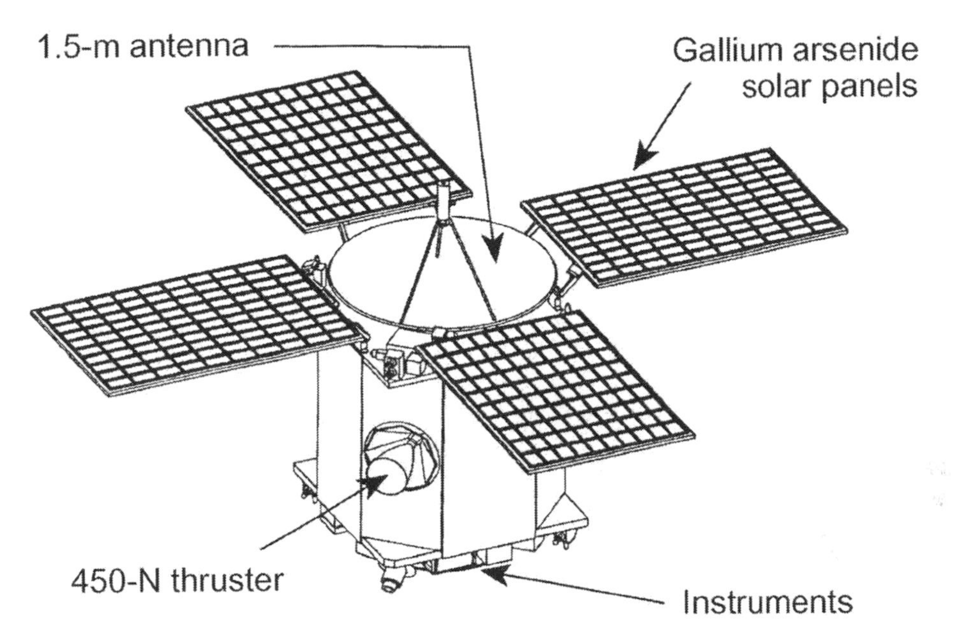

The Near-Earth Asteroid Rendezvous (NEAR) spacecraft, which was the first mission of the Discovery program to be launched.

and how this interacted with the target. The scientific suite accounted for 56 kg of the 805-kg launch mass.

The NEAR spacecraft was delivered to Cape Canaveral in December 1995. On 16 February 1996 the first opportunity of the 16-day window was lost due to winds at high altitude and malfunctioning equipment, but at 20:45 UTC on 17 February the Delta II lifted off. After a few minutes in parking orbit, the upper stage inserted the spacecraft into a 0.99 × 2.18-AU orbit. There was a moment of concern when a new receiver installed at the Canberra station of the Deep Space Network failed to receive telemetry at the end of the escape maneuver, but when the antenna was switched to an older receiver the telemetry appeared, establishing the fault to be in the ground equipment.[43] Soon thereafter, the multispectral camera was calibrated by taking 40 images of the Moon from a distance of 1.5 million km. On 24 March it imaged the 'bright comet' C/1996B2 Hyakutake, which was at that time prominent in the northern skies of Earth. The characteristic long thin plasma tail was not seen, but the inner coma, 16.7 million km away, was easily spotted. These observations confirmed the ability of the spacecraft to image a rapidly moving target. During the next year and a half, NEAR performed five course corrections and more instrument calibrations. On 19 February 1997 it was occulted by the disk of the Sun as viewed from Earth, and engineering and scientific experiments were undertaken to assist in the planning of future missions which were under study not only at JPL but also at APL, including the descendants of the 1980s Starprobe that was to have had a very small perihelion.[44]

Comet Hyakutake as seen by the NEAR spacecraft.

The first real burst of activity came with the encounter of Mathilde in late June 1997. Starting 42 hours prior to the time of closest approach, the spacecraft took a sequence of images in the predicted direction of the target to identify it against the background of stars for optical navigation. But because the approach was from the asteroid's night-side, the dim speck was almost lost in the Sun's glare. Because the heliocentric distance was 1.99 AU, the solar panels provided only sufficient power to operate the imager; the other instruments remained inactive. The actual imaging sequence did not start until 5 minutes before closest approach, with the first images showing an irregular crescent at a resolution of 500 meters. At 12:56 UTC on 27 June 1997 NEAR flew by Mathilde at a range of 1,212 km and a relative speed of 9.93 km/s. The highest resolution images obtained at that time recorded details as small as 160 meters. The imaging continued for 20 minutes after closest approach, and concluded with 200 frames of the space surrounding the asteroid to search for a satellite, but nothing as large as 40 meters was found. The flyby produced a total of 534 images. As a result of the asteroid's extremely slow rotation, it was possible to view only one of its hemispheres, and since no shadow migration was detected it was impossible to infer the spin axis. Based on the observed hemisphere, Mathilde proved to be 66 × 48 × 46 km in size and to have a very dark surface that reflected only 3.5 to 5 per cent of the light that it received from the Sun. It had at least five large craters with diameters greater than 19 km. The largest crater, which was also the best imaged, had a diameter of 33 km (i.e. half the asteroid's major axis) and a depth of between 5 and 6 km – this was difficult to determine because its floor was in shadow. This was partly overlapped by another crater with a diameter of 26 km. Despite the severity of these impacts, except for possessing a few polygonal craters the asteroid did not show any evidence of fracturing or grooving resulting from the shock of such major events, which suggested that it was a 'rubble pile'. This was supported by the Doppler tracking data, which gave an estimate of the mass of the asteroid, and thereby its mean density, showing this to be very low. Inspired by the dark surface, the International Astronomical Union named the features on Mathilde after terrestrial coal fields, with the largest crater becoming Karoo.[45,46,47]

On 3 July 1997 NEAR executed a large trajectory correction of 269 m/s to lower its perihelion from 0.99 to 0.95 AU in order to target the gravity-assist that it was to receive from the Earth flyby. In August, the magnetometer detected a magnetic cloud which took almost 6 hours to wash over the spacecraft. It is possible that this was ejected by the coronal mass ejection observed the previous week by the SOHO spacecraft. Several other magnetic field disturbances were later observed that could be traced back to solar phenomena.[48] On returning to Earth in January 1998, NEAR calibrated the magnetometer and laser altimeter. Moreover, to test the accuracy of its attitude control system, it oriented itself so that its solar panels reflected sunlight towards Earth to enable amateur astronomers to monitor its passage across the sky. The closest point of approach was at 07:23 UTC on 23 January, at an altitude of 540 km above southwestern Iran. It performed two sequences of observations with the camera and infrared spectrometer, the second of which covered Antarctica. It also took images of the Moon's southern hemisphere for data on scattered light. Further imaging of Antarctica on the outbound leg was used to produce an amazing 'Earth spin movie'.[49] It was later discovered that NEAR gained an unaccountable excess velocity increment of 13 millimeters per second from the flyby. A similar, although lesser, excess had been observed for the Galileo spacecraft. The NEAR flyby is the largest such anomaly seen to date.[50] More than 2 months later, when the spacecraft was 33,650,000 km from Earth, sunlight again briefly reflected off its solar panels and telescopes were able to see it, thereby establishing a record for the most distant spacecraft ever imaged from Earth.

Systematic optical navigation of Eros began in November 1998 and continued at weekly intervals until 14 December, and then at daily intervals until 19 December. At 22:03 UTC on 20 December the bipropellant engine was to start for a 15-minute burn to increase the spacecraft's velocity by 650 m/s to rendezvous with Eros in its orbit. However, after a series of low-thrust burns designed to settle the propellants in the tanks, the main engine cut off within a fraction of a second of igniting. Some 37 seconds later, communications ceased. In this circumstance, NEAR should have entered a spin-stabilized 'safe mode' that maintained its solar panels facing the Sun and swept its fan-beam antenna by Earth at 3-hourly intervals. Nothing was heard. One possibility was that the batteries had discharged, and that while the spacecraft was recharging them it had switched off every system, including the transmitter. It

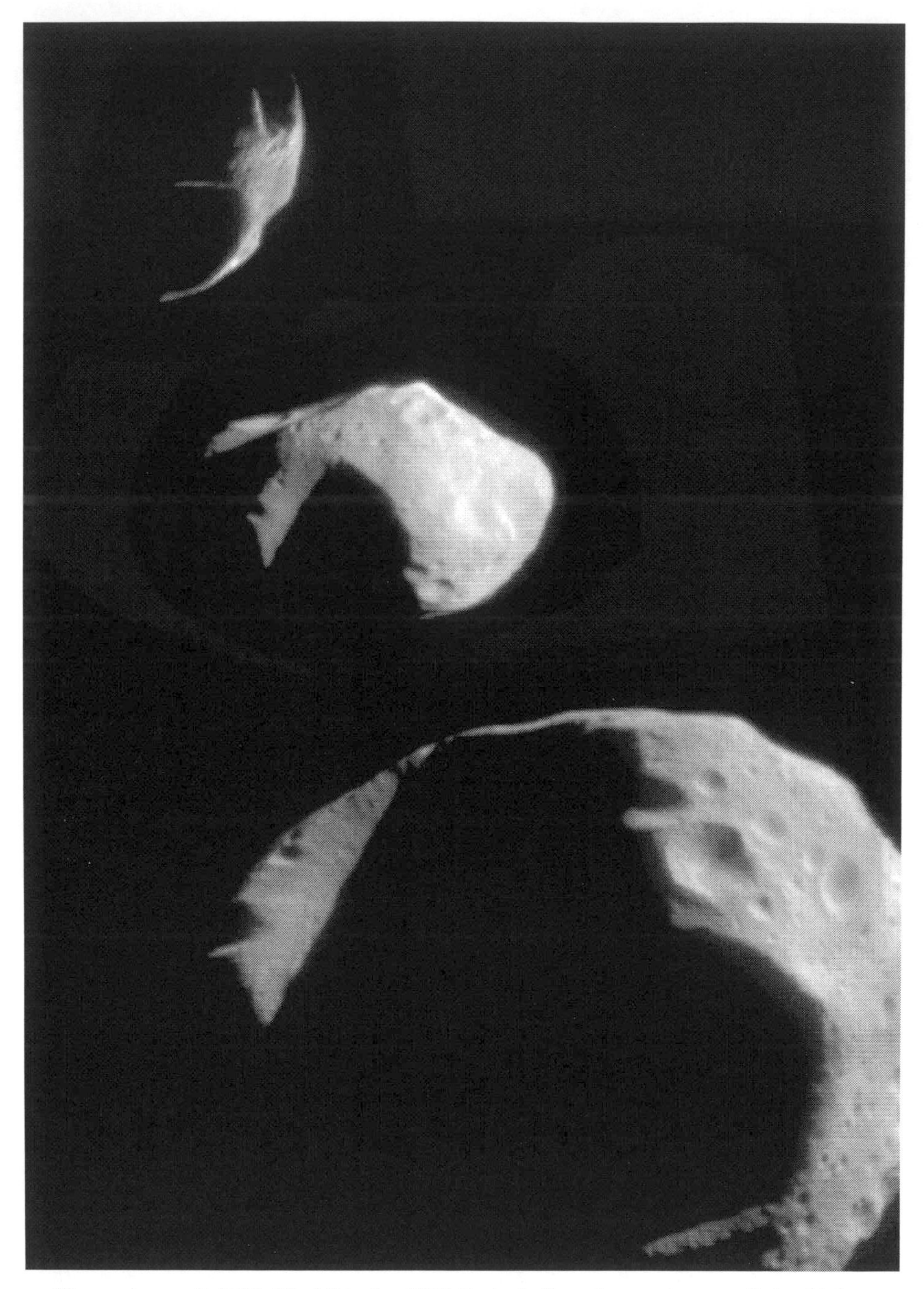

Three views of (253) Mathilde by NEAR, including (bottom) one of the highest resolution frames.

The south polar regions of Earth and the Moon viewed by NEAR after its flyby.

was an anxious wait, but after 27 hours of silence contact with NEAR was restored. In the meantime, NEAR had tumbled, at one point facing its solar panels directly away from the Sun, and used about 29 kg of hydrazine trying to stop this rotation. In addition, a residue of efflux had coated the optics, with the result that from this point on computer processing would be required to overcome a slight degradation of the image quality.[51] A major concern was that the spacecraft, having failed to make the rendezvous maneuver, was rapidly closing on Eros and within 24 hours would pass it by at a range of less than 4,000 km. To recover some science from the situation, a contingency flyby imaging sequence was prepared. The flyby at 18:41 UTC on 23 December was at a range of 3,827 km and a relative speed of just 965 m/s. The imaging started 3 hours prior to closest approach, and over an interval of 6.7 hours obtained 1,026 frames.[52] However, owing to the uncertainty in the position of the asteroid, only 222 frames captured all or part of it. The best resolution was about 400 meters. As inferred from ground-based observations, Eros is a highly irregular body. The imagery revealed it to be 33 × 13 × 13 km in size, with a convex and a concave hemisphere giving it a distinctive 'bean' or 'kidney' shape. Despite its elongated form, the presence of five craters about a kilometer in size and linear markings 20 km in length suggested a solid body, and this was consistent with the mass and density inferred from Doppler tracking. Near-infrared spectra indicated the presence of olivine and pyroxene on the surface of the asteroid. The imaging on the inbound and outbound legs covered the entire realm of gravitational influence, but there was no evidence of a satellite larger than 50 meters.[53,54]

The results from the impromptu flyby fell far short of what had been expected from rendezvousing with the asteroid to conduct a detailed investigation. However,

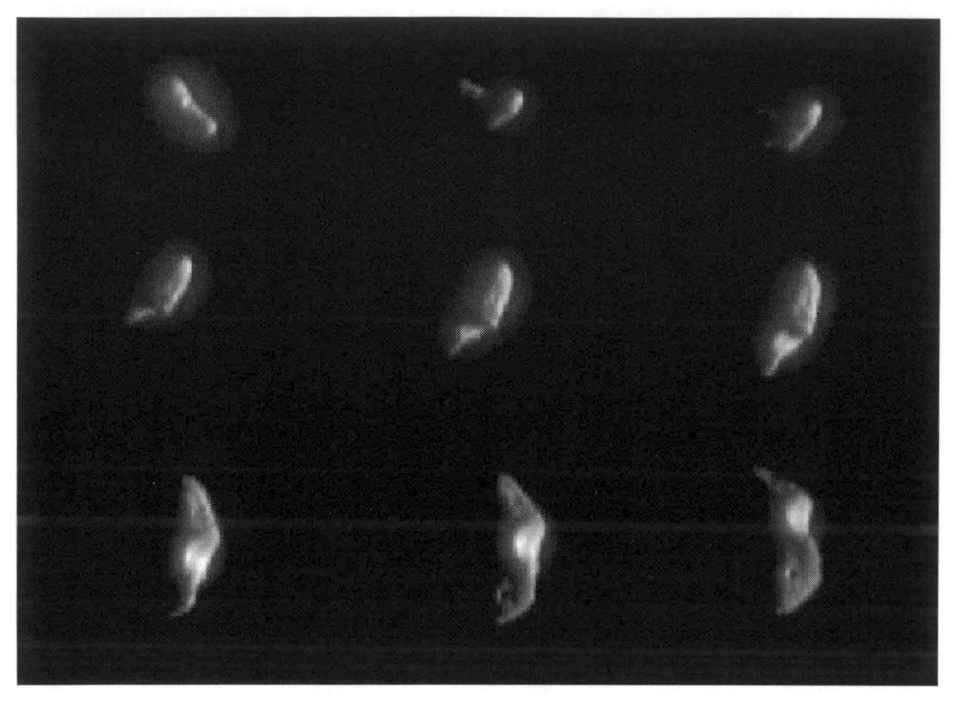

Images of (433) Eros taken by NEAR during its tantalizing December 1998 flyby.

there were options to recover this science. If a large burn were to be executed in the next few days this would permit a return to Eros in the near future, but would leave little propellant for the maneuvers planned for when in orbit of the asteroid. On the other hand, a maneuver early in January 1999 could set up an approach to Eros one year later, at which time the rendezvous maneuver would leave sufficient propellant to conduct the orbital phase of the mission. But first the cause of the bipropellant engine's misfire of 20 December had to be determined. It was found that the threshold of an accelerometer to detect uncontrolled thrust during engine burns had been set too low and had been exceeded by the engine-start, which led the guidance software to cancel the burn and enter safe mode. Once this value had been changed, it should be possible to fire the engine.[55,56] On 3 January 1999 the spacecraft accomplished a 24-minute burn to increase its heliocentric velocity by 932 m/s and thereby set up a return to Eros on 14 February 2000 – this date being a compromise to save several months of operating costs and to mark Valentine's Day.

Regular observations of Eros resumed when the range reduced to 100,000 km in December 1999. An extensive search for small satellites down to 20 meters in size was performed, but none were found. On 3 February 2000 a maneuver set up the desired rendezvous. At 15:33 UTC on 14 February, when NEAR was some 200 km ahead of Eros as viewed from the Sun, as dictated by the requirements of the scientific observations, it fired its engine to slow its heliocentric velocity by 10 m/s and so be captured by the asteroid's weak gravity – the first time that this had been done. A series of six correction maneuvers gradually lowered the 321 × 366-km initial orbit to a 50-km near-circular polar orbit on 30 April. At about that time the spacecraft was dedicated to the memory of Eugene M. Shoemaker, who died in 1997; making it the NEAR–Shoemaker mission. He had led the field in understanding the effects of small asteroids impacting on Earth.

A spacecraft in orbit of Eros had peculiar characteristics owing to the asteroid's irregular shape and small mass. In the highest orbits the spacecraft's speed was no greater than walking speed. The solar radiation pressure was sufficient to displace the focus of the orbit up to 900 meters from the center of mass of Eros, and an orbit below 35 km was unstable. Doppler tracking and landmark sightings provided an accurate orbit determination, which in turn provided the mass of the asteroid. By interpolating laser altimetry to compute the shape of the asteroid, the mean density could be determined. This proved that unlike Mathilde, Eros is a solid body. Owing to the elongated shape, the escape speed from the surface of Eros varies by a factor of five depending on the position.[57]

A mosaic of Eros taken on 3 March 2000 showing details as small as 20 meters. Note the large boulders on the floor of the crater Psyche. (JPL/NASA/Caltech)

The imagery showed Eros to be peppered by craters with diameters in the range 0.5 to 1 km. The largest individual features identified were a 5.5-km and a 7.6-km crater, and an irregular depression 10 km across which gave the body its distinctive 'saddle' shape. The saddle is believed to have been formed by an impact at a time when Eros was part of a larger parent asteroid, or perhaps as this progenitor broke up. The IAU nomenclature for Eros drew upon mythology and lovers in literature. The 5.5-km crater was named Psyche, and the saddle Himeros. As an exception to this rule, the 7.6-km crater was named in honor of Shoemaker. Various grooves, ridges and depressions were also evident. A ridge seen during the 1998 flyby rises up to 200 meters, varies between 1 and 2 km in width and extends along half of the asteroid. A complex ridge that spans much of the northern hemisphere consists of a series of segments several tens of meters tall and less than 300 meters wide which may represent failures induced along lines of weakness by a large impact. Millions of boulders were scattered around, with a strong concentration in a depression west of Himeros. However, there was no clear indication that these had been tossed out by impacts, or that they had rolled downslope. Perhaps they are debris accreted in low-energy impacts. Hints were found of the presence of regolith, which appears to have obliterated all but the largest craters, creating a crater-size distribution that is unique to Eros. At least one area was found (the saddle) in which the crater density was much lower than the average, suggesting that the regolith was unusually deep in this location or that the area had been resurfaced in some way. Many mysterious bright patches of soil might indicate recent exposures. It was evident that Eros had not been ejected from the main belt of asteroids by a collision, as such an impact would have erased most of the surface features. Perhaps its orbit was perturbed by an orbital resonance with Jupiter, or by the slow but progressive Yarkovsky effect related to solar heating.

The color, infrared spectra and mean density of Eros were all consistent with the composition of ordinary chondrite meteorites. The infrared spectrometer failed on 13 May, before its observations reached the southern hemisphere.[58,59] The X-ray and gamma-ray spectrometer had difficulty obtaining more than a few spectra from such a small body, but a series of solar flares starting in May bathed the asteroid in charged particles, enabling better data to be obtained which indicated that Eros was not a thermally differentiated body, and confirmed it to have a surface composition similar to ordinary chondrites – albeit with some differences.[60]

After imagery of both hemispheres had been obtained, it was decided to further investigate the regolith. The orbit was adjusted to skim over the surface at a height of 6,430 meters on 26 October. During this flyover at a 'cycling speed' of 22 km/h about 250 images with resolutions as good as 1 meter were obtained, together with fine-scale laser altimetry. The gently undulating surface confirmed the paucity of small craters down to 3.5 meters in diameter. The presence of an extensive layer of regolith was inferred from observations of partially buried rocky debris, and by the discovery of smooth deposits no more than a few meters deep on crater floors.[61,62] These 'ponds' appeared to be concentrated near the equator, and various processes have been suggested for their creation, including fine ejecta from the formation of Shoemaker crater, downslope movement caused by seismic shock, and electrostatic

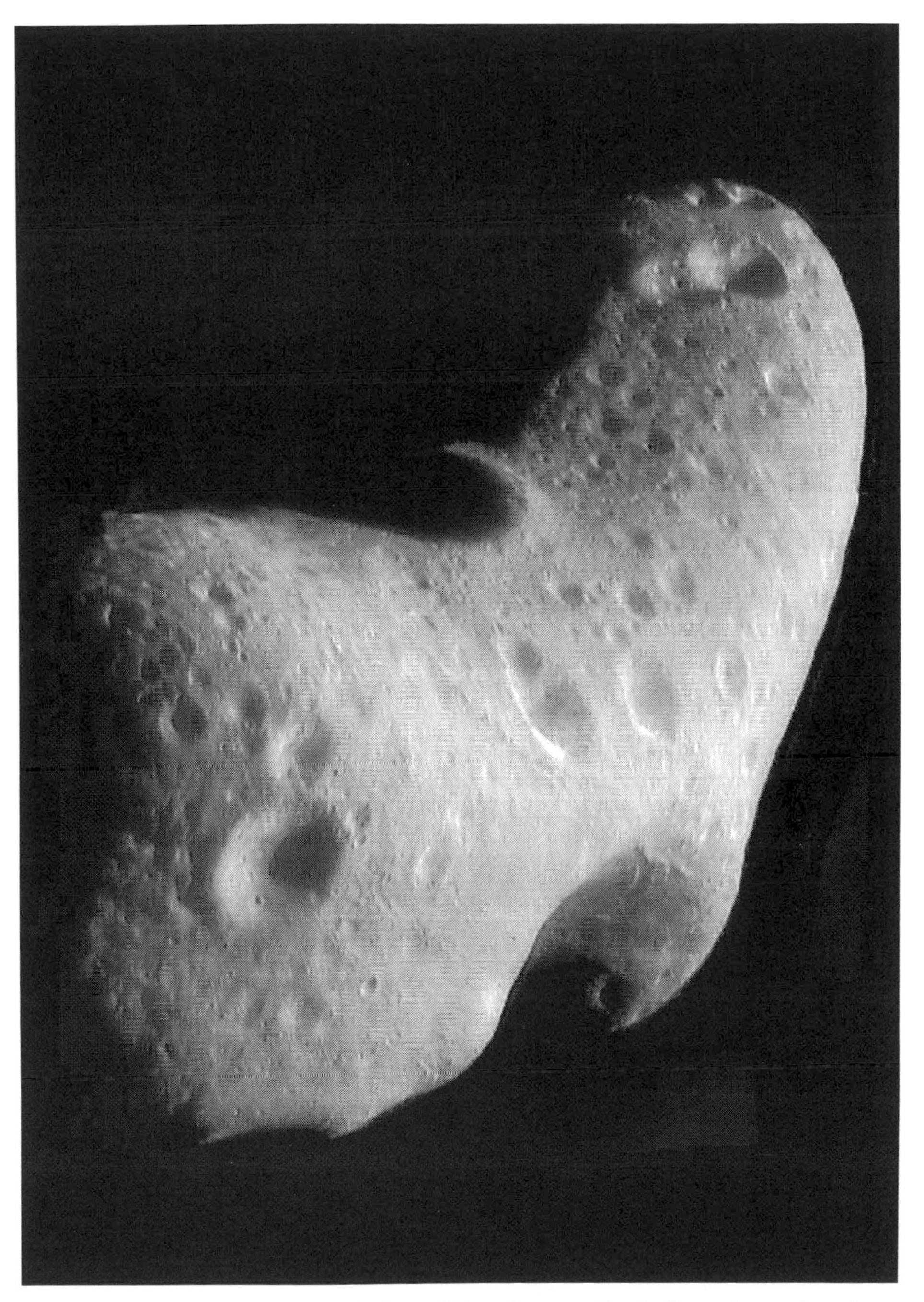

In this view of the northern hemisphere of Eros the crater Psyche forms the notch at the top, and Himeros, the saddle, is on the opposite side. (JPL/NASA/Caltech)

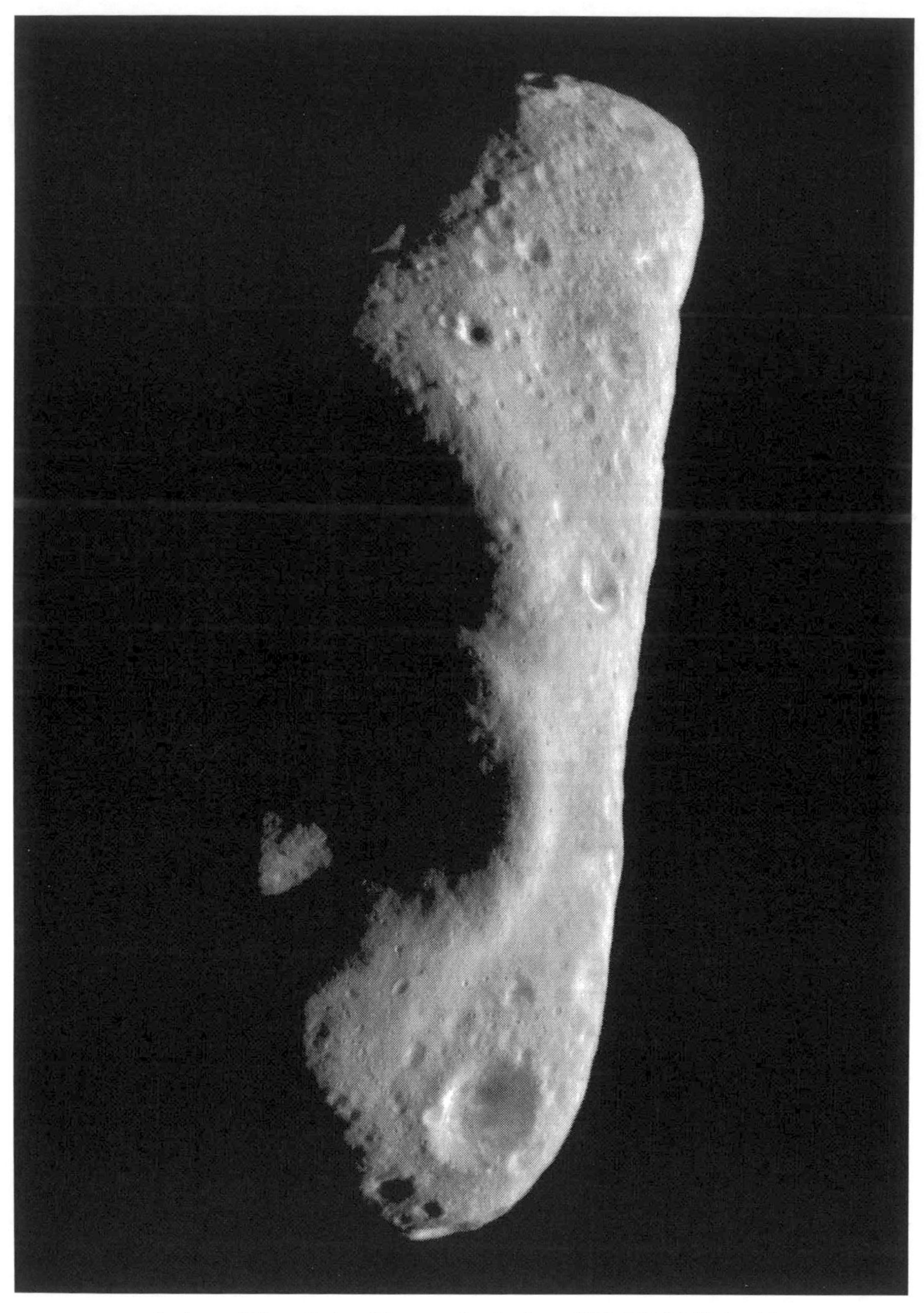

A view of Eros facing Himeros in shadow. (JPL/NASA/Caltech)

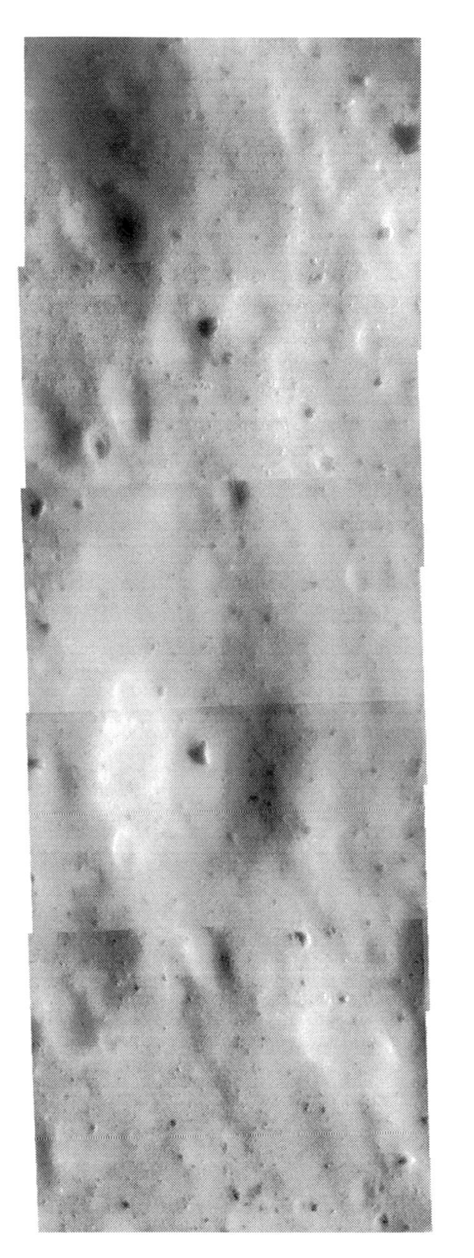

A mosaic taken during NEAR's low-altitude flyover on 26 October 2000. Note the smooth terrain and the paucity of impact craters. The laser altimeter showed the boulder at top center to be 20 meters across and to stand 7 meters high, meaning that most of it is probably buried in fine regolith. The altimeter path also crossed the flat 'ponded' deposit inside the 190-meter-diameter crater at the center.

levitation.[63] After this low pass, the spacecraft was in an unstable orbit that could have ended in an impact, so on approaching apoapsis it maneuvered into a 200-km circular orbit.

In all, a total of 25 orbit correction maneuvers were executed, with the orbital radius ranging from an initial 365 km down to 19 km between 25 and 28 January 2001, when the spacecraft made a series of low passes over the 'ends' of Eros at an altitude of 2.7 km.[64] The plan had been to conclude the orbital mission a year after orbit insertion simply by commanding NEAR to switch off, but a more flamboyant conclusion was devised in which it would be steered towards a slow-speed impact. On 12 February 2001 the orbit was first lowered to 36 × 7 km, and at periapsis the spacecraft performed a deorbit burn to establish a vertical descent trajectory. Five further burns were made during the 4.5-hour fall in the asteroid's weak gravity to direct the descent towards the saddle. In the final 37 minutes, the spacecraft took about 70 images of the landing area. The view from altitudes above 1 km showed a blocky surface, but below this altitude it was of a relatively block-free area which probably corresponded to a 'pond' of dust. The final image was taken at an altitude of 129 meters and showed the floor of a 100-meter crater, with two small collapse pits in the dust, several centimeters deep. When the spacecraft came into contact with the surface its orientation was altered, preventing the antenna from pointing at Earth, with the result that the last quarter of the final image was lost. NEAR gently touched down at a speed of 1.6 m/s, bounced slightly, and at 19:44 UTC came to rest at an estimated position of 35.7°S, 279.5°W, almost certainly on one side and on the corners of two of its solar panels. For the first time, a US spacecraft was the first to land on another celestial body.[65,66,67,68,69] A beacon signal was immediately received, followed by telemetry which showed that NEAR was in excellent health. If the spacecraft had settled with the imager facing towards the horizon, a stunning new perspective on the asteroid would have been obtained, but the instrument was buried in the dust. On the other hand, because the gamma-ray spectrometer was facing down it would be able to yield high-quality data on the composition of the surface. Consequently, a two-week extension was granted to the mission in order to integrate a gamma-ray analysis. During this time, magnetometer data was returned at slow speed using the low-gain antenna. Finally, on 28 February, as 'winter' loomed at the landing site, the final data was downloaded and the spacecraft was ordered into hibernation.[70,71]

When the Sun rose over the landing site in late 2002, a single attempt was made on 10 December to revive NEAR, but there was no response, possibly because the spacecraft had succumbed to the extreme cold during the months spent in darkness. The mission had been an outstanding success, returning over 160,000 images and 11 million laser 'echoes', making a brief flyby of one asteroid and then characterizing another in detail for the first time. Furthermore, it had established the feasibility of low-cost planetary missions.

The last four images of NEAR's descent sequence on 12 February 2001. Note the transition from the rocky terrain at right to the smooth 'ponded' terrain at left. The large boulder in NEAR's last frame at left is about 5 meters wide. The spacecraft landed some 7 meters beyond the lefthand edge of the mosaic.

COMPLETING THE CENSUS

Little was learned of Pluto in the 40 years following its discovery apart from the fact that it traveled around the Sun in an inclined and eccentric orbit. Estimates of its mass and size, and therefore its mean density and composition, were extremely uncertain. Only in the early 1970s was it discovered that the spin axis lies almost in the plane of its orbit, as is the case with Uranus; and this explained a trend toward dimming which had been observed since the 1950s – it was a seasonal effect due to the reduced exposure of a bright polar cap as the planet approached equinox. The first crude spectra were also obtained in the early 1970s, and indicated the presence of ice, particularly water ice, on the surface. However, the discovery which would revolutionize our understanding of Pluto came in 1978, when James W. Christy of the US Naval Observatory noticed that in telescopic images Pluto seemed to have a bulge that moved in a regular manner from night to night – it was a satellite! This made possible the first accurate measurement of the mass of Pluto, proving it to be just 0.2 per cent that of Earth. The mass of Charon, as Christy named the moon, is 1/11th that of Pluto. In relation to its parent, this made Charon the solar system's largest moon.*

* The next closest match, our own Moon, is 1/81st the mass of Earth.

Soon other peculiarities were noted. Charon orbits Pluto in the same time that the planet takes to complete an axial revolution, causing the moon to hover above the same spot on the planet. It was also realized that starting in about 1988 the orbit of Charon would be viewed edge-on from Earth, with the result that for a few years before and afterwards Pluto and Charon would undergo a series of mutual eclipses and occultations. As this geometry occurs only twice during Pluto's 248-year orbit of the Sun, the discovery of Charon was timely. By observing such events it would be possible to calculate the diameters of the bodies and, with their masses known, also their densities. Furthermore, by measuring the spectrum of Pluto when Charon was hidden and then subtracting this from the combined spectrum, it would be possible to infer details of their respective surfaces. Such observations over the next decade showed Pluto and Charon to be quite different. Whereas the surface of Pluto was rich in methane, nitrogen and carbon monoxide ice, Charon appeared to be covered by crystalline water ice. Since methane ice at very low temperature has a structural behavior quite different from that of water ice, the surface of Charon could record impacts dating back billions of years whilst the surface of Pluto will have tended to relax and erase all but the most recent impacts. Observations of Pluto occulting a star in 1988 showed an anomalous dimming that revealed the presence of a tenuous atmosphere with a surface pressure one-millionth that of Earth, and also the possibility of hazes near to the ground. Similar observations of Charon showed no detectable indication of an envelope. However, Pluto's atmosphere was probably due to the sublimation of ices while the planet was passing perihelion in 1989, making it a short-lived feature. Nevertheless, in a certain sense Pluto is a huge comet![72]

In view of the fact that until recently our knowledge of this frigid world was so paltry, it is hardly surprising that it attracted little interest from those who planned solar system exploration. Only the original concept for the Grand Tour included a reconnaissance of Pluto; and in the actual case, that of Voyager 1, this opportunity was sacrificed in favor of a close inspection of Saturn's large moon, Titan, which in retrospect was a sensible decision.[73] A study of a dedicated Pluto mission began in 1989, just 60 days after Voyager 2 flew by Neptune, whose amazing moon, Triton, is believed to closely resemble Pluto. Because deep-space missions were becoming ever more complex and expensive, it was felt that a costly reconnaissance mission would be rejected, so the goal of the study, called 'Pluto 350', was to develop a spacecraft with a total mass, including payload and RTG power supply, of 350 kg, which was half that of Voyager. The minimized payload would be only a camera, an ultraviolet spectrometer, a plasma and a radio-science package. It would address some basic questions about this unique system, characterize the surfaces of both bodies, study the structure of the planet's atmosphere and haze layers, and find out whether the perceived similarities between Triton and Pluto are real. Of particular interest would be observations to understand the process which prevents the bright south polar cap from weathering and growing dark – such as a seasonal effect due to the eccentric orbit and highly inclined spin axis, or some kind of icy volcanism that periodically resurfaces the poles. In addition, the mission would search for smaller moons orbiting Pluto. It was soon noted that the nature of the Pluto system meant that if a mission were not to be launched soon, then the science it could undertake would be diminished. In particular, whereas at equinox in the late 1980s almost all of Pluto's and Charon's surfaces had been illuminated, by 2015 almost half of their southern hemispheres would be in darkness. Furthermore, since the planet would chill as it drew away from the Sun after perihelion, if the spacecraft were to arrive after about 2020 the atmosphere would have collapsed onto the surface. To save on the cost of a large launcher to send the spacecraft directly to Pluto, it would use the gravity-assist technique to build up the energy required to reach its destination. The key was a Jovian flyby, but if encounters with Venus and Earth could be arranged this might enable the standard Delta II to be used. Ironically, the Pluto 350 proposal was criticised for its low mass and high perceived risk. In response, NASA studied a costly Mariner Mark II flyby mission with a spacecraft based on the architecture developed for Cassini. On approaching Pluto, the spacecraft would release a small battery powered probe on a trajectory timed to arrive either half a rotation before or half a rotation after its mothership, in order to image the hemisphere that would be in darkness for the main event. Nevertheless, in a time of shrinking budgets for scientific missions, Pluto 350 was recommended over the more expensive but less risky Mariner Mark II proposal.[74] Meanwhile, thanks to the lobbying of the 'Pluto Underground' group of US planetary scientists, support for a mission had grown to the point that NASA's Solar System Exploration Committee recommended it as a priority mission for the next decade.

At that point, history took an unexpected turn. JPL engineers Robert Staehle and Stacy Weinstein took inspiration from a series of American stamps dedicated to the exploration of the solar system which depicted Pluto as a fuzzy ball with the words "not yet explored", and they called on JPL to study the smallest possible spacecraft that could accomplish three key scientific objectives of mapping the surfaces of Pluto and Charon at a resolution of 1 km; making temperature maps of their visible hemispheres; and profiling the atmosphere of the planet. As one of the authors of a 1981 study of a 20–30-kg spacecraft to study the solar corona (a sort of 'Sunblazer Mark II'), Staehle had already been involved in 'microspacecraft' and was familiar with the challenges. The results of this Pluto mission study were surprising, and, in hindsight, in line with the "faster, cheaper, better" mantra that came to dominate the development of spacecraft of all types during that decade. With a mass of less than 150 kg, the Pluto Fast Flyby (PFF) mission raised the prospect of truly revolutionizing spacecraft design.[75,76] In fact, it was so small that a Titan IV launcher would be able to place it on a direct solar system escape trajectory to Pluto that would cross the 5-billion-km gulf in 7 years. On the original plan, a pair of spacecraft would be launched on trajectories timed to reach Pluto half a revolution apart in order to inspect as much as possible of the surfaces of the planet and its satellite. The payload would consist of a camera, an infrared spectrometer and an ultraviolet spectrometer. Of course, the trajectories would be arranged to provide radio-occultations of both Pluto and Charon. To prove that the entire payload would not exceed 7 kg, and to choose the instruments for the spacecraft, NASA funded seven engineering studies, the most interesting of which were two highly integrated systems which combined the three instruments in a single unit with a common structure, electronics and (in one case) even optics.[77] By 1994, it seemed likely that Pluto Fast Flyby, whose mass had meanwhile increased to the 150-kg limit, would be built and launched in the first decade of the new millennium. In the meantime, an even more innovative (and riskier) Pluto mission was proposed to the Discovery program. This called for a battery powered spacecraft equipped with only a camera left over from Mars Observer.

At this point, NASA made the unprecedented change in policy of requiring cost estimates for all future missions to include the cost of launch. A pair of Titan IVs would cost in excess of $800 million, and thereby push the overall cost of the Pluto Fast Flyby mission over the billion-dollar mark. Daniel Goldin stressed that the mission would never be approved unless a way was found to reduce the cost of launch.[78] There were three options. One was to stick with the Titan IV but send only a single spacecraft, accepting the loss of redundancy and the loss of some coverage. The second option was to use a smaller launch vehicle and use the gravity-assist technique to build up the energy required to reach Pluto. The remaining option was to invite international collaboration. In fact, if slingshots at Venus and Jupiter were added, then it would be possible to dispatch the mission using either the Molniya or the Proton, both of which were 'workhorses' for planetary exploration.[79] In exchange for a free ride on a Proton (valued at $30 million), the Russians sought participation by, for example, designing small (6-kg) probes that would be released a month before the encounter in order to directly sample Pluto's atmosphere. Referred to as 'Drop Zonds' by the Russians, these could carry either a miniaturized mass spectrometer to measure the composition of the atmosphere, a fish-eye camera to characterize the limb hazes, or another relevant experiment. In this manner, the mission to complete mankind's initial reconnaissance of all the major bodies of the solar system would see its two main players, once rivals, finally cooperating. In fact, Pluto Fast Flyby became part of a larger study for joint US/Russian planetary missions that included a 'Mars Together' mission and a Solar Probe to venture within three solar radii of the photosphere, which forms the visible 'surface' of the Sun. When in late 1994 the Russians said they wished to be paid for the launch, Germany's Max Planck Institute for Planetary Physics said it would cover this in exchange for the mission dropping a probe on Io while flying through the Jovian system. Meanwhile, proposals were being made to exploit Pluto Fast Flyby spares, including fast flybys within 4 years of launch of Uranus, Neptune or (2060) Chiron – which is either an asteroid that develops a coma at perihelion, or a vast and almost dormant cometary nucleus that was discovered in 1977 by Charles T. Kowal as the initial member of a class of Centaur objects traveling in eccentric orbits between Saturn and Uranus.[80,81,82,83]

In 1995, as proposals for cooperating with Russia faltered, largely as a result of the dire finances of that country's space institutions, development of Pluto Fast Flyby halted. It was replaced by Pluto Express, which maintained the planned payload mass but halved the overall mass to 75 kg. It was believed that if Pluto Express was approved, then it could be launched in the first years of the new millennium and reach Pluto in the first half of the 2010s, before conditions there diminished the opportunities for science. Studies were made into alternative power sources to a heavy RTG, and of using a less powerful but cheaper launcher.[84] The mission soon picked up a second (but no less significant) objective which prompted its name being changed to Pluto–Kuiper Express (PKE). When sensitive electronic detectors were introduced at the world's largest astronomical observatories in the late 1980s and early 1990s, this facilitated the discovery by David C. Jewitt and Jane X. Luu in 1992 of the first object in the Kuiper Belt, which forms a reservoir of bodies made mostly of water ice in orbits beyond the orbit of Neptune and near the plane of the ecliptic (hence the alternative name of Trans-Neptunian Objects). This belt, whose existence was postulated by Kenneth E. Edgeworth in 1949 and independently by Gerard P. Kuiper in 1951, is believed to be the main source of short-period comets. In retrospect, it was evident that Pluto and Charon are large members of this class. Object (15760) 1992QB1, which Jewitt and Luu found, is estimated to be one-tenth the size of Pluto. There may well be 100,000 'Kuiperoids' larger than 100 km. By 1995 they were being discovered at a rate of dozens per year. As with Pluto, many of these objects appear to be binary.[85] It is noteworthy that (136199) Eris, which was found by Michael E. Brown in 2005 on images taken two years previously (hence its initial designation 2003UB313) is actually larger than Pluto. The mission planners therefore proposed that if Pluto–Kuiper Express had sufficient fuel after its Pluto flyby, it should go on to investigate at least one of these other objects.

The Pluto mission then became part of the Outer Planets/Solar Probe Project established in 1998 as part of NASA's Origins Program. This called for three 'Ice and Fire' missions to some of the solar system's most difficult targets. In addition to Pluto–Kuiper Express, it included the Europa Orbiter to ascertain whether there really is an ocean beneath the icy crust of the eponymous Jovian moon, and an

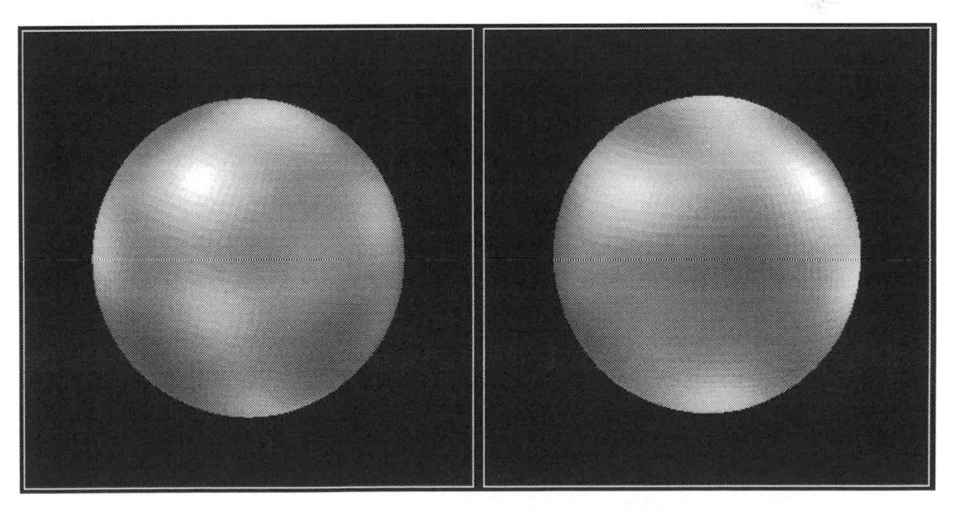

These synthetic views of Pluto, obtained in 1996 by the Hubble Space Telescope are about the best that can be attained from Earth. (Courtesy of Alan Stern of the Southwest Research Institute, Marc Buie of the Lowell Observatory, NASA and ESA)

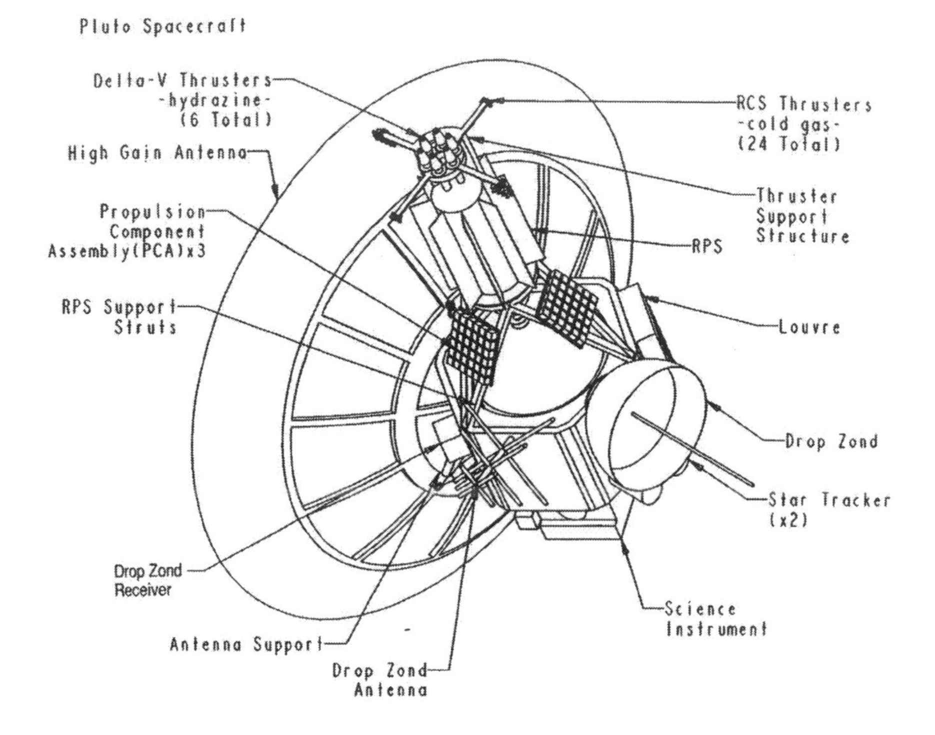

The small Pluto–Kuiper Express spacecraft, with a Russian 'Drop Zond' attached.

ambitious Solar Probe to pass within three solar radii of the photosphere. The tentative schedule was to launch the Europa Orbiter in 2003, Pluto–Kuiper Express in 2004 and the Solar Probe in 2007, but if necessary the Europa Orbiter would be held back in order to allow Pluto–Kuiper Express to meet its launch window. All these missions were to make use of autonomous technologies for self-monitoring and self-commanding in order to minimize ground intervention. They were also to deeply integrate the scientific payload into the spacecraft as a so-called 'sciencecraft'. The core software would be common to all three spacecraft, as would be some of the communication and propulsion components. It was hoped that a high degree of commonality, and JPL's advanced development programs, would enable the cost of each mission to be reduced to less than the already 'cheap' Mars Pathfinder. Further cost savings would be made by using relatively unskilled staff drawn from university students for tracking and other routine activities. Of course, each spacecraft would need systems specific to its own mission. For example, the Europa Orbiter would require extensive radiation shielding because Europa orbits well within Jupiter's radiation belts, and the Solar Probe would require shielding both against the heat of the Sun at perihelion and against the deep cold at aphelion near Jupiter's orbit. Moreover, while the Solar Probe would use solar panels and RTGs, both Pluto–Kuiper Express and the Europa Orbiter would use the advanced new RTGs which were being developed by the US Department of Energy. The baseline launcher for Pluto–Kuiper Express was now the Delta II which, although low-cost, left a shortfall in performance. The option was left open of using the Shuttle/IUS combination, in which case the spacecraft would be augmented by either a deep-atmospheric probe for Jupiter or a microlander for Pluto similar to the Drop Zonds proposed by the Russians.[86,87]

Just as Pluto–Kuiper Express's future appeared to be more or less assured, NASA announced in September 2000 that it had ordered a halt to work in order to concentrate resources on the higher priority Europa Orbiter, but within a matter of months this was canceled. NASA then said that Pluto–Kuiper Express had been put on hold because its projected cost had soared to over $800 million. With no firm date for the resumption of work or a new launch opportunity, Jovian gravity-assisted trajectories to Pluto becoming impractical after the middle of the decade, and the prospect of reaching the planet only after its atmosphere had collapsed, the prospects did not look good. However, scientists, engineers and space enthusiasts joined together in a global campaign to pressure the US Congress to compel NASA to reinstate the mission. It eventually did so, but in the form of a completely new project named New Horizons whose fortunes will be described in a later volume in this series.

LOW-COST MASTERPIECE

After the loss of Mars Observer, NASA administrator Daniel Goldin directed JPL to establish three teams to explore the possibility of recovering the lost science in 1994 or 1996 by either developing a 'Mars Observer 2' rectified of the fault that crippled its predecessor; using miniaturized technology akin to the Department of Defense's Clementine mission; or inviting Russian participation. After digesting the results of these investigations, in February 1994 NASA requested $77 million in its 1995 budget to start the Mars Surveyor program, which would involve a series of missions. The program would be managed by JPL, and its overall objectives would be to explore Mars; to seek evidence of life, whether past or present; to understand the current climate, and investigate how this has changed over time; and to map its resources. The core theme of the investigation was the role of water in the planet's history. The program envisaged launching an orbiter in 1996 and pairs of orbiters and landers at each launch window thereafter, through to a low-cost sample-return mission at some point between 2005 and 2010. The budget was capped at $100 million per year, with an additional $20 million for operations.

The name was reminiscent of the Surveyors that NASA had sent to the Moon in the 1960s to prepare for the Apollo landings, and as such hinted at robotic pioneers clearing the way for human missions to the Red Planet. In fact, the long-term goals included the development of a means of manufacturing propellant on the surface of the planet as part of a cost-effective robotic sample-return mission and, ultimately, a human landing.

After an extremely fast 2-month selection process, in July 1994 JPL announced that the Mars Global Surveyor (MGS) spacecraft to be launched in 1996 would be designed and built by Martin Marietta in Denver, Colorado, which had made Mars Observer. The 28-month development phase was one of the briefest in recent US space exploration. It was estimated that spacecraft development, launch, cruise and operations in orbit around Mars for one local year would cost about $250 million. A key innovation was to use the aerobraking technique which Magellan had tested in Venus orbit. Mars Global Surveyor would enter into a highly elliptical capture orbit with a period of about 2 days, and then employ aerobraking to turn this into the low circular 2-hour mapping orbit which Mars Observer would have achieved by a series of propulsive maneuvers. The benefit of aerobraking was evident in a comparison of the amount of propellant required to attain the same orbit: Mars Observer had carried 1,536 kg, but Mars Global Surveyor would need only 393 kg! Whereas Mars Observer, with a launch mass of 2,500 kg, had required a Titan III launch vehicle costing in excess of $500 million, Mars Global Surveyor, weighing a mere 1,062 kg, could be launched on a Delta II costing only $55 million, with a PAM-D solid-rocket setting it on course for Mars.

The penalty paid for utilizing this strategy to achieve the operating orbit was a postponement of mapping, but the aerobraking phase of the mission would last just a few months. The mapping orbit would be 378 km high with a period of just under 118 minutes, and by virtue of being inclined at 92.9 degrees to the equator it would be sun-synchronous, and therefore precess at the same rate as Mars traveled around its orbit. This meant that it would always view the surface in the same illumination conditions. As a compromise between the differing requirements of the instrument teams, the spacecraft was to cross the equator at 2 p.m. local time. As the ground track would migrate eastwards by a mere 59 km every 7 days or 88 revolutions, the spacecraft would be able to make a series of observations of any site which was of particular interest.

As conditions in the Martian atmosphere at a height of about 115 km, at which aerobraking would occur, were not precisely known, the spacecraft was built with sufficient margin to ensure that it would not overheat or suffer damage in any other way. In fact, the aerobraking strategy was made flexible enough to cope with large variations in the density of the upper atmosphere – such as if it were to inflate in response to being heated by a dust storm, since the mission would occur at the peak of the storm season – and the uneven gravity field. A significant portion of the project's contingency fund was allocated to improving the model of the upper atmosphere. The radio tracking would be almost continuous during an aerobraking pass, in order to monitor the spacecraft's progress and provide a timely response to any anomaly or unplanned event, and 4 hours on each orbit would be dedicated to determining the parameters of the orbit to measure the progress of the process. This would continue until the apoapsis was reduced from the initial altitude of about 54,000 km down to 1,800 km, and the period was 2.4 hours. Then the spacecraft would become more active, firing its thrusters at apoapsis to gradually lift its periapsis to about 140 km at a pace which was consistent with the diminishing effects of aerobraking acting to put the apoapsis at a height of 450 km.

This 'walk-out' was to ensure that in the event of a loss of communication, JPL would have several days in which to re-establish control of the spacecraft before it succumbed to drag. Aerobraking would finish with engine burns to circularize the orbit. A significant complication in this strategy was that the final orbit had to be established in such a manner that the spacecraft would cross the equator within a few minutes of 2 p.m. local solar time on each orbit. The plan was to initiate mapping in March 1998. This phase of the mission would last a local year of 687 Earth days. In the ensuing 6 months, Mars Global Surveyor was to serve as a relay satellite for any lander that might be on the surface. As its mission drew to an end, the spacecraft was to raise its orbit in order not to fall into the atmosphere and rain debris onto the surface, because it had not been sterilized.

To reduce costs and to speed development, Martin Marietta (later Lockheed Martin Astronautics) proposed that 75 per cent of the hardware spares and software which had been written for Mars Observer should be used for the first orbiter of the new program. This would save not only on the procurement of components, but also by reducing qualification testing. In fact, although the spacecraft had a new composite structure and propulsion module, its electronics were all spares left over from Mars Observer.[88] Unlike previous JPL-controlled missions, Mars Global Surveyor, and indeed the subsequent missions of the Mars Surveyor program, were to be 'flown' from a mission support area at Lockheed Martin's facility in Denver.

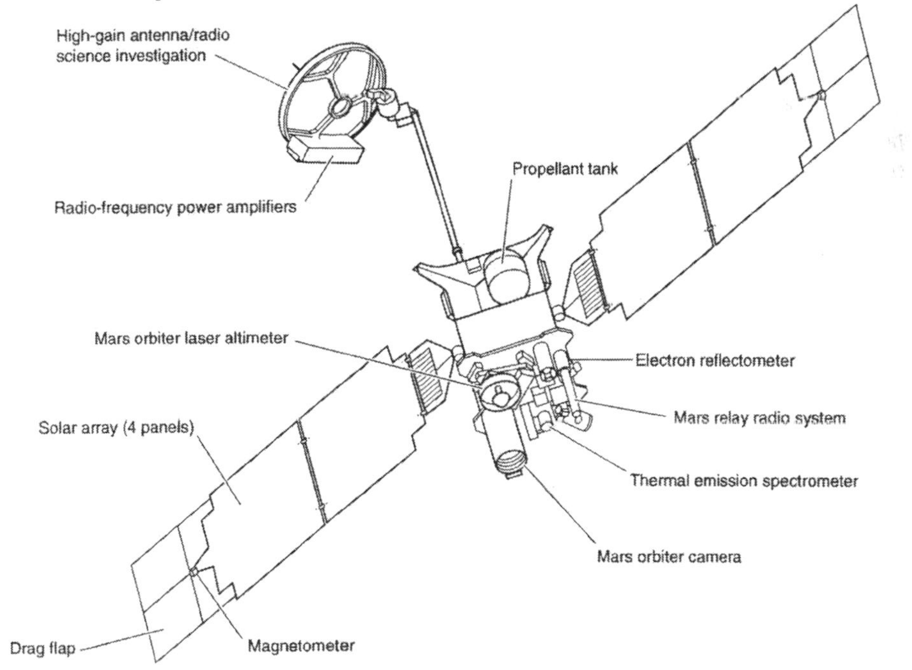

Mars Global Surveyor in its mapping configuration. (Reprinted from Surkov, Yu. A., "Exploration of Terrestrial Planets from Spacecraft", Chichester, Wiley–Praxis, 1997)

Mars Global Surveyor was essentially a box 1.22 × 1.22 × 0.76 meters in size. An equipment module housed the electronics and most of the instruments and was connected to the module of batteries, propellant and pressurant tanks and all of the propulsion elements. The design of the propulsion system specifically avoided all of the failure scenarios suggested as the cause of the loss of Mars Observer, including propellant mixing. In fact, parts of the system were of Cassini heritage. There was a bipropellant engine for major maneuvers. This was a UK-built Royal Ordnance Leros-1B with a thrust of 596 N, and was an improved version of the Leros-1 used by the NEAR spacecraft. Attitude determination was by Sun and stellar sensors, a horizon sensor, an inertial platform with four accelerometers, four gyroscopes and four reaction wheels as actuators. There were a dozen 4.45-N hydrazine thrusters for minor course changes and for unloading the momentum wheels of the attitude control system.[89] Power was provided by two wings of solar panels. Each wing had two 1.51-meter-wide 1.85-meter-long panels, and a small rectangular flap at its tip. With both panels in line, the wing was 3.88 meters long. The inboard panel of each wing was connected by hinges and a damper to a Y-shaped yoke that was in turn mounted on the bus by a 2-axis gimbal. The wings could be individually gimbaled. The total collection area was 6 square meters. The inner panel had gallium arsenide cells and the outer panel had silicon cells. The output was 980 W near Earth, and 660 W in Mars orbit. The rear of the panels was to serve as a drag surface during the aerobraking passes. In fact, from the very beginning the design of the solar wings took into account their dual purpose. For example, their area was selected to create the required drag, and the solar cell adhesives were tailored to the temperatures expected from atmospheric heating. The small drag flaps at the tips were added late in the development. A pair of nickel-hydrogen buffer batteries were to sustain the spacecraft in the planet's shadow. The overall span with the solar panels and flaps all in line was 12 meters. For aerobraking passes, the spacecraft would be oriented with its solar wings canted in a 'Vee' configuration to serve as 'air plows', with the instruments side of the bus facing back in the 'protected' direction. The downlink would use a 25-W transmitter operating in the X-Band, and a 1.5-meter-diameter high-gain antenna having 2 degrees of freedom of articulation that was mounted on a 2-meter-long mast. A Ka-Band engineering experiment was also carried. The high-gain antenna would be maintained stowed against the body of the spacecraft during the cruise and aerobraking, and deployed for mapping. The telemetry rates would vary between 10 and 85 kbps, depending on the distance between Mars and Earth. Four low-gain antennas, two for receiving and two for transmitting, were mounted on the body for low-rate telemetry and emergencies that denied the use of the high-gain antenna. There were two redundant solid-state recorders, each with a capacity of 1.5 Gbits, to store instrument data for later replay to Earth.

The role of the early orbiters of the Mars Surveyor program was to recover the science lost with Mars Observer. Mars Global Surveyor had six of its predecessor's instruments, with a total mass of 75 kg. The infrared radiometer of Mars Observer weighing 44 kg was assigned to the 1998 orbiter, together with a new instrument; and the 23-kg gamma-ray spectrometer was assigned to the 2001 orbiter.

Like Mars Observer, the remote-sensing instruments on Mars Global Surveyor were mounted on the Mars-facing side of the body. This required the spacecraft to make a 360-degree rotation once per orbit, while maintaining the high-gain antenna pointing at Earth and the solar panels face-on to the Sun.

The camera inherited from Mars Observer had two independent subsystems. A narrow-angle camera used f/10 Ritchey–Chretien optics with a 350-cm focal length and a linear CCD array of 2,048 pixels. In the wide-angle camera piggy-backed on the baffle of the narrow-angle camera, a pair of fish-eye lenses formed an image at a single focal plane, where there was a linear CCD array of 3,456 pixels. The pixel arrays were oriented perpendicular to the direction of travel to operate in the 'push broom' mode by which the spacecraft's motion built up images of arbitrary length. In the mapping orbit the narrow-angle camera had a resolution of 1.5 meters. The wide-angle camera was to produce horizon-to-horizon views through red and blue filters with a resolution at the nadir of 250 meters. There were no moving parts for focusing; this was achieved by accurately controlling the temperature of the optics. Data could be returned in real-time by reading the CCD directly onto the telemetry link, or be stored on the solid-state recorder for later replay.[90,91]

The laser altimeter measured the duration of flight of laser pulses fired from the spacecraft to the surface of the planet and back, with the reflection being observed by a Cassegrain telescope with 0.5-meter-diameter optics. The instrument was to fire

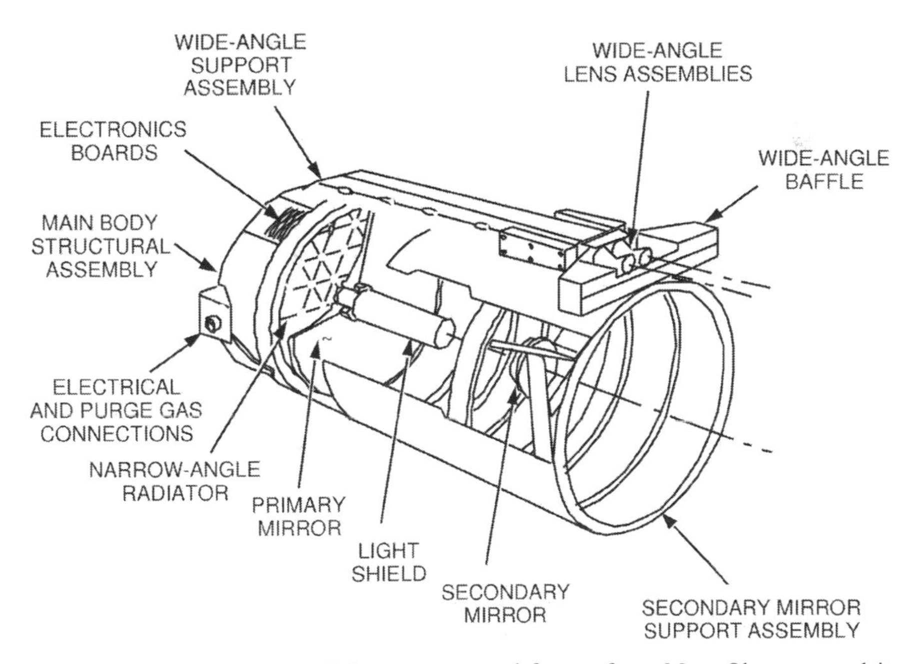

The camera for Mars Global Surveyor was a left-over from Mars Observer, and it consisted of a narrow-angle camera and (mounted on the tube of the telescope) a wide-angle camera that used a pair of fisheye lenses with fixed red and blue filters.

ten pulses per second, illuminate a 160-meter footprint and measure the altitude to within several tens of meters. The data from the first altimeter ever to operate in orbit of the planet was expected to greatly enhance the topographic map based on terrestrial radars, the radio-occultations of Mariner 9 and the Viking orbiters, stereo imagery and atmospheric column-depth data.[92] The thermal emission spectrometer comprised an interferometer and a thermal radiometer, each with its own optics. Its resolution was at best about 1 km. By using a scanning mirror, it could also obtain data from the atmosphere at the limb. It was to map the composition of minerals, rocks and surface ice; determine the composition and distribution of airborne dust; measure the thermal characteristics of the surface and atmosphere – characterizing the dynamics of the latter; measure the temperature and composition of clouds; and investigate the growth and retreat of the polar caps.[93] Mars Observer was the first American spacecraft since Mariner 4 to be sent to Mars with a magnetometer. This was carried over to Mars Global Surveyor to exploit the low periapsis during the aerobraking phase of the mission to study the interaction of the solar wind with the planet's ionosphere and atmosphere at altitudes never previously explored, in order to clarify the nature of the magnetic field that had been inferred from the scant data returned by Soviet orbiters. Mariner 4 had not detected a magnetic field during its flyby, but the Soviet orbiters had flown well within the ionosphere, and hence had been better positioned to study a weak field. The instrument consisted of a pair of 3-axis fluxgate magnetometers and an electron reflectometer, to map the strength of the magnetic field down to the top of the atmosphere and also the local distribution of electrons and electric fields. On Mars Observer the magnetometer had been mounted on a boom, but on Mars Global Surveyor the sensors were at the tips of the solar wings, and the electron reflectometer was on the instrument panel of the body. In order to accommodate the magnetometer, the solar wings had been made as magnetically 'clean' as possible.[94] Extensive gravity and radio-occultation data was also to be obtained using the ultrastable oscillator of radio-science system. The accelerometers and horizon sensors were to be used during aerobraking passes to collect data on the upper atmosphere.

Finally, like Mars Observer, Mars Global Surveyor carried a French-supplied radio relay experiment to serve as a downlink for landers. It would store data in the camera's memory, and then this would be transferred to the solid-state recorder for later replay to Earth. Apart from providing a beacon signal to instruct the lander to start transmitting data, the link was one-way. The 1-meter helical antenna provided a data rate of 128 kbps at a range of 1,300 km and 8 kbps at 5,000 km. Its first use was to be to relay data from the landers that the Russian Mars 96 spacecraft would release. In 1999 the first lander of the Mars Surveyor program was to release a pair of Deep Space 2 microprobe penetrators, whose transmissions would be relayed by Mars Global Surveyor.[95,96,97,98]

In August 1996, a few weeks after the 20th anniversary of the Viking 1 landing and three months before Mars Global Surveyor was to be launched, a NASA team made the startling announcement that the Red Planet had once hosted life. The key to this discovery was a 1.9-kg stone which was found on the Allan Hills ice field in Antarctica on 27 December 1984, and hence designated ALH84001. In contrast to

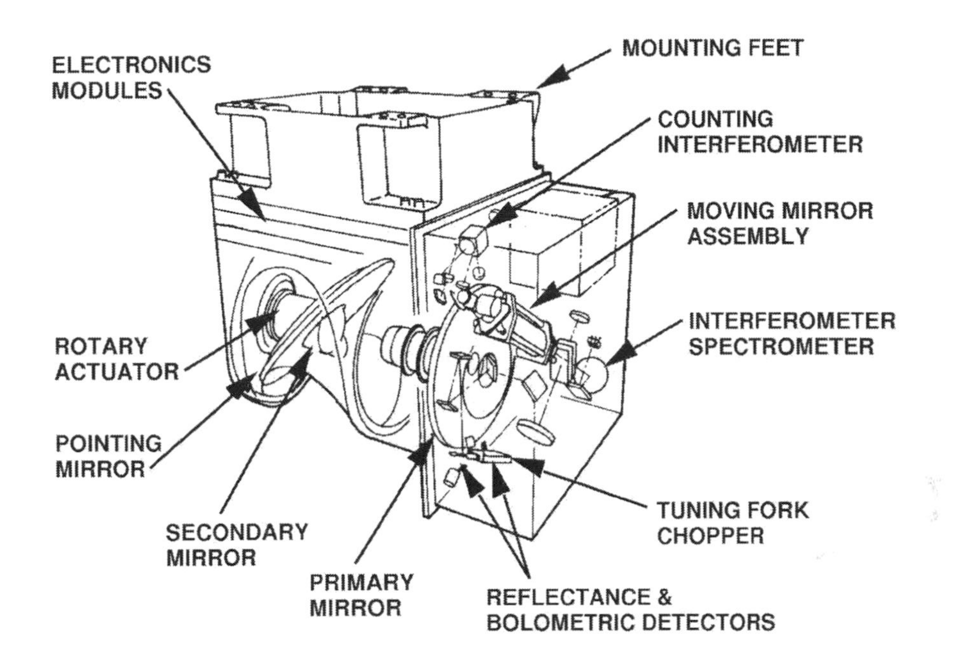

Like the other instruments on Mars Global Surveyor, the thermal emission spectrometer was a reflight of a Mars Observer instrument.

elsewhere on Earth, where meteorites are difficult to spot among the litter of rocks, in Antarctica they are immediately recognizable lying on the ice. Many thousands have been found there since the 1960s. As one researcher put it, Antarctica "is the poor man's space probe".[99] At the time of its discovery ALH84001 was classified as an igneous fragment of the asteroid (4) Vesta, but in 1993 it was realized that its composition more closely resembled the SNC (Shergottite–Nakhlite–Chassignite) meteorites known to have originated on Mars.[100] Yet ALH84001 differed from the SNCs. In particular, whereas none of this group exceeded 1.3 billion years in age, ALH84001 crystallized about 4.5 billion years ago. A detailed study revealed that it had been subjected to intense shock by the formation of an impact crater nearby. It was ejected into space by another impact some 15 million years ago. Then it fell on Antarctica roughly 13,000 years ago, where it was protected by the ice until being exposed at the surface.

Scientists analyzing meteorites at the Johnson Space Center announced that, in their opinion, ALH84001 contained "evidence" of Martian life. However, as they readily admitted, their investigation was biased by the fact that (as had the Viking experimenters) they were "searching for Martian biomarkers on the basis of what [was known] about life on Earth". A characteristic of the stone was the presence of small orange-hued globules of carbonate minerals that formed by the simultaneous presence of water and carbon dioxide with Martian characteristics. These carbonate 'blobs' were rimmed by alternating dark and bright layers. The dark rims proved to

Left to right: Wesley T. Huntress (NASA's Associate Administrator for Space Science), Daniel Goldin (NASA's Administrator), David S. McKay and Everett K. Gibson (co-investigators) pose with a chip of meteorite ALH84001 at the 7 August 1996 press conference for the controversial 'life on Mars' discovery. (NASA/Bill Ingalls)

have minuscule embedded grains of magnetic materials like magnetite and iron sulfide, whose simultaneous presence in such close proximity would have required alternating oxidizing and reducing environments which, whilst unlikely in natural conditions, are characteristic of microbial systems. If magnetite crystals displaying so few structural defects and impurities were found native to Earth, they would be immediately classified as having been produced by biological processes. What was interesting, was that terrestrial organisms produce magnetite crystals as a means of orienting themselves relative to Earth's magnetic field, and the discovery of similar crystals in ALH84001 might hint that Mars once had a global magnetic field. The carbonate globules proved to be infused with complex organic molecules known as polycyclic aromatic hydrocarbons. Although these are made by biological systems, they can also be created by other chemical processes and have been found in some 'ordinary' meteorites. It was concluded that in this case they originated from Mars, since terrestrial analogs result primarily from human industrial activities and have different compositions and characteristics. In fact, the most abundant 'pollutant' of this type was completely absent in all of the samples. Furthermore, the distribution of the polycyclic aromatic hydrocarbons in ALH84001 mimicked that which could be expected from the decay of in-situ organic matter. The weakest line of evidence (but the most visually stunning) took the form of clusters of elongated shapes in or near the carbonates which were no more than 100 nanometers in length. Although these were reminiscent of fossil bacteria, the argument for their being so was weak because the process of slicing and preparing the meteorite samples for examination could have created such tubular structures. Furthermore, being at least an order of magnitude smaller than the smallest known terrestrial bacteria, it would be difficult to fit ordinary DNA inside them. As the team leader David S. McKay concluded at a hastily arranged press conference at NASA Headquarters on 7 August 1996 to present the results of the study, "we have these lines of evidence and none of them in itself is definitive. But taken together, the simplest explanation to us is that they are the remains of Martian life."[101]

One aspect of the vigorous scientific debate which ensued focused on when the carbonate globules were made, and the conditions prevailing at that time. Although some analysts said they were produced 3.6 billion years ago, others dated them at a mere 1.3 billion years – at which time Mars may have been undergoing volcanic and hydrothermal activity. Moreover, scientists could not decide the temperature at which the carbonates formed. Some people said they formed out of a fluid that was no hotter than 80°C, whilst others argued for temperatures of at least 650°C which no known form of life could have endured. Although most of the characteristics of ALH84001 which suggested life proved to have alternative non-biological origins, a significant percentage of the magnetite crystals have shapes and morphologies which would seem to be compatible only with a biological origin. If the magnetite crystals had been subjected to great heat, they would have oriented their magnetic fields in a single direction as they cooled, which was not the case.[102,103] The debate, which was protracted and bitter, failed to deliver a definitive consensus on whether the characteristics of ALH84001 were indicative of Martian life. This issue served to highlight how difficult it is to define what is biological and what is not on the nanometer scale, and the lessons learned will inform the analysis of material from a sample-return mission. If indeed life did develop on Mars billions of years ago, the question became whether this had survived to the present day.[104,105]

While the debate about ALH84001 was underway, JPL investigated whether the sample-return mission of the Mars Surveyor program could be advanced to 2005 or even as early as 2003. Some scientists said that to maximize the chances of a single mission finding evidence of life, a precursor mission in 2001 should scout potential landing sites to determine whether the soils could contain evidence of life forms. In another option, precursors in 2001 and 2003 could deliver rovers to collect and pile up interesting rocks for collection by a human mission in the 2010s.[106] President Clinton reportedly found the ALH84001 findings to be so compelling that he asked Al Gore to convene a 'space summit' to assess the planning for exploring Mars.[107] Dick Morris, Clinton's political adviser, pushed for an early human Mars mission, no doubt to reap the political possibilities of announcing it. Many scientists thought that this would be a "calamitous mistake". As Thomas Gold of Cornell University, with characteristic bluntness, put it, "the last thing you would want is to have men there, copiously contaminating Mars and ruining it forever".[108] Fortunately, Goldin knew that such an early human mission made no sense technically, financially or politically.[109]

One week after the ALH84001 announcement, Mars Global Surveyor arrived at Cape Canaveral, having been airlifted during the night to avoid imposing the stress of daylight turbulence on its structure. The original plan had been to transport it by truck, but when the failure of Galileo's high-gain antenna was attributed in part to its repeated trips across the country by truck, it was decided to deliver Mars Global Surveyor by air.[110] The launch window extended from 6 to 25 November. Clouds and high-altitude winds ruled out the first day, but it lifted off on 7 November. The 7925 configuration of the Delta II rocket utilized nine strap-on solid boosters. US Air Force instrumentation aircraft introduced in the Apollo era were stationed over the Indian Ocean in order to monitor the second stage of the Delta leaving parking orbit and then the Star 48B solid-motor as the spacecraft was inserted into a 0.98 × 1.49-AU heliocentric orbit.[111] About an hour after launch, the spacecraft deployed its solar panels. However, the early telemetry indicated that one of the two wings had failed to latch into position. An early reconstruction suggested that the arm that turned the damper to prevent the wing from being damaged on halting suddenly had wedged between the inboard panel and the yoke, penetrating a short distance into the panel and preventing it from reaching its latching point. Although this would not affect power generation, the fact that the wing was not latched required a reassessment of the aerobraking plan. Both wings were to have been canted back at 30 degrees with their solar cells on the protected side, but the drag would force the damaged wing against the side of the bus, rendering the spacecraft unstable. It was decided instead to set the secure wing at 34 degrees, rotate the damaged wing so that its solar cells faced forward, then set its yoke at 51 degrees so that its panel was at 31 degrees, to even up the aerodynamic drag. The interplanetary transfer was one which subtended an

Mars Global Surveyor during ground preparations. This is the high-gain antenna side of the spacecraft. The main engine is visible at the bottom of the 'box'. The only instrument on view is the antenna of the relay experiment which is the white 'rod' at the top.

arc of more than 180 degrees around the Sun, in order to minimize the energy of the insertion maneuver at Mars. The initial trajectory had incorporated an offset to ensure that the unsterilized Star 48B stage could not strike Mars. On 21 November the spacecraft made a 43-second burn for a 27-m/s change in velocity to eliminate this precautionary offset. The trajectory was further refined on 20 March 1997 and 25 August. The option of a correction on 21 April was judged to be unnecessary. In the meantime, the spacecraft calibrated its instruments. An attempt to image Earth and the Moon had been made several weeks into the cruise, but this was frustrated by an error of several degrees in the attitude determination.

On 9 January 1997 the spacecraft's attitude was such that Earth finally entered the beam of the stowed high-gain antenna, and communicating became easier and faster. During the cruise, it flew spin-stabilized with its antenna pointed at Earth and solar wings canted back. Images of the Pleiades star cluster were taken in order to check the alignment and focus of the camera's optics and CCD. Two sets of observations of Mars were made during the cruise. The imaging in early July was in support of the Mars Pathfinder landing, and was done in cooperation with the Hubble Space Telescope which, although operating at longer range, had higher resolution. The imaging on 19 and 20 August was a contingency against the loss of the spacecraft on arrival. Although the resolution was no better than 20 km per pixel because the spacecraft was still some 5.5 million km from the planet, the images showed some changes in the albedo markings compared to the time of the Viking missions.[112]

On 12 September 1997, Mars Global Surveyor fired its bipropellant engine for 1,340 seconds to change its velocity relative to Mars by 973 m/s. During the burn, the spacecraft used a novel technique whereby it constantly adjusted its attitude in order to hold the thrust vector as close as possible to the velocity, because this was more efficient in terms of propellant. Nevertheless, the maneuver consumed 77 per cent of the propellant load. The spacecraft achieved an orbit which ranged between 263 km (the objective was 250 km) and 54,026 km, inclined at 93.3 degrees to the planet's equator and with a period that was a mere 25 seconds short of the 45-hour target. In fact, because one of the solar wings had not latched, the parameters of the orbit had been revised slightly to reduce the aerodynamic drag and heating rates in the early aerobraking. After just three orbits, the 'walk-in' phase of the aerobraking process commenced. First, a burn reduced the periapsis to 150 km. Once the drag at this altitude had been confirmed to match the model prediction, the spacecraft was allowed to continue. The telemetry indicated that there was some deflection in the damaged wing, evidently due to the pressure compressing the damper arm. Five days after the spacecraft's arrival, the spring began in the southern hemisphere, and with it the prospect of dust storms that would heat the atmosphere and cause it to inflate. The main aerobraking phase began on the 12th revolution, on 2 October, at which time the periapsis was 110 km and the mean dynamic pressure imposed on the spacecraft of 0.53 Pa was reducing the period by 75 minutes with each pass. The pressure was deflecting the damaged wing by as much as 13 degrees, which took it beyond the latching point, but it did not lock into position. In fact, it seemed that the damaged wing was shedding small pieces of graphite from its face sheets, and as these reflected

The launch of Mars Global Surveyor on a Delta II rocket. Owing to its relatively low cost, this rocket became the standard launcher for the Discovery and Mars Surveyor missions.

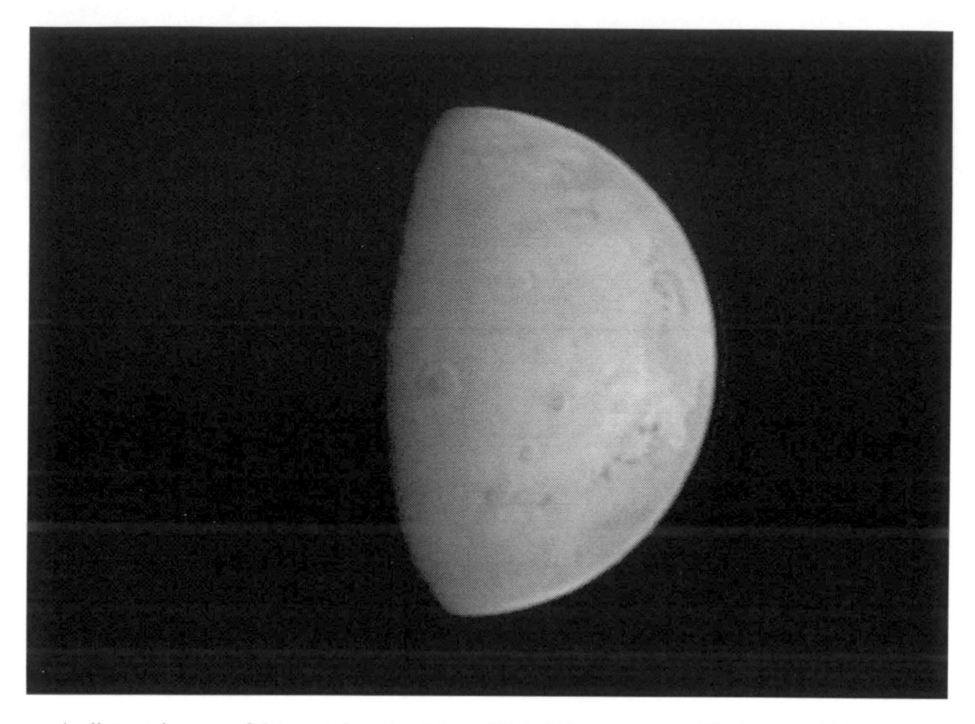

A distant image of Mars taken by Mars Global Surveyor on 21 August 1997. It is centered on Olympus Mons, and shows the line of volcanoes on the Tharsis bulge. (NASA/JPL/MSSS)

sunlight the glint was noted by the star sensors.[113] There was concern that if the pressure were to significantly increase, then a clevis might break and seriously damage the solar cells. It was evident from the response of the loose wing that its failure to deploy properly was poorly understood. When the pressure rose to 0.9 Pa on the 15th revolution, JPL decided to call a halt to gain time to analyse the state of the spacecraft and the prospects for the mission. Accordingly, on 12 October the periapsis was raised to a more benign 172 km. Although the atmospheric passes had reduced the period to just under 35 hours, it was clear that a hiatus would make it impossible to achieve a circular orbit in the desired orientation within the time available.

After analyzing all the telemetry from the first 15 orbits, engineers realized that not only had the loose wing failed to latch, but that the damage was more serious than thought. A carbon composite and aluminum honeycomb yoke face sheet had probably cracked near the gimbal point when the undamped panel had abruptly stopped during deployment, crushing one side of the yoke. However, tests of a deliberately damaged yoke defined a new, lower, dynamic pressure limit which the yoke should be able to sustain for 1,000 aerobraking cycles. The 3-week review thus established that Mars Global Surveyor could resume aerobraking, providing that the dynamic pressure did not exceed 0.20 Pa – a value that included a safety margin sufficient to accommodate likely fluctuations in the density of the atmosphere. A

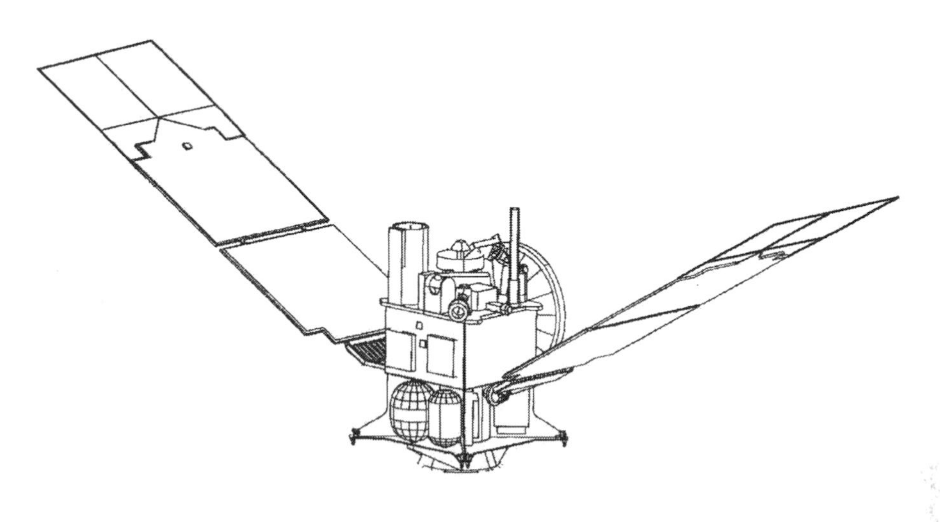

The nominal aerobraking configuration of Mars Global Surveyor.

complete revision of the circularization plan was ordered. One option was to angle the damaged wing at 85 degrees so that it rested on its support bracket. This would minimize the load on the hinge, but require the spacecraft to penetrate deeper into the atmosphere, which would in turn present thermal and control problems, so this option was rejected.[114] The new aerobraking plan divided the circularization into three phases. The first phase, which would last from November 1997 to April 1998, would trim the period to 11.64 hours. There would then be a hiatus of several months, during which the spacecraft would make preliminary scientific observations and pass through solar conjunction. Aerobraking would resume in September and continue until February 1999. The intervening hiatus, which was referred to as the 'science phasing orbit', was to restore the planned orientation of the orbital plane relative to the Sun, but in the opposite sense – i.e. instead of the spacecraft crossing the equator northbound at 2 p.m. local time, it would do so southbound. There would be some risks in this aerobraking scheme. For example, when the spacecraft passed through the planet's shadow while near apoapsis the period of darkness would greatly exceed the design maximum of the batteries and at the same time cause the solar cells to chill almost to their qualification limit. Moreover, the propellant margin was so narrow that there was enough for only one additional walk-out and walk-in sequence in the event that the aerobraking had to be interrupted again.

The walk-in began on 7 November by lowering the periapsis to 135 km, and the proper aerobraking resumed six orbits later, on 15 November. Several mechanical characteristics of the damaged wing were to be monitored to determine whether the crack in the yoke was stable or was continuing to propagate. On 23 November the mean dynamic pressure increased to 0.32 Pa because of a regional dust storm, and Mars Global Surveyor had to raise its periapsis in order to maintain a safe pressure; only to lower it again after the storm had abated in December. And so the process continued.[115,116]

A number of scientific observations were being made during periapsis passes in the aerobraking process. By careful planning, the narrow-angle camera was able to take images, albeit at reduced resolution, as Mars Global Surveyor slewed from its aerobraking orientation in order to aim its high-gain antenna (whose boom had not yet been deployed and therefore was still held fixed against the bus) at Earth. Features that had been shaped by wind erosion were almost ubiquitous. In fact, most regions had meter-scale dunes, ripples, ridges, drifts and grooves which were in one way or another related to the wind. Images of the walls of Valles Marineris at resolutions of 5 to 10 meters showed extensive layering up to 50 meters thick, but it was not possible to determine whether this was superposition of volcanic flows or wind or water deposits. Images of the southern polar terrain were obtained during the local spring, just as the carbon dioxide cap was starting to sublimate, and then again as it retreated. Meanwhile, the wide-angle camera monitored the formation of cloud and haze over elevated terrain such as the Tharsis bulge, and also the evolution of the regional dust storm that developed in Noachis Terra in the southern hemisphere in late November and early December 1997. Interestingly, no hazes and clouds were seen anywhere on the planet within 4 days of the storm breaking out, probably owing to a changed temperature distribution in the atmosphere.[117] In December 1997 and January 1998 targeted observations were made to map the landing sites that were being considered for the first lander of the Mars Surveyor program. When periapsis passes occurred over dark areas like Syrtis Major and Meridiani Sinus, the thermal emission spectrometer showed them to be covered with similar minerals, including pyroxene, as well as an extensive layer of dust. These early results were also used to place constraints on the presence of carbonates. The absence of carbonates was one of the outstanding mysteries of the planet's geological history. The landscape appeared to have been extensively shaped by water, yet carbonates, which form in the presence of carbon dioxide and water, seemed to be rare. This instrument also monitored the November regional storm, observing an increase in both the temperature and the opacity of the lower atmosphere.[118]

The laser altimeter took its first strip of data on 15 September 1997, lasting about 20 minutes during which the spacecraft was below an altitude of 800 km. The line of data points stretched across the northern plains and the Elysium rise, a volcanic region with peaks standing 5 km above the plain. On this elevated terrain, the line crossed a narrow chasm which proved to be more than three times deeper than the Grand Canyon.[119] Many more such strips in October and November confirmed that the northern plains are extremely flat, with the slopes rarely exceeding 3 degrees. The reason for this smoothness was not immediately evident, but the resemblance to the abyssal plains of the deep ocean floor on Earth suggested the possibility that this low-lying part of Mars once held an ocean. Despite the fact that this terrain spans most of the northern hemisphere, the volume of water needed to fill it would have been less than that which the planet is believed to have lost by accumulation of ice at the poles and by atmospheric erosion to space. Data points over the layered polar terrain, which stands more than 1 km over the adjacent terrain, scattered the laser in a way which indicated

This early image from Mars Global Surveyor's narrow-angle camera shows layering in the wall of western Candor Chasma. (NASA/JPL/MSSS)

that this mass was composed primarily of ice. Profiles of crater ejecta in the polar regions were in many cases consistent with impacting into permafrost. Some of the altimetry lines crossed the Tharsis rise, and even some of the volcanoes atop its summit, revealing clues to their geological history. A pass over Valles Marineris found terraces dating from the last flood of the channel, and others that might represent an earlier water line. Slope measurements gave new estimates of the flood velocity and discharge in the canyons.[120]

The accelerometers returned extensive data on the atmosphere above an altitude of 110 km, and how this varied with the time of day; for which the only previous measurements had been readings by the two Viking landers and Mars Pathfinder – the latter having reached Mars only a few months before Mars Global Surveyor. In late November 1997 there was a 130 per cent increase in the atmospheric density during aerobraking due to the dust storm in Noachis Terra heating and inflating the atmosphere. Other smaller increases may well have been due to regional storms elsewhere, but no such activity was observed by either Mars Global Surveyor or the Hubble Space Telescope. The accelerometry and other engineering data such as the torques acting on the aerobraking spacecraft yielded data of scientific interest, in particular the zonal wind speeds (i.e. parallel to the lines of latitude) in the high atmosphere. Models had suggested that the winds at this altitude would be 100 m/s on average, but at the time of the November dust storm the spacecraft encountered winds of 300 m/s. In accordance with predictions, the wind was westerly most of the time; but there were some exceptions, in particular an easterly wind at 200 m/s late in the aerobraking activity. Owing to the fact that aerobraking was drawn out beyond the three months originally intended, it was possible to see seasonal trends in the data.[121,122,123,124]

The magnetometer reported on the interaction between the solar wind and the planet's ionosphere. The spacecraft's crossings of the interfacing bow shock made a map of where the bow shock was located and how its shape varied. Overall, this confirmed the results of the Fobos 2 mission, and showed that Mars interacts with the solar wind in a manner akin to the other unmagnetized terrestrial planet, Venus. All the Soviet reports of an intrinsic magnetic field had been made from within the ionosphere, but by virtue of the fact that it was aerobraking Mars Global Surveyor was able to collect data from beneath the ionosphere. When a field was detected on 15 September, shortly after orbit insertion, this led researchers to speculate that it was a weak global magnetic field. However, as later periapsis passes yielded more data it became evident that this magnetism was confined to isolated areas. It would be a while before a pattern became evident (see below). Overall, the results proved that Mars does not possess a global field.[125]

On 27 March 1998, after the 201st orbit, aerobraking was halted in accordance with the revised schedule.[126] For the next six months, Mars Global Surveyor would make preliminary scientific observations. On 5 April it inspected the infamous 'Face on Mars' which had so fascinated UFO-fans since being found on a Viking picture. When viewed through a clearing in the clouds which masked the Cydonia region in winter, this feature proved to be just one of a number of mesas. Its resemblance to a face was evident only at low resolution and a particular angle

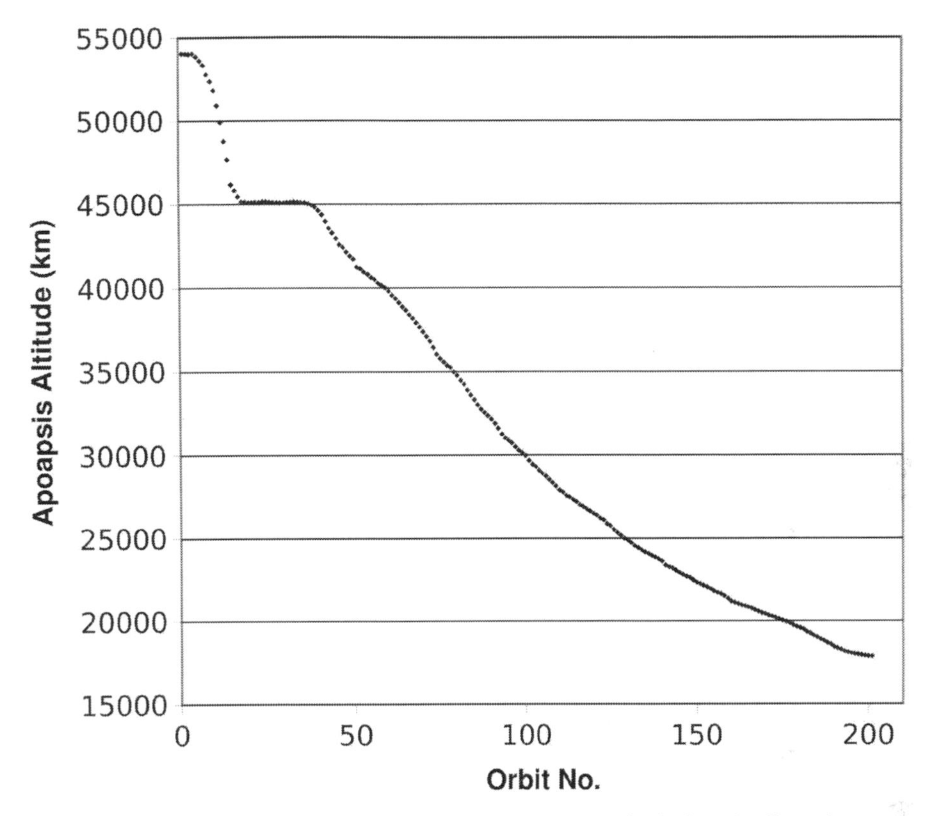

A plot of the altitude of Mars Global Surveyor's apoapsis during the first phase of aerobraking. Note the fast reduction in altitude at the beginning, then a halt during investigations into the damaged solar panel before a resumption at a slower pace.

of illumination.[127] Although communications were impaired as Mars approached the Sun in the sky as viewed from Earth, and ceased on 12 May when the planet was occulted, this was an opportunity for solar science and for a 24-day period centered on the conjunction the spacecraft's X-Band and Ka-Band transmitters were used to 'sound' the corona and near-solar environment.[128] Extensive data was obtained during the aerobraking and science phasing orbits, including 2,140 images, millions of spectra, hundreds of radio-occultation profiles, laser altimetry, and extensive magnetometer sensing. And, of course, the engineering sensors gave atmospheric data. The camera issued good results despite the handicap of not being able to be precisely focused owing to the fact that the elliptical orbit imposed varying thermal loads which were difficult to predict. The imaging prior to the mapping mission was intended to cover as wide a range of geological landforms and local times as possible as a safeguard against the spacecraft being lost during its final aerobraking. One picture of a 50-km crater in Noachis Terra showed dark wedge-shaped depressions on its rim that some people thought might be evidence of groundwater seepage.

Furthermore, the crater had a dark floor with islands, bays and peninsulas which suggested that this water had ponded and subsequently either evaporated or frozen. A picture of Nanedi Vallis in Xante Terra showed for the first time a 200-meter-wide channel nestled within the broader canyon. This seemed to some people to be a 'smoking gun' indicating that the channel was formed by water running across the surface, but ground collapse could have been a contributing factor. Images of a mysterious fretted terrain at the boundary between the elevated southern region and the low-lying northern plains showed cliffs, valleys, grooves and parallel ridges which were strongly suggestive of glacial activity, but with significant differences to analogous terrestrial terrain.[129]

During the hiatus in aerobraking, Mars Global Surveyor had four occasions to observe Phobos, in particular on 19 August when it flew by the moon at a range of 1,080 km. High-resolution images and spectra were taken in an effort to clarify the surface composition and to shed light on the moon's origin. In addition, high-resolution images were obtained of the Viking and Mars Pathfinder landing sites. As the spacecraft sailed to apoapsis, spin-stabilized and with its axis maintained toward the Sun, the thermal emission spectrometer swept the planet's far southern latitudes. In periapsis passes over the northern latitudes, the laser altimeter revealed that this ice cap resides in a 5-km-deep depression and rises about 3 km above its surroundings. The volume of the cap equated to about half that of the Greenland ice cap, but the lay of the land suggested that it had been much more extensive in the past.[130]

Aerobraking resumed on orbit 573, with the periapsis being lowered to 120 km by a series of three maneuvers. In fact, this aerobraking phase started a week later than planned owing to a software problem which developed just minutes before the walk-in maneuver and placed the spacecraft into safe mode. No images were taken, but the magnetometer continued to collect measurements. The latitude of periapsis was allowed to drift towards the south during this aerobraking, until it was almost over the south pole. This showed that the magnetic anomalies were concentrated in the ancient terrain of the southern hemisphere. Thus was revealed one of the most interesting discoveries of the mission. The fact that the strength of the field was not correlated with individual craters or other features indicated that the sources of the magnetism were very ancient. Indeed, the magnetism predated the formation of the Hellas and Argyre impact basins an estimated 3.9 billion years ago. There were no significant anomalies over Tharsis, Elysium, the Valles Marineris complex or the low-lying northern plains. Unfortunately, because the aerobraking would last only a few months, it was possible to compile only a partial and rudimentary map of the anomalies in the southern hemisphere. Terra Cimmeria showed parallel bands 200 km wide by up to 2,000 km in length with large magnetic moments (larger in fact that any known on Earth) and alternating polarities. As lava solidifies, the iron which it contains 'locks in' the magnetic field prevailing at that time. From time to time the polarity of the Earth's magnetic field 'flips', and this imprints a pattern of magnetic stripes on the newly extruded lava at the margins between the tectonic plates of the ocean floor. Whilst the existence of remanent magnetism on Mars implied that the planet once had a

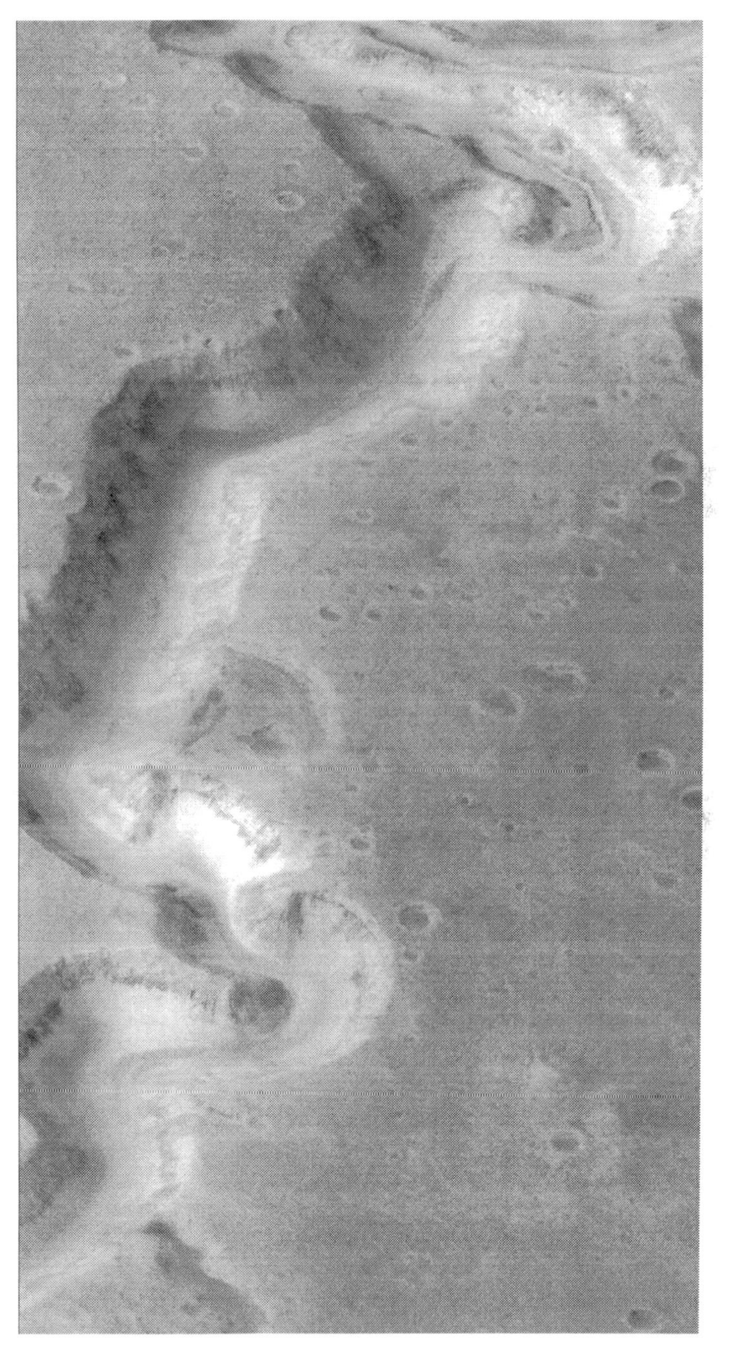

A view of Nanedi Vallis showing a narrow channel on the floor of the canyon. This is particularly evident at the bend at the top of the image. (NASA/JPL/MSSS)

A close-up of Phobos taken by Mars Global Surveyor during the hiatus in aerobraking. (NASA/JPL/MSSS)

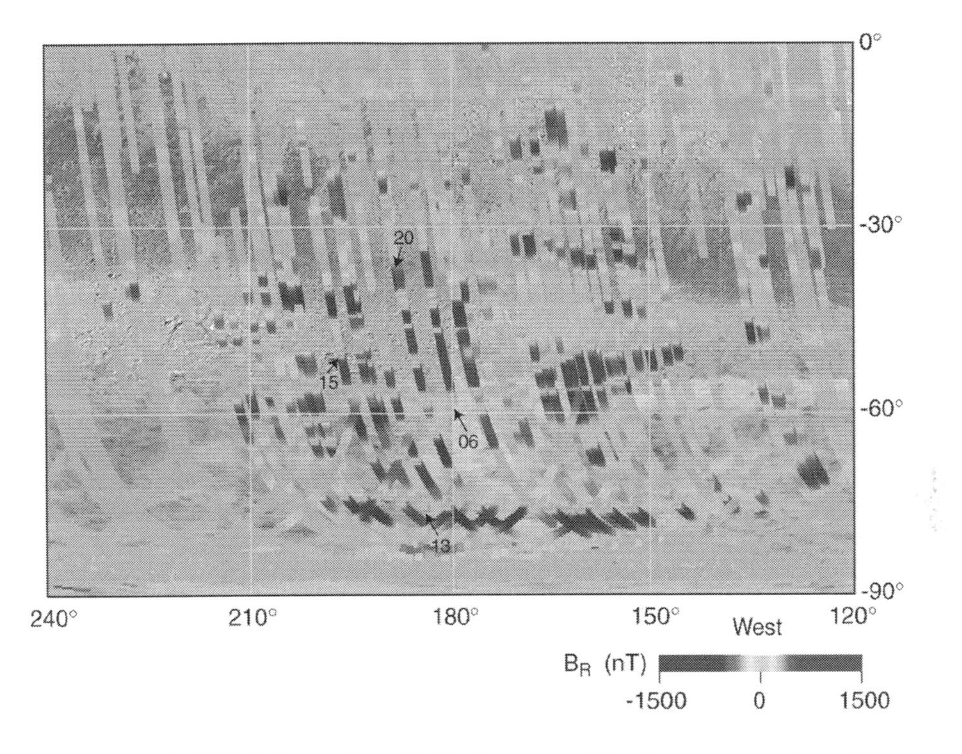

The magnetic field of the southern highlands measured by the magnetometer on Mars Global Surveyor made a pattern of stripes of opposite polarity (which are difficult to tell apart in this black-and-white view). (JPL/NASA/Caltech)

global field, this evidently faded after a few hundred million years. The Earth's magnetic field is created by a dynamo effect in its liquid-iron core. If a dynamo once operated on Mars and then shut down, this implied that the core was inactive. The magnetic fields provided an intriguing insight into the planet's early history. At an altitude of 400 km the Martian crustal magnetic field measured up to 200 nT, as against 26,000 nT for Earth; very weak on average. Nevertheless, the intensity of the field over Terra Sirenum was sufficient to create a small bow shock when exposed to the solar wind on the day-side, and a variety of other Earth-like magnetospheric phenomena. This may well have been the source of field periodicities recorded by Fobos 2. The discovery that Mars may have had a global field early in its history strengthened the case for the Martian origin of the magnetite crystals in the meteorite ALH84001.[131,132]

The 62-m/s walk-out maneuver on 4 February 1999, on orbit 1,284, ended the orbit cicularization phase by lifting the periapsis to 377 km. There had been a scare when the control center in Denver was flooded the previous day, but it was brought back into use just in time. A total of 92 trajectory control maneuvers had facilitated 891 atmospheric passes. The ground track was close to the desired 2 p.m. equator crossing, and the period was just short of 1.97 hours. The spacecraft

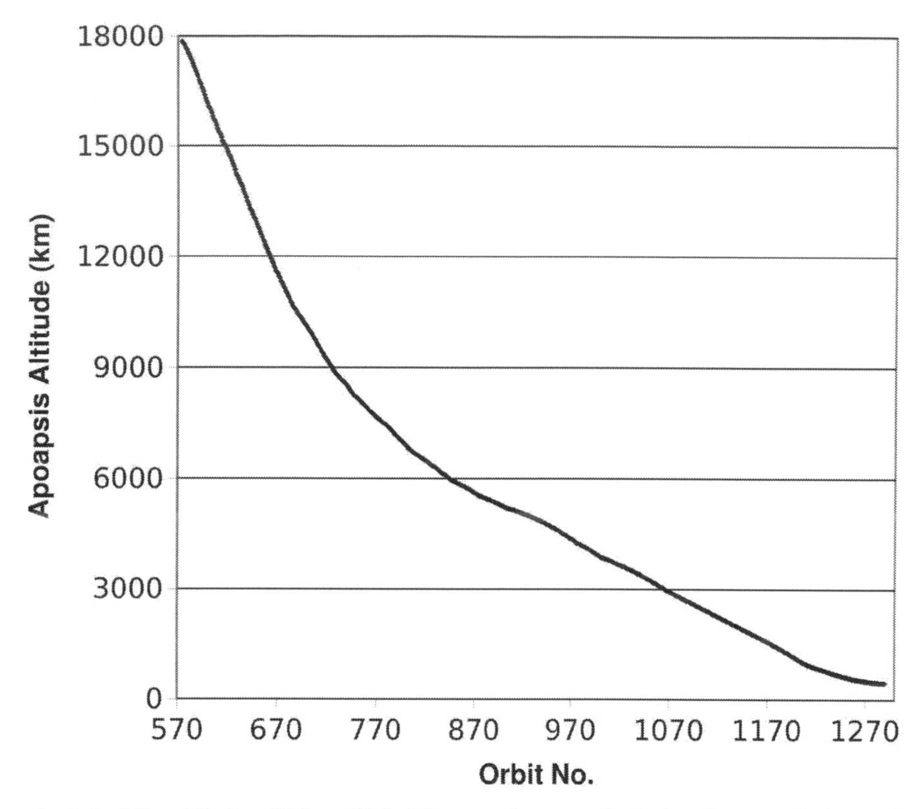

A plot of the altitude of Mars Global Surveyor's apoapsis during the second phase of aerobraking.

was left in this orbit for a fortnight, and continuously tracked through 88 revolutions to obtain data to improve the model of the planet's gravity field. The spacecraft was then maneuvered precisely into the desired mapping orbit on 19 February, and officially started its mapping mission on 9 March. The first 20 days of mapping were carried out with the high-gain antenna stowed. In this 'fixed high-gain antenna' phase, only the gimbal was exercised, with data collected during 7 or 8 consecutive orbits being returned over the next 4 or 5 orbits. This was done as a precaution against a recurrence of the solar wing deployment problem, because the spring, damper and latch architecture of the mast was similar to that of the solar wing. Moreover, there was concern that the mast might have been damaged by having remained stowed for so long.[133] It was successfully deployed on 29 March, but as the antenna was changing from its parking azimuth to Earth-pointing on 15 April the azimuth gimbal jammed. This put the spacecraft into safe mode. Tests over the next few days confirmed the presence of an obstacle at the 41.35-degree position. Further tests over a 2-week period, during which the instruments were powered off, led engineers to conclude that the gimbal assembly was being obstructed by a bolt that was creating a hard stop. (Other possible causes investigated early on included a bent or broken tooth on the gear drive, and interference from a cable or thermal blanket.) Fortunately, for 9 months starting in early May the geometry would allow the antenna to articulate without hitting the obstruction. Thereafter, the 2-degree-of-freedom gimbal could flip the antenna 'upside down' in order to reverse the azimuth motion. Nevertheless, this would restrict the ability of the spacecraft to return data in real-time, and it would rule out taking radio-occultation data for 14 months starting in February 2000.[134]

Mapping resumed on 6 May 1999. Shortly thereafter, the spacecraft performed a 4-week geodesy campaign. The opportunity for this had been noted in planning the Mars Observer mission, and carried over to Mars Global Surveyor. The wide-angle camera took in excess of 2,000 images covering most of the planet's surface from 70°S to the north pole (the northern summer solstice having occurred in January) at two viewing angles for stereoscopic coverage. Moreover, precise tracking allowed the positions of features to be accurately identified. This dataset at a resolution of 250 meters then provided the basis for a new map, on which the laser altimetry was superimposed. When the coverage was extended to the south pole in December the map was completed.[135] Also, in June more narrow-angle pictures were taken of the candidate sites for Mars Polar Lander, and in August follow-up observations were made of the final and backup sites. In addition, it surveyed the equatorial sites that were being considered for the lander planned for 2001. Unfortunately, when Mars Polar Lander arrived on 3 December 1999, nothing was heard from either it or its pair of Deep Space 2 microprobes. Much of Mars Global Surveyor's time from then until February 2000 was devoted initially to trying to make contact using the relay experiment, then to making specific searches for the lander or its parachute. The orbiter had to adopt attitudes offset from nadir-viewing in order to cover the entire landing ellipse at high-resolution; unfortunately in vain. In an effort to clarify whether the narrow-angle camera was capable of resolving a lander sitting on the surface, the Viking 1 and Mars Pathfinder sites were re-examined, but the vehicles proved too small to be distinguished at a resolution of 1.5 meters. However, the camera had sufficient resolution to identify some of the larger rocks viewed by Pathfinder, including the large dark one named 'Couch'. Interestingly, these pictures showed a ripple texture at this site that had not been evident at ground level, and that could be attributed to deposits left by the floods which swept through the region at some indeterminate time in the past. Operations were suspended between 21 June and 13 July 2000 as Mars again passed through solar conjunction. The primary mission was concluded on 1 February 2001, on the 8,505th mapping orbit, having covered a full local year. By this time, over 2 terabits of data and 83,000 pictures had been returned, which was more than all the previous missions to Mars combined. It had been possible to obtain more than 300 images per day when Mars was closest to Earth, and 60 when the planets were farthest apart. The fact that the images were usually compressed for transmission meant that more data could be returned than would otherwise have been the case (70 per cent being estimated), but if parts of the data stream were lost this could result in a corrupted frames, and during the primary mission some 12,000 pictures were spoiled in some way, although in many cases the impairment was minor.

The results were exceptional. As an article summarizing the imagery put it, the "results generally [indicated] a planet far more diverse than previously known and in some cases substantially unlike the broad consensus views developed during the past 20–25 years of analyses of Mariner 9 and Viking orbiter data". Furthermore, "every [narrow-angle] image can, in and of itself, tell a story about the nature of Martian geomorphic and geologic history".

In fact, even the earliest observations proved that many of the impressions from the Mariner 9 and Viking missions were wrong. Peering down on craters, chasms, mesas, valleys, etc, Mars Global Surveyor provided a considerable insight into the first kilometer or so of the crust, showing it to be extensively layered, which was an unexpected result because the surface, particularly in the southern hemisphere, had been generally thought to be similar to the non-stratified lunar highlands. For example, there was layering in the northern part of the Hellas basin. And most impressively, layers in the walls of Valles Marineris seemed to extend down to a depth of at least 10 km below the adjacent surface, most of them very likely dating back billions of years. In other places, such as Meridiani Sinus, layered terrain extended for hundreds of kilometers. There was layered terrain or stairs-stepped terrain displaying regular and repetitive layers of uniform thickness in the craters Holden, Gale and other parts of Terra Tyrrhena. Layering was found to occur preferentially in confined sites where surface water would have tended to collect, such as in canyons and craters. At none of these sites was there evidence of how the water arrived, or how it drained away; and nor were there any salts left behind if it evaporated in situ. One possibility was that all such flow features had been buried. Another was that the water had been of artesian origin. In a case like Meridiani Sinus, the layering covered such a broad and unconfined area that a vast ocean would have been required to submerge it! Or perhaps the layers were formed during a brief period when the chaotic oscillations of the planet's spin axis produced a different climate. On the other hand, perhaps the layering was made by a process that did not involve water.[136,137] As Michael C. Malin, chief scientist for the camera team, pointed out, "the presence of these layers is 'smoking gun' evidence of there being things that have changed on the planet".

Studies of other seemingly water-carved features mostly failed to yield evidence of flows that persisted over long periods of time, and the lack of tributary networks suggested a groundwater source, rather than surface water. That said, the structure near the crater Holden showed clear signs of sedimentation and ancient, persistent flows of surface liquid. For example, one valley had steep-walled, flat-floored and uniformly box-shaped sections similar to a drainage basin. At the end of the valley, where it entered an unnamed crater that may once have harbored a lake similar in size to the Sea of Galilee, there was a fan-shaped feature that resembled sediment dumped at a river delta.[138] This contrasted with Ma'adim Vallis, which at 900 km long and some 10 km wide is one of the largest valleys in the southern hemisphere. As revealed by laser altimetry, this seemed to have been formed by a catastrophic flood resulting from the overflow of a lake in the highlands that covered an area of a million square kilometers. The flow breached the southern wall of the large crater Gusev, in which the water would have ponded.[139]

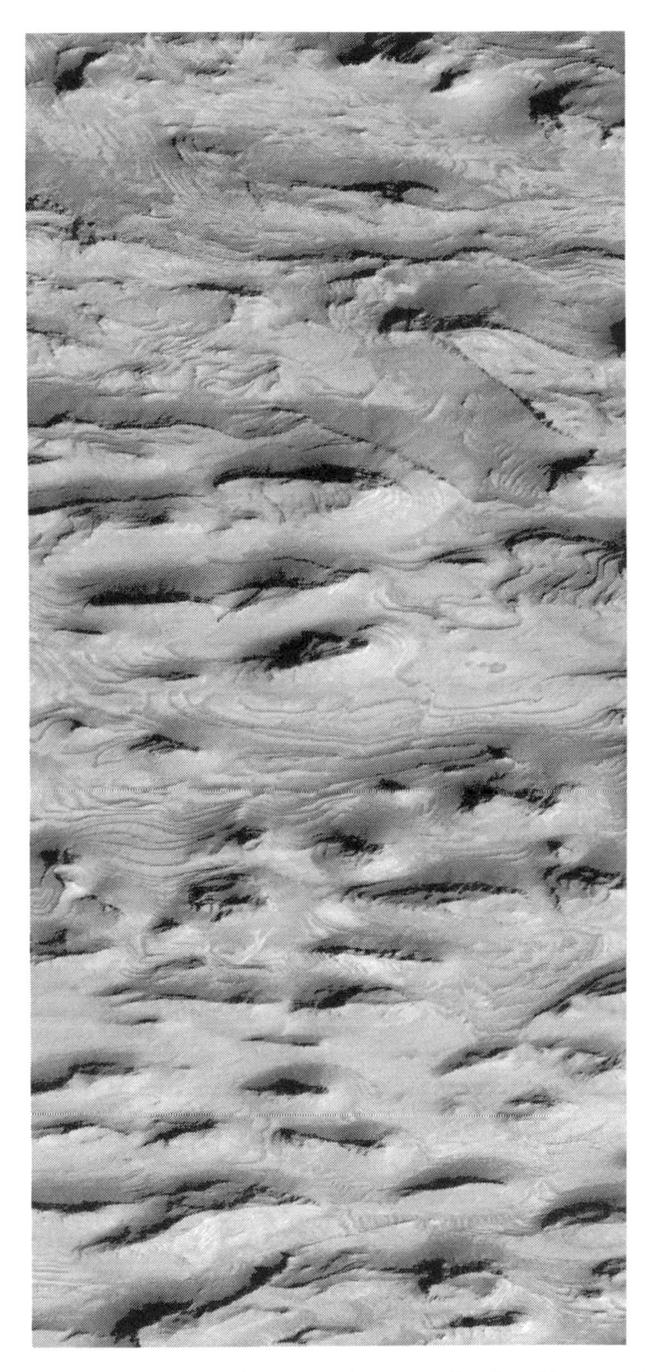

The remarkable cliff-bench terrain in southwestern Candor Chasma. (NASA/JPL/MSSS)

Intriguing gullies on the wall (top) of a small unnamed crater in the southern hemisphere of Mars. (NASA/JPL/MSSS)

But the most striking discovery by Mars Global Surveyor relating to water was announced in June 2000. Over a period of a few months, images taken to provide context for laser altimetry had revealed over 100 examples of small-scale features. In each case, one or more branched channels emerged from a wide alcove and then tapered downslope for some distance towards broad triangular aprons of material that had clearly been transported down the channels. In about one-third of the cases, these features were on the walls or central peaks of impact craters, and the others were on the walls of the Nirgal and Dao valleys. The majority were on poleward-facing slopes in the ancient southern terrain greater than latitude 30°S. The youth of these features was indicated by that fact that there were generally no impacts on the gullies, and in some cases they were superimposed on dunes and other short-lived wind-formed features. Intriguingly, they appeared in regional clusters, and within a given cluster appeared to emerge from the same geological layer. On Earth, these characteristics are shared by landslides and the seepage of water. Although water cannot remain in the liquid state on the Martian surface owing to the pressure and temperature, if groundwater were to emerge from a rock exposure and drain downslope this could cut a gully before it evaporated to leave behind an apron of rock. On the other hand, it could simply have been a dry landslide. However, the correlation with latitude and poleward orientation, and indeed other characteristics, argued for an insulation-dependent origin. Perhaps a reservoir of liquid groundwater became trapped by a dam of ice, rock and debris until the pressure breached the barrier on exposed slopes, creating a short-lived flow of liquid water, perhaps laced with chemicals that would lower its freezing-point. A significant volume of rapidly flowing water would be required to form a gully: a volume of liquid equivalent to an Olympic-sized swimming pool would be required to cut the smaller gullies, and 100 times this for the larger ones. In the polar regions, where carbon dioxide could freeze during the winter, gullies could be formed by the thaw of carbon dioxide snow in the spring.[140,141,142]

As regards windblown deposits, small dune fields first observed by Mariner 9 had shifted. It was found that there were no light-colored large dunes; they were all darker than their surroundings. But small ripples were either the same as, or lighter than, their surroundings. This was probably due to differences in the size and/or the composition of the particles. In some places there were fields of 'petrified' dunes whose age was indicated by cratering. In fact, one of the original objectives for the narrow-angle camera when this was proposed for Mars Observer in 1985 was to search for such ancient features. Thousands of narrow-angle images showed a large number of thin streaks several meters wide that often crisscrossed each other. They were found at all elevations, ranging from the summit of the highest volcano to the floor of the deepest basin. Their cause was revealed by images which were taken as dust devils were sweeping across the surface. The existence of dust devils had been suspected from earlier orbital imagery, and also from meteorological data provided by the Viking and Mars Pathfinder landers. Being comprised of very thin layers of dust, these streaks were extremely short-lived. Their abundance indicated that dust devils must be very common.

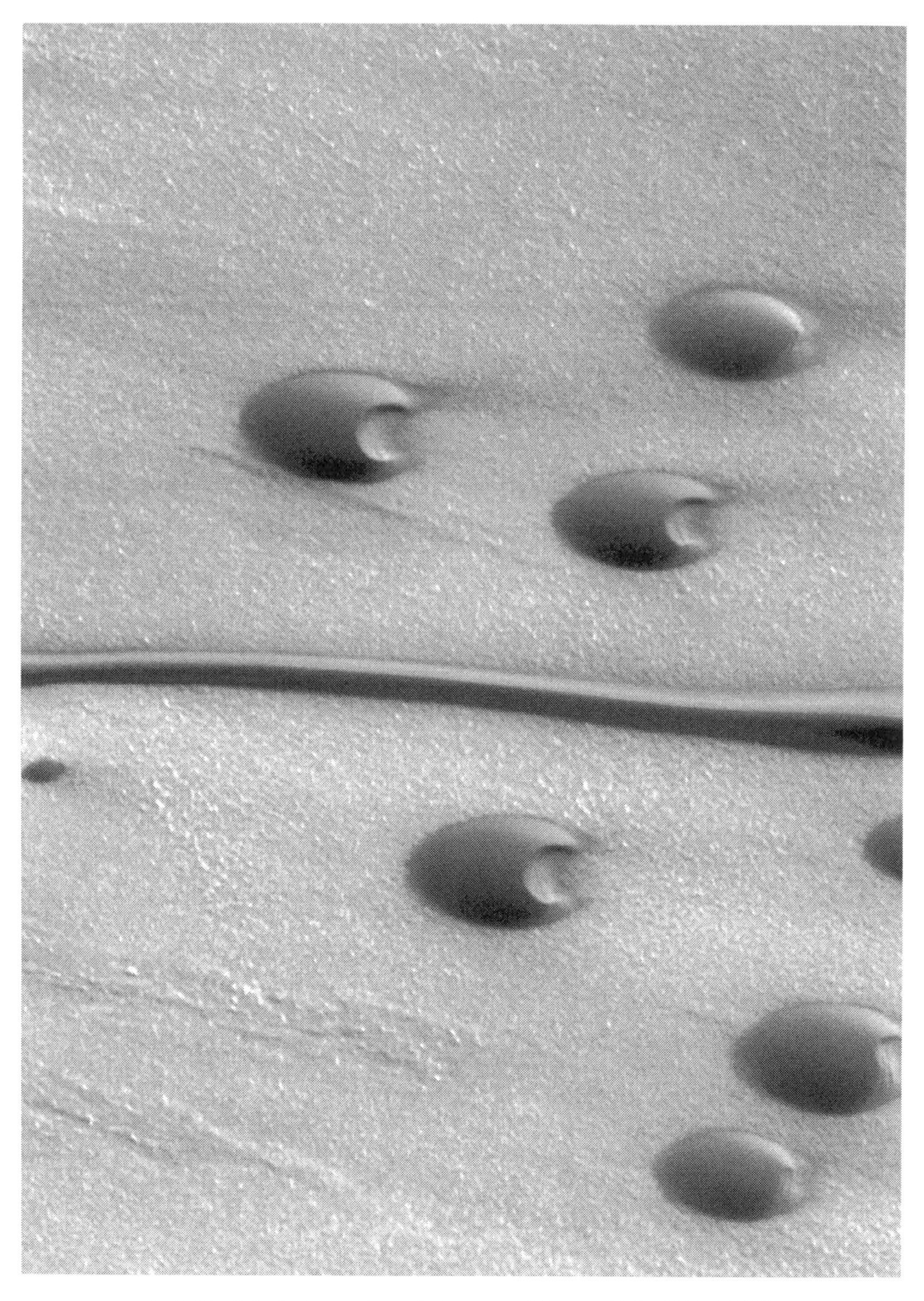

A field of dark, horn-shaped dunes on Mars. (NASA/JPL/MSSS)

Streaks left by dust devils on Argyre Planitia. (NASA/JPL/MSSS)

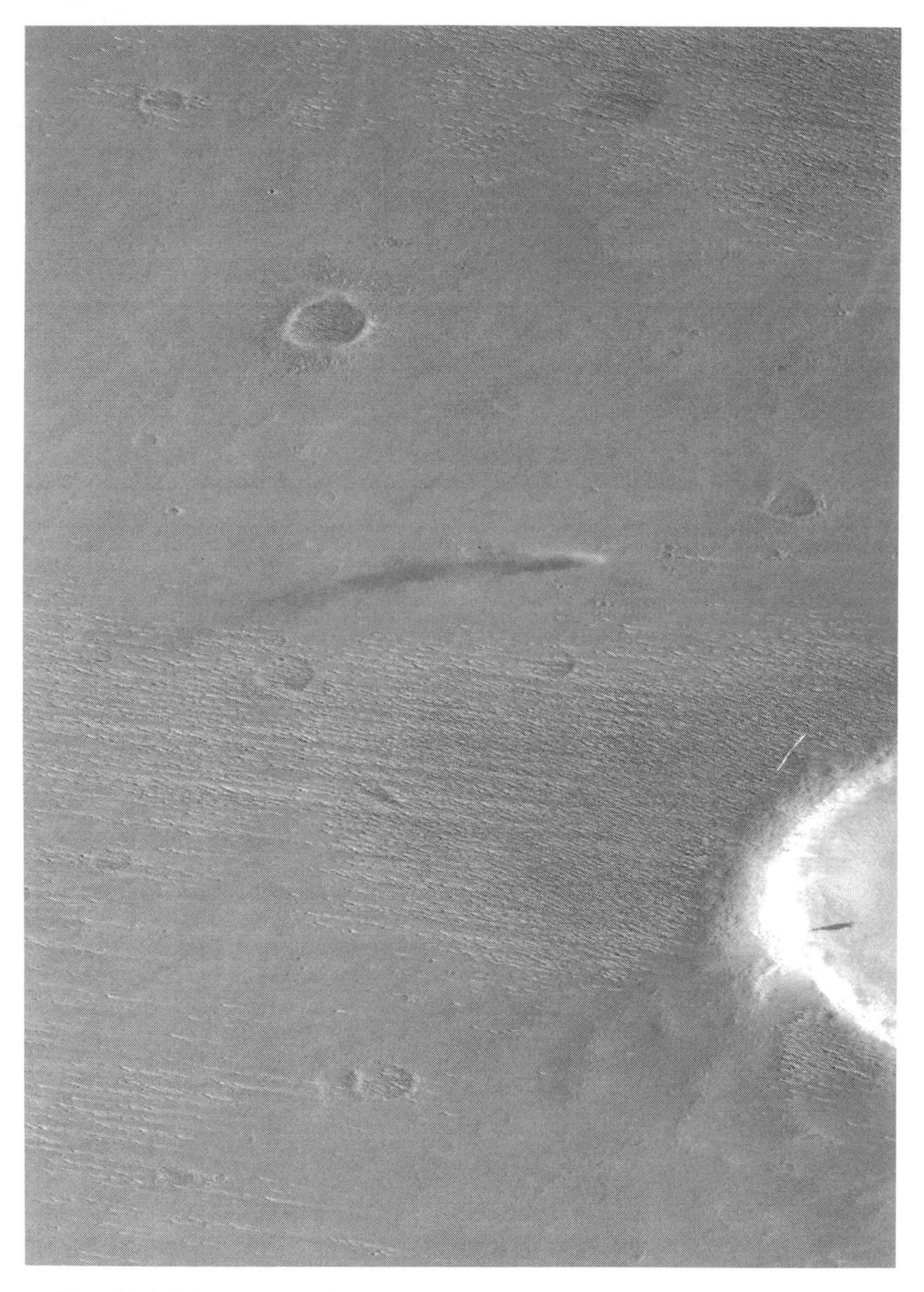

Mars Global Surveyor caught a dust devil in the act on western Daedalia Planum. (NASA/JPL/MSSS)

Because the mapping mission lasted an entire local year, Mars Global Surveyor was able to monitor, in turn, the retreat of the polar ice in both hemispheres. As the southern cap withdrew, a remarkable pattern of circular depressions reminiscent of Swiss cheese was exposed on the residual cap. In the north, the surface was largely flat, disrupted by small pits or buttes only a few meters in size. Images of the polar layered terrains showed layers which were thinner than were resolved by previous orbiters. Such thin layers were evidence of cycles of dust deposition and erosion on timescales much shorter than could be accounted for by climate changes resulting purely from shifts in the planet's spin axis. Dune fields in the polar regions showed dark spots that radiated streaks. These were initially thought to have been caused by small explosions of defrosting vapor, but they later proved to be exposed dark sand spread about by the wind.

Crater counts suggested that the floor of the caldera of Arsia Mons was at most 130 million years old, and the plains south of Cerberus were no older than about 10 million years old. The youth of the lava flows around Elysium Mons was indicated by the fact that they have crater densities 1,000 times less than the lunar maria. But because erosive processes on Mars could bias age estimates based on cratering, the issue of whether the planet is volcanically active today remains open.[143] Other dark features in Viking imagery which were suspected to be due to volcanism in recent times showed no evidence of such an origin. The slopes of the three large Tharsis volcanoes had young-looking deposits on their northwestern flanks, but these gave the impression of having been left by glaciers. Perhaps in geologically recent times the chaotic wandering of the planet's spin axis had imposed polar conditions on the tropics.[144]

One of Mars Global Surveyor's objectives had been to document the boulders in outflow channels in order to evaluate discharge rates and flow formation models, but these proved to be too small to be resolved. This was confirmed when the Mars Pathfinder set down in such a channel. Nor was it possible to find evidence that it had rained at some point in the planet's history. After studying the Viking orbital imagery of the northern plains, Timothy J. Parker suggested that several short-lived oceans had formed in this low-lying area.[145,146] Mars Global Surveyor took dozens of images in search of landforms such as beaches, barriers, tombolos and fluvial deltas to test this idea. Features were found that could not be easily explained, but the evidence for shorelines was disputed. Although one of the putative boundaries of the ocean, the outer 'Arabia shoreline', was found by the laser altimeter to have variations in elevation of over 5 km, the inner 'Deuteronilus shoreline' remained within 280 meters of its average elevation. Whilst the hypothesis was certainly not proven, it remains quite popular. In fact, proponents of the ocean contended that the early afternoon illumination of the images was ill-suited to discerning the delicate features of a shoreline, which in any case may have been short-lived. For example, the rim of Lake Bonneville, which covered large parts of Utah, Nevada and Idaho, has almost completely disappeared in less than 100,000 years.[147,148,149]

Throughout the mapping mission, the wide-angle camera produced global maps on almost every orbit at a resolution of 7.5 km in order to monitor the atmosphere and seasonal surface changes.

'Swiss cheese' terrain near the south polar cap of Mars, showing mesas only a few meters high and circular pits. (NASA/JPL/MSSS)

A remarkable heart-shaped pit in Acheron Catena. (NASA/JPL/MSSS)

The thermal emission spectrometer showed the surface of Mars to be dominated by just two types of volcanic rock that had compositions similar to their terrestrial counterparts. Apart from patches of hematite, there were no large concentrations of other minerals. For this analysis, the instrument had targeted the large dark regions that were believed to represent dust-free terrain – these were the 'seas' on the maps drawn up prior to the space age. The ancient southern hemisphere seemed to have a basaltic composition, while the northern hemisphere resembled andesite, which is an igneous rock rich in aluminum and silicon that (on Earth) is indicative of crustal material which has been remelted by later volcanism. This was consistent with the soil analyses performed by the Viking and Mars Pathfinder landers on the northern plains. But there were some remarkable anomalies. In addition, no region had the same composition as the basaltic silicon-poor meteorites thought to have originated from Mars – the best-matching composition was in a region that accounted for just 15 per cent of the surface, and it seemed rather unlikely that all the meteorites apart from ALH84001 derived from one region; even if it did include the planet's largest volcanoes. One striking finding by this instrument was of gray crystalline hematite in Meridiani Sinus, Aram Chaos and the Ophir Candor part of Valles Marineris, all near the equator. To some extent these areas correlated with finely layered terrains. Hematite comes in two forms. Because the fine-grained form is mixed into the dust on Mars, it has been distributed globally. Gray hematite is much more granular. Its presence on Mars was remarkable because on Earth most hematite mineralization processes require a long-term supply of water that is rich in dissolved iron-bearing minerals. It is most commonly associated with hydrothermal activity. On the other hand, the instrument found the greenish iron-magnesium silicate mineral olivine to be common, and because this is easily weathered by water, its presence argued for the surface of the planet having been arid for billions of years. The data could also distinguish low-iron and high-iron forms of olivine.[150] One of the main objectives of the instrument had been to look for carbonates which (it was thought) must have formed early in the planet's history when the atmosphere was much denser and the climate was wetter. It was expected that carbonates would have precipitated out of solution as bodies of standing water evaporated. But such large-scale deposits were absent. Later analyses found that the infrared spectra showed the presence of small amounts of carbonate, in particular as magnesite in the dust.[151,152,153] The mystery of the missing carbonate evaporites was resolved several years later when a Mars Exploration Rover sampled the hematite at Meridiani Sinus (as will be explained in a future volume in this series).

Ionospheric radio-occultations and measurements by the two particles and fields instruments provided a detailed map of the obstruction that Mars represented in the solar wind. This was generally Venus-like, except over regions where the remanent magnetic field was intense, where it was Earth-like. Contrary to expectations, there appeared to be no direct influence by the magnetic field on the location or altitude of the planetary bow shock, since it was too high to be influenced in this way. The magnetopause had been detected by Fobos 2, but Mars Global Surveyor extended the dataset from a few dozen crossings to many hundreds, and yielded new insights into the physics of this boundary. In the process of collecting this data, the electron reflectometer became only the second instrument to sample the ionosphere in situ – the only previous studies having been made by the Viking landers in the process of entering the atmosphere. Mars Global Surveyor was also the first mission to return data on how the Martian ionosphere responded to solar flares over a long period of time. This was done by using radio-occultations to measure the density profiles of electrons in the ionosphere. In fact, by 2001 some 1,867 such measurements had been made, which was over four times as many as were made by all the previous missions combined.[154,155]

The laser altimeter mapped almost all of the planet. The firing and receiving of a total of 500 billion pulses was in itself a notable achievement, because no such instrument had previously been operated for so long. The data was so detailed that Mars became better known than the topography of some of Earth's continents. For example, Tharsis was resolved into one isolated rise to the north, dominated by the low shield of Alba Patera, and a larger rise on which stood the volcanoes of Arsia, Pavonis and Ascraeus Mons. Tharsis produced stress faults and radial fractures that contributed, amongst other things, to the formation of the Valles Marineris canyon system. Olympus Mons seemed to be physically isolated from Tharsis.[156] Tracking of Mars Global Surveyor in its low mapping orbit provided a much more detailed map of the gravity field than had been possible with previous missions. The gravity data confirmed the hemispheric dichotomy. When this data was combined with the laser altimetry, it yielded the first reliable estimates of the thickness of the planet's crust and upper mantle. The crust in the southern hemisphere was thick, with few gravitational anomalies. In the north the crust was thinner, and showed a variety of anomalies. There were distinct anomalies associated with Tharsis, Olympus Mons, Valles Marineris and the Isidis basin. Tracking of the Viking orbiters had revealed there to be a large mass concentration associated with the Tharsis volcanoes, and a mass deficit associated with Valles Marineris. Furthermore, the two hemispheres seemed to compensate for the overlying topography in different ways. A number of anomalies in the north that did not correlate with topographic features could mark impact basins occupied by volcanic deposits or other sediments. One such site was Utopia, which had long been thought to be a 1,500-km-diameter impact cavity. Chryse was also thought to be an ancient basin, and was marked by an anomaly, but it had been completely filled by sediments from several large outflow channels. The poles had different gravitational signatures, possibly as a result of differences in structure, composition or even age. The Mars Global Surveyor data indicated that the crust thinned from the southern regions toward the north, with the trend continuing through the Tharsis region. The altimetry revealed there to be an abrupt boundary between the old and cratered southern highlands and the smooth northern lowlands. The fact that there was no crustal discontinuity to match this topographic boundary implied that this is not a manifestation of the planet's internal structure. This cast doubt on the hypothesis that the northern lowlands are the remnant of a vast impact basin produced very early in the planet's history. Interestingly, gravity profiles of the northern regions revealed structures which might be buried channels that transported water and sediments from the southern highlands into the putative northern-hemisphere ocean.[157,158]

The tracking of Mars Global Surveyor revealed for the first time how the shape of Mars (and hence its gravity field) changed in response to the 'tidal' attraction of the Sun as the planet pursued its somewhat elliptical orbit. Because the shape of a planet depends on its internal structure, it was possible to study the characteristics and state (liquid or otherwise) of the core. The results ruled out a completely solid iron core, and instead implied a large core which was at least partially liquid at its periphery and rich in light elements such as sulfur. Analysis of Martian meteorites and Mars Global Surveyor's discovery of a fossil magnetic field had already indicated the core to be rich in iron.[159,160]

In bistatic radar experiments Mars Global Surveyor reflected its radio signal off the Martian surface directly back to Earth, or bounced it off the surface at grazing angles during radio-occultations. Such experiments had been conducted in 1969 by Mariners 6 and 7, and later by Viking 2.[161] In particular, these observations were to characterize the terrain where Mars Polar Lander had been lost; showing this to be typical of the polar regions.[162,163]

Since Mars Global Surveyor was so healthy at the end of its primary mission, it was granted an extension. During this time engineers tested a novel way to acquire high-resolution images in which the spacecraft was made to roll more rapidly than usual. It normally made a full 360-degree roll once per orbit, to hold its science deck facing the ground for nadir-viewing. By rolling at a faster rate, it was possible to keep a target in the field of view of the push-broom CCD sensor for longer and thereby attain a higher resolution in the along-track direction. Indeed, it produced a resolution of 30 cm. (This technique would be dramatically used in 2004 to image the Mars Exploration Rovers, discerning details as small as their wheel tracks.) The spacecraft was also made to view to the side of its track for stereoscopic imaging.

It was during its extended mission that Mars Global Surveyor became the first orbiter to see most of the phases of a global dust storm. Such a storm had already masked the planet when Mariner 9 arrived in 1971. There had been major storms in 1973 and 1977, but not global in extent. Mars Global Surveyor had seen a number of regional storms, but this was the first one that spanned the entire planet. It began when a small orange cloud appeared in Hellas on 15 June 2001. After giving rise to local and regional storms over the next 16 days in Hellas and along the boundary of the south polar cap, it doubled its extent in 2 days by simultaneously expanding south to the polar cap, east through Hesperia and north right across Terra Thyrrena and over the equator to Syrtis Major. At its peak rate of expansion, the front of the storm was advancing at about 30 m/s. After 13 days the southern hemisphere was obscured, and much of the northern hemisphere with it, but strong air currents had halted the storm's northward progress at 60°N. There was dust at an altitude of 60 km, as there had been in 1971. By 4 July the storm was truly global and, as in 1971, the planet had become a featureless ball apart from a few high points that poked through. Contradicting the belief that a global dust storm constituted a single event which progressively engulfed the planet, these observations showed that a number of regional storms formed along the leading edge of the main disturbance, and that these both sustained and helped to propagate a high-altitude dust cloud around the globe. The thermal emission spectrometer recorded an increase in the atmospheric

A series of images centered at 270°W (the longitude of Syrtis Major and the Hellas basin) showing thermal emission spectrometer measurements of the dustiness of the Martian atmosphere during the planet-encircling dust storm of 2001 (although the difference between clear and very dusty is difficult to see in this black-and-white view). (Courtesy of Philip Christensen of ASU, and NASA)

temperature of 30–40°C when the storm was at its peak, and revealed how layers of warm and cold air interacted to make the fierce winds which powered the storm. Slight signs of atmospheric clearing 43 local days after the storm erupted were the first indication that the global storm had started to abate, but the regional storms remained as active as ever. The northern hemisphere was clear in less than 20 days, but low-lying regions like Valles Marineris on the equator and Hellas in the south were still cloaked by pinkish clouds. The spacecraft was unable to observe the final phases of the storm, because an attitude control problem on 6 September put it into safe mode. In fact, the attitude control system had been experiencing difficulty in recognizing reference stars for a some while, each time interrupting observations. When monitoring was resumed a fortnight later, a few local storms were still active but the atmosphere was much clearer. However, it would be several months before the opacity of the atmosphere returned to its average value. Follow-on observations revealed that the storm had altered the boundaries of the dark and bright areas, and the anomalous temperatures had affected the rate at which the polar cap receded. A number of altered meteorological phenomena persisted for some time. In fact, a dust storm has a similar effect on the Martian atmosphere as a major volcanic eruption does on Earth – a recent example being Mount Pinatubo in the Philippines in 1991. Other large dust storms covered part or all of the southern hemisphere in the next two Martian years, but the they were not monitored in such detail. Remarkably, during three local years of operations, Mars Global Surveyor spotted over 5,700 local dust storms.[164,165] On 30 June 2001, as the dust storm was raging, the laser diode of the altimeter failed, and thereafter the instrument's telescope was able to operate only as a radiometer.

Mars Global Surveyor's longevity after completing its own (delayed) primary mission enabled it to assist with later missions. In late 2001 the camera and thermal emission spectrometer provided daily reports on the state of the upper atmosphere for the just-arrived Mars Odyssey, which was in the process of aerobraking. Then, by sheer coincidence, the plane of Mars Global Surveyor's orbit enabled it to assist with the arrival of the Mars Exploration Rovers in early 2004. Its orbit was refined to place it overhead to record telemetry during a critical 70-second period as each lander made its descent. In transmitting this engineering data to Earth, it confirmed that the landings had gone to plan. In each case, Mars Global Surveyor then served as the main relay for returning scientific data from the surface to Earth – a role it shared with Mars Odyssey.[166]

Mars Global Surveyor was also used for unprecedented exercises. In May 2003 it imaged a conjunction of the Earth and Jupiter in which it resolved details as fine as the coastline of South America. The idea of imaging another spacecraft orbiting Mars was first proposed in 2003, but because the process would require the active spacecraft to perform a series of unusual maneuvers this was deferred. On 20 April 2005 Mars Global Surveyor took an image of the European Mars Express orbiter at a range of 250 km, resolving it as a blurry streak. The next day, controllers took advantage of the fact that Mars Odyssey would pass within 15 km of Mars Global Surveyor to snap two images during a single slewing maneuver, the first inbound at a range of 90 km and the second after the flyby at a range of 135 km. It managed to resolve the solar panels, antennas and booms of the target, thereby demonstrating that in favorable circumstances this technique offered a means of diagnosing the health of other satellites.[167] Mars Global Surveyor collected a great quantity of data on the Martian environment that had rarely been sampled previously, and at such times only briefly. An unusual result from electron profiles measured during radio-occultations was the identification on 26 April 2003 of a short-lived plasma layer at an altitude of about 90 km that appeared to have been produced by a shower of meteors just as the planet crossed the orbit of the short-period comet du Toit–Hartley. Similar plasma layers are created by meteors disintegrating in the Earth's atmosphere, and were one of the early rationales for developing radio-telescopes.[168] In October 2003 the spacecraft recorded how the Martian magnetosphere responded to the arrival of the mass of plasma ejected by the Sun in its 'Halloween' flares. The results implied an increased rate of loss of atmospheric gases that probably mimicked conditions much earlier in the planet's history. On 40 occasions between January 2004 and February 2006, Mars Global Surveyor happened to pass within 400 km of Mars Express. On the closest pass, on 10 February 2006, the two were a mere 26 km apart. These opportunities were exploited to make simultaneous observations of the Martian plasma.[169] Another unusual test was to measure the minuscule precession of the orbit by the Lense–Thirring 'frame dragging' effect of General Relativity. An analysis of tracking during a 5-year period provided a controversial result that is believed to be rather more precise than previous tests in orbit around Earth.[170,171] In August 2003 the camera team began to solicit and accept imaging requests from members of the public. The first narrow-angle shot of this 'public target program' was taken on 4 September and showed details of the caldera of Pavonis Mons.

The real benefit of the extended mission was the ability to monitor polar winter and summer cycles, seasonal phenomena like dust storms and medium- and small-scale surface changes. The results were impressive. For example, laser altimetry of the topography of the polar caps enabled the seasonal change in cap thickness to be correlated with the cycle of exchanging carbon dioxide between the ice and the atmosphere. Most of the mass exchange occurred at latitudes of less than 73 to 75 degrees, while the greatest change in elevation, which was of the order of several meters, occurred at latitudes above 80 degrees. The anomalous rate of sublimation during atmosphere-heating dust storms was also recorded. Simultaneous Doppler tracking served to measure the mass (and hence the density) of the sublimating ice. The results suggested that the caps were covered by hard, dense carbon dioxide ice rather than fluffy snow or frost. In fact, the thermal emission spectrometer was able to discern three forms of carbon dioxide on the surface: fine-grained frost, coarse-grained frost and slab ice. A remarkable 50-billionth of the planet's total mass took part in the exchange cycle, and the tracking data documented the variation in the gravity field with the annual thawing and freezing of the caps. Observations of the southern polar terrain made one local year apart were used to look for year-to-year changes in morphology. Pits, mesas and other features, most of which were tens of meters across but only a few meters high, appeared to be shaped by a combination of collapse and erosion. The scarps retreated at a rate that was consistent with the

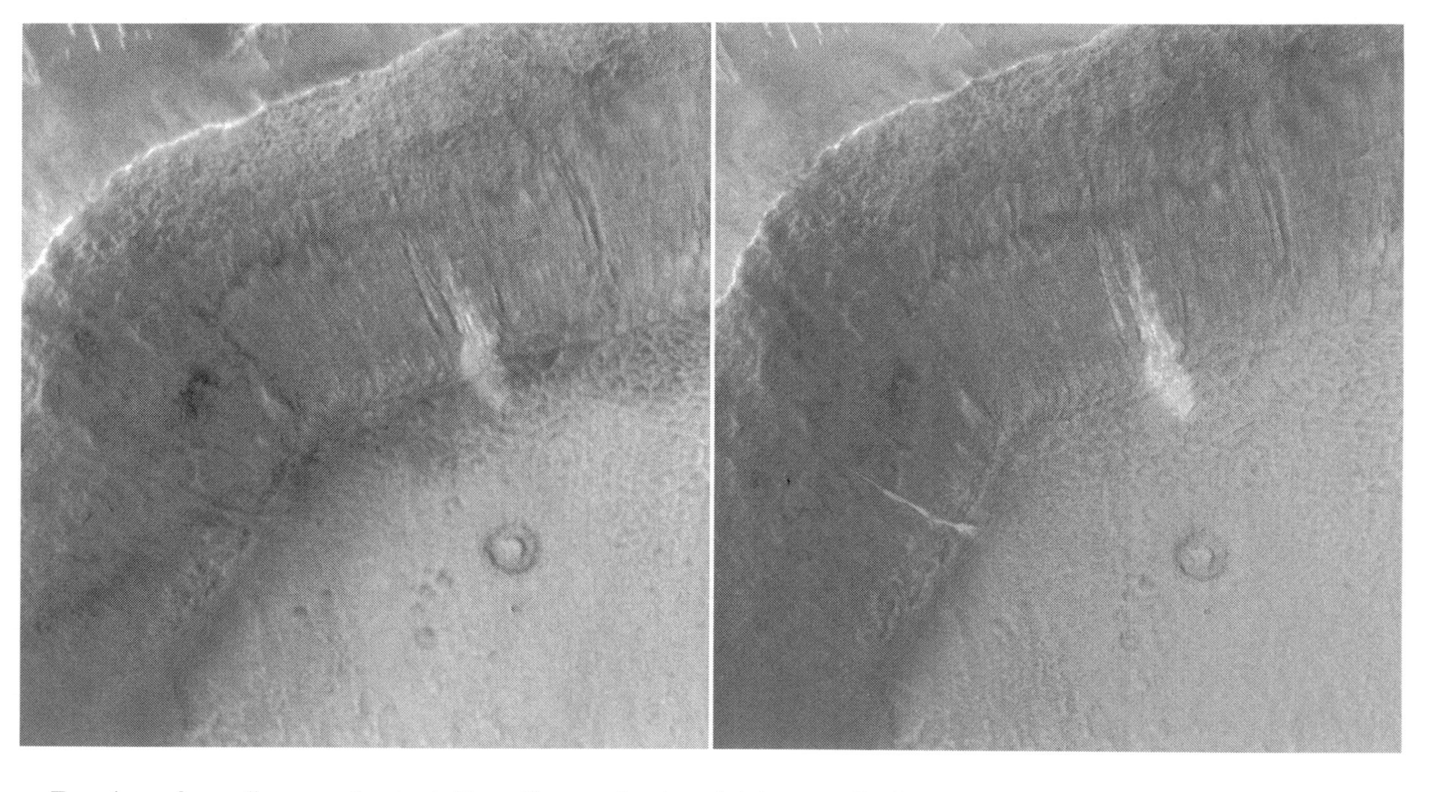

Two views of a small unnamed crater in Terra Sirenum showing a bright new gully that appeared between 2001 (left) and 2005 (right). (NASA/JPL/MSSS)

sublimation of carbon dioxide ice. Furthermore, the erosion rate seemed to indicate that the amount of carbon dioxide which the caps released into the atmosphere was increasing by as much as 1 per cent per local decade. These observations served to indicate the dynamic nature of the Martian climate.[172,173,174]

Further dramatic results were reported in 2006. The small gullies on slopes had been revisited to look for changes. There was a new feature at a site in the Centauri Montes region that had not been there 5 years earlier, and another in Terra Sirenum that had formed in the last 4 years. In both cases the feature was long, branched and flowed around obstacles. Of the various mechanisms proposed, the most likely was a low-viscosity fluid such as water. The light tone of the deposits argued for salty water possibly transporting dust and/or silt. However, if the liquid was indeed water, this still posed major questions. How was water maintained in a liquid state just beneath the surface – as it had to be in order to be able to leak out from slopes? Where did it originate from? How widespread was it? It is remarkable to note that most of the gullies appear to be in ice-free locations. Another change was found on 6 January 2006 in a wide-angle image at a resolution of 230 meters that was taken as context for a narrow-angle view – it showed a dark 'spot' about 1 km in size that had not been there previously. The following month the narrow-angle camera was aimed right at this spot, and showed there to be a fresh crater located at its center. A comparison with earlier images of this site taken by Mars Global Surveyor and other orbiters established that the impact had occurred some time after November 2004. The camera team promptly scheduled new narrow-angle views of sites which had already been imaged – and discovered no fewer than 39 new spots, of which at least 20 proved to be new craters ranging in size from just over 2 meters (i.e. to the limiting resolution) to about 150 meters, implying that the impactors were at most a few meters in size. Most of the craters were at the center of dark areas. In many cases there were multiple pits. Often there were extensive rays and other patterns. The importance of this study was that it provided a direct measurement of the current cratering rate, and indirectly confirmed that some craterless regions must indeed be very young.[175,176,177]

By now, Mars Global Surveyor was beginning to suffer from old age. Many of its components had long-since exceeded their design lifetimes. In July and August 2005 it had its first serious computer problems, with a back-to-back malfunction of both the primary and backup computers putting it into safe mode; but the problem was cured. With sufficient attitude control propellant to sustain operations well into the 2010s, it continued its highly varied extended mission. On 10 July 2006 the camera was pointed at Deimos for the first time. Although the resolution was lower than provided by Viking 2 because the range was 23,000 km, the pictures helped in refining the orbit of this tiny moon and in unraveling its geology.

On 2 November 2006, just after the mission was granted its fourth extension, routine commands were sent to Mars Global Surveyor directing it to move its solar wings off Sun-pointing in order to improve the thermal control. The next telemetry indicated that the drive seemed to have stuck and the spacecraft had activated the backup drive; which appeared to be working as intended. Contact should have been re-established 2 hours later, as the spacecraft emerged from occultation on the next

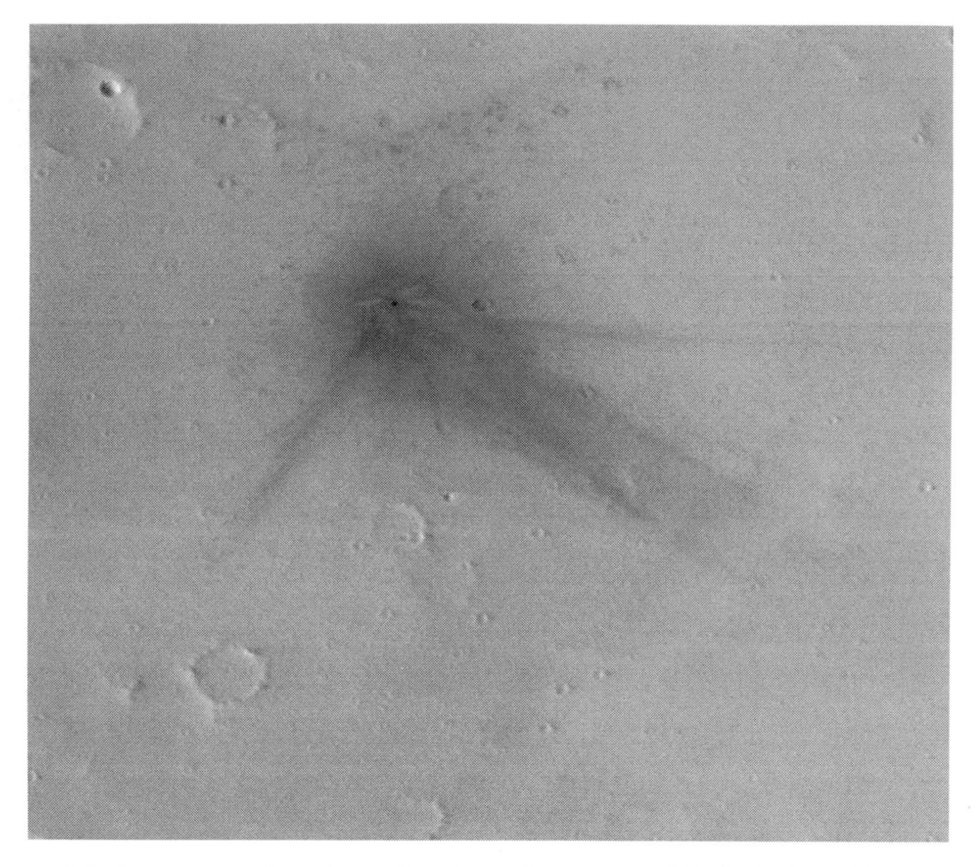

A fresh crater some 20 meters in diameter on the upper north flank of Ulysses Patera. (NASA/JPL/MSSS)

orbit, but it was never heard from again. (It was later discovered that four sporadic signals were fortuitously recorded on 5 November.) If the backup drive system had jammed while on the night-side of the planet, the spacecraft should have adopted a strategy of alternating between a power-optimum attitude of facing the solar wings at the Sun and a communication-optimum attitude of aiming the high-gain antenna at Earth. It was decided to enroll the assistance of some of the spacecraft that were either orbiting or on the surface of Mars. On 14 November Mars Express and the recently arrived Mars Reconnaissance Orbiter were told to try to image the ailing spacecraft – or if the damaged yoke had finally failed, its debris. Commands were also sent out in the blind directing Mars Global Surveyor to switch on its relay link and communicate with the Opportunity rover, which, if it received a signal, would relay the diagnostic data from that test to Earth through Mars Odyssey.[178] All these attempts were in vain, however, and on 28 January 2007 Mars Global Surveyor was declared lost. An internal board set up to investigate the incident and formulate recommendations for future missions identified the origin of the problem: in a sort of repeat of the memory-location error that had doomed the Viking 1 lander more than 25 years earlier, in June 2006 a command to position the high-gain antenna in the event of an emergency had been written into the wrong address of Mars Global Surveyor's computer.[179] This corruption had prompted the spacecraft to drive the solar wing to its hard stop, which it mistook for a stuck gimbal. Furthermore, when the emergency began, the spacecraft's battery was directly exposed to the Sun and this caused a false signal indicating that the battery was heating as a result of an overcharge. With one battery being recharged and the other not, within 12 hours both batteries had depleted. Worst of all, the corrupted pointing parameters caused the high-gain antenna to point away from Earth, precluding communications at the crucial time. The spacecraft's orbit will decay in about 40 years, and it will burn up in the atmosphere.[180]

In the years since the start of the Mars Global Surveyor project it had cost a total of only $377 million, making it one of the most cost-effective spacecraft ever. By any measure, it was an amazing success. Having spent 9 years in Martian orbit and exceeded its design life by five times, it far surpassed the 4-year orbital life of Viking 1. The camera, in particular, performed magnificently. Its acquisition rate exceeded the transmission rate of the high-gain antenna by a factor of at least 1,000, and as a consequence its 240,000 narrow-angle images represented an area corresponding to just 5.2 per cent of the planet's surface. The narrow-angle camera could never have mapped the entire planet, and had never actually been expected to, but Mars Global Surveyor had provided an excellent preliminary high-resolution view of the planet, and had survived long enough to pass the baton to Mars Reconnaissance Orbiter, whose downlink bandwidth was considerably greater.

SINKING THE HERITAGE

The Fobos missions of 1988 were intended to be only the first step in the renewed Soviet program to explore Mars in which at least eight missions involving orbiters, rovers and balloons would pave the way for a sample-return mission.

In collaboration with its traditional space exploration partners, the Soviet Union was developing the technologies for this ambitious effort. After research by VNII Transmash in the 1970s on designs for the rover to be delivered to Mars by the 4M mission, which was canceled, studies of Marsokhods resumed in 1988 with the Mir (Peace) six-wheeled 'wheel-walking' prototype. This consisted of three hinged cabs that could be mutually rotated to negotiate large obstacles, with each cab having a pair of cone-cylinder-shaped wheels that occupied most of the chassis bottom and allowed almost no ground clearance – that is, the area of the wheels that came into contact with the ground was maximized to ensure traction on a variety of loose and compact soils. In order to make the vehicle bottom-heavy (and thereby maximize stability) most of the motors, electronics and other equipment were installed inside the wheels. The cross-country performance was increased by wheel-walking, which involved lengthening or shortening the articulation joints between the cabs to vary the separation between the pairs of wheels. After successfully testing the 200-kg prototype, a smaller version using many of the same systems was developed for a mission in the 1990s. It would use panoramic cameras for navigation and scientific observations; employ a sampling device or robotic manipulator to feed material to instruments for geochemical investigations; have an antenna for communication with an orbiter, but no direct-to-Earth link; and be powered by an RTG. Engineers conducted extensive tests of the Marsokhod and its traction on the volcanic soils of Kamchatka. Upon the collapse of the Soviet Union, the Planetary Society, which is an American lobby for planetary exploration, provided the Russians with $150,000 to conduct additional tests in Death Valley. Fitted with US-made instruments and a robotic arm, it was again put through its paces in 1995 to demonstrate both lunar and Mars sorties on the volcano Kilauea in Hawaii, in a collaborative venture by the Lavochkin Association, VNII Transmash, the Planetary Society, McDonnell Douglas and NASA's Ames Research Center. Other cooperation agreements were signed with Italian industries and institutes.[181,182,183,184,185,186,187]

Following the lead of the Vega missions, another payload under consideration for Mars in the 1990s was the use of balloons. Jacques Blamont, the Frenchman who, amongst other things, was one of the proponents in the 1970s and 1980s of aerostats for Venus, had conceived of an arrangement consisting of an upper helium-filled balloon and a lower solar montgolfier. With a volume of 2,000 cubic meters the helium balloon would provide buoyancy in darkness, while the aerostat remained stationary with a drag-rope payload resting on the surface. At dawn the heat of the Sun would cause the 3,800-cubic-meter montgolfier to rise several kilometers. The winds would carry it as much as 500 km per day. At sunset it would settle to enable the instruments to sample another location. It would use an inflation system similar to that of the Vega balloons and the total mass was to be about 30 kg.[188] In the end, it was decided to use a simpler configuration consisting only of a helium balloon. To refine the technology, test instruments and evaluate payload concepts IKI and the Planetary Society jointly sponsored piloted flights in a hot-air balloon from the Prienai airfield in Lithuania in August 1988.[189] More tests of the definitive robotic configuration were carried out by Soviet, French and US engineers in the Mojave desert in 1990.

Penetrators were also under consideration. These would be dropped from orbit, and on impact would dig several meters into the surface. The intention was to use a gas-filled airbrake to slow the descent to limit the deceleration on impact to 500 g. The power supply, radio system and data management equipment would be based on the Vega designs, suitably modified to survive the impact.[190]

As for the orbiters, the Soviets gave some consideration to using an aerocapture method to achieve an initial orbit around the planet by penetrating the atmosphere protected by an aerodynamic nose heat shield – as was envisaged by NASA for its ambitious Mars Rover and Sample Return proposal – but if this was pursued at all, it was soon abandoned.

In the framework for Mars exploration presented at an IKI seminar in July 1988, the robotic sample-return mission would fly in 2005, and be followed by a human mission around 2010. Architectures for both missions were presented, based on the heavyweight Energiya launch vehicle and the use of nuclear-electric propulsion in

A VNII Transmash prototype Marsokhod with conical-cylindrical wheels which would provide traction in loose soil similar to that of a tracked vehicle.

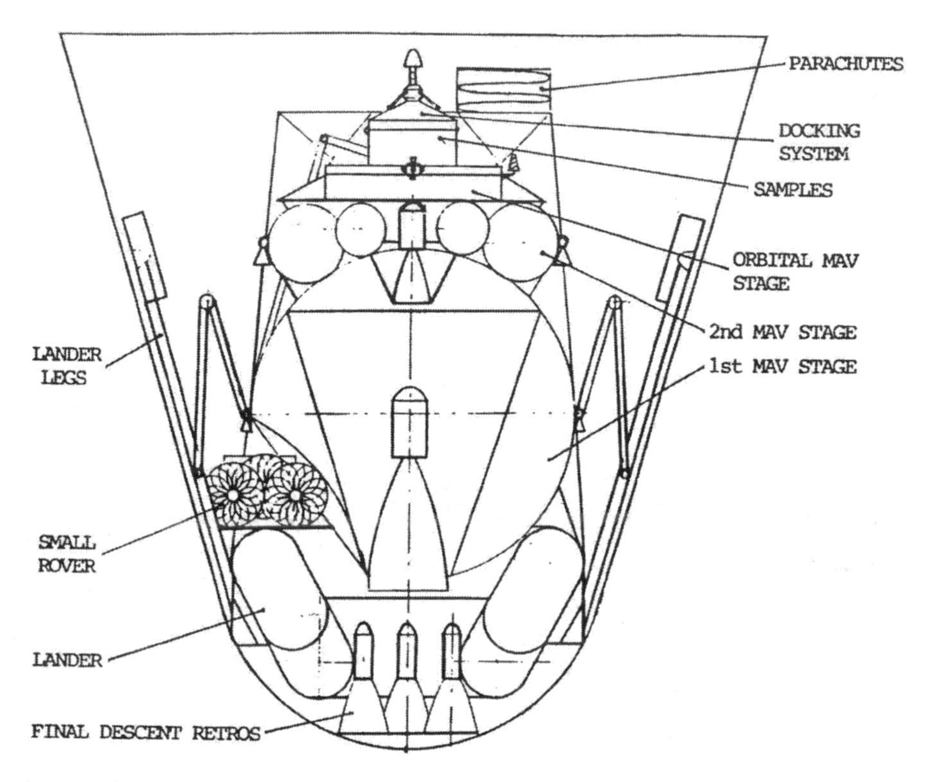

One possible configuration of the Soviet Mars Sample Return lander. Note that the ascent stage has a docking system. (Reprinted from Lusignan, B., et al., (ed.), "The Stanford US–USSR Mars Exploration Initiative", Stanford University, 1991)

space.[191] The Soviets faced the same trade-offs as their US counterparts: whether the ascent vehicle should lift off directly for Earth, or rendezvous with an escape stage in orbit around Mars; and upon reaching Earth whether it should dive into the atmosphere or brake into orbit to be recovered by cosmonauts.[192] The Mars landing vehicle would employ a combination of aerodynamic decelerators, parachutes and braking engines to effect a soft landing. Samples could either be collected from the immediate vicinity by a robotic arm, or be collected by a rover and transferred to the ascent stage. One novel idea was to incorporate the ascent stage into the rover. This would eliminate the need to navigate a return to the lander, but the additional mass would affect mobility and greatly increase the power required for locomotion.

In the profile involving rendezvous, the 2-stage ascent vehicle would deliver the samples to a 200-km Mars orbit, and the orbiter would be the active partner in the rendezvous. Unlike the Americans, the Soviets had a great deal of experience in automatic dockings. In keeping with their tradition of rugged technology, once the vehicles had maneuvered into close proximity a mechanical system would perform the final alignment and coupling; thus minimizing the need for complex electronics and control systems. Then the sample container would be passed to the return stage

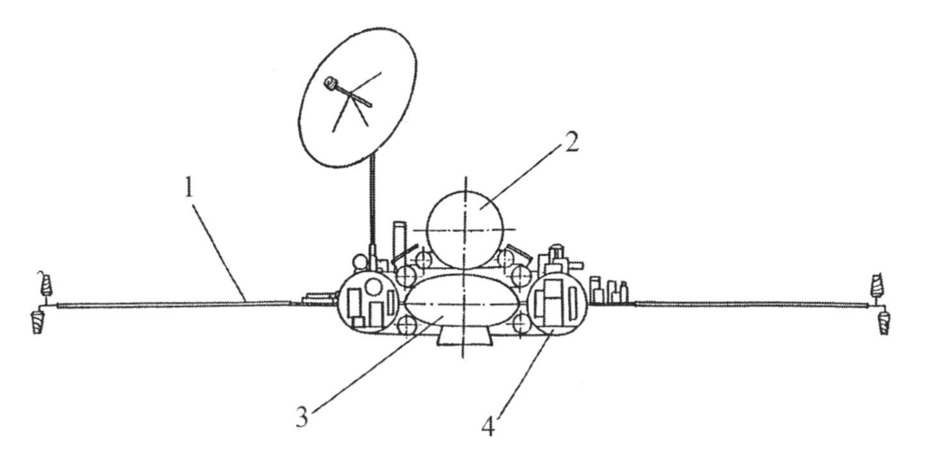

A UMVL Mars orbiter carrying a pod for returning high-resolution photographic film to Earth: (1) solar array; (2) film pod; (3) Earth-return engine; (4) scientific instruments. (Reprinted from Lusignan, B., et al., (ed.), "The Stanford US–USSR Mars Exploration Initiative", Stanford University, 1991)

of the orbiter and held in a hermetic, thermally controlled and vibration-suppressed environment. If the Energiya launch vehicle were used, the heavier stack would comprise a cryogenic Earth-escape stage, a course correction and braking engine, and a Mars exploration complex consisting of an orbiter, a variety of small landers and penetrators, and the main lander and direct-to-Earth ascent vehicle.[193]

In the original plan, extensive exploration would begin in 1992 and 1994, with dual orbiters carrying aerostats and small rovers capable of traveling 200 km which would survey sites for future landers and larger rovers. Because it would coincide with the 500th anniversary of Columbus's voyage, the Soviets attributed his name to the 1992 mission. Other possible payloads included penetrators, small battery- or RTG-powered landers (up to ten of which could be carried by a single spacecraft) and subsatellites for gravity field studies. A UMVL bus carrying a high-resolution film Earth-return pod was also considered, as a means of gaining experience which would be essential for a sample-return mission. But in 1988 the first mission was postponed to 1994 to give the scientists and engineers more 'breathing room' – in fact, the detailed design had yet to start.[194]

The detailed Mars program presented early in 1989 envisaged the objectives of the Mars 94 mission as being to provide high-resolution imagery of the surface and to deliver both balloons and landers. The second phase, which had been delayed to 1996, would see orbiters and rovers paving the way for the sample-return mission in 1998 or 2001. The imagery from Mars 94 was to be used to select landing sites for the rovers. In particular, detailed information on the surface 'rockiness' was a key factor in the selection of a safe site.[195] The Mars 94 mission was approved later in 1989, with the first 20 million of the 500 million roubles that the project would cost being granted a few months later. It was to be launched around 20 October 1994, and arrive at Mars 11 months later. At this point, the space program became tightly intertwined with Soviet and Russian politics. First, in 1990, in the wake of the Fobos failures, the government made drastic cuts to the space budget, and then the Minister of General Machine Building, Oleg Baklanov, who oversaw the space industry, joined the ranks of the failed coup of August 1991 that sought to prevent the break up of the Soviet Union. By 1 January 1992 the Soviet Union had ceased to exist, and been superseded by the Confederation of Independent States, the main member of which was the Russian Federation. One early Russian decision was to overcome the lack of management of the Soviet space program by establishing the Russian Space Agency (RKA; Rossiyskoye Kosmisheskoye Agenstvo) to supervise the national program. Mars 94 was endorsed as a cornerstone scientific mission of the space program by both the Russian Space Agency and the Russian Academy of Sciences. However, the economic crisis that accompanied the decline of the Soviet Union had taken its toll with the decision in 1991 that the plan would have to be revised so that Mars 94 would comprise a single orbiter carrying small landers and penetrators, and Mars 96 would deliver the balloons and rovers. At about this time, plans for follow-on missions began to lose definition. By 1992 the rampant inflation affecting the Russian economy had driven the Mars program to the verge of cancellation. The only hope was international funding. Germany and France had already invested $120 million in the project. The Russians invited American help, and Mars 94 became part of the 1992 Bush–Yeltsin agreement which called for an increase in collaboration in space, and which eventually resulted in missions by the Space Shuttle to the Mir space station. On that occasion, it was proposed that the payload of Mars 94 be revised to ferry an American 'hard' lander in addition to the Russian small stations and penetrators, but US scientists were not particularly keen on this idea, and eventually the American contribution became just one experiment to be carried by the Russian small stations.

Not surprisingly, work progressed at a snail's pace. It was 1993 before IKI was able to announce the final selection of scientific instruments. The fact that many of the manufacturers of space hardware were now beyond Moscow's influence meant that Lavochkin had unprecedented difficulty getting the parts that it required. The Russian Space Agency estimated that one-fourth of its total budget was assigned to Mars 94, but even with launch less than a year away work continued to fall behind schedule.[196] In July 1993 Viatcheslav M. Kovtunenko died. As Lavochkin's chief designer he had been the driving force behind the development of the UMVL bus. He was superseded by Stanislav Kulikov.[197] Another serious blow was dealt by the loss of Mars Observer the following month, because it was to have relayed for the Russian surface probes. The Russian orbiters would serve in this role, but their elliptical orbits would provide fewer opportunities for contact than Mars Observer would have in its low circular orbit.[198]

The version of the UMVL bus developed for Mars 94 was designated the M1 by Lavochkin. The 'dry' mass of the spacecraft was about 1,750 kg, of which 600 kg was attributed to its suite of over 20 instruments. With its solar panels and other appendages deployed, it was some 3 meters tall and had a span of about 9 meters. With two small stations, two penetrators and 3,142 kg of propellant for the Fregat engine, the total mass was 6,180 kg (some sources say 6,825 kg); only slightly less than for a Fobos. The cylindrical equipment tower of the Fobos was replaced by a flat top-deck on which were mounted two fixed solar panels and a pair of mating points for landers. To the equipment deck were then attached a variety of antennas and platforms. Using the non-steerable parabolic high-gain antenna, data rates of between 64 and 128 kbps would enable about 0.5 Gbits per day to be returned from Mars. In a departure from the Fobos architecture, the M1 was given an articulated deployable scan platform called Argus for precise 3-axis pointing. The 115-kg scan platform had thermal control and electrical interfaces for an 84-kg suite of remote sensing instruments that included German high-resolution and wide-angle cameras, a French visible and infrared spectrometer, and a Russian navigation camera that would assist in precise pointing.[199] Argus was developed by VNII Transmash, and was the cause of much concern because it proved difficult for Russian engineers to develop a sufficiently accurate stabilization system for it. At one point Lavochkin suggested scrapping it and mounting the remote-sensing instruments directly on the body, as on Fobos, but to have done this would have so curtailed their observation time as to question their usefulness. As a result, German engineers were dispatched to Saint Petersburg to assist in debugging the platform. A second, much simpler platform with 2 degrees of freedom carried a multichannel optical spectrometer, a stellar photometer and a gamma-ray spectrometer.

Sponsorship for Mars 94 came from 20 countries, including the US, Hungary, the newly reunified Germany, France, Finland, Greece (its first planetary mission) and the European Space Agency. There had also been talks with Chinese scientists, but they did not participate in this mission.[200,201]

The Argus scan platform of the Russian Mars 94/96 spacecraft. (VNII Transmash)

Germany's involvement began with the Institute for Optoelectronics of the West German space agency opting to build a high-resolution camera and the Institute of Cosmic Research of the East German Academy of Sciences a wide-angle camera. After national reunification, these instruments were combined into a single project led by Dornier with Jenoptik Carl Zeiss Jena as its subcontractor. The objectives included geomorphological studies of volcanic, fluvial, glacial and other features; characterization of the polar caps; 3-dimensional topographic studies; stratigraphic analyses; photometry and mineralogy of the terrain; geodesy and cartography over varying scales; high-resolution imaging of landing sites; the evolution of clouds and dust storms; and global coverage of Phobos and Deimos. Both cameras were designed as linear 5,184-pixel push-broom scanners with multiple CCD sensors in parallel. There were three sensors in the wide-angle camera to allow stereoscopic imaging, and nine in the high-resolution camera to facilitate multispectral imaging, stereoscopic imaging and photometric surveying. The resolution would depend on the position of the spacecraft in its eccentric orbit, but it was expected at best to be 12 meters for high-resolution imaging and 100 meters for wide-angle imaging. The length of a scan could be selected to suit the requirements of observation, with the integration time of the CCD being tailored to the orbital velocity. Extensive data compression and encoding would be applied on board to reduce the volume of data to a manageable level for transmission. Integrated with a 12-kg 1.5-Gb solid-state memory that was also to serve other instruments, the two cameras had a total mass of 48 kg.[202,203]

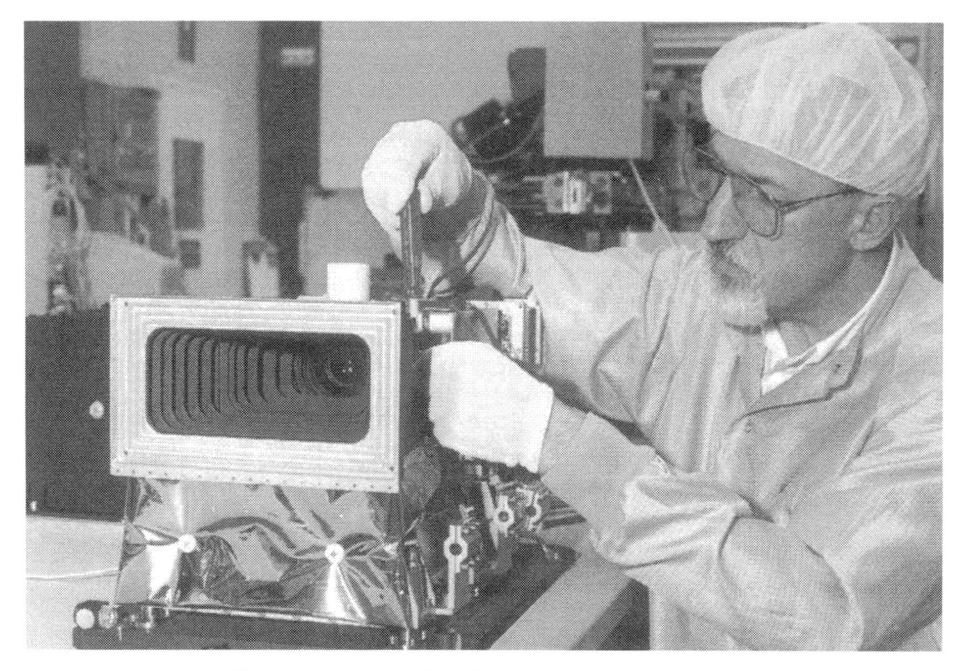

The German Mars 94/96 wide-angle camera. (DLR)

A visible and infrared spectrometer shared the Argus platform with the cameras. It was to map the surface composition and the main constituents of the atmosphere, ices, frosts, etc. The Termoskan of the Fobos missions was to be reflown to obtain data on the thermal inertia of the surface, temperature fields and thermal dynamics, and heat sources. A planetary Fourier spectrometer was to profile carbon dioxide to determine the distribution of temperatures, winds and aerosols. A high-resolution mapping spectrophotometer was to analyse the spectra of rocks and atmospheric aerosols. A multichannel optical spectrometer was to determine vertical profiles of ozone, water, carbon monoxide and oxygen by limb observations during solar and stellar occultations. This instrument had two detectors: the solar one was mounted on the body in a fixed position and the stellar one was on the small scan platform. An ultraviolet spectrophotometer was to map hydrogen, helium and oxygen in the upper atmosphere, measure the deuterium in order to estimate the rate at which the atmosphere was losing water to space, and make observations of the interplanetary medium and interstellar medium. A gamma-ray spectrometer was to map naturally radioactive minerals on the surface to infer aspects of its chemistry. Humidity, ice and water deposits were to be mapped by a neutron spectrometer. A radar which operated at a long wavelength was to investigate the near-surface substructure and the presence of subterranean deposits of ice.

Like Fobos, Mars 94 had a number of instruments to study the Martian plasma environment, ionosphere and magnetosphere. A quadrupole mass spectrometer was to map the composition and profiles of the upper atmosphere and ionosphere. The ASPERA ion spectrometer and neutral particle imager of Fobos was to be carried. An omnidirectional plasma energy and mass analyzer was to measure the structure, dynamics and origin of plasma in near-Mars space. Two similar instruments were to study plasma of ionospheric origin, and the main parameters of the ionosphere. In passive mode, the long-wavelength radar would be able to chart the distribution of electrons in the ionosphere and the dynamics of its interaction with the solar wind. A low-energy charged-particle spectrometer was to study low-energy cosmic rays in the interplanetary cruise and then investigate the near-Mars environment. The plasma wave instrument included three Langmuir probes and three search-coil magnetometers. An instrument was to map the distribution of electron velocities. A pair of fluxgate magnetometers were to study the magnetic field in interplanetary space, and then near Mars. Other magnetometers were mounted on the landers and penetrators in the hope that they would settle the question of whether Mars had an intrinsic magnetic field. In addition, gamma-ray spectrometers were to study solar flares and, in collaboration with Ulysses and several near-Earth satellites, precisely locate celestial bursts. The occultation of celestial X-ray sources by the Martian atmosphere would also be studied. During the cruise an oscillation photometer was to monitor the Sun and other stars in order to collect helioseismology and astroseismology data. An instrument was to measure the radiation dosage during the cruise and in Mars orbit to gain data relevant to planning a human mission to Mars. In fact, this instrument consisted of several units monitoring photons, charged particles, cosmic rays, X-rays and micrometeoroids.[204]

The M1 was to ferry two landers on its top deck. Like the M-71 landers, these

A mockup of the Russian 'small station' lander for Mars. Its design was strongly influenced by the landers previously built by Lavochkin for the Moon and Mars.

'small stations' were egg-shaped and after landing were to open four spring-loaded petals in order to level and stabilize the capsule. With the petals closed, the capsule was about 60 cm wide. A heat shield of 1-meter diameter would be used for entry and the initial atmospheric flight. After a single parachute on a 130-meter harness had slowed its rate of descent to 26 m/s, a spherical airbag in two hemispheres that gave the appearance of a huge beach ball would inflate for landing; no retrorockets would be used. (The use of airbags was pioneered by the Soviets in the early 1960s with their lunar landers, long before being introduced to the public by JPL's Mars Pathfinder.) Approximately 10 minutes after landing the airbag would deflate and be jettisoned. Over the next few minutes the petals would open. Three of the petals had spring-loaded booms at their tips to place instruments either in contact with the ground or as far as possible from the metallic structure. The total mass of such a capsule was just 33.5 kg, of which fully 7 kg was scientific instruments. Two small coffee-cup sized RTGs in hardened shells (never before used on a Soviet planetary mission) were augmented by buffer batteries.

The small station's payload included a Finnish–French–Russian imaging system which had a 500 × 400-pixel descent camera to place the lander's observations into context, and a camera with a linear 1,048-pixel CCD and a rotating head to return a 6,000-pixel-wide 360-degree panorama with a 60-degree coverage in elevation. An alpha-particle, proton and X-ray spectrometer that was built by Germany with substantial Russian and American input was mounted on a boom to determine the

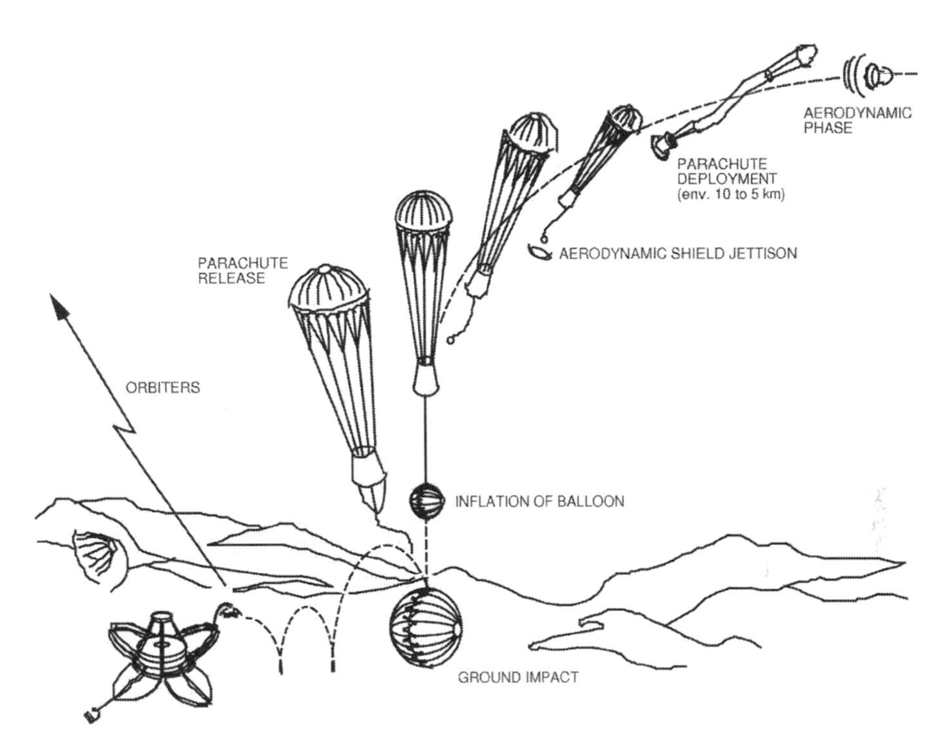

The descent profile of a Russian 'small station' lander. Unlike Mars Pathfinder, it would not use retrorockets to halt its fall immediately before reaching the surface.

abundances of elements in the surface. Single-axis seismometers were mounted on the landers and penetrators to determine whether the planet was seismically active; this was something that the Vikings had been unable to establish. The meteorology package on an extensible pod on top of the capsule had a thermometer, barometer, relative-humidity sensor, anemometer and an optical-depth sensor to measure the quantity of airborne dust. Accelerometers were also to measure the structure of the atmosphere during entry and landing. The US contribution was an instrument to characterize peroxides and other oxidants in order to test the hypothesis implied by the Viking biology experiments that the Martian soil supports chemical oxidation processes that mimic metabolic reactions. The instrument consisted of seven cells, each of which had nine spots of reactants which were sensitive to different types of oxidants. Two cells on a sensor head would be exposed to the Martian air, and five would be put in contact with the ground by a spring-loaded boom. The instrument was intended to be as self-contained as possible, since the lander would not be able to provide either power or telemetry bandwidth until about a month after landing. Even although its total mass budget was just 0.85 kg, it had its own power supply and data storage capability. It was developed by JPL in about a year, and at a cost of just $4 million.[205] A US instrument to study the magnetic properties of the soil had been considered, but was not flown.[206] Each lander also contained a 'Visions of

Mars' compact disk sponsored by the Planetary Society which contained novels, stories, articles, audio clips and images attesting to mankind's fascination with the Red Planet, as a gift to any future human explorers (or indeed other creatures) that might recover it.[207] The small stations were expected to operate on the surface for a local year.

Mars 94 also carried a pair of 2-meter-long spear-shaped titanium penetrators. On being released and spun up longitudinally for stability, a penetrator would fire a solid-fuel braking engine at its rear to adopt a trajectory which would cause it to hit the atmosphere about 20 hours later. The forebody had a diameter of 12 cm, but the aftbody started 17 cm in diameter and near the tail flared out as a 78-cm-diameter cone that housed a decelerator and its inflation system. On entering the atmosphere at 4.6 km/s, the decelerator would open to slow the penetrator's flight and stabilize it by limiting its angle of attack to a narrow range of values. Within a few minutes of entering the atmosphere, the dart would hit the ground at a speed of 60–90 m/s and split into two parts linked by a cable rope. The aftbody would remain on the surface with the meteorology sensors and camera, as well as the antennas and communication systems, while the forebody carrying the other instruments and the command system dug to a depth of 6 meters. The ability of a penetrator to survive such an impact was demonstrated initially in lift-shaft drop tests and later by drops from a helicopter. The scientific payload accounted for 8 kg of the total mass of about 120 kg. A British–Russian accelerometer was to measure the mechanical characteristics of the ground as the forebody dug itself in. A Russian 2,048-pixel linear camera would return a panorama of the emplacement site, and a meteorology package similar to that of the small stations would report on the environment. A German–Russian–Romanian gamma-ray, alpha-particle, proton, neutron and X-ray spectrometer would not only analyze the soil chemistry but also measure the water content, if any, at depth. A magnetometer would measure the local magnetic field. Another Russian instrument was to measure the thermal diffusivity, heat capacity and conductivity of the soil, as well as the heat flow from the interior.[208] An RTG would sustain a penetrator for half a local year, during which time a seismometer was to listen for activity. The data would be periodically uplinked to an orbiter for relay to Earth.

The landers and penetrators were subjected to a planetary protection effort with the goal of meeting the maximum biological risk specified by the United Nations Committee of Space Research of 300 'spores' per square meter on the external and internal exposed surfaces. A number of techniques were used, including gamma-ray, dry heat and hydrogen peroxide sterilization, and alcohol and sporicide cleansing rounded out with ultraviolet exposure, microbiological control, very sterile clean rooms, etc.[209]

The follow-on mission would also use a UMVL orbiter. It would carry a conical entry capsule whose base was 3.5 meters in diameter, containing a Marsokhod built by VNII Transmash and a large balloon supplied by the CNES French space agency. After the orbiter had completed a 10-day imaging reconnaissance, it would release the capsule. After the initial aerodynamic braking, the sections housing the rover and the balloon were to disengage and descend separately by parachute. The rover

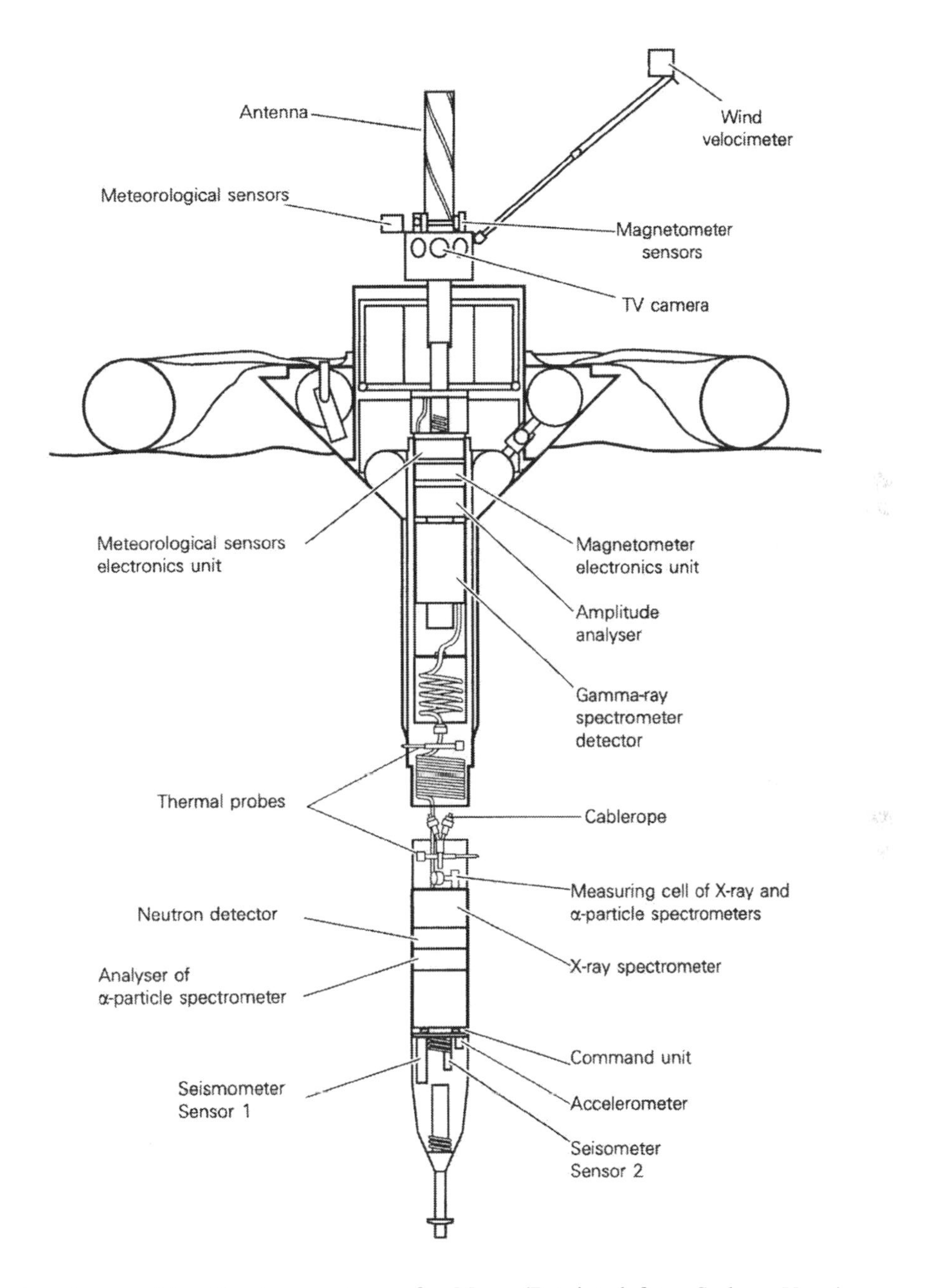

A cutaway of the Russian penetrator for Mars. (Reprinted from Surkov, Yu. A., "Exploration of Terrestrial Planets from Spacecraft", Chichester, Wiley–Praxis, 1997)

would reach the surface cocooned in an impact-absorbing airbag. After shedding the bag, the rover would conduct its 2-year mission, moving at a top speed of some 500 meters per hour.[210] A number of instruments would be mounted on the chassis within reach of a manipulator arm. A television system with four cameras was to provide panoramic views of the surface for navigational and scientific purposes. A laser aerosol spectrometer would measure the presence of dust suspended in the atmosphere. A quadrupole mass analyzer would measure the composition of the air at ground level. A meteorology package would monitor the environment. A visible and infrared spectrometer would measure the mineralogy of the surface. Magnets emplaced within sight of the TV would indicate the magnetic characteristics of the soil and windblown dust. A low-frequency radio would 'sound' the structure of the uppermost 150 meters of the ground. The manipulator, which was to be capable of digging 10 cm into the regolith and of grinding small rocks, would feed samples to a pyrolytic gas chromatograph to analyze the material. Four instruments were to be mounted on the arm itself: a camera for closely observing rocks, an alpha-particle proton and X-ray spectrometer, a Mossbauer spectrometer to study the state of iron in the rocks, and a gas analyzer to detect minority constituents of the atmosphere. It was even possible that a minirover (perhaps even JPL's Rocky 4, of which more later) would be carried to scout the terrain ahead when the Marsokhod experienced difficulties.

The inflation of the balloon by a compressed helium system would begin at an altitude of about 8 km. During this operation, which would take several minutes, the heat shield would serve as ballast. The inflated balloon would be released at its cruising altitude of 3 km. In contrast to Venus, the thin atmosphere of Mars would require the balloon to be large, and hence made of lightweight material capable of maintaining its helium pressure during the chilly night. A mylar film was proved to

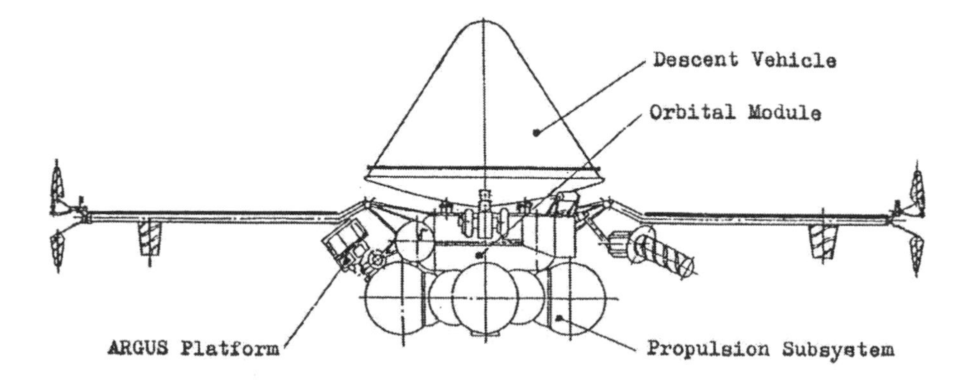

The Russian Mars 96/98 orbiter, carrying a rover and balloon in the entry capsule. (Babakin Center image; reproduced with permission of Cépaduès Editions from Eremenko, A, Martinov, B., Pitchkhadze, K., "Rover in the Mars 96 Mission". In: "Missions, Technologies et Conception des Vehicules Mobiles Planetaires", Toulouse, Cépaduès, 1993)

have most of the properties required. The 64-kg probe consisted of a cylindrical envelope 13.2 meters across and 42 meters tall which held 5,000 cubic meters of helium, suspended from which was a 15-kg gondola with batteries, a transmitter to communicate with an orbiter, and most of the scientific suite. The payload included four cameras with focal lengths ranging from 350 mm for the very-high-resolution camera, to just 6 mm in the case of the panoramic camera that was to provide 'fish-eye' pictures. A Russian–French infrared spectrometer would provide multispectral mineralogy data along the ground track, and a magnetometer would determine how the magnetic field varied. An integrated altimeter and reflectometer would measure the altitude of the balloon, and the albedo and variation in elevation of the surface. The meteorology instruments included pressure, temperature and humidity sensors. A 7-meter-long rope dangled from the gondola to provide stabilization during the night, and in particular to avoid either the gondola or balloon coming into contact with the ground. The rope was actually a number of articulated titanium segments, inside which were scientific and engineering sensors and associated electronics. It would host a gamma-ray spectrometer to analyze the surface, a thermometer, and a French–Latvian–Russian subsurface radar which would use the rope as its antenna to seek ice deposits in the uppermost 100 meters of the surface. The total length of the balloon, including the tethers between its various sections and the tail rope was about 100 meters. The plan was for the balloon to fly during the day, and settle on its rope as the air temperature and the pressure in the envelope fell during the night. There was some concern that if it were to land in bad weather its envelope might rip. The nominal mission duration was 10 days, during which it would travel hundreds if not thousands of kilometers.[211,212,213,214]

In April 1994 the US partners were told that owing to financial difficulties and the slow pace of ground testing of the spacecraft and instruments, the launch of the first mission would have to be postponed to November 1996. It would use the same window as Mars Pathfinder and Mars Global Surveyor. This raised the prospect of flying both Russian spacecraft simultaneously, but the budget would not permit this and the first mission was renamed Mars 96 and the second Mars 98. The fact that the 1996 window was not as favorable as 1994 meant that the spacecraft would have to be lightened. It is a testimony to the ability of the Russian engineers that this was achieved without seriously impacting the scientific payload. By late 1995, the Russian Space Agency realized that it could no longer afford to launch Mars 98 on a Proton rocket, and transferred it to the less powerful Molniya rocket, which had not been used for a planetary mission since 1972. Although work began to reduce the mass of the spacecraft, and in particular the balloon by shortening the envelope and moving the entire payload to the tail rope, Mars 98 was canceled. All the effort was switched to preparing Mars 96, and the Lavochkin engineers worked around the clock.[215]

As with Fobos, Mars 96 was too heavy for the Proton to inject directly into the interplanetary transfer orbit. After stage D had placed its payload into an elliptical orbit with an apogee of 314,000 km, the Fregat engine would supply the additional 575 m/s required to head for Mars. After this burn, the spacecraft would deploy its solar panels and the smaller of the scan platforms. As with Mars Global Surveyor,

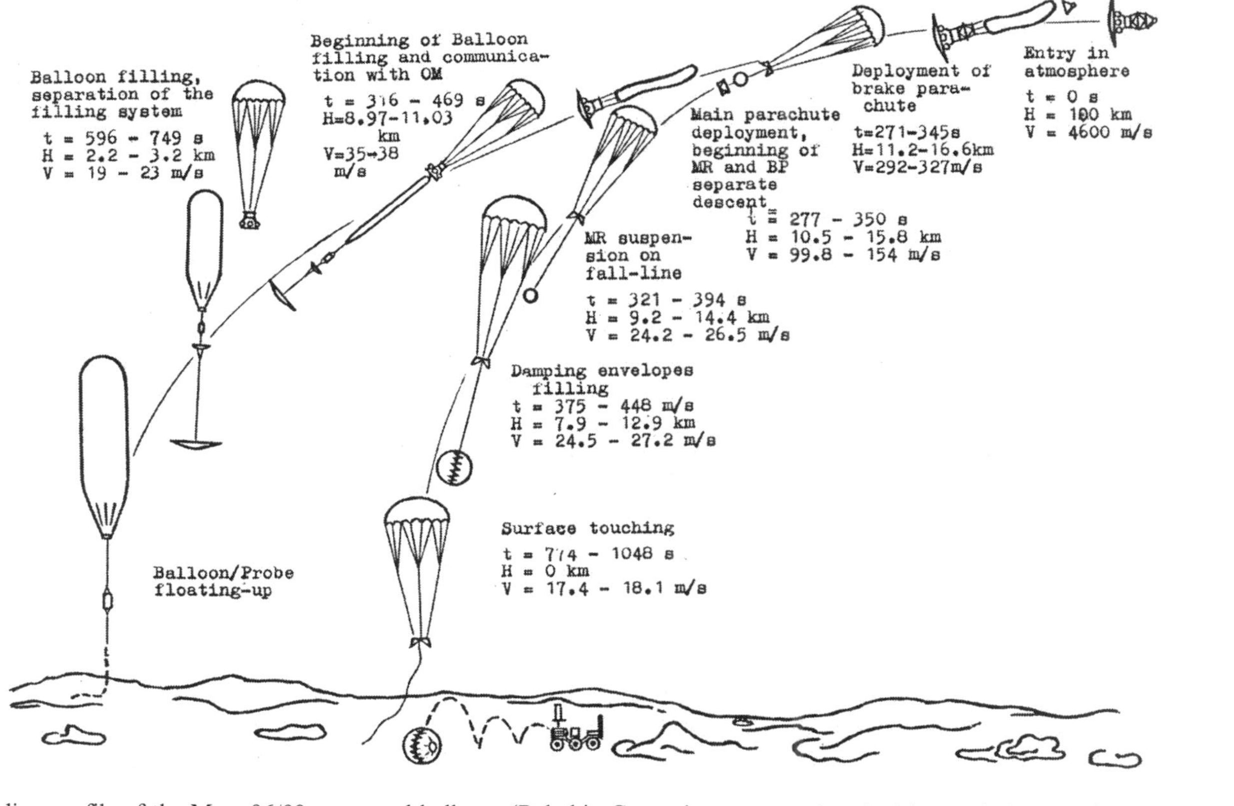

The landing profile of the Mars 96/98 rover and balloon. (Babakin Center image; reproduced with permission of Cépaduès Editions from Eremenko, A, Martinov, B., Pitchkhadze, K., "Rover in the Mars 96 Mission". In: "Missions, Technologies et Conception des Vehicules Mobiles Planetaires", Toulouse, Cépaduès, 1993)

Mars 96 was to pursue an interplanetary transfer orbit that would sweep more than 180 degrees around the Sun and arrive at the Red Planet in September 1997. There were to be three course corrections: the first two burns were to establish a collision course with the planet; then, immediately after the small stations had been released about 5 days from Mars, the final maneuver of 35 m/s would deflect off to the side in preparation for orbit insertion. The small stations would remain passive in flight, and enter the atmosphere at 5.75 km/s on a trajectory depressed below the local horizontal at an angle of about 16.5 degrees. The primary targets were at 41.31°N, 153.77°W and 32.48°N, 163.32°W, both in the low-lying Arcadia Planitia, with a backup at 3.65°N, 193°W, but because the dispersion ellipse for an entry directly from the interplanetary approach trajectory is much larger than from orbit these were 'area targets' rather than specific features.

On 23 September 1997 the main spacecraft was to perform a 1,020-m/s braking burn to enter an initial 500 × 52,000-km orbit inclined at 100 degrees to the equator with a period of 43 hours. In stages it would then lower the periapsis to 300 km and the apoapsis to 22,000 km, in the process reducing the period to about 14.77 hours. The penetrators were to be released 7–28 days after arrival. In order to provide a good baseline for seismometry, one penetrator was to be aimed at Arcadia, close to the small stations, and the other at least 90 degrees away in Utopia Planitia. After the penetrators had been released, the orbiter would jettison the unit on which they had been carried so that it could deploy the Argus scan platform. This would mark the start of the orbital science mission, which would last a local year. After that, the Russians had in mind an attempt to use aerobraking to trim the period of the orbit to about 10 hours.[216,217]

Testing of the completed spacecraft began in January 1996, and it was delivered to Baikonur in mid-October. A funding delay had caused a slip in the production of the Proton rocket earmarked for the mission, but the military agreed to allow one of its rockets to be used, providing it received a replacement by the end of the year.[218] Although the launch window opened on 12 November, it had been decided not to attempt the launch until four days later, on the date that would minimize the Fregat engine's escape burn. On being launched on 16 November, the mission was named Mars 8.

Because Russia could not afford to operate the fleet of tracking ships which the Soviet Union had deployed around the world to monitor the operations of its space missions, when stage D restarted its engine over the South Atlantic to establish the preliminary elliptical orbit it was out of communication. The investigation of what happened during this maneuver was made difficult by the absence of telemetry. In particular, it is unclear whether the engine failed to restart, or it did so and then cut off several seconds later. In any event, the orbit was not drawn out into an ellipse with a high apogee. Unfortunately, Mars 8's sequencer then caused it to separate from stage D, which precluded any opportunity to diagnose the fault with stage D and reschedule the maneuver. At the appointed time, the spacecraft fired its Fregat engine, but because it was still in the initial orbit this burn achieved an apogee of only 1,500 km. The fact that the perigee was still low (certainly under 100 km; some reports say 87 km, although this seems implausible) meant it would soon fall into the

The Russian Mars 8 spacecraft in Lavochkin's integration hall. The cones housing the 'small station' landers are on the top, between the folded solar panels, and one of the penetrators is in view attached to the Fregat stage at bottom left. (DLR)

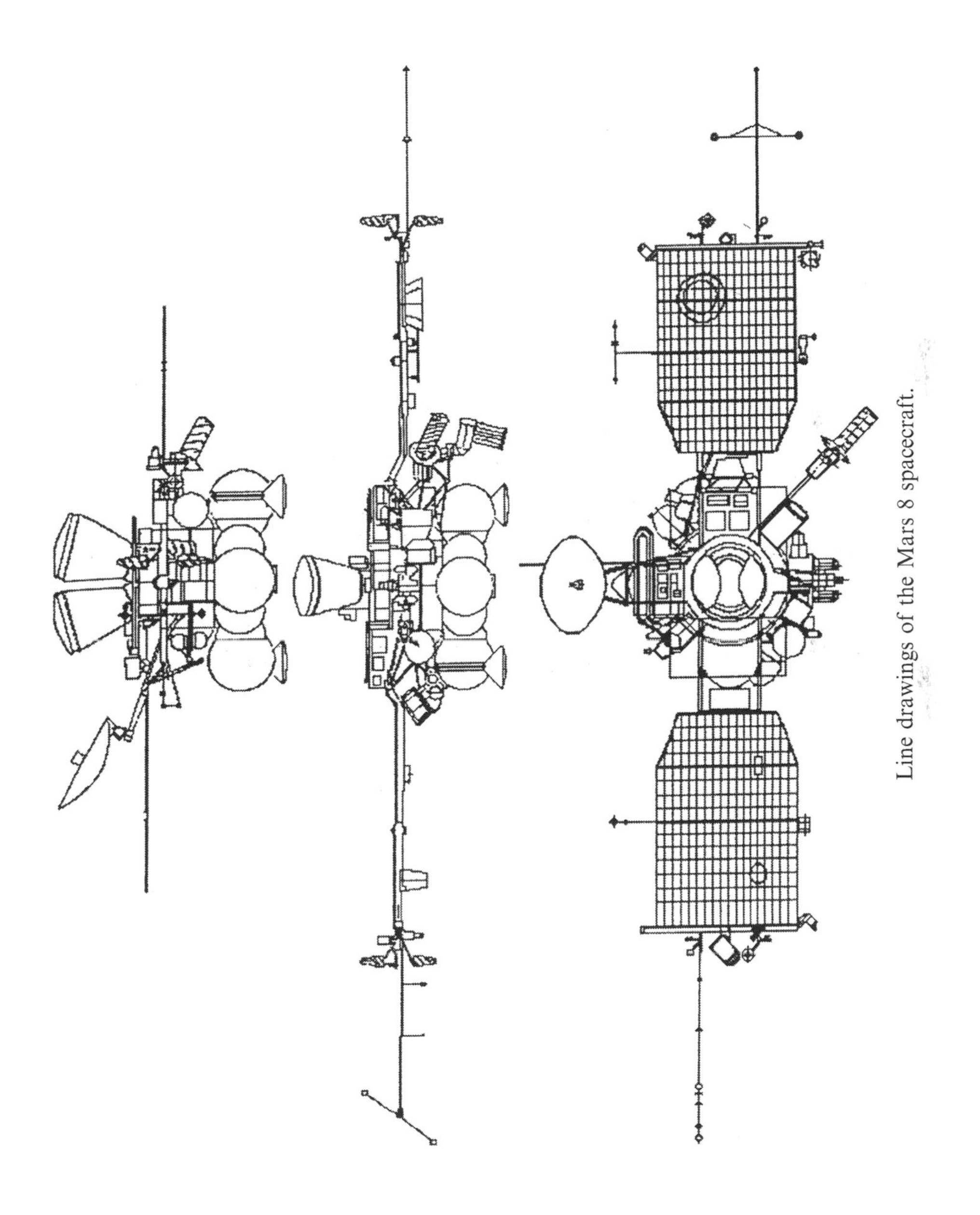

Line drawings of the Mars 8 spacecraft.

atmosphere. Ironically, the first transmission received by the deep-space communication center at Yevpatoria was interpreted as confirming that the mission was safely on course for Mars. The US military had only limited radar coverage of the South Atlantic, but they reported that a large object associated with this mission was still in low orbit. Owing to the dilapidated state of their tracking facilities, the Russians were unable to confirm that the spacecraft had separated from stage D. It was clear that one way or another the spacecraft would soon enter the atmosphere carrying the 18 pellets (each of which contained 15 grams of plutonium dioxide) of the RTGs in the surface probes. The concern was relieved on 18 November when the large object burned up over the Pacific Ocean some 560 km from Easter Island. In fact, Mars 8 had re-entered over southern Chile the previous day, being seen by a number of observers, some of them professional astronomers, as a bright, slowly moving meteor in the night sky. Debris is reported to have been found south of the Oyuni salt plateau in Bolivia at an elevation of over 3,000 meters, 100 km beyond the Chilean border, but as the Russians did not have the funds to mount a recovery operation the RTGs were abandoned. Thus the first (and so far only) Russian deep-space mission came to an inglorious end. Owing to the absence of telemetry during the critical maneuver, it was not possible to determine precisely what went wrong. The fact that the spacecraft's attitude control system directed stage D in this part of the mission raised the possibility of there having been a communications issue that foiled the maneuver.[219,220,221,222,223,224]

Nevertheless, a significant portion of the orbital science was able to be salvaged by sending some of the instruments on later missions. In particular, the European Space Agency funded the Mars Express orbiter specifically for this purpose.

The longest-lived heritage of the ill-fated UMVL bus whose unenviable record of achievement was one partially successful mission out of three attempts, was the Fregat ADU stage, which was adapted for use by many Russian launch vehicles, including the Soyuz rocket. The Soyuz–Fregat combination has deployed a number of missions, including ESA's Mars Express and Venus Express planetary orbiters and the Cluster satellites to study the Earth's magnetosphere.

WHEELS ON MARS

In the early 1990s several NASA centers were studying strategies for renewing the exploration of the surface of Mars. The Ames Research Center's MESUR (Mars Environmental Survey) envisaged an incremental approach in which a number of small landers of a given configuration would be delivered over several consecutive launch windows. Each Delta II payload would involve a carrier stage ferrying four autonomous entry vehicles. On nearing the planet, the landers would be released in series and independently targeted. In order to cut costs, there would be no precision radar or variable-throttle engines; the landing system would employ a combination of heat shield, retrorockets, parachutes and airbags. Once on the surface, the RTG-powered landers would image the site and provide meteorological and seismometry data. Ames envisaged delivering four landers in 1998, and eight in each of 2001 and

2003. The total cost of $1 billion would be spread over almost a full decade. A rival proposal from the Langley Research Center called for a Mars Polar Penetrator network. In this scheme, a subscale Viking aeroshell would make a direct entry and emplace a number of CRAF-heritage penetrators across the south polar region. JPL was also investigating using a technique similar to MESUR to deploy a network of small penetrators and landers.[225,226,227,228]

On abandoning Mars Rover and Sample Return, JPL asked its robotics group to devise smaller and more efficient rovers than Robby – the one which served as the basis of the studies for a sample-return rover. Whereas previously the imperative had been to design a rover that would operate for years and travel for hundreds of kilometers, the new ones were to be much more modest. Don Bickler, a mechanical engineer at JPL who experimented with suspension architectures for rovers, came up with an elegant six-wheeled rocker-bogie system which placed a small bogie at the end of a master bogie and used independent wheel steering to turn left or right. This not only maintained wheel contact with the ground in almost any attitude, but also gave a better distribution of weight than would an articulated-body system like Robby or the Russian Marsokhods. A preliminary prototype, named Rocky, was built in 1989 and subjected to thorough testing. A design for Rocky 2 was not built, but Rocky 3 was built and proved even better at negotiating obstacles.[229,230,231] On hearing of the MESUR proposal, JPL developed a 7.1-kg prototype Mars Science Microrover (MSM) named Rocky 4, and suggested that such a vehicle should form the payload for one of the Ames landers. At the Mars Science Working Group held at NASA Headquarters in October 1991, Ames proposed that that a SLIM (Surface Lander Investigation of Mars) be launched as early as 1996 to prove the concept of the MESUR lander. The Mars Science Working Group not only endorsed SLIM, it also urged that the payload should be JPL's rover. NASA agreed, and transferred both MESUR and SLIM to JPL as the MESUR Pathfinder project.

When NASA introduced the low-cost Discovery program in late 1992, MESUR Pathfinder, whose development was by then well underway, was announced as one of the early missions. When NASA decided not to pursue the MESUR multiprobes, MESUR Pathfinder was renamed Mars Pathfinder.

In fact, Mars Pathfinder did not fit the Discovery program's model very well, in that it was not a PI-led mission, and the engineering goals of proving that airbags could cushion the impact of a small planetary lander and that a small rover was a viable tool were ranked above the scientific objectives. The rover was cost-capped at $25 million, and the overall development budget was $171 million. Including launch and flight operations, the mission would cost about $265 million. Although JPL had led the field in planetary exploration, Mars Pathfinder would be its first lander.[232] Adopting the 'skunk works' philosophy used by Lockheed in developing some of the most advanced US military aircraft, JPL created a highly focused team in which experienced managers led young engineers, encouraging unconventional thinking with few constraints on creativity, allowing informal reporting and a relief from the usual bureaucratic paperwork.[233] Being a technology demonstrator, Mars Pathfinder would have only a limited number of scientific instruments. The lander would have only a suite of meteorological sensors, a camera and small magnets in the camera's

The little rover that could: Don Bickler's first Rocky prototype of 1989, from which Sojourner and its successors were derived. (JPL/NASA/Caltech)

field of view. The rover was to have carried a neutron spectrometer, a rock chipper and spectrometer to analyze the interior of rocks, and was to deploy a seismometer away from the lander.[234] However, mass constraints obliged JPL to install only one instrument; namely an alpha, proton and X-ray spectrometer identical to that developed for the Mars 96 small landers, which a simple mechanism would place into contact with individual rocks. This would require many hours of integration time to build up a high-quality spectrum, and so it would be operated through the Martian night.

The 10.6-kg rover was 65 cm long, 48 cm wide and 30 cm tall. Its boxy body held the electronic systems, insulated by tailored panels of aerogel and warmed by small radioisotope pellets. The rocker-bogie suspension system was mounted on the sides of the body. The master bogie on each side was designed to lay straight during the cruise in order to reduce the vehicle's height inside the folded up lander to 18 cm. After landing, a spring and latch mechanism would restore the master bogie to its operating configuration in an 'inverted-vee' shape. There were six 13-cm-diameter aluminum wheels, each with its own electric drive motor and independent steering, with stainless steel treads and cleats for improved grip in many types of soil. The rover was powered by a solar array with a collection area of 0.2 square meters on the upper deck, augmented by non-rechargeable batteries. It had a pair of imagers for stereoscopic vision. To minimize costs it was decided not to install a dedicated computer and instead to read the CCDs directly from the main computer, whose 80C85 CPU was a space-qualified form of the 1976-vintage Intel 8085 chip that was used on the very first 'home computers'. This architecture was so slow, in fact, that it took several tens of seconds to read a single image. It would have been prohibitive on Earth, where the sensor would become fogged by thermal noise, but on Mars the temperatures were sufficiently low that a usable image would result. The cameras were on a frame on the front of the body, together with five lasers that projected a fixed pattern on the terrain ahead. By measuring how this pattern was deformed, the software was able to detect obstacles.[235] The vehicle could either be 'guided' from Earth by Computer-Aided Remote Driving in which it was provided a sequence of motions to be executed, or in an autonomous Behavior Control mode in which it would 'decide' on a route to reach a specified point, adjusting its path to avoid unexpected obstacles. Its top speed was 1 cm/s. Its safety features included a tilt meter that would intervene if there was a risk of tipping over. The rocker-bogie would be able to negotiate rocks 20 cm tall. It was the most autonomous planetary vehicle yet deployed 'in the field'. It would communicate with the lander using a small rod antenna, on which it would rely because it had no direct link to Earth. The nominal duration for the rover's mission was fixed at one week, during which it would remain within the immediate vicinity of the lander, but if it survived much longer than this it would explore further out. Alongside the alpha, proton and X-ray spectrometer on the rear, the rover had a crude color camera. It also carried several technological experiments including sensors to measure the wear and abrasion of wheel material, and a solar cell to study the stickiness of windblown dust. After a competition, JPL announced in 1995 that the rover would be called 'Sojourner', after Sojourner Truth, an anti-slavery and women's rights activist of the American Civil War. The engineering model, which was identical but not flight-worthy, was called Marie Curie in honor of the famous French scientist.[236,237]

The lander had a triangular base and side petals, giving it a tetrahedral shape in flight. One of the technological innovations for JPL was the airbag system that was to cushion the high-speed landing. The Soviets had introduced airbags for its lunar landers in the early 1960s, but this was the first time that NASA had used them. An airbag was mounted under each side of the tetrahedron. Each airbag consisted of six 1.8-meter lobes in a 'billiard rack' arrangement made from four layers of high-strength Vectran fabric (manufactured by the same company that made spacesuits for astronauts). They would be inflated by quick-action gas generators which were essentially solid-fuel motors. The airbags completely surrounded the lander. With its airbags deployed, the lander was about 5.3 meters wide, 4.3 meters high and 4.8 meters deep. The Lewis Research Center conducted a series of drop tests in various situations in simulated Martian conditions and, after initial problems, demonstrated that the fabric could withstand vertical velocities of 14 m/s, horizontal velocities of 20 m/s and impacts with rocks as tall as 0.5 meter.[238] The equipment case within the tetrahedron was a truncated pyramid, and housed the electronics, batteries and the communication systems – which included a modem to communicate with the

A view of the front of Sojourner during ground preparations.

The Mars Pathfinder airbags during ground tests, showing the 'billiard rack' arrangement of the lobes.

rover. Mounted on its exterior there was a short low-gain antenna and a small motorized high-gain antenna in the shape of a hockey puck. On top was the Mars Pathfinder imager. This incorporated 23-mm f/10 optics and a pair of 256 × 256-pixel CCDs, each having a 12-color filter wheel. It was mounted on a head that could rotate in azimuth and elevation, in turn on a spring-coiled telescoping mast which deployed to a height of 1.5 meters. The camera was the most fragile part of the lander, which was otherwise designed to accommodate very rough terrains and rocks. In fact, if the lander were to come to rest too steeply inclined, the telescopic mast would not be able to be deployed and the mission would have to be conducted with the camera held in its stowed position.[239] Images from Pathfinder were to be processed on the ground, and within half an hour incorporated into a 3-dimensional virtual-reality rendition to be used in planning traverse routes for the rover.[240] With its petals deployed, the lander spanned 2.75 meters. It had arrays of solar cells on the upper surfaces of the petals, with a collection area of 2.8 square meters. In fact, this was the first solar-powered lander sent to Mars. Whip booms installed at the ends of two of the solar panels had thermocouples installed at various heights and a hot-wire anemometer at the tip, in addition to having small windsocks for imaging by the camera. As in the case of Viking, this lander had magnets in view of the camera to monitor the build up of windblown dust. The third panel accommodated the rover, whose only interface to the lander was its wheel locks. Once the rover was configured to leave the lander, it would run down a spring-deployed coiled metal slide. There were two slides, one in front of the rover and the other behind it, just in case a rock obstructed egress in one direction. The baseline mission duration for the lander was one month, but the atmospheric scientists hoped to receive weather reports for an entire local year.

The lander was to be protected during atmospheric entry by an aeroshell which was 2.65 meters in diameter and 1.5 meters tall. The development cost was reduced by exploiting the materials research invested in the Viking aeroshell. As a subscale version of the Viking aeroshell, it was a 70-degree half-angle cone with a spherical nose cap. One modernization in the heat shield was that instead of aluminum spars and stringers, a carbon-epoxy composite structure was used. To this was bonded a reinforcing honeycomb for a 1.91-cm layer of ablator material. As for Viking, this was a mix of ground cork and silica and phenolic microspheres within a matrix of silicone binder. The Viking heritage enabled the validation testing to be limited to computational fluid dynamics simulations and arc-jet testing to prove the aeroshell would survive the greater heat loads of a direct entry from the interplanetary cruise. The structural and thermal requirements of the conical backshell were less severe; so much so, in fact, that the composite fiber lay-up was less than half the thickness of the main shield, and in this case the 0.48-cm layer of ablator was simply able to be sprayed on. Inside the backshell there was a plate with the attachments and interface for the lander. This was the only part of the aeroshell to be made by JPL; the remainder was built by Martin Marietta. Six thermocouples were built into the heat shield at various depths on the stagnation point (i.e. the tip of the nose, which would endure the greatest thermal load), at mid-span and at the edge to measure the heating.[241] In addition, pressure, temperature and density profiles would be inferred

Sojourner about to be sealed inside Mars Pathfinder. The white cylinder on the lander is the camera head.

from measurements made by a set of accelerometers. Other accelerometers would provide information to the parachute deployment system.

The apex of the backshell mated to the 2.65-meter-diameter disk-shaped cruise stage, on the other side of which was an array of solar cells. It was spin stabilized, and carried 94 kg of hydrazine for trajectory control. For once, the urge to load up the spacecraft with instruments had been resisted, and no scientific observations were to be made in flight. On approaching Mars, the cruise stage would release the lander. The initial braking would be borne by the heat shield, but several minutes later, still traveling supersonically, a parachute would open to provide stabilization and additional braking. Then the forward heat shield would be released to fall away and the backshell would unreel the lander on a 20-meter-long bridle and the airbags would inflate. When a radar altimeter reported an altitude of 80 meters, three solid-fuel rockets on the backshell were to fire for about 2.2 seconds in order to bring the lander to a halt about 13 meters above the ground, at which point the lander would be cut loose. The residual thrust would carry the backshell and parachute clear. The lander would fall freely and then undergo a series of diminishing bounces, in the process traveling as much as several hundred meters. The only transmission during the descent would be a carrier signal from an omnidirectional antenna, the Doppler of which would provide a measure of the deceleration. This would be monitored by the Deep Space Network. On coming to rest, the lander would send a 'semaphore' signal to indicate that it had survived. However, this would not be effective unless the lander happened to be orientated with its antenna pointing upwards. After it was at rest, the airbags were to deflate and their fabric retracted by cables on motorized spools. Then other motors would deploy the three movable petals. These had been made sufficiently powerful for the process of opening the petals to flip the lander upright if it had come to rest on its side.[242] The landing would be tense for the team at JPL, because whilst they would be able to infer from the Doppler that the lander had reached the surface, it might be several minutes before the semaphore was able to be received.

Unlike the Viking orbiters, which had been able to survey landing site options prior to releasing their landers, the site for Mars Pathfinder had to be selected prior to launch, with the interplanetary cruise being refined to aim for this point. There were fewer constraints, since the airbag system would accommodate terrain much rougher than that of the Vikings. But because the lander was solar powered the site had to be near 15°N, the latitude at which the Sun would be overhead at the time of arrival in July 1997. A workshop of planetary scientists was convened in 1994 to review the shortlist of over 20 candidates. They selected a 100 × 200-km ellipse in Ares Vallis. This was not far from the original landing site for Viking 1, which had been rejected as too rough; and 850 km from where that lander actually set down. In 50-meter-resolution Viking imagery, the site appeared to be a flood plain at the mouth of a major outflow channel. Radar and thermal inertia data implied that Ares Vallis was rockier than 90 per cent of the surface. A major part of the attraction of this site to the scientists was that it would host rocks transported from the southern highlands – an area that makes up two-thirds of the surface but had not previously been sampled. The craters on the alluvial terrain suggested that the floods occurred

The cruise stage and aeroshell of Mars Pathfinder about to be mated with the final stage of the launch vehicle. Note the separation springs.

over 2 billion years ago. The duration and volume of water were not known, but it was expected that an examination at ground level, and in particular the distribution of the rocks and their state of erosion, would shed light on this epoch.[243,244] But not everyone was satisfied. In a report on the renewed program for exploring Mars the US National Research Council criticized the mission, citing the lack of a descent camera and the rover's limited mobility as major drawbacks of the architecture, but this assessment seems not to have fully appreciated that the mission was primarily a technology demonstrator.[245]

With its airbags and rover payload, the lander had a mass of 360 kg. The total mass of the airbags and their inflation system, together with the solid rockets, was 104 kg. The remainder of the 896-kg launch mass comprised the cruise stage and aeroshell – the forward heat shield was 64.4 kg, and the backshell and parachute system was 56.9 kg.

Major elements of Mars Pathfinder arrived at Cape Canaveral by truck one day ahead of Mars Global Surveyor, and Sojourner was flown in a few weeks later. As was the case for most of the Discovery missions, Pathfinder was to be launched by a Delta II equipped with an additional stage for the escape maneuver. The window opened on 2 December 1996 and ran to the end of the month. On the first day the weather ruled out a launch. The next day the countdown was abandoned with one minute remaining owing to a fault on the ground. But Pathfinder was successfully launched on 4 December, and injected into a 0.95 × 1.60-AU transfer which would enable it to reach the Red Planet two months before Mars Global Surveyor. Apart from initially being unable to determine the orientation of its spin axis as a result of a small piece of dislodged insulation masking a Sun sensor, Pathfinder was in good condition. The rover received its first health check on 17 December. During the cruise, Pathfinder maneuvered on 9 January, 3 February, 6 May and 25 June 1997 to refine its aim for the landing site. Meanwhile, the Hubble Space Telescope monitored the atmosphere of the planet in support of the mission. It spotted a small dust storm in Valles Marineris a week prior to Pathfinder's arrival, but because this was 1,000 km to the south of Ares Vallis it was not expected to pose a problem for the lander's descent.

The cruise stage was jettisoned 30 minutes from Mars. From now on, there was no possibility of altering the trajectory, nor the rate of spin or angle of attack. The entry interface altitude was arbitrarily designated as 3,522.2 km from the center of the planet. Pathfinder hit the atmosphere at a relative speed of 7.26 km/s. The peak deceleration load of 15.9 g occurred about 78 seconds later, at an altitude of 33 km. When the aerodynamic braking had slowed the rate to a Mach number of 1.8, at an altitude of 9 km, the parachute deployed. The forward heat shield was discarded 20 seconds later, at an altitude of 7,600 meters. Some 20 seconds after that, the bridle lowered the lander and the airbags inflated. The altimeter locked onto the surface at an altitude of 1,591 meters; a little higher than planned. Post-flight analysis showed that when the rockets fired the lander was at 88 meters, significantly higher than predicted, and descending at 63 m/s. The most likely explanation was that the drag of the parachute had been poorly modeled. The result was that most of the ensuing events occurred at heights other than planned. In particular, the bridle was cut at a height of 21 meters. Four seconds later, Pathfinder hit the ground at 18 m/s with a load of 16 g, and bounced back to a height of 15 meters. Over the next two and a half minutes the accelerometers recorded no fewer than 14 bounces of diminishing magnitude. Despite the greater than predicted dynamic loads, the airbags retained their integrity. The accelerometers were switched off 6 minutes after entry, but it is likely that Pathfinder continued to bounce for several minutes more, in the process traveling about 1 km, prior to rolling up and over a gentle rise and coming to rest at 16:56:55 UTC on 4 July at 19.28°N, 33.52°W, only 19 km from the target point. This position was reconstructed from tracking data, and was slightly different from that determined later from landmark recognition.[246,247,248,249] The prompt receipt of the 'semaphore' by the Madrid Deep Space Network station indicated that Pathfinder had come to rest on its base. It was only the third probe to survive a landing on the Red Planet. After retracting the airbags and deploying the petals to expose its solar panels, the lander powered down to await sunrise.

Stable communications via the low-gain antenna were established about 4 hours after landing and, after an engineering status check, the data stored during the entry and descent was replayed. Pathfinder provided the first profile of the atmosphere in darkness. The results were both interesting and controversial. In particular, above 60 km it was much colder than measured by Viking 1, although a comparison was difficult because the latter entered in the late afternoon. In contrast, below 60 km it was about 20°C warmer than measured by Viking 1. This contradicted telescopic observations suggesting that the lower atmosphere had been cooling for years. All but one of the thermocouples buried in the heat shield returned data; the one which failed unfortunately being the one mounted near the surface at the stagnation point. The results were in reasonable agreement with expectation. This was an important confirmation of the soundness of the heat shield design, which was to be used for future landers.[250,251,252] After assessing Pathfinder's state of health, the camera head was unlatched. Once it had located the Sun in the sky, this information, in addition to the local time, enabled the lander to compute the location of Earth in the sky and aim the high-gain antenna. Telemetry indicated that the lander had settled almost level, tilted by only 2 degrees. Ironically, this discouraged some of the scientists, as it raised the possibility that the site would prove to be flat and uninteresting! After the first few images had shown details of the hardware, the field of view revealed a multitude of rocks with a variety of shapes, sizes and textures. Moreover, while the immediate vicinity was fairly flat, there were rolling ridges a few meters away and, on the horizon, two hills, one conical and the other flat-topped, which, although no more than 50 meters tall, were promptly named the 'Twin Peaks'. The only issue identified from these early images was that a portion of an airbag posed an obstacle to the deployment of the rover's ramp. In a maneuver rehearsed in simulations, the petal was partially raised and then the spool motor commanded to reel in the fabric a little further.

The Pathfinder site was the roughest and most varied yet visited. On the western horizon the flanks of the Twin Peaks showed remarkable traces of deposits, gullies and flood-etched terraces. To the south, a low rise might be the rim of a moderately large crater whose ejecta would have been intermixed with rocks and debris left by the flooding. Given the plentiful topographical references, it was relatively easy to pinpoint the site, initially in Viking imagery and later in higher resolution pictures taken by follow-on orbiters. Pathfinder had landed 860 meters from the northern of the Twin Peaks (the flat one) and within 1 km of the other. A little over 2 km to the south was the 'Big Crater', which was 1.5 km in diameter. Several knobs were also glimpsed through the haze, the farthest being some 40 km away. Groupings of rocks and boulders near the lander were often tilted or aligned to the northeast. The overall impression was consistent with flood debris. There were small dune fields nearby, and layers of sand and dust covering most of the rocks. The rocks showed a variety of textures and shapes ranging from rounded to highly angular. Several of the rocks showed traces of wind erosion which had never before been seen on the planet, and downwind tails were common. The plentiful pebbles could have been rounded by a flood, by waves breaking on a shore, or by the motion of a glacier, or indeed could be solidified drops of lava or impact melt. The appearance of the sky was a surprise. Because there had been no major dust storms for years, it had been expected that the sky would be blueish but it was pinkish, as it was for the Vikings; which indicated that dust must be perennially present in the atmosphere. The act of drawing in the airbags had revealed a darker subsoil. Similar patches were seen in

A mosaic taken by Mars Pathfinder shortly after landing. The Twin Peaks are on the horizon, and Sojourner is still in its undeployed configuration on one of the side petals. Note the airbag at right that is obstructing one of the rolled-up exit ramps. (JPL/NASA/Caltech)

Mars Global Surveyor imaged the Mars Pathfinder landing site. The spacecraft landed near the center of the image. The two knobs at center left are the Twin Peaks. Big Crater is at the bottom of the picture, and North Peak is at the top. (NASA/JPL/MSSS)

panoramic images, some of them probably indicating were the lander had bounced. It was not long before the engineers and scientists began to assign names to rocks, drawing their inspiration either from the appearance of the rock or from fictional characters, in particular cartoon characters. As a result, the press conferences (and indeed the ensuing scientific papers) were replete with references to 'Yogi', 'Moe', 'Scooby Doo', 'Couch', 'the Dices', 'Ginger', and even 'Darth Vader'. In contrast, the Pathfinder lander was designated the Carl Sagan Memorial Station in honor of the recently deceased planetologist and science popularizer.[253,254,255,256,257]

The only major issue identified on the first local day, or sol, was that Pathfinder and Sojourner, after exchanging several corrupted frames of data, had completely ceased to communicate between themselves. However, the problem healed itself and on Sol 2, one day later than planned, Sojourner was finally able to 'stand up'. It was subsequently found that the communication problem had arisen because the radio systems of the lander and rover, which used a modified form of a commercial radio modem, had been operating at different temperatures, and this had shifted their frequencies by different amounts. In fact, the radio systems included small electric heaters to overcome this. Meanwhile, the exit ramps had been deployed. The front one ended suspended above the ground, but the rear one led to a flat patch that was clear of rocks yet would give ready access to a mottled rock that had been named 'Barnacle Bill'. Watched by the lander's camera, Sojourner drove slowly down the ramp and halted with all six of its wheels on the dirt. It was 27 years since the first automated lunar rover, and only one year less since the unsuccessful landing of the first Marsokhodik.[258] The first thing that Sojourner did was put its spectrometer on the ground in order to collect data overnight to determine the chemistry of the soil. After Pathfinder had finished a preliminary panorama with Sojourner safely on the ground, it was commanded to release the coiled camera boom, which rapidly rose to its deployed height. That night, the camera was used to locate Deimos as a two-pixel dot shining by reflected sunlight in the darkness.

On Sol 3, Sojourner was to reach its first rock, Barnacle Bill, but prior to this it was commanded to lock five wheels and scrape the soil by rotating the sixth wheel in place. The patch of disturbed soil this produced was then imaged by Pathfinder. This done, it assumed its 'ram' position and drove backwards, successfully placing the spectrometer in contact with Barnacle Bill on the first try. From this position, Sojourner took its first picture of Pathfinder, but owing to the low perspective the airbags obstructed the view of the structure. The analysis of Barnacle Bill, the first Martian rock ever analyzed in situ – the Vikings had analyzed only sand and dirt – was a surprise. It was so rich in silicon that it required the presence of quartz, much like a terrestrial andesite.[259,260,261,262] There was some concern, however, that what was being analysed was a weathered 'rind' that was not typical of the rock within. The second target was a large rounded rock named Yogi which had an intriguing two-colored appearance. When Sojourner approached this rock on Sol 6, it bumped against it. Because the spectrometer was on the rear, where there were no guidance sensors, this was not a significant issue – all that would be necessary would be to order the rover to move away, adjust its orientation and try again. Unfortunately, owing to a series of glitches, four sols elapsed before the instrument was able to be

As Sojourner drove away from Mars Pathfinder it provided this view back towards the lander's exit ramp. Most of the lander's structure is obscured by the retracted airbags. (JPL/NASA/Caltech)

A Sojourner view of the Yogi boulder that dramatically shows the low perspective of the rover's camera. (JPL/NASA/Caltech)

emplaced against the rock – which was a significant delay in terms of the nominal 1-week mission duration. After Yogi, Sojourner headed for Scooby Doo, a white rock, pausing on the way to abrade a sandy area named the 'Cabbage Patch'. For this move, the autonomous navigation system was used for the first time. Next, the soil near the rock 'Lamb' was analyzed. Meanwhile, lander operations were focused on building panoramas to assist the rover team with their navigation, taking pictures to record the position of the rover at the end of each day, and regular monitoring of the windsocks and dust magnets. As the lander monitored the rover's progress, the imagery showed a number of patches of bright pink material that were exposed by the rover's wheels but were otherwise not disturbed by the cleats. After a rock near Yogi named 'Souffle' had been examined, it was decided to undertake a clockwise drive around Pathfinder to reach the only safe path into an interesting grouping of boulders dubbed the 'Rock Garden'.

On Sol 38, now well into its extended mission, Sojourner reached the point of access to the Rock Garden. It began by inspecting 'Wedge', which had been named for its shape. In fact, this was only the third rock to be studied by the spectrometer. Then as the rover attempted to enter the Rock Garden a guidance sensor which had been experiencing some drift caused it to climb up onto Wedge, prompting the tilt sensor to abort the maneuver. In order to make progress, it was decided to turn off the protection logic. Over the next 20 sols, the rover analyzed Shark, Moe and Half Dome in turn. After the loss of its batteries on Sol 58, the rover could accumulate spectrometer data only during the day. On Sol 76, still in the Rock Garden, it spied a small field of dunes in a trough beyond the cluster of rocks that was not visible to Pathfinder's camera.[263]

Images of the twilight sky were taken on several sols at dusk and dawn, often in conjunction with observations by the Hubble Space Telescope, to evaluate the dust and hazes. The rate at which the output of the solar panels fell day by day provided a measure of dust accumulation. On some occasions blue clouds of water ice were seen, and observations of the Sun at suitable wavelengths measured the abundance of water in the atmosphere. The dust on the magnets appeared to consist at least in part of iron oxides formed in a wet environment. The thermocouples 0.65, 0.9 and 1.4 meters above the surface recorded temperatures as warm as –10°C during the afternoon, and as chilly as –76°C immediately before dawn. They also provided the interesting observation that the surface was warmer than the air during the day, and cooler at night. In fact, during the day the top sensor was as much as 15°C cooler than the bottom sensor, with the difference inverted in darkness. These temperature gradients can trigger small vortices which suck up dust. These dust devils might, in fact, be the mechanism that continuously refills the atmosphere with dust. During the first month Pathfinder's anemometer and barometer noted four of these vortices sweeping by. Since it was late summer in the northern hemisphere, Pathfinder (as had the Vikings) recorded the dwindling pressure as the southern polar cap drew carbon dioxide from the atmosphere.

By Sol 58, Sojourner had used its wheels to perform 12 abrasion and scraping

On Sol 8 Sojourner took this picture showing one of its hazard-detection laser stripes. (JPL/NASA/Caltech)

A panorama of the Mars Pathfinder landing site. Sojourner is alongside the Yogi boulder. (JPL/NASA/Caltech)

tests on soils, dunes and surface crusts, and its spectrometer had analyzed six soils and five rocks. The fact that the soil was chemically similar to that seen by both of the Vikings raised the prospect that the dust storms had homogenized the material on a global basis. Some rocks appeared to be impact breccias, composed of smaller angular fragments cemented in a fine-grained matrix of powdered rock. Others had a layered appearance suggestive of a sedimentary origin, possibly in the presence of a stable body of water. If silicon and potassium-rich rocks such as Barnacle Bill were typical of the highlands, this would be difficult to reconcile with what was known of the planet's early geological history. Overall, Pathfinder and Sojourner provided strong hints that whilst Mars was once warmer and wetter, its climate is probably stable over long periods.

Other results were obtained from radio tracking of the lander on the surface. In particular, by combining Viking and Pathfinder data it was possible to estimate the angle by which the spin axis of the planet had precessed over a 21-year interval. This gave the first measurement of the planet's moment of inertia, and because this depends on the configuration of the interior, it was possible to confirm that there is a metallic core.[264,265,266,267,268,269,270,271,272,273]

After Sojourner had finished in the Rock Garden, the plan was to put it through its paces by sending it on a true traverse to evaluate its capabilities and assist in the development of the next generation of rovers. However, Pathfinder was ailing. The fact that the batteries were no longer recharging properly meant it would soon have to cease nocturnal observations. On the morning of Sol 85 (28 September) it failed to report. It is believed that the batteries failed during the night, stopping its clock. Although it would have revived at sunrise, the fact that it no longer knew the time meant it could not calculate the position of Earth in the sky to direct the high-gain antenna. A link was briefly re-established on 1 October, and again 7 October, but without data being returned. Without battery power to sustain it through the night, Pathfinder would have become progressively colder, until the electronics were no longer capable of reviving at sunrise. Attempts to send 'wake up' commands were abandoned on 10 March 1998. Just days before Pathfinder developed this problem, an emergency program had been uploaded to Sojourner. If it were to find itself out of contact with Pathfinder for five days during its traverse, it was to head home. On the other hand, a 'zone of avoidance' was defined around Pathfinder to preclude a collision. These conflicting orders – to return to the lander, yet maneuver around it – were designed to ensured that the rover would end up circling around the lander until either it met an obstacle it could not negotiate or it, too, failed.[274]

Pathfinder returned a total of 300 Mb of data, including 16,661 lander and 550 rover images, 15 chemical analyses and much meteorological data. It produced five panoramas of ever better spatial and spectral resolution, although when contact was lost only 83 per cent of the last one had been transmitted. The best views included a pair of bright spots to the southwest and southeast that were initially interpreted as the heat shield and backshell, which implied that wind had been blowing from the southwest at the time of the descent. The site was subsequently imaged by orbiters at progressively greater resolution – most recently in December 2006 by Mars Reconnaissance Orbiter. This resolved small details of Pathfinder, such as its

On Sol 76 Sojourner took this view of dunes beyond the Rock Garden. This area could not be seen by the stationary lander. (JPL/NASA/Caltech)

airbags, most of the rocks that were analyzed by Sojourner, and what appeared to be the rover itself some 6 meters to the south; a little closer to the lander than when contact was lost. No tracks were seen, quite possibly because they had been eroded by the wind. The backshell and parachute were clearly seen, although from the lander's point of view they were both beyond a ridge. The object which had been misidentified in Pathfinder's imagery as the backshell was deemed to be a piece of aluminized airbag thermal blanket, marking possibly the site of the first bounce. The heat shield, which was about 1 km away, had shattered into at least three pieces on impact.[275,276]

Pathfinder had surpassed its 'life expectancy' by a factor of three. Sojourner had operated for twelve times the duration of its quite conservative target. The fact that the rover actually outlasted the lander was remarkable. Sojourner drove a distance of 100 meters, although it never strayed more than about 12 meters from its lander. Although this was the second Discovery mission to be launched, it was the first to reach fruition, and showed that the 'faster, cheaper, better' strategy could indeed be effective. Mars Pathfinder was also a milestone in the era of the World Wide Web, with the mission's website receiving a record 46 million visits on 8 July, just as the rover began its investigations.

MARTIANS WORLDWIDE

America was not alone in the early 1990s in studying the possibility of creating a network of scientific stations on Mars. At about this time, ESA was studying the MARSNET project as one of the medium-sized missions by which it intended to participate in exploring the Red Planet. The idea to land three or four small probes was conceived in the aftermath of the proposal for the Kepler orbiter being refused, and industrial study contracts were issued in 1992. Together with MESUR and the Mars 94 and 96 missions, MARSNET was to complete an international network of about twenty surface stations. The principal instrument on the MARSNET landers would be a seismometer, but they would also have panoramic and descent cameras, atmospheric and meteorology instruments, and a suite of geochemical experiments which could be deployed and moved from one rock to another by an updated form of the 1971-vintage PrOP-M Marsokhodik supplied by Russia. The landers could either be flown individually or together on a Mariner Mark II-class bus. They were to establish a seismometry network in the Tharsis region, with the sites separated on average by 3,500 km in Tempe Terra, Candor Chasma and Daedalia Planum. An alternative scheme called for three landers to be positioned at the vertices of an equilateral triangle, and a fourth lander on the opposite hemisphere to improve the ability to investigate the planet's core. A combined penetrator and lander was also studied, and subjected to extensive laboratory tests in the 1990s. It was envisaged that MARSNET would be launched by a Delta rocket in 2001 or 2003.[277,278,279]

Mars missions were studied by both of Japan's space agencies, with a particular emphasis being placed on large rovers. Starting in 1988 NASDA investigated five possible missions, including an orbiter, a sample-return from one of the satellites, a MESUR-like lander, a large rover, and a planetary sample-return. One requirement was that a mission be compatible with the H-II launch vehicle, which was then in development. Interest focused on a 435-kg rover capable of driving up to 1,600 km over a 3-year period. Three potential routes were drawn up: one through the mesas of Cydonia, one along Valles Marineris, and the other starting at Margaritifer Sinus and ending at Meridiani Sinus. Meanwhile, engineers at ISAS were studying a drill for a rover which would be capable of penetrating to a depth of 1.5 meters into the Martian surface.[280,281]

All the major space agencies of the world met in Wiesbaden, Germany, in May 1993 to coordinate their plans. An International Mars Exploration Working Group was created to define a common strategy. It strongly recommended a network like MARSNET as the next logical step. ESA renamed its plan INTERMARSNET and invited broader international participation. Although NASA was moving towards a series of low-cost missions with its Discovery program, the European strategy was to develop several expensive 'cornerstone missions' and supplement this program with medium-class missions which, whilst considerably cheaper, were still nearly twice the price cap of a Discovery mission. In 1992 ESA had invited medium-class proposals and INTERMARSNET was submitted. The plan called for a spacecraft similar to that intended for the Rosetta comet orbiter. It would carry four 40-kg landers, have a total mass of 2,500 kg and be launched in 2003 by an Ariane 5. The payloads for the landers would be similar to those of the small stations of Mars 94, with each lander having a descent camera, a panoramic camera, a small meteorological suite, a seismometer, a magnetometer and an alpha-particle, proton and X-ray spectrometer. They were to operate on the surface for a full local year. In addition, scientific data was to be obtained during the entry and descent.[282] Unfortunately, ESA rejected INTERMARSNET. It was revived as a joint project with America in the context of the Mars Surveyor program, this time as a simpler orbiter carrying three landers similar to one that NASA was developing, which would be sent to the crater Gusev, the plains of Uranius Patera and the highlands of Coprates, but once again it failed to gain approval.[283]

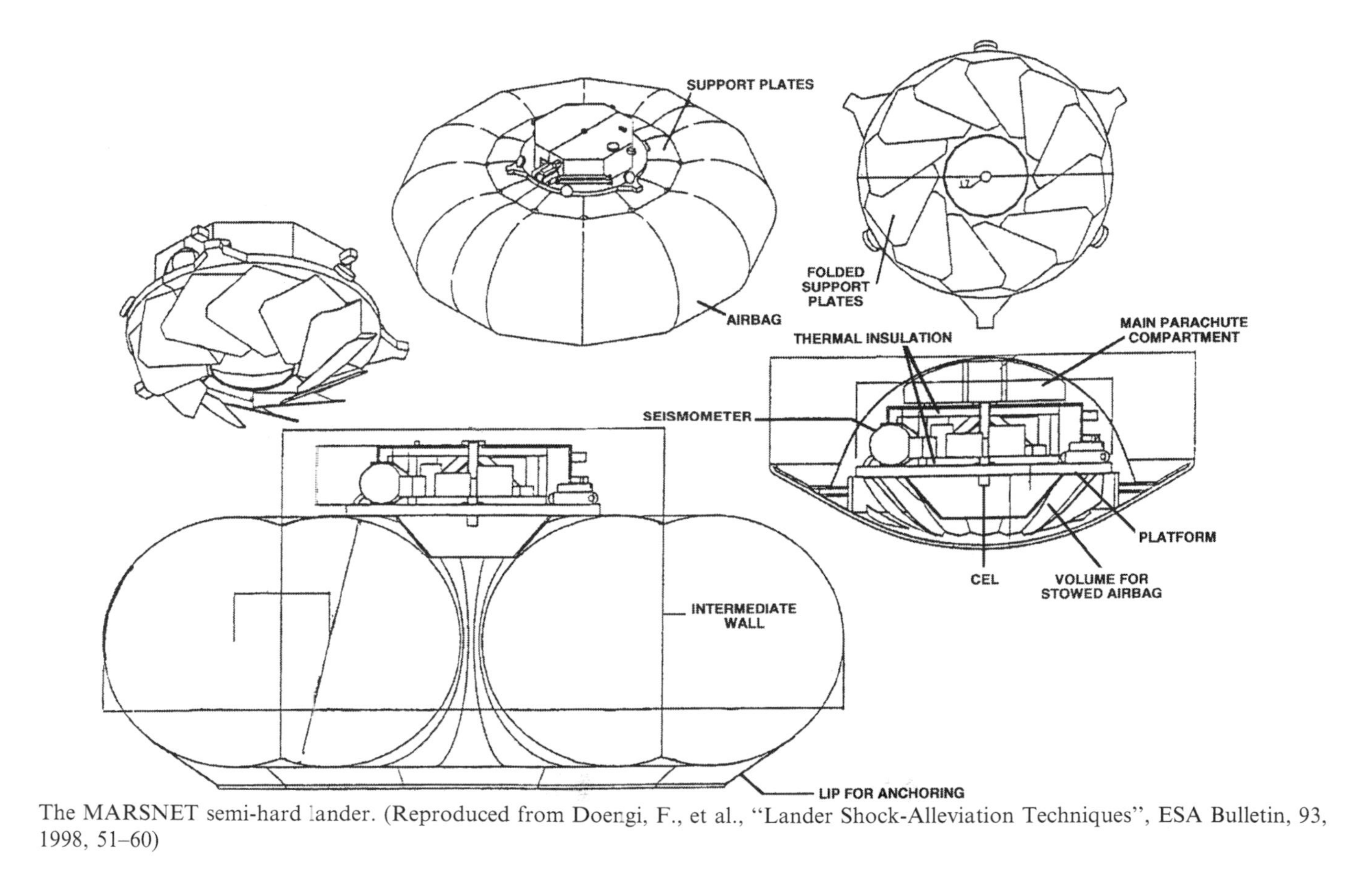

The MARSNET semi-hard lander. (Reproduced from Doengi, F., et al., "Lander Shock-Alleviation Techniques", ESA Bulletin, 93, 1998, 51–60)

MEANWHILE IN AMERICA

Nevertheless NASA was exuberant. The Mars Surveyor and Discovery programs had built up a welcome momentum in the spirit of 'faster, cheaper, better' and the development of the next missions, including an orbiter and a lander for Mars, was progressing nicely. At last, after almost two decades, planetary exploration seemed to be entering a second golden age.

REFERENCES

1 Flight-1990
2 Bevilacqua-1994
3 Gianvanni-1990
4 Goy-1990
5 Avanesov-1991
6 Avanesov-1989b
7 Luttmann-1992
8 Volare-1989
9 Westwick-2007o
10 McCurdy-2005b
11 Krimigis-1995
12 Surkov-1993
13 NASA-1993c
14 Carroll-1993a
15 Furniss-1993
16 Spudis-1994
17 Nelson-1995
18 Nelson-1997
19 Carroll-1995
20 Divsalar-1995
21 Killeen-1995
22 NASA-1995
23 Goldman-1994
24 Ulivi-2004
25 Chapman-1995
26 Ostro-1996
27 Westwick-2007p
28 Lenorovitz-1993
29 Lenorovitz-1994
30 Robertson-1994
31 Scott-1996a
32 Boain-1993
33 Butrica-1996f
34 Duxbury-1997
35 Hudson-2000
36 Jaroff-1997
37 Freese-2000
38 For the discovery of Eros see Part 1, pages lii–liii
39 McCurdy-2005c
40 Veverka-1997a
41 Farquhar-1995
42 Santo-1995
43 McCurdy-2005d
44 Bokulic-1997
45 Veverka-1997a
46 Veverka-1997b
47 Yeomans-1997
48 Rust-2005
49 Izenberg-1998
50 Antreasian-1998
51 Li-2002
52 Dunham-2000
53 Yeomans-1999
54 Veverka-1999
55 APL-1999
56 Dunham-2000
57 Yeomans-2000
58 Veverka-2000
59 Zuber-2000a
60 Trombka-2000
61 Veverka-2001a
62 Cheng-2001
63 Robinson-2001
64 Cheng-2002
65 Nelson-2001
66 Veverka-2001b
67 Kelly Beatty-2001
68 Asker-2001a
69 Asker-2001b
70 McCurdy-2005e
71 Evans-2001

72 Binzel-1990
73 For Voyager 1's trajectory see Part 1, pages 310–311
74 Stern-2007
75 Sobel-1993
76 Westwick-2007p
77 Staehle-1994
78 Stern-2007
79 Bojor-1996
80 Weinstein-1993
81 Stern-1993
82 NASA-1994
83 Stern-2007
84 Stern-1998
85 Weissman-1999
86 Staehle-1999
87 Woerner-1998
88 Smith-1994
89 Furniss-1997
90 Malin-1991
91 Malin-2001a
92 Smith-1998
93 Christensen-1998
94 Acuña-1998
95 Cunningham-1996
96 Palluconi-1997
97 Albee-2001
98 Lee-1996
99 Sawyer-2006a
100 For Martian meteorites see Part 1, pages 233–234
101 Sawyer-2006b
102 McKay-1996
103 Kelly Beatty-1996b
104 Thomas-Keprta-2002
105 Treiman-1999
106 Anselmo-1996a
107 Anselmo-1996b
108 Rogers-1996
109 Sawyer-2006c
110 Scott-1996b
111 Covault-1996a
112 Malin-2001a
113 Smith-1997a
114 Wilmoth-1999
115 Johnston-1998
116 Albee-1998
117 Malin-1998
118 Christensen-1998
119 Parker-1998
120 Smith-1998
121 Keating-1998
122 Baird-2007
123 Crowley-2007
124 Tolson-1999
125 Acuña-1998
126 Lyons-1999
127 ST-1998
128 Morabito-2000
129 Malin-1999
130 Zuber-1998
131 Acuña-1999
132 Connerney-1999
133 Esposito-1999
134 Smith-1999a
135 Caplinger-2001
136 Malin-2000a
137 Tytell-2001a
138 Malin-2003
139 Rossman-2002
140 Malin-2000b
141 Tytell-2000
142 Tytell-2001b
143 ST-2000b
144 Tytell-2004
145 Parker-1989
146 Parker-1993
147 Malin-2001a
148 Kelly Beatty-1999
149 Head-1999
150 Tytell-2001c
151 Bandfield-2000
152 Christensen-2001
153 Bandfield-2003
154 Brain-2006
155 Mendillo-2006
156 Smith-1999b
157 Smith-1999c
158 Zuber-2000b
159 Yoder-2003
160 Dehant-2003
161 For Viking bistatic radar observations see Part 1, pages 252–253
162 Simpson-2000
163 Simpson-2002
164 Cantor-2007
165 Tytell-2001d
166 Edwards-2004

167 Dornheim-2005
168 Christou-2007
169 Brain-2006
170 Iorio-2007
171 Krogh-2007
172 Smith-2001
173 Malin-2001b
174 Paige-2001
175 Malin-2006
176 Naeye-2007
177 Chaikin-2007
178 Covault-2006
179 For the Viking 1 memory location error see Part 1, page 254
180 NASA-2007
181 Kemurdjian-1992
182 Eremenko-1993
183 VnIITransmash-1999
184 Bogatchev-2000
185 Carroll-1993b
186 McDonnell Douglas-1995
187 Caprara-1992
188 Lenorovitz-1987a
189 AWST-1988c
190 Vorontsov-1989
191 AWST-1988d
192 Lenorovitz-1987b
193 Lusignan-1991
194 AWST-1988d
195 Flight-1989c
196 Burnham-1996
197 Harvey-2007c
198 Friedman-1993
199 VnIITransmash-1999
200 Flight-1989d
201 Spaceflight-1992h
202 Neukum-1996
203 Flight-1991
204 MSSS-1996
205 Lande-1995
206 Flight-1992b
207 Lomberg-1996
208 Galeev-1995
209 Debus-2002
210 Eremenko-1993
211 Laplace-1993
212 Carroll-1993b
213 Carlier-1995
214 Galeev-1995
215 Burnham-1996
216 Surkov-1997f
217 MSSS-1996
218 AWST-1996
219 Oberg-1999
220 Furniss-1996
221 Covault-1996b
222 Flight-1997
223 Harvey-2007c
224 Clark-2000
225 Hubbard-1992
226 Mishkin-2003c
227 Willcockson-1999
228 Burke-1990
229 Bickler-1993
230 Bickler-1989
231 Mishkin-2003d
232 Mishkin-2003e
233 Westwick-2007r
234 Mishkin-2003f
235 Mishkin-2003g
236 Mishkin-2003h
237 NASA-1997
238 Cadogan-1998
239 Mishkin-2003i
240 Becker-2005
241 Willcockson-1999
242 NASA-1997
243 Golombek-1997
244 Bell-1998
245 Asker-1996
246 Spencer-1999b
247 Golombek-1997
248 Schofield-1997
249 Spencer-1998
250 Schofield-1997
251 Kahn-1997
252 Willcockson-1999
253 Golombek-1997
254 Smith-1997b
255 Golombek-1998
256 Collins Petersen-1997
257 Stooke-2000
258 For the PrOP-M Marsokhodik see Part 1, page 105–108
259 Golombek-1997
260 Golombek-1998
261 Collins Petersen-1997
262 Rieder-1997

263 Mishkin-2003j
264 Golombek-1997
265 Smith-1997b
266 Golombek-1998
267 Schofield-1997
268 Rover Team-1997
269 Rieder-1997
270 Folkner-1997b
271 Hviid-1997
272 Goldman-1997
273 Bell-1998
274 Mishkin-2003k
275 Parker-2007a
276 Parker-2007b
277 Chicarro-1993
278 Scoon-1993
279 Doengi-1998
280 Maeda-1993
281 Kawashima-1993
282 Chicarro-1994
283 Surkov-1997g

Glossary

ADU: Avtonomnaya Dvigatel'naya Ustanovka; Autonomous Engine Unit

Aerobraking: A maneuver where a spacecraft's orbit is changed by reducing its energy by repeated passages through a planet's upper atmosphere.

Aerocapture: A maneuver where a spacecraft enters into orbit around a planet by slowing it down by a passage through the upper levels of a planet's atmosphere.

Aerogel: A silicon-based foam in which the liquid component of a gel has been replaced with gas or, for use in space, effectively with vacuum, to produce a solid with a very low density.

AGORA: Asteroidal Gravity Optical and Radar Analysis

ALH: Allan Hills (meteorites)

AMPTE: Active Magnetospheric Particle Tracer Explorer

AMSAT: The Radio Amateur Satellite Corporation.

Aphelion: The point of maximum distance from the Sun of a heliocentric orbit. Its contrary is perihelion.

APL: The Applied Physics Laboratory of Johns Hopkins University.

Apoapsis: The point of maximum distance from the central body of any elliptical orbit. This word has been used to avoid complicating the nomenclature, but a term tailored to the central body is often used. The only exceptions used herein owing to their importance were for Earth (apogee) and the Sun (aphelion). The contrary of apoapsis is periapsis.

Apogee: The point of maximum distance from the Earth of a satellite orbit. Its contrary is perigee.

AS: Aerostatnaya Stantsiya; Aerostatic probe

ASI: Agenzia Spaziale Italiana; Italian Space Agency

ASLV: Advanced Satellite Launch Vehicle

ASPERA: Automatic Space Plasma Experiment with a Rotating Analyzer

Astronomical Unit: To a first approximation the average distance between the Earth and the Sun is 149,597,870,691 (± 30) meters.

AU: Astronomical Unit

BMDO: Ballistic Missile Defense Organization

Booster: Auxiliary rockets used to boost the lift-off thrust of a launch vehicle.

Bus: A structural part common to several spacecraft.

C4: Comet Coma Chemical Composition

CAESAR: Comet Atmosphere Encounter and Sample Return, or Comet Atmosphere and Earth Sample Return

CFD: Computational Fluid Dynamics

CHON: Carbon, Hydrogen, Oxygen and Nitrogen-rich molecules

CIA: Central Intelligence Agency

CISR: Comet Intercept and Sample Return

CNES: Centre National d'Etudes Spatiales; the French National Space Studies Center

CNRS: Centre National de la Recherche Scientifique; the French National Scientific Research Center

CNSR: Comet Nucleus Sample Return

CNUCE: Centro Nazionale Universitario di Calcolo Elettronico; the Italian National University Center for Electronic Computation

Conjunction: The time when a solar system object appears close to the Sun as seen by an observer. A conjunction where the Sun is between the observer and the object is called 'superior conjunction'. A conjunction where the object is between the observer and the Sun is called 'inferior conjunction'. See also opposition.

CONSCAN: Conical Scan

CONTOUR: Comet Nucleus Tour

Cosmic velocities: Three characteristic velocities of spaceflight:

First cosmic velocity: Minimum velocity to put a satellite in a low Earth orbit. This amounts to some 8 km/s.

Second cosmic velocity: The velocity required to exit the terrestrial sphere of attraction for good. Starting from the ground, this amounts to some 11 km/s. It is also called 'escape' speed.

Third cosmic velocity: The velocity required to exit the solar system for good.

CRAF: Comet Rendezvous/Asteroid Flyby

Cryogenic propellants: These can be stored in their liquid state under atmospheric pressure at very low temperature; e.g. oxygen is a liquid below –183°C.

DAS: Dolgozhivushaya Avtonomnaya Stanziya; Long-Duration Autonomous Station

Deep Space Network: A global network built by NASA to provide round-the-clock communications with robotic missions in deep space.

Direct ascent: A trajectory on which a deep-space probe is launched directly from the Earth's surface to another celestial body without entering parking orbit.

DMSP: Defense Meteorological Satellite Program

DSN: Deep Space Network

DSPSE: Deep Space Program Science Experiment

DZhVS: Dolgozhivushaya Veneryanskaya Stanziya; long-duration Venusian probe

ECAM: Earth-Crossing Asteroid Mission

Ecliptic: The plane of the Earth's orbit around the Sun.

Ejecta: Material from a volcanic eruption or a cratering impact that is deposited all around the source.

EOS: Eole–Venus

EPONA: Energetic Particle Onset Admonitor

ESA: European Space Agency

Escape speed: See Cosmic velocities

ESO: European Southern Observatory

ESRO: European Space Research Organization (incorporated into ESA)

Flyby: A high relative speed and short-duration close encounter between a spacecraft and a celestial body.

GEM: Galileo Europa Mission

GEM: Giotto Extended Mission

GMM: Galileo Millennium Mission

GRB: Gamma-Ray Bursts

GSFC: Goddard Space Flight Center

GSLV: Geostationary Satellite Launch Vehicle

HAPPEN: Halley Post-Perihelion Encounter

HEOS: Highly Eccentric Orbit Satellite

HER: Halley Earth Return

HIM: Halley Intercept Mission

HMC: Halley Multicolor Camera

HST: Hubble Space Telescope

Hypergolic propellants: Two liquid propellants that ignite spontaneously on coming into contact, without requiring an ignition system. Typical hypergolics for a spacecraft are hydrazine and nitrogen tetroxide.

IACG: Inter-Agency Consultative Group

ICM: International Comet Mission

IHW: International Halley Watch

IKI: Institut Kosmicheskikh Isledovanii; the Russian Institute for Cosmic Research

IMEWG: International Mars Exploration Working Group

IMP: Interplanetary Monitoring Platform

IRAS: InfraRed Astronomical Satellite

IRIS: Italian Research Interim Stage

ISAS: Institute of Space and Astronautical Sciences

ISEE: International Sun–Earth Explorer

ISO: Infrared Space Observatory

ISPM: International Solar Polar Mission

ISPP: In-Situ Propellant Production

ISRO: Indian Space Research Organization

IUE: International Ultraviolet Explorer

IUS: Inertial Upper Stage (previously: Interim Upper Stage)

JOP: Jupiter Orbiter with Probe

JPA: Johnstone Plasma Analyzer

JPL: Jet Propulsion Laboratory; a Caltech laboratory under contract to NASA

KGB: Komityet Gosudarstvennoy Bezapasnosti; Committee for the State Security

KTDU: Korrektiruyushaya Tormoznaya Dvigatelnaya Ustanovka; course correction and braking engine

Lander: A spacecraft designed to land on another celestial body.

LaRC: Langley Research Center

Launch window: A time interval during which it is possible to launch a spacecraft to ensure that it attains the desired trajectory.

LEAP: Light ExoAtmospheric Projectile

LESS: Low-cost Exploration of the Solar System

Lyman-alpha: The emission line corresponding to the first energy level transition of an electron in a hydrogen atom.

MAGE: Moteur d'Apogée Geostationnaire Européen; European Geostationary Apogee Motor

MAOSEP: Multiple Asteroid Orbiter with Solar Electric Propulsion

MAV: Mars Ascent Vehicle

MBB: Messerschmitt Bölkov Blohm

MEI: Moskovskiy Energeticheskiy Institut; Moscow's Power Institute

MER: Mars Exploration Rover

MESUR: Mars Environmental Survey

MGCO: Mars Geoscience/Climatology Orbiter

MGS: Mars Global Surveyor

MIT: Massachusetts Institute of Technology

MORO: Moon Orbiting Observatory

MPF: Mars Pathfinder, MESUR Pathfinder

MPO: Mercury Polar Orbiter

MRSR: Mars Rover and Sample Return

MSM: Mars Science Microrover

MSX: Midcourse Space Experiment

MS-T5: Mu Satellite-Test 5

MUADEE: Mars Upper Atmosphere Dynamics, Energetics and Evolution Mission

MUSES: MU [rocket] Space Engineering Satellite

NAS: National Academy of Sciences

NASA: National Aeronautics and Space Administration

NASDA: National Space Development Agency

NEAR: Near Earth Asteroid Rendezvous

NEARS: Near Earth Asteroid Returned Sample

NEP: Nuclear Electric Propulsion

Occultation: When one object passes in front of and occults another, at least from the point of view of the observer.

OOE: Out-Of-Ecliptic mission

Orbit: The trajectory on which a celestial body or spacecraft is traveling with respect to its central body. There are three possible cases:

Elliptical orbit: A closed orbit where the body passes from minimum distance to maximum distance from its central body every semiperiod. This is the orbit of natural and artificial satellites around planets and of planets around the Sun.

Parabolic orbit: An open orbit where the body passes through minimum distance from its central body and reaches infinity at zero velocity in infinite time. This is a pure abstraction, but the orbits of many comets around the Sun can be described adequately this way.

Hyperbolic orbit: An open orbit where the body passes through minimum distance from its central body and reaches infinity at non-zero speed. This describes adequately the trajectory of spacecraft with respect to planets during flyby manoeuvres.

Opposition: The time when a solar system object appears opposite to the Sun as seen by an observer.

Orbiter: A spacecraft designed to orbit a celestial body.

PAH: Polycyclic Aromatic Hydrocarbons

PAM: Payload Assist Module

Parking orbit: A low Earth orbit used by deep-space probes before heading to their targets. This relaxes the constraints on launch windows and eliminates launch vehicle trajectory errors. Its contrary is direct ascent.

Periapsis: The minimum distance point from the central body of any orbit. See also apoapsis.

Perigee: The minimum distance point from the Earth of a satellite. Its contrary is apogee.

Perihelion: The minimum distance point from the Sun of a heliocentric orbit. Its contrary is aphelion.

PFF: Pluto Fast Flyby

PKE: Pluto Kuiper Express

POLO: Polar Orbiting Lunar Observatory

PrOP: Pribori Otchenki Prokhodimosti; instrument for cross-country characteristics evaluation

PSLV: Polar Satellite Launch Vehicle

PVM: Pioneer Venus Multiprobe

PVO: Pioneer Venus Orbiter

'Push-broom' camera: A digital camera consisting of a single row of pixels, with the second dimension created by the motion of the camera itself.

Re: Earth radii (6,371 km)

Rendezvous: A low relative speed encounter between two spacecraft or celestial bodies.

Retrorocket: A rocket whose thrust is directed opposite to the motion of a spacecraft in order to brake it.

Rj: Jupiter radii (approximately 71,200 km)

RKA: Rossiyskoye Kosmisheskoye Agenstvo; Russian Space Agency

Rover: A mobile spacecraft to explore the surface of another celestial body.

RPA: Rème Plasma Analyzer

RTG: Radioisotope Thermal Generator

RTH: Radioisotope Thermal Heater

SAR: Synthetic Aperture Radar

SEI: Space Exploration Initiative

SETI: Search for Extraterrestrial Intelligence

SLIM: Surface Lander Investigation of Mars

SLV: Satellite Launch Vehicle

SMACS: Small Missions to Asteroids/Comets

SNAP: System for Nuclear Auxiliary Power

SNC: Shergottites–Nakhlites–Chassignites meteorites

SOCCER: Sample of Comet Coma Earth Return

SOHO: Solar and Heliospheric Observatory

Solar flare: A solar chromospheric explosion creating a powerful source of high energy particles.

Space probe: A spacecraft designed to investigate other celestial bodies from a short range.

Spectrometer: An instrument to measure the energy of radiation as a function of wavelenght in a portion of the electromagnetic spectrum. Depending on the wavelength the instrument is called, e.g. ultraviolet, infrared, gamma-ray spectrometer etc.

Spin stabilization: A spacecraft stabilization system where the attitude is maintained by spinning the spacecraft around one of its main inertia axes.

SSEC: Solar System Exploration Committee

SSED: Solar System Exploration Division

STS: Space Transportation System; the Space Shuttle

Synodic period: The period of time between two consecutive superior or inferior conjunctions or oppositions of a solar system body.

TAU: Thousand Astronomical Units mission

TDRS: Tracking and Data Relay Satellite

Telemetry: Transmission by a spacecraft via a radio system of engineering and scientific data.

3-axis stabilization: A spacecraft stabilization system where the axes of the spacecraft are kept in a fixed attitude with respect to the stars and other references (the Sun, the Earth, a target planet etc.)

TOPS: Thermoelectric Outer Planet Spacecraft

TOS: Transfer Orbit Stage

UDMH: Unsymmetrical DiMethyl Hydrazine

UMVL: Universalnyi Mars, Venera, Luna; Universal for Mars, Venus and the Moon

UTC: Universal Time Coordinated; essentially Greenwich Mean Time

V2: Vergeltungswaffe 2 (vengeance weapon 2)

Vidicon: A television system based on resistance changes of some substances when exposed to light. It has been replaced by the CCD.

VLA: Very Large Array

VLBI: Very Long Baseline Interferometry

VMPM: Venus Multiprobe Mission

VOIR: Venus Orbiting Imaging Radar

VRM: Venus Radar Mapper

Yarkovsky effect: A force acting on a rotating body in space caused by the emission of thermal photons, carrying momentum. Small asteroids and meteoroids are known to be perturbed by this effect.

Appendix 1

CHRONOLOGY OF SOLAR SYSTEM EXPLORATION 1983–1996

Date	Event
16 October 1983	The first orbital radar images of Venus are returned by Venera 15
7 January 1985	Sakigake, the first non-US, non-Soviet deep-space probe is launched
11 June 1985	The Vega 1 balloon is the first 'aircraft' to fly in the atmosphere of another planet (Venus)
11 September 1985	The International Comet Explorer flies by comet Giacobini–Zinner
28 January 1986	Space Shuttle Challenger explodes, derailing US planetary exploration initiatives
14 March 1986	The European Giotto flies by comet Halley
10 August 1990	Magellan enters orbit around Venus to deliver a complete radar map of the planet
29 October 1991	Galileo flies by asteroid Gaspra
13 September 1994	Ulysses makes the first pass over one of the poles of the Sun
7 December 1995	Galileo enters orbit around Jupiter as its atmospheric capsule plunges into the planet
17 February 1996	NEAR-Shoemaker, the first low-cost Discovery mission is launched
Related milestones	
4 July 1997	Mars Pathfinder lands on Mars carrying the first working planetary rover
14 February 2000	NEAR-Shoemaker enters orbit around asteroid Eros
12 February 2001	NEAR-Shoemaker lands on asteroid Eros

Appendix 2

PLANETARY LAUNCHES 1983–1996

Launch Date	Name	Main Target	Launcher	Nation
12 August 1978	International Cometary Explorer	P/Giacobini –Zinner	Delta 2914	USA
2 June 1983	Venera 15	Venus	8K82K Proton K/D-1	USSR
7 June 1983	Venera 16	Venus	8K82K Proton K/D-1	USSR
15 December 1984	Vega 1	Venus/Halley	8K82K Proton K/D-1	USSR
21 December 1984	Vega 2	Venus/Halley	8K82K Proton K/D-1	USSR
7 January 1985	Sakigake	P/Halley	Mu-3SII	Japan
2 July 1985	Giotto	P/Halley	Ariane 1	ESA
18 August 1985	Suisei	P/Halley	Mu-3SII	Japan
7 July 1988	(Fobos 1)	Mars/Phobos	8K82K Proton K/D-1	USSR
12 July 1988	(Fobos 2)	Mars/Phobos	8K82K Proton K/D-1	USSR
4 May 1989	Magellan	Venus	OV 104 + IUS	USA
18 October 1989	Galileo	Jupiter	OV 104 + IUS	USA
6 October 1990	Ulysses	Solar orbiter	OV 103 + IUS	ESA/USA
25 September 1992	(Mars Observer)	Mars	Commercial Titan III	USA
25 January 1994	(Clementine)	Moon/Asteroid	Titan IIG SLV	USA
17 February 1996	NEAR-Shoemaker	Asteroid	Delta 7925-8	USA
7 November 1996	Mars Global Surveyor	Mars	Delta 7925A	USA
16 November 1996	(Mars 8)	Mars	8K82K Proton K/D-1	Russia
4 December 1996	Mars Pathfinder	Mars	Delta 7925A	USA

Missions in parentheses are missions that failed, but the status of Fobos 2 and Clementine is disputed. Clementine successfully completed its lunar orbiter mission but not its asteroid flyby.

Appendix 3

GALILEO ORBITS AND ENCOUNTERS

Orbit	Satellite	Date	Minimum Distance (km)
J0	Europa	7 December 1995	32,958
J0	Io	7 December 1995	898
G1	Ganymede	27 June 1996	835
G2	Ganymede	6 September 1996	260
C3	Callisto	4 November 1996	1,136
E4	Europa	19 December 1996	692
J5	*No Encounters*	20 January 1997	–
E6	Europa	20 February 1997	586
G7	Ganymede	5 April 1997	3,102
G8	Ganymede	7 May 1997	1,603
C9	Callisto	25 June 1997	418
C10	Callisto	17 September 1997	539
E11	Europa	6 November 1997	2,042
E12	Europa	16 December 1997	205
E13	Europa	10 February 1998	3,562
E14	Europa	29 March 1998	1,645
E15	Europa	31 May 1998	2,515
E16	Europa	20 July 1998	1,837
E17	Europa	26 September 1998	3,582
E18	Europa	22 November 1998	2,273
E19	Europa	1 February 1999	1,439
C20	Callisto	5 May 1999	1,315
C21	Callisto	30 June 1999	1,047
C22	Callisto	14 August 1999	2,296
C23	Callisto	16 September 1999	1,057
I24	Io	11 October 1999	612

Orbit	Satellite	Date	Minimum Distance (km)
I25	Io	26 November 1999	300
E26	Europa	3 January 2000	351
I27	Io	22 February 2000	198
G28	Ganymede	20 May 2000	809
G29	Ganymede	28 December 2000	2,337
C30	Callisto	25 May 2001	138
I31	Io	6 August 2001	194
I32	Io	16 October 2001	184
I33	Io	17 January 2002	101.5
A34	Amalthea	5 November 2002	244
J35	*Jupiter Impact*	21 September 2003	–

Galileo orbits were named according to the satellite that was the main target of the periapsis pass: A = Amalthea, C = Callisto, E = Europa, G = Ganymede, I = Io, J = no encounter.

Chapter references

[Acuña-1998] Acuña, M.H., et al., "Magnetic Field and Plasma Observations at Mars: Initial Results of the Mars Global Surveyor Mission", Science, 279, 1998, 1676–1680

[Acuña-1999] Acuña, M.H., et al., "Global Distribution of Crustal Magnetization Discovered by the Mars Global Surveyor MAG/ER Experiment", Science, 284, 1999, 790–793

[Adams-1981] Adams, R.E.W., Brown, W.E. Jr., Patrick Culbert, T., "Radar Mapping, Archeology, and Ancient Maya Land Use", Science, 213, 1981, 1457–1463

[Akiba-1980] Akiba, R., et al., "Orbital Design and Technological Feasibility of Halley Mission", Acta Astronautica, 7, 1980, 797–805

[Albee-1994] Albee, A.L., Uesugi, K.T., Tsou, P., "SOCCER: Comet Coma Sample Return Mission". In: Lunar and Planetary Institute, Workshop on Particle Capture, Recovery and Velocity/Trajectory Measurement Technologies, 1994, 7–11

[Albee-1998] Albee, A.L., Palluconi, F.D:, Arvidson, R.E., "Mars Global Surveyor Mission: Overview and Status", Science, 279, 1998, 1671–1672

[Albee-2001] Albee, A.L., et al., "Overview of the Mars Global Surveyor Mission", Journal of Geophysical Research, 106, 2001, 23,291–23,316

[Alekseev-1986] Alekseev, V.A., et al., "The Plasma Near the Sun Sounded by Venera 15 Radio Signals: a VLBI Experiment", Soviet Astronomy Letters, 12, 1986, 204–207

[Alexandrov-1989] Alexandrov, Yu.N., Krivtsov, A.P., Rzhiga, O.N., "Venera 15 and 16 Spacecraft: Some Results on Venus Surface Reflectivity Measurements", paper presented at the XXI Lunar and Planetary Science Conference, Houston, 1989

[Alfvèn-1970] Alfvèn, H., "Exploring the Origin of the Solar System by Space Missions to the Asteroids", paper presented at the Third Conference on Planetology and Space Mission Planning, New York, October 1970

[Alvarez-1997] Alvarez, W., "T.Rex and the Crater of Doom", Princeton University Press, 1997

[Anderson-1977] Anderson, J.D., et al., "An Arrow to the Sun". In: "Proceedings of the International Meeting on Experimental Gravitation, Pavia, September 17–20, 1976", Rome, Accademia Nazionale dei Lincei, 1977, 393–422

[Anderson-1994] Anderson, J.D., Vessot, R.F.C., Mattison, E.M. , "Gravitational Experiments for Solar Probe", paper presented at the Memorial Conference "Ideas for Space Research after the Year 2000", Padua, 18–19 February 1994

[Anderson-1996a] Anderson, J.D., Sjogren, W.L., Schubert, G., "Galileo Gravity Results and the Internal Structure of Io", Science, 272, 1996, 709–712

[Anderson-1996b] Anderson, J.D., et al., "Gravitational Constrains on the Internal Structure of Ganymede", Nature, 384, 1996, 541–543

[Anderson-1997a] Anderson, J.D., et al., "Gravitational Evidence for an Undifferenciated Callisto", Nature, 387, 1997, 264–266
[Anderson-1997b] Anderson, J.D., et al., "Europa's Differentiated Internal Structure: Inferences from Two Galileo Encounters", Science, 276, 1997, 1236–1239
[Anderson-1998] Anderson, J.D., et al., "Europa's Differentiated Internal Structure: Inferences from Four Galileo Encounters", Science, 281, 1998, 2019–2022
[Anderson-2004] Anderson, J.D., et al., "Discovery of Mass Anomalies on Ganymede", Science, 305, 2004, 989–991
[Anderson-2005] Anderson, J.D., et al., "Amalthea's Density is Less than that of Water", Science, 308, 2005, 1291–1293
[Andreichikov-1986] Andreichikov, B.M., et al., "Element Abundances in Venus Aerosols by X-Ray Radiometry: Preliminary Results" Soviet Astronomy Letters, 12, 1986, 48–49
[Angold-1992] Angold, N., et al., "Ulysses Operations at Jupiter – Planning for the Unknown", ESA Bulletin, 72, 1992, 44–51
[Angold-2008] Angold, N., "Ulysses, the Over Achiever", presentation at "The Ulysses Legacy" Press Conference, Paris, ESA Headquarters, 12 June 2008
[Anselmo-1987a] Anselmo, L., Trumpy, S., "Low Cost Mission to Near-Earth Asteroids", Paper AAS 87–405
[Anselmo-1987b] Anselmo, L., "Proposta di Missione Spaziale agli Asteroidi Apollo-Amor-Aten" (A proposed space mission to Apollo-Amor-Aten asteroids), CNUCE Internal Report C87–11, 17 March 1987 (in Italian)
[Anselmo-1990] Anselmo, L., Pardini, C., "Piazzi: a Probe to the Apollo-Amor Asteroids", in: "Proceedings of the 17th International Symposium on Space Technology and Science", Tokyo, 1990, 1879–1884
[Anselmo-1991] Anselmo, L., Milani, A., "Reconnaissance Mission to Near-Earth Asteroids", The Journal of the Astronautical Sciences, 39, 1991, 469–485
[Anselmo-1996a] Anselmo, J.C., "Mars Sample Return Still Years Away", Aviation Week & Space Technology, 28 October 1996, 69
[Anselmo-1996b] Anselmo, J.C., "Life on Mars? Evidence Emerges", Aviation Week & Space Technology, 12 August 1996, 24–25
[Anselmo-2007] Anselmo, L., Personal communication with the author, 8 September 2004
[Antreasian-1998] Antreasian, P.G., Guinn, J.R., "Investigations into the unexpected Delta-V Increases during the Earth Gravity Assists of Galileo and NEAR", Paper AIAA-98-4287
[APL-1999] "The NEAR Rendezvous Burn Anomaly of December 1998: Final Report of the NEAR Anomaly Review Board", Laurel, The Johns Hopkins University Applied Physics Laboratory, November 1999
[Aran-2007] Aran, A., et al., "Modeling and Forecasting Solar Energetic Particle Events at Mars: the event on 6 March 1989", Astronomy & Astrophysics, 469, 2007, 1123–1134
[Armstrong-1997] Armstrong, J.W., et al., "The Galileo/Mars Observer/Ulysses Coincidence Experiment". Paper presented at the 2nd Edoardo Amaldi Conference on Gravitational Waves, Geneva, 1997
[Armstrong-2002] Armstrong, J.C., Wells, L.E., Gonzalez, G., "Rummaging through Earth's Attic for Remains of Ancient Life", Arxiv pre-print astro-ph/0207316
[Asker-1996] Asker, J.R., "Next Missions to Mars May Prove Too Small", Aviation Week & Space Technology, 19 August 1996, 26–27
[Asker-2001a] Asker, J.A., "Attempt at Hard Landing Set For Asteroid Spacecraft" Aviation Week & Space Technology, 5 February 2001, 42–43
[Asker-2001b] Asker, J.A., "Cheating Death, NEAR Lands, Operates on Eros", Aviation Week & Space Technology, 19 February 2001, 24–25

[Aston-1986] Aston, G., "Electric Propulsion: A Far Reaching Technology", Journal of the British Interplanetary Society, 39, 1986, 503–507
[Atkinson-1996] Atkinson, D.H., Pollack, J.B., Seiff, A., "Galileo Doppler Measurements of the Deep Zonal Winds at Jupiter", Science, 272, 1996, 842–843
[Atkinson-1997] Atkinson, D.H., Ingersoll, A.P., Seiff, A., "Deep Winds on Jupiter as Measured by the Galileo Probe", Nature, 388, 1997, 649–650
[Atzei-1989] Atzei, A., et al., "Rosetta/CNSR – ESA's Planetary Cornerstone Mission", ESA Bulletin, 59, 1989, 18–29
[Avanesov-1989a] Avanesov, G.A., et al., "Television Observations of Phobos", Nature, 341, 1989, 585–587
[Avanesov-1989b] Avanesov, G., et al., "Regatta-Astro Project: Astrometric Studies from Small Space Laboratory". In: Preceedings of the 141st IAU Symposium on the Inertial Coordinate System on the Sky, Leningrad, 1989, 361–366
[Avanesov-1991] Avanesov, G.A., Kostenko, V.I., "Regatta v Kosmicheskiy Polyet pod Solnetsnim Parusom" (Regatta for flight in space with a solar sail), Zemlya I Vselennaya, January 1991, 3 (in Russian)
[AWST-1979] "OMB Kills Halley/Tempel 2 Mission", Aviation Week & Space Technology, 26 November 1979, 20
[AWST-1980] "Venus Orbiting Imaging Radar Design Proposals Due This Year", Aviation Week & Space Technology, 24 March 1980
[AWST-1983] "Hardware From Past Programs Will Cut Venus Mapper Costs", Aviation Week & Space Technology, 14 February 1983, 20
[AWST-1985a] "Industry Observer", Aviation Week & Space Technology, 29 July 1985, 11
[AWST-1985b] "NASA Chief Favors ISTP, Ocean Mapping Over Comet Mission", Aviation Week & Space Technology, 16 September 1985, 18
[AWST-1986a] "New Vega Flyby", Aviation Week & Space Technology, 24 March 1986, 22
[AWST-1986b] "Glavcosmos Formed by Soviets to Help Run Space Program", Aviation Week & Space Technology, 24 March 1986, 77
[AWST-1987a] "Soviet Mars Mission Will Use Modular Propulsion System", Aviation Week & Space Technology, 2 November 1987, 81
[AWST-1987b] "JPL Studying Aeroshell Structure for Mars Rover, Spacecraft", Aviation Week & Space Technology, 3 August 1987, 61
[AWST-1988a] "NASA Board Nears End of Fire Review", Aviation Week & Space Technology, 24 October 1988, 24
[AWST-1988b] "Magellan Fire Inquiry Board Urges New Test Procedures", Aviation Week & Space Technology, 14 November 1988, 35
[AWST-1988c] "U.S./Soviet Balloon Flights May Aid Future Mars Missions", Aviation Week & Space Technology, 29August 1988, 49
[AWST-1988d] "Soviets Consider Varied Concepts for 1994 Mars Exploration Flight", Aviation Week & Space Technology, 18 July 1988, 19
[AWST-1989a] "CRAF Will Be First in Series of Missions Using Mariner Mk.2", Aviation Week and Space Technology, 9 October 1989, 99–109
[AWST-1989b] "Magellan's Radar Images of Venus to Unmask Cloud-Shrouded Planet", Aviation Week & Space Technology, 9 October 1989,113–115
[AWST-1989c] "Galileo Represents Peak in Design Complexity", Aviation Week & Space Technology, 9 October 1989, 77–78
[AWST-1990a] "Magellan Switched to Safer Mode; Computer Faults Still Puzzle Controllers", Aviation Week & Space Technology, 10 September 1990, 30

[AWST-1990b] "Magellan Spacecraft Regains High-Data Rate Communications", Aviation Week & Space Technology, 17 September 1990, 41
[AWST-1996] "Crunch Time for Mars 96", Aviation Week & Space Technology, 7 October 1996, 70
[AWST-2002] "Quit Fiddling and Image Jupiter Moon", Aviation Week & Space Technology, 8 July 2002, 70
[Baird-2007] Baird, D.T., et al., "Zonal Wind Calculations from Mars Global Surveyor Accelerometer and Rate Data", Journal of Spacecraft and Rockets, 44, 2007, 1180–1187
[Balogh-1984] Balogh, A., "AGORA: Asteroid Rendezvous", Spaceflight, June 1984, 242–245
[Balsiger-1986] Balsiger, H., et al., "Ion Composition and Dynamics at Comet Halley", Nature, 321, 1986, 330–334
[Balsiger-1988] Balsiger, H., Fechtig, H., Geiss, J., "A Close Look at Halley's Comet", Scientific American, September 1988, 62–69
[Bandfield-2000] Bandfield, J.L., Hamilton, V.E., Christensen, P.R., "A Global View of Martian Surface Compsition from MGS-TES". Science, 287, 2000, 1626–1630
[Bandfield-2003] Bandfield, Glotch, T.D., V.E., Christensen, P.R., "Spectroscopic Identification of Carbonate Minerals in the Martian Dust", Science, 301, 2003, 1084–1087
[Banerdt-1994] Banerdt, W.B., et al., "Gravity Studies of Mead Crater, Venus", paper presented at the Lunar and Planetary Science Conference XXV, Houston, 1994
[Barbieri-1985] Barbieri, C., et al., "La Halley Multicolour Camera: Contributo Italiano alla sua Realizzazione" (The Italian Contribute to the Halley Multicolour Camera). In: "Le Comete nell'Astronomia Moderna: Il Prossimo Incontro con la Cometa di Halley" (Comets in Modern Astronomy: The Forthcoming Encounter with Halley's Comet), Naples, Guida, 1985, 229–250 (in Italian)
[Barbosa-1992] Barbosa, D.D., Kivelson, M.G., "Ulysses Spacecraft Rendezvous with Jupiter", Science, 257, 1992, 1487–1489
[Basilevsky-1988] Basilevsky, A.T., "Northern Beta: Photogeologic Analysis of Venera 15/16 Images and Maps", paper presented at the XIX Lunar and Planetary Science Conference, Houston, 1988
[Basilevsky-1992] Basilevsky, A.T., Weitz, C.M., "The Geology of the Venera/Vega Landing Sites", paper presented to the International Colloquium on Venus, Pasadena, 10–12 August 1992
[Baumgärtel-1998] Baumgärtel, K., et al., "'Phobos Events' – Signature of Solar Wind Interaction with a Gas Torus?", Earth Planets Space, 50, 1998, 453–462
[Becker-2005] Becker, S.C., "Astro Projection: Virtual Reality, Telepresence, and the Evolving Human Space Experience", Quest, 12 No.3, 2005, 34–55
[Beech-1990] Beech, P., Meyer, D., "Post-Launch Operations and Data Production", ESA Bulletin, 63, 1990, 60–63
[Bell-1998] Bell, J.. "Mars Pathfinder: Better Science?", Sky & Telescope, July 1998, 36–43
[Belton-1991] Belton, M.J.S., et al., "Images from Galileo of the Venus Cloud Deck" Science, 253, 1991, 1531–1536
[Belton-1992] Belton, M.J.S., et al., "Galileo Encounter with 951 Gaspra: First Pictures of an Asteroid", Science, 257, 1992, 1647–1652
[Belton-1994] Belton, M.J.S., et al., "First Images of Asteroid 243 Ida", Science, 265, 1994, 1543–1547
[Belton-1996] Belton, M.J.S., et al., "Galileo's First Images of Jupiter and the Galilean Satellites", Science, 274, 1996, 377–385
[Bender-1978] Bender, D.F., "Ballistic Trajectories". In: Neugebauer, M., Davies, R.W., "A Close-Up of the Sun", Pasadena, JPL, 1978, 535–543

[Bertaux-1986] Bertaux, J.L., et al., "Active Spectrometry of the Ultraviolet Absorption within the Venus Atmosphere", Soviet Astronomy Letters, 12, 1986, 33–36
[Bertotti-1992] Bertotti, B., et al., "The Gravitational Wave Experiment", Astronomy & Astrophysics Supplement Series, 92, 1992, 431–440
[Bertotti-1995] Bertotti, B., et al., "Search for Gravitational Wave Trains with the Spacecraft Ulysses", Astronomy & Astrophysics, 296, 1995, 13–25
[Bertotti-2007] Bertotti, B., interview with the author, Pavia, 27 April 2007
[Bevilacqua-1994] Bevilacqua, F., Cesare, S., "A Project for a Solar Sail Propelled Spaceship", Journal of the British Interplanetary Society, 47, 1994, 57–66
[Bibing-1989] Bibing, J.-P., et al., "Results from the ISM Experiment", Nature, 341, 1989, 591–593
[Bickler-1989] Bickler, D.B., "Articulated Suspension System", United States Patent No. 4,840,394, 20 June 1989
[Bickler-1993] Bickler, D.B., "The New Family of JPL Planetary Surface Vehicles". In: "Missions, Technologies et Conception des Vehicules Mobiles Planetaires", Toulouse, Cépaduès, 1993
[Bindschadler-2003] Bindschadler, D.L., et al., "Project Galileo: Final Mission Status", paper presented at the LIV Congress of the International Astronautical Federation, Bremen, 2003
[Binzel-1990] Binzel, R.P., "Pluto", Scientific American, June 1990, 50–58
[Bird-1992a] Bird, M.K., et al., "The Coronal-Sounding Experiment", Astronomy & Astrophysics Supplement Series, 92, 1992, 425–430
[Bird-1992b] Bird, M.K., et al., "Ulysses Radio Occultation Observations of the Io Plasma Torus During the Jupiter Encounter", Science, 257, 1992, 1531–1535
[Blamont-1987a] Blamont, J., "Venus Devoilée" (Venus Unveiled), Paris, Editions Odile Jacob, 1987, 251 (in French)
[Blamont-1987b] ibid 285
[Blamont-1987c] ibid., 173–215
[Blamont-1987d] ibid. 248
[Blamont-1987e] ibid., 247
[Blamont-1987f] ibid., 250
[Blamont-1987g] ibid., 232–238
[Blamont-1987h] ibid., 302–304
[Blamont-1987i] ibid., 249–250
[Blamont-1987j] ibid., 312
[Blamont-1987k] ibid., 317
[Blamont-1987l] ibid., 320
[Blamont-1987m] ibid., 335
[Blamont-1989] Blamont, J.E., et al., "Vertical Profiles of Dust and Ozone in the Martian Atmosphere Deduced from Solar Occultation Measurements", Nature, 341, 1989, 600–603
[Blume-1984] Blume, W.H., et al., "Overview of the Planetary Observer Program", Paper AIAA-84-0454
[Boain-1993] Boain, R.J., "Clementine II: a Double Impact Asteroid Flyby and Impact Mission", paper presented at the Workshop on Advanced Technology for Planetary Instruments, 28–30 April 1993
[Bockstein-1988] Bockstein, I., Chochia, P., Kronrod, M,, "Methods of Venus Radiolocation Map Sythesis using Strip Images of Venera-15 and Venera-16 Space Station", Earth, Moon and Planets, 43, 1988, 233–259
[Bogatchev-2000] Bogatchev, A., et al., "Walking and wheel-walking robots", paper presented at the 3rd International Conference on Climbing and Walking Robots, Madrid, 2000

[Bojor-1996] Bojor, Yu., et al., "Analysis of the Mission Profiles and Means for the US-Russian Project of the Mission to Pluto", paper presented at the First IAA Symposium on Realistic Near-Term Advanced Scientific Space Missions, Aosta, 25–27 June 1996

[Bokulic-1997] Bokulic, R.S., Moore, W.V., "The NEAR Solar Conjunction Experiment", paper dated 1997

[Bond-1993] Bond, P., "Close Encounter with a Comet", Astronomy, November 1993, 42–47

[Bonnet-1994] Bonnet, R.M., "The Influence of Giuseppe Colombo on the ESA Science Programme", paper presented at the Memorial Conference "Ideas for Space Research after the Year 2000", Padua, 18–19 February 1994

[Bonnet-2002] Bonnet, R.M., "History of the Giotto Mission", Space Chronicle, 55, 2002, 5–11

[Borg-1994] Borg, J., Bribing, J.-P., Maag, C., "Main Characteristics of the COMET/COMRADE Experiments", Paper presented at the Workshop on Particle Capture, Recovery and Velocity/Trajectory Measurement Technologies, 1994

[Bortle-1996] Bortle, J.E., "Winter's Express Comet", Sky & Telescope, February 1996, 94–95

[Brain-2006] Brain, D.A., "Mars Global Surveyor Measurements of the Martian Solar Wind Interaction", Space Science Reviews, 126, 2006, 77–112

[Bromberg-1966] Bromberg, J.L., Gordon, T.J.: "Extensions of Saturn", paper presented at the XVII International Astronautical Congress, Madrid, 1966

[Brownlee-2003] Brownlee, D.E., et al., "Stardust: Comet and Interstellar Dust Sample Return Mission", Journal of Geophysical Research, 108, 2003, 1–1 to 1–15

[Bruns-1990] Bruns, A.V., et al., "Solar Brightness Oscillations: Phobos 2 Observations", Soviet Astronomy Letters, 16, 1990, 140–145

[Burke-1984] Burke, J.D., "The Missing Link Revealed", Studies in Intelligence, Spring 1984, 27–34

[Burke-1990] Burke, J.D., Mostert, R.N., "A Network of Small Landers on Mars", Paper AIAA-90-3577-CP

[Burnham-1996] Burnham, D., Salmon, A., "Mars '96: Russia's Return to the Forbidden Planet", Spaceflight, August 1996, 272–274

[Burns-1999] Burns, J.A., et al., "The Formation of Jupiter's Faint Rings", Science, 284, 1999, 1146–1150

[Butrica-1996a] Butrica, A.J., "To See the Unseen – A History of Planetary Radar Astronomy", Washington, NASA, 1996, 177–187

[Butrica-1996b] ibid., 194

[Butrica-1996c] ibid., 187–188

[Butrica-1996d] ibid., 193

[Butrica-1996e] ibid., 204

[Butrica-1996f] ibid., 252–254

[Cadogan-1998] Cadogan, D., Sandy, C., Grahne, M., "Development and Evaluation of the Mars Pathfinder Inflatable Airbag Landing System", paper IAF-98-I.6.02

[Calder-1992a] Calder, N., "Giotto to the Comets", London, Presswork, 1992, 20–28

[Calder-1992b] ibid., 29–38

[Calder-1992c] ibid., 65

[Calder-1992d] ibid., 69–70

[Calder-1992e] ibid., 64

[Calder-1992f] ibid., 37 and 45

[Calder-1992g] ibid., 114–115

[Calder-1992h] ibid., 82

[Calder-1992i] ibid., 64

[Calder-1992j] ibid., 96–97

[Calder-1992k] ibid., 98–99
[Calder-1992l] ibid., 100–101
[Calder-1992m] ibid., 105–106
[Calder-1992n] ibid., 109–110
[Calder-1992o] ibid., 118–119
[Calder-1992p] ibid., 135–136
[Calder-1992q] ibid., 128
[Calder-1992r] ibid., 123
[Calder-1992s] ibid., 148
[Calder-1992t] ibid., 147–163
[Calder-1992u] ibid., 164–197
[Caldwell-1992] Caldwell, J., Turgeon, B., Hua, X.-M., "Hubble Space Telescope Imaging of the North Polar Aurora on Jupiter", Science, 257, 1992, 1512–1515
[Canby-1986] Canby, T.Y., "Are The Soviets Ahead in Space?", National Geographic, October 1986, 420–459
[Cantor-2007] Cantor, B.A., "MOC Observations of the 2001 Mars Planet-Encircling Dust Storm", Icarus, 186, 2007, 60–96
[Caplinger-2001] Caplinger, M.A., Malin, M.C., "Mars Orbiter Camera Geodesy Campaign", Journal of Geophysical Research, 168, 2001, 23,595–23,606
[Caprara-1992] Caprara, G., "L'Italia nello Spazio" (Italy in Space), Rome, Valerio Levi, 1992, 202 (in Italian)
[Carlier-1993] Carlier, C., Gilli, M., Laidet, L., "CNES: The French Space Agency 1962–1992", Paper IAA-2-1-93-669, presented at the XLIV Congress of the International Astronautical Federation, Graz, 1993
[Carlier-1995] Carlier, C., Gilli, M., "The First Thirty Years at CNES: the French Space Agency 1962–1992", Paris, CNES/La Documentation Française, 1995, 141 and 210
[Carlson-1991] Carlson, R.W., et al., "Galileo Infrared Imaging Spectroscopy Measurements at Venus", Science, 253, 1991, 1541–1548
[Carlson-1992] Carlson, R.W., et al., "Near-Infrared Mapping Spectrometer Experiment on Galileo", Space Science Reviews, 60, 1992, 457–502
[Carlson-1996] Carlson, R., et al., "Near-Infrared Spectroscopy and Spectral Mapping of Jupiter and the Galilean Satellites: Results from Galileo's Initial Orbit", Science, 274, 1996, 385–388
[Carlson-1999a] Carlson, R.W., "A Tenuous Carbon Dioxide Atmosphere on Jupiter's Moon Callisto", Science, 283, 1999, 820–821
[Carlson-1999b] Carlson, R.W., et al., "Sulfuric Acid on Europa and the Radiolytic Sulfur Cycle", Science, 286, 1999, 97–99
[Carlson-1999c] Carlson, R.W., et al., "Hydrogen Peroxide on the Surface of Europa", Science, 283, 1999, 2062–2064
[Carlson-2007] Carlson, R.W., Personal communication with the author, 3 December 2007
[Carr-1998] Carr, M.H., et al., "Evidence for a Subsurface Ocean on Europa", Nature, 391, 1998, 363–365
[Carroll-1993a] Carroll, M.W., "Cheap Shots", Astronomy, 21, August 1993, 38–47
[Carroll-1993b] Carroll, M., "Mars: The Russians are Going! The Russians are Going!", Astronomy, October 1993, 26–33
[Carroll-1995] Carroll, M., "New Discoveries on the Horizon: NASA's Next Missions", Astronomy, 23, November 1995, 36–43
[Carroll-1997] Carroll, M., "Europa: Distant Ocean, Hidden Life?", Sky & Telescope, December 1997, 50–55

[Caseley-1990] Caseley, P.J., Marsden, R.G., "The Ulysses Scientific Payload", ESA Bulletin, 63, 1990, 29–38
[Cattermole-1997] Cattermole, P, Moore, P., "Atlas of Venus", Cambridge University Press, 1997, 50–103
[Chaikin-2007] Chaikin, A., "Global Surveyor's Last Hurrah", Sky & Telescope, April 2007, 38–41
[Chapman-1995] Chapman, R.J., Regeon, P.A., "The Clementine Lunar Orbiter Project", paper presented at the Austrian Space Agency Summer School 1995, Alpbach, Germany, 26 July–3 August 1995
[Cheng-2001] Cheng, A.J., et al., "Laser Altimetry of Small Scale Features on 433 Eros from NEAR–Shoemaker", Science, 2001, 292, 488–491
[Cheng-2002] Cheng, A.F., et al., "Small Scale Topography from Laser Altimetry and Imaging", Icarus, 155, 2002, 51–74
[Chicarro-1993] Chicarro, A., Scoon, G., Coradini, M., "MARSNET – A Network of Stations on the Surface of Mars", ESA Journal, 17, 1993, 225–237
[Chicarro-1994] Chicarro, A., Scoon, G., Coradini, M., "INTERMARSNET – An International Network of Stations on Mars for Global Martian Characterization", ESA Journal, 18, 1994, 207–218
[Christensen-1998] Christensen, P.R., et al., "Results from the Mars Global Surveyor Thermal Emission Spectrometer", Science, 279, 1998, 1692–1698
[Christensen-2001] Christensen, P.R., et al., "Mars Global Surveyor Thermal Emission Spectrometer Experiment: Investigation Description and Surface Science Results", Journal of Geophysical Research, 106, 2001, 23,823–23,871
[Christou-2007] Christou, A.A., Vaubaillon, J., Withers, P., "The Dust Trail Complex of Comet 79P/du Toit–Hartley and Meteor Outbursts on Mars", Astronomy & Astrophysics, 471, 2007, 321–329
[CIA-1988] "Soviet Scientific Space Program: Gaining Prestige", Langley, CIA, January 1988, 10
[Clark-2000] Clark, P.S., "Launch Profiles Used by the Four-Stage Proton-K", Journal of the British Interplanetary Society, 53, 2000, 197–214
[Collins-1986] Collins, D.H., Miller, S.L., "Comet Rendezvous: The Next Stage in Cometary Exploration", Journal of the British Interplanetary Society, 39, 1986, 263–272
[Collins Petersen-1997] Collins Petersen, C., "Welcome to Mars", Sky & Telescope, October 1997, 34–37
[Combes-1986] Combes, M., et al., "Infrared Sounding of Comet Halley from Vega 1", Nature, 321, 1986, 266–268
[Connerney-1999] Connerney, J.E.P., et al., "Magnetic Lineations in the Ancient Crust of Mars", Science, 284, 1999, 794–798
[Cosmovici-1983] Cosmovici, C.B., Schmidt, E., Stanggassinger, U., "ASTIS: Infrared Spectrometer for the German Asteroids Mission". In: "Asteroids, comets, meteors; Proceedings of the Meeting", Uppsala University, 1983, p. 187–191
[Covault-1979] Covault, C., "Funds Cut Forces Comet Strategy Shift", Aviation Week & Space Technology, 3 December 1979, 61–65
[Covault-1985a] Covault, C., "U.S. Plans Soviet Talks on Joint Manned Mission", Aviation Week & Space Technology, 7 January 1985, 16–18
[Covault-1985b] Covault, C., "First Comet Probe Reveals Structure of Great Complexity", Aviation Week & Space Technology, 16 September 1985, 16–19
[Covault-1985c] Covault, C., "NASA Defining Mission to Return Cometary Matter to Earth in 1990s", Aviation Week & Space Technology, 9 December 1985, 115–117

[Covault-1985d] Covault, C., "Soviets in Houston Reveal New Lunar, Mars, Asteroid Flights", Aviation Week & Space Technology, 1 April 1985, 18–20
[Covault-1989a] Covault, C., "Magellan Prepared for Course Correction as Astronauts Land Atlantis in Crosswind", Aviation Week & Space Technology, 15 May 1989, 25
[Covault-1989b] Covault, C., "Galileo Launch to Jupiter by Atlantis Culminates Difficult Effort with Shuttle", Aviation Week & Space Technology, 9 October 1989, 58–67
[Covault-1996a] Covault, C., "Mars Surveyor Leads New Era of Exploration", Aviation Week & Space Technology, 11 November 1996, 22–24
[Covault-1996b] Covault, C., "Confusion Marks Mars 96 Failure", Aviation Week & Space Technology, 25 November 1996, 71–72
[Covault-2006] Covault, C., "Rescue Ops Over Mars", Aviation Week & Space Technology, 27 November 2006, 53–55
[Cowley-1985] Cowley, S.W, "ICE Encounters Giacobini–Zinner", Nature, 317, 1985, 381
[Crowley-2007] Crowley, G., Tolson, R.H., "Mars Thermospheric Winds from Mars Global Surveyor and Mars Odyssey Accelerometers", Journal of Spacecraft and Rockets, 44, 2007, 1188–1194
[Cunningham-1983] Cunningham, C., "European Satellite Studies of Minor Planets", Minor Planets Bulletin, 10, 1983, 26–27
[Cunningham-1985] Cunningham, C., "European Satellite Studies of Minor Planets II", Minor Planets Bulletin, 12, 1985, 29–30
[Cunningham-1988a] Asteroid diameters are from IRAS data published in: Cunningham, C.J., "Introduction to Asteroids", Richmond, Willmann-Bell, 1988, 148–164
[Cunningham-1988b] ibid., 135
[Cunningham-1988c] ibid., 132
[Cunningham-1988d] ibid., 93–123
[Cunningham-1988e] ibid., 132–133
[Cunningham-1988f] ibid., 135
[Cunningham-1988g] ibid., 134
[Cunningham-1988h] ibid., 136–138
[Cunningham-1988i] ibid., 136
[Cunningham-1988j] ibid., 133–134
[Cunningham-1988k] ibid., 89–92
[Cunningham-1989] Cunningham, C., "European Satellite Studies of Minor Planets III", Minor Planets Bulletin, 16, 1989, 20–21
[Cunningham-1996] Cunningham, G.E., "Mars Global Surveyor Mission", Acta Astronautica, 38, 1996, 367–375
[Dale-1986] Dale, D., Felici, F., Lo Galbo, P., "The Giotto Project: From Early Concepts to Flight Model", ESA Bulletin, 46, 1986, 22–33
[Davies-1988a] Davies, J.K., "Satellite Astronomy: The Principles and Practice of Astronomy from Space", Chichester, Ellis Horwood, 1988, 119–120
[Davies-1988b] ibid., 115–116
[Davies-2003] Davies, A.G., "Temperature, Age and Crust Thickness Distribution of Loki Patera on Io from Galileo NIMS Data: Implications for Resurfacing Mechanism", Geophysical Research Letters, 30, 2003, 2133–2136
[Dawson-2004] Dawson, V.P. Bowles, M.D., "Taming Liquid Hydrogen: The Centaur Upper Stage Rocket 1958–2002", Washington, NASA, 2004, 202–207
[Day-2006] Day, D.A., "The Heat of a Burning Atom", Spaceflight, April 2006, 145–150
[Debus-2002] Debus, A., et al., "Landers Sterile Integration Implementations: Example of Mars 96 Mission", Acta Astronautica, 50, 2002. 385–392

[Dehant-2003] Dehant, V., "A Liquid Core for Mars?", Science, 300, 2003, 260–261
[Divsalar-1995] Divsalar, D., Simon, M.K., "CDMA With Interference Cancellation for Multiprobe Missions", JPL TDA Progress Report 42-120, 1995, 40–53
[Doengi-1998] Doengi, F., et al., "Lander Shock-Alleviation Techniques", ESA Bulletin, 93, 1998, 51–60
[Doody-1993] Doody, D.F., "Grappling for Gravity", Sky & Telescope, August 1993, 20
[Doody-1995] Doody, D.F., "Aerobraking the Magellan Spacecraft in Venus Orbit", Acta Astronautica, 35, 1995, 475–480
[Dornheim-1985] Dornheim, M.A., "Soviets' Vega 2 Balloon, Lander Transmit Data from Venus", Aviation Week & Space Technology, 24 June 1985, 22–24
[Dornheim-1988] Dornheim, M.A., "Magellan Probe Signifies Renewed Interest in Planetary Programs", Aviation Week & Space Technology, 6 June 1988, 38–41
[Dornheim-1989] Dornheim, M.A., "Galileo Thrusters Approved for Flight, But Mission Plan May be Abbreviated", Aviation Week & Space Technology, 10 April 1989. 23
[Dornheim-1990a] Dornheim, M.A., "Magellan Begins Systems Checkouts After Entering Orbit Around Venus", Aviation Week & Space Technology, 20 August 1990, 30–31
[Dornheim-1990b] Dornheim, M.A., "Magellan Radar Produces Sharp Images, But Computer Problems Vex Controllers", Aviation Week & Space Technology, 27 August 1990, 29
[Dornheim-1991] Dornheim, M.A., "Improper Antenna Deployment Treathens Galileo Jupiter Mission", Aviation Week & Space Technology, 22 April 1991, 25
[Dornheim-2005] Dornheim, M.A., "Sat-to-Sat Photos May Help Diagnose Ills of Other Spacecraft", Aviation Week & Space Technology, 30 May 2005, 47
[Dubinin-1998] Dubinin, E., et al., "Multiple Shocks near Mars", Earth Planets Space, 50, 1988, 279–287
[Dunham-1990] Dunham, D.W., Jen, S.-C., Farquhar, R.W., "Trajectories for Spacecraft Encounters with Comet Honda–Mrkos–Pajdušáková in 1996", Acta Astronautica, 22, 1990, 161–171
[Dunham-2000] Dunham, D.W., et al., "Recovery of NEAR's mission to Eros", Acta Astronautica, 47, 2000, 503–512
[D'Uston-1989] D'Uston, C., et al., "Observation of the Gamma-Ray Emission from the Martian Surface by the APEX Experiment", Nature, 341 1989, 598–600
[Duxbury-1997] Duxbury, T.C., "Proposed Clementine II Mission", paper dated October 1997
[Eaton-1990] Eaton, D., "The Ulysses Storage and Recertification Activities: The Managerial Problems", ESA Bulletin, 63, 1990, 73–77
[Eberhart-1985] Eberhart, J., "The ICE Plan Cometh", Science News, 31 August 1985, 138–139
[Eberhardt-1986] Eberhardt, P. et al., "The CAESAR Project – A Comet Atmosphere Encounter and Sample Return", In: ESA Proceedings of the 20th ESLAB Symposium on the Exploration of Halley's Comet. Volume 2: Dust and Nucleus, 1986, 243–248
[Edwards-2004] Edwards, C.D. Jr., et al., "A Martian Telecommunications Network: UHF Relay Support of the Mars Exploration Rovers by the Mars Global Surveyor, Mars Odyssey, and Mars Express Orbiters", paper presented at the LV Congress of the International Astronautical Federation, Vancouver, 2004
[Efimov-2008a] Efimov, A.I., et al., "Coronal Radio-Sounding Detection of a CME During the 1997 Galileo Solar Conjunction", Advances in Space Research, 42, 2008, 110–116
[Efimov-2008b] Efimov, A.I., et al., "Solar Wind Turbulence During the Solar Cycle Deduced from Galileo Coronal Radio-Sounding Experiments", Advances in Space Research, 42, 2008, 117–123

[Elachi-1980] Elachi, C., "Spaceborne Imaging Radar: Geologic and Oceanographic Applications", Science, 209, 1980, 1073–1082

[Elfving-1993] Elfving, A., "Automation Technology for Remote Sample Acquisition". In: "Missions, Technologies et Conception des Vehicules Mobiles Planetaires", Toulouse, Cépaduès, 1993

[Eremenko-1993] Eremenko, A, Martinov, B., Pitchkhadze, K., "Rover in 'the Mars 96' Mission". In: "Missions, Technologies et Conception des Vehicules Mobiles Planetaires", Toulouse, Cépaduès, 1993

[Erickson-1999] Erickson, J.K., et al., "Project Galileo: Completing Europa, Preparing for Io", paper presented at the L Congress of the International Astronautical Federation, Amsterdam, 1999

[Erickson-2000] Erickson, J.K., et al., "Project Galileo: Surviving Io, Meeting Cassini", paper presented at the LI Congress of the International Astronautical Federation, Rio de Janeiro, 2000

[ESA-1975] "14th SOL Meeting", ESA document 4164, 1 September 1975

[ESA-1976] "Out-of-Ecliptic Mission – Progress Report", ESA document 4327, 3 December 1976

[ESA-1979a] "Report on Studies for a Comet Mission to Halley and Tempel-2", document ESA 4214, 12 January 1979

[ESA-1979b] "International Comet Mission", document ESA 8047, containing correspondence, reports etc. dated between 22 March 1979 and 19 April 1980

[ESA-1979c] "31st SOL Meeting: Paris from 3 May to 4 May 1979", document ESA 4218, 25 June 1979

[ESA-1979d] "Ad-hoc Panel on polar orbiters of the Moon and Mars", document ESA 4743, July 1979

[ESA-1979e] "Exploration of Mars", document ESA 4712, 22 November 1979

[ESA-1980] "32nd SOL meeting : Noordwijk on 22/11/1979", document ESA 4221, 31 January 1980

[Esposito-1994] Esposito, P.B., et al., "Navigating Mars Observer: Launch Through Encounter and Response to the Spacecraft's Pre-Encounter Anomaly", Paper AAS 94-119

[Esposito-1999] Esposito, P. et al., "Navigating Mars Global Surveyor Through the Martian Atmosphere: Aerobraking 2", paper AAS 99-443

[Etchegaray-1987] Etchegaray, M.I. (ed.), "Preliminary Scientific Rationale for a Voyage to a Thousand Astronomical Units", Pasadena, JPL, 1987

[Evans-2001] Evans, L.G., et al., "Elemental Composition from Gamma-Ray Spectroscopy of the NEAR–Shoemaker Landing Site on 433 Eros", Meteoritics & Planetary Science, 36, 2001, 1639–1660

[Farquhar-1976] Farquhar, R.W., Muhoen, D.P., Richardson, D.L., "Mission Design for a Halo Orbiter of the Earth", Paper AIAA 76–810

[Farquhar-1983] Farquhar, R., "ISEE-3 A Late Entry in the Great Comet Chase", Astronautics & Aeronautics, September 1983, 50–55

[Farquhar-1995] Farquahr, R.W, Dunham, D.W., McAdams, J.V., "NEAR Mission Overview and Trajectory Design", Paper AAS 95-378

[Farquhar-1999] Farquhar, R.W., "The use of Earth-return trajectories for missions to comets", Acta Astronautica, 44, 1999, 607–623

[Farquhar-2001] Farquhar, R.W., "The Flight of ISEE-3/ICE: Origins, Mission History, and a Legacy", The Journal of the Astronautical Sciences, 49, 2001, 23–73

[Ferrin-1988] Ferrin, I., Gil, C., "The Aging of Comets Halley and Encke", Astronomy and Astrophysics, 194, 1988, 288–296

[Festou-1986] Festou, M.C., et al., "IUE Observations of Comet Halley During the Vega and Giotto Encounters", Nature, 321, 1986, 361–363

[Fischer-1996] Fischer, H.M., et al., "High-Energy Charged Particles in the Innermost Jovian Magnetosphere", Science, 272, 1996, 856–858

[Flight-1987] "Mars Observer Delayed", Flight International, 28 March 1987, 135

[Flight-1988] "ESA Nears Science Mission Decision", Flight International, 19 November 1988, 21

[Flight-1989a] "Phobos 2 in Trouble", Flight International, 14 January 1989, 13

[Flight-1989b] "Dress Rehearsal Proves Magellan", Flight International, 20 June 1990, 21

[Flight-1989c] "Soviets Turn from Mars to Moon", Flight International, 7 January 1989, 7

[Flight-1989d] "Greece in Space", Flight International, 14 October 1989, 32

[Flight-1990] "Mars-Race Spacecraft Needs $10 Million Funding", Flight International, 4 April 1990, 28

[Flight-1991a] "Galileo High-Gain Antenna Failure Blurs Jupiter", Flight International, 8 May 1991, 10

[Flight-1991] "Dornier Wins Mars Camera Contract", Flight International, 4 December 1991, 16

[Flight-1992a] "Successful Giotto Set for Third Comet", Flight International, 22 July 1992, 17

[Flight-1992b] "NASA/Russian Mars Agreement", Flight International, 21 October 1992, 23

[Flight-1993] "India Aims at Mercury", Flight International, 31 March 1993, 19

[Flight-1994] "Martin Marietta Drops Observer Claim", Flight International, 19 January 1994, 23

[Flight-1997] "Mars Find", Flight International, 22 January 1997, 22

[Folkner-1997a] Folkner, W.M., et al., "Earth-Based Radio Tracking of the Galileo Probe for Jupiter Wind Estimation", Science, 275, 1997, 644–646

[Folkner-1997b] Folkner, W.M., et al., "Interior Structure and Seasonal Mass Redistribution of Mars from Radio Tracking of Mars Pathfinder", Science, 278, 1997, 1749–1751

[Forward-1986] Forward, R.L., "Feasibility of Interstellar Travel: A Review", Journal of the British Interplanetary Society, 39, 1986, 379–384

[Frank-1996] Frank, L.A., et al., "Plasma Observations at Io with the Galileo Spacecraft", Science, 274, 1996, 394–395

[Freese-2000] Freese, J.J., "The Viability of U.S. Anti-Satellite (ASAT) Policy: Moving Toward Space Control", USAF Institute of National Security Studies, INSS Occasional Paper 30, January 2000, 21–22

[Friedlander-1971] Friedlander, A.L., Niehoff, J.C., Waters, J.I., "Trajectory Requirements for Comet Rendezvous", Journal of Spacecraft, 8, 1971, 858–866

[Friedman-1980] Friedman, L.D., "A Proposal for a U.S. Initiative: The International Halley Watch", Paper AIAA-80-0113

[Friedman-1988] Friedman, L., "Starsailing: Solar Sails and Interstellar Travel", New York, Wiley Science, 1988

[Friedman-1993a] Friedman, L.D., "What Happened to Mars Observer?", The Planetary Report, November/December 1993, 4

[Friedman-1993b] Friedman, L.D., "Loss of Mars Balloon Relay Will Affect Mars '94 and '96 Missions", The Planetary Report, November/December 1993, 5

[Friedman-1994] Friedman, L.D., "Cleverness: Colombo's Legacy to Mission Design", paper presented at the Memorial Conference "Ideas for Space Research after the Year 2000", Padua, 18–19 February 1994

[Furniss-1987a] Furniss, T., "Countdown to Co-operation", Flight International, 5 December 1987, 30–33

[Furniss-1987b] Furniss, T., "Soviets Plan 1992 Mars Rover" Flight International, 24 October 1987, 35
[Furniss-1987c] Furniss, T., "Phobos – The Most Ambitious Mission", Flight International, 27 June 1987, 43–45
[Furniss-1990a] Furniss, T., "Infernal Device", Flight International, 26 September 1990, 40–42
[Furniss-1990b] Furniss, T., "Shuttle Clear to Launch Ulysses", Flight International, 19 September 1990, 17
[Furniss-1990c] Furniss, T., "Discovery Gives a Boost to Ulysses", Flight International, 17 October 1990, 10
[Furniss-1990d] Furniss, T., "Aerobraking Development Begins", Flight International, 31 January 1990, 17
[Furniss-1992a] Furniss, T., "Return to the Red Planet", Flight International, 2 September 1992, 149–150
[Furniss-1992b] Furniss, T., "Mars Observer en Route after Scars", Flight International, 7 October 1992, 21
[Furniss-1993] Furniss, T., "Low Cost Discoveries", Flight International, 17 March 1993, 27
[Furniss-1996] Furniss, T., "Mars Probe May Be to Blame for Failure", Flight International, 27 November 1996, 26
[Furniss-1997] Furniss, T., "The Mars Burn", Flight International, 26 November 1997, 49
[Galeev-1990] Galeev, A.A., et al., "K Solntsu!" (To the Sun!), Nauka v SSSR, No.1, 1990, page unknown. (in Russian)
[Galeev-1995] Galeev, A.A., et al., "Russian Programs of Planetary Exploration: Mars-94/98 Missions", Acta Astronautica, 35, 1995, 9–33
[Garcia-1991] Garcia, H.A., Fárník, F., "Stereoscopic Measurements of Flares from Phobos and GOES", Solar Physics, 131, 1991, 137–148
[Garvin-1988] Garvin, J.B., Ulaby, F.T., "Dielectric Properties of Meteorites: Implications for Radar Observations of Phobos", paper presented at the Lunar and Planetary Science Conference XIX, Houston, 1988
[Geenty-2005] Geenty, J., "Flights of Fancy. The Lost Space Shuttle Missions of 1986", Spaceflight, 47, January 2005, 26–32
[Geissler-1998] Geissler, P.E., et al., "Evidence for Non-Synchronous Rotation of Europa", Nature, 391, 1998, 368–370
[Geissler-1999] Geissler, P.E., et al., "Galileo Imaging of Atmospheric Emissions from Io", Science, 285, 1999, 870–874
[Gel'man-1986] Gel'man, B.G., et al., "Reaction Gas Chromatography of Venus Cloud Aerosols" Soviet Astronomy Letters, 12, 1986, 42–43
[Gianvanni-1990] Gianvanni, P., "Capitana Italica verso Marte?" (Italian Flagship to Mars?), JP4, January 1990, 10 (in Italian)
[Gierasch-2000] Gierasch, P.J., et al., "Observation of Moist Convection in Jupiter's Atmosphere", Nature, 403, 2000, 628–630
[Giorgini-1995] Giorgini, J., et al., "Magellan Aerobrake Navigation", Journal of the British Interplanetary Society, 48, 1995, 111–122
[Goldman-1994] Goldman, S. J.: "Clementine Maps the Moon", Sky & Telescope, August 1994, 20–24
[Goldman-1997] Goldman, S.J., "A Sol in the Life of Pathfinder", Sky & Telescope, November 1997, 32–34
[Golombek-1997] Golombek, M.P., et al., "Overview of the Mars Pathfinder Mission and Assessment of Landing Site Predictions", Science, 278, 1997, 1743–1748

[Golombek-1998] Golombek, M.P., "The Mars Pathfinder Mission", Scientific American, July 1998, 40–49
[Gore-1986] Gore, R., "Halley's Comet 1986 – More than Met the Eye", National Geographics, December 1986, 758–785
[Goy-1990] Goy, F., "Regata per Marte" (Mars Regatta), Volare, June 1990, 42–45 (in Italian)
[Graps-2000] Graps, A.L., et al., "Io as a Source of the Jovian Dust Streams", Nature, 405, 2000, 48–49
[Grard-1982] Grard, R., "Kepler – A Mission to the Planet Mars", ESA Bulletin, 32, 1982, 22–24
[Grard-1986] Grard, R., et al. "Observations of Waves and Plasma in the Environment of Comet Halley", Nature, 321, 1986, 290–291
[Grard-1988] Grard, R., "The Vesta Mission – A Visit to the Small Bodies of the Solar System", ESA Bulletin, No. 55, 1988, 36–40
[Grard-1989a] Grard, R., et al., "First Measurements of Plasma Waves near Mars", Nature, 341, 1989, 607–609
[Grard-1989b] Grard, R.J.L., Marsden, R.G., "The Phobos Mission: First Results from the Plasma Wave System and Low-Energy Telescope", ESA Bulletin, 56, 1989, 81–82
[Grard-1994] Grard, R., Scoon, G., Coradini, M., "Mercury Orbiter – An Interdisciplinary Mission", ESA Journal, 18, 1994, 197–205
[Gringauz-1986] Gringauz, K.I., et al. "First In Situ Plasma and Neutral Gas Masurements at Comet Halley", Nature, 321, 1986, 282–285
[Grün-1993] Grün, E., et al., "Discovery of Jovian Dust Streams and Interstellar Grains by the Ulysses Spacecraft", Nature, 362, 1993, 428–430
[Grün-1994] Grün, E., et al., "Interstellar Dust in the Heliosphere", Astronomy and Astrophysics, 286, 1994, 915–924
[Grün-1996] Grün, E., et al., "Dust Measurements During Galileo's Approach to Jupiter and Io Encounter", Science, 274, 1996, 399–401
[Guernsey-2001] Guernsey, C.S., "Propulsion Lessons Learned from the Loss of Mars Observer", Paper AIAA-2001-3630
[Gurnett-1991] Gurnett, D.A., et al., "Lightning and Plasma Wave Observations from the Galileo Flyby of Venus", Science, 253, 1991, 1522–1525
[Gurnett-1996a] Gurnett, D.A., et al., "Galileo Plasma Wave Observations in the Io Plasma Torus and Near Io", Science, 274, 1996, 391–392
[Gurnett-1996b] Gurnett, D.A., et al., "Evidence for a Magnetosphere at Ganymede from Plasma-Wave Observations by the Galileo Spacecraft", Nature, 384, 1996, 535–537
[Gurnett-1997] Gurnett, D.A., et al., "Absence of a Magnetic-Field Signature in Plasma-Wave Observations at Callisto", Nature, 387, 1997, 261–262
[Gurnett-2002] Gurnett, D.A., et al., "Control of Jupiter's Radio Emission and Aurorae by the Solar Wind", Nature, 415, 2002, 985–987
[Hainaut-2004] Hainaut, O.R., et al., "Post-Perihelion Observations of Comet 1P/Halley. V: rh = 28.1 AU", Astronomy and Astrophysics, 417, 2004, 1159–1164
[Hainaut-2007] Hainaut, O.R., Personal communication with the author, 9 June 2007
[Harland-2000a] Harland, D.M., "Jupiter Odissey: The Story of NASA's Galileo Mission", Chichester, Springer–Praxis, 2000, 45–49
[Harland-2000b] ibid., 57–62
[Harland-2000c] ibid., 72–78
[Harland-2000d] ibid., 111–125
[Harland-2000e] ibid., 138–143, 186–189, 250–255 and 301–306

[Harland-2000f] ibid., 190–196
[Harland-2000g] ibid., 265
[Harland-2000h] ibid., 204–207
[Harland-2000i] ibid., 265–268
[Harland-2000j] ibid., 155–160
[Harland-2000k] ibid., 268
[Harland-2000l] ibid., 313
[Harland-2000m] ibid., 173–174
[Harland-2000n] ibid., 313–315
[Harland-2000o] ibid., 285–288
[Harland-2000p] ibid., 208–211
[Harland-2000q] ibid., 219–222
[Harland-2000r] ibid., 223–226
[Harland-2000s] ibid., 226–227
[Harland-2000t] ibid., 330–348
[Harvey-2000a] Harvey, B., "The Japanese and Indian Space programs", Chichester, Springer–Praxis, 2000, 3–37
[Harvey-2000b] ibid., 127–189
[Harvey-2007a] Harvey, B., personal communication with the author, 15 November 2007
[Harvey-2007b] Harvey, B., "Russian Planetary Exploration: History, Development, Legacy and Prospects", Chichester, Springer–Praxis, 2007, 251–252
[Harvey-2007b] ibid., 266–275
[Harvey-2007c] ibid., 281–284
[Hawkyard-1990] Hawkyard, A., Buia, P., "The Ulysses Spacecraft", ESA Bulletin, 63, 1990, 40–49
[Head-1991] Head, J.W., et al., "Venus Volcanism: Initial Analysis from Magellan Data", Science, 252, 1991, 276–288
[Head-1999] Head, J.W. III, et al., "Possible Ancient Oceans on Mars: Evidence from Mars Orbiter Laser Altimeter Data", Science, 286, 1999, 2134–2137
[Head-2001] Head, J., et al., "Ganymede: Very High Resolution Data from G28 Reveal New Perspectives on Processes and History", paper presented at the XXXII Lunar and Planetary Science Conference, Houston, 2001
[Henderson-1989] Henderson, B.W., "NASA Scientists Hope Mars Rover Will be Precursor to Manned Flight", Aviation Week & Space Technology, 9 October 1989, 85–94
[Hengeveld-2005] Hengeveld, E., "The Reluctant Space Shuttle", Spaceflight, December 2005, 460–464
[Hill-2002] Hill, T.W., "Magnetic Moments at Jupiter", Nature, 415, 2002, 965–966
[Hirao-1984] Hirao, K., "The Suisei/Sakigake (Planet-A/MS-T5) Missions". In: "Space Missions to Halley's Comet", Noordwijk, ESA SP-1066, 1984
[Hirao-1986] Hirao, K., Itoh, T. "The Planet-A Halley Encounters", Nature, 321, 1986, 294–297
[Hirao-1987] Hirao, K., Itoh, T., "The Sakigake/Suisei Encounter with Comet P/Halley", Astronomy & Astrophysics, 187, 1987, 39–46
[Hoppa-1999] Hoppa, G.V., et al., "Formation of Cycloidal Features on Europa", Science, 285, 1999, 1899–1902
[Hoppa-2000] Hoppa, G.V., et al., "Europa's Sub-Jovian Hemisphere from Galileo I25: Tectonic and Chaotic Surface Features", paper presented at the Lunar and Planetary Science Conference XXXI, Houston, 2000
[Hoppa-2001] Hoppa, G.V., et al., "Europa's Rate of Rotation Derived from the Tectonic Sequence in the Astypalaea Region", Icarus, 153, 2001, 208–213

[Houpis-1986] Houpis, H.L.F., Gombosi, T.I., "An Icy-Glue Nucleus Model of Comet Halley", In: ESA Proceedings of the 20th ESLAB Symposium on the Exploration of Halley's Comet. Volume 2: Dust and Nucleus, 1986, 397–401

[Hubbard-1992] Hubbard, G.S., et al., "Mars Environmental Survey (MESUR): Science objectives and mission description", paper presented at the Lunar and Planetary Institute Workshop on the Martian Surface and Atmosphere Through Time, 1992

[Hudson-2000] Hudson, R.S., et al., "Radar Observations and Physical Model of Asteroid 6489 Golevka", Icarus, 148, 2000, 37–51

[Hufbauer-1992] Hufbauer, K., "European Space Scientists and the Genesis of the Ulysses Mission, 1965–1979". In: Russo, A. (ed.), "Science Beyond the Atmosphere: the History of Space Research in Europe", ESA, Proceedings of a Symposium held in Palermo, 5–7 November 1992

[Hughes-1980] Hughes, D., "Mission to the Comets", New Scientist, 10 January 1980, 66–69

[Hviid-1997] Hviid, S.F., et al., "Magnetic Properties Experiments on the Mars Pathfinder Lander: Preliminary Results", Science, 278, 1997, 1768–1770

[IAUC-3737] "International Astronomical Unit Circular No. 3737", 21 October 1982

[IAUC-3937] "International Astronomical Unit Circular No. 3937", 12 April 1984

[IAUC-7243] "International Astronomical Unit Circular No. 7243", 23 August 1999

[IAUC-8107] "International Astronomical Unit Circular No. 8107", 4 April 2003

[Iorio-2007] Iorio, L., "High-Precision Measurement of Frame-Dragging with the Mars Global Surveyor Spacecraft in the Gravitational Field of Mars", Arxiv pre-print gr-qc/0701042

[Ivanov-1988] Ivanov, M.A., "The Results of Morphometric Study of the Tessera Terrain of Venus from Venera 15/16 Data", paper presented at the XIX Lunar and Planetary Science Conference, Houston, 1988

[Ivanov-1990] Ivanov, B.A., "Venusian Impact Craters on Magellan Images: View from Venera 15/16", Earth Moon and Planets, 50/51, 1990, 159–173

[Izenberg-1998] Izenberg, N.R., Anderson, B.J., "NEAR Swings by Earth en Route to Eros", Eos Transaction American Geophysical Union, 79, 1998, 289–295

[Jaffe-1980] Jaffe, L.D., et al., "An Interstellar Precursor Mission", Journal of the British Interplanetary Society, 33, 1980, 3–26

[James-1982] James, W., "Unveiling Venus with VOIR", Sky & Telescope, February 1982, 141–144

[Janin-1984] Janin, G., "Towards the Halley Comet". In: "Mathématiques spatiales pour la préparation et la réalisation de l'exploitation des satellites/Space mathematics for the preparation and the development of satellites exploration", Toulouse, Cépaduès, 1984, 1051–1071

[Jaroff-1997] Jaroff, L., "Dreadful Sorry, Clementine", Time Magazine, 27 October 1997, page unknown

[Jenkins-2002] Jenkins, R.M., "The Giotto Spacecraft plus 'Why was Giotto Special?'", Space Chronicle, 55, 2002, 12–30

[Johnson-1991] Johnson, T.V., et al., "The Galileo Venus Encounter", Science, 253, 1991, 1516–1518

[Johnston-1998] Johnston, M.D., et al., "Mars Global Surveyor Aerobraking at Mars", Paper AAS 98-112

[Johnstone-1986] Johnstone, A., et al., "Ion Flow at Comet Halley", Nature, 321, 1986, 344–347

[Jones-2002] Jones, G.H., "Ulysses's Encounter with Comet Hyakutake", paper presented at the Asteroids, Comets, Meteors – ACM 2002 International Conference, 29 July–2 August 2002

[Jones-2003] Jones, G.H., et al., "Possible Distortion of the Interplanetary Magnetic Field by the Dust Trail of Comet 122P/De Vico", The Astrophysical Journal, 597, 2003, L61–L64
[JP4-1992] "Tecnospazio e le Comete" (Tecnospazio and Comets), JP4, April 1992, 11 (in Italian)
[JPL-1978] "Solar Polar Fact Sheet", Pasadena, JPL, 26 July 1978
[JPL-1991] "Outward to the Beginning: The CRAF and Cassini Missions", JPL Brochure 400-341, June 1991
[JPL-1993] "Mars Observer Loss of Signal: Special Review Board Final Report", JPL Publication 93–28, November 1993
[JWG-1986a] Joint Working Group on Cooperation in Planetary Exploration, "United States and Western Europe Cooperation in Planetary Exploration", Washington, National Academic Press, 1986
[JWG-1986b] ibid., 146–147
[JWG-1986c] ibid., 149–157
[JWG-1986d] ibid., 59–64
[Kahn-1997] Kahn, R., "A Martian Mystery", Sky & Telescope, October 1997, 38–39
[Kamoun-1982] Kamoun, P., et al., "Comet Grigg–Skjellerup: Radar Detection of the Nucleus", Bulletin of the American Astronomical Society, 14, 1982, 753
[Kaneda-1986] Kaneda, E., et al., "Strong Breathing of the Hydrogen Coma of Comet Halley", Nature, 320, 1986, 140–141
[Kargel-1997] Kargel, J.S., "The Rivers of Venus", Sky & Telescope, August 1997, 32–37
[Kawashima-1993] Kawashima, N., et al., "Development/Drilling Test of Auger Boring Machine on Board Mars Rover for Mars Exploration". In: "Missions, Technologies et Conception des Vehicules Mobiles Planetaires", Toulouse, Cépaduès, 1993
[Keating-1998] Keating, G.M., et al., "The Structure of the Upper Atmosphere of Mars: In Situ Accelerometer Measurements from Mars Global Surveyor", Science, 279, 1998, 1672–1676
[Keller-1986] Keller, H.U., et al., "First Halley Multicolour Camera Imaging Results from Giotto", Nature, 321, 1986, 320–326
[Keller-1988] Keller, H.U., Kramm, R., Thomas, N.., "Surface Features on the Nucleus of Comet Halley", Nature, 331, 1988, 227–231
[Kelly Beatty-1984] Kelly Beatty, J., "Radar Views of Venus", Sky & Telescope, February 1984, 110–112
[Kelly Beatty-1985a] Kelly Beatty, J., "A Radar Tour of Venus", Sky & Telescope, June 1985, 507–510
[Kelly Beatty-1985b] Kelly Beatty, J., "Comet G–Z: The Inside Story", Sky & Telescope, November 1985, 426–427
[Kelly Beatty-1993] Kelly Beatty, J., "Working Magellan's Magic", Sky & Telescope, August 1993, 16–20
[Kelly Beatty-1993b] Kelly Beatty, J., "The Long Road to Jupiter", Sky & Telescope, April 1993, 18–21
[Kelly Beatty-1995] Kelly Beatty, J., "Ida & Company", Sky & Telescope, January 1995, 20–23
[Kelly Beatty-1996a] Kelly Beatty, J., "Into the Giant", Sky & Telescope, April 1996, 20–22
[Kelly Beatty-1996b] Kelly Beatty, J., "Life from Ancient Mars?", Sky & Telescope, October 1996, 18–19
[Kelly Beatty-1999] Kelly Beatty, J., "In Search of Martian Seas", Sky & Telescope, November 1999, 38–41
[Kelly Beatty-2001] Kelly Beatty, J., "NEAR Falls for Eros", Sky & Telescope, May 2001, 34–37

[Kemurdjian-1992] Kemurdjian, A.L., et al., "Soviet Developments of Planet Rovers in Period of 1964–1990". In: "Missions, Technologies et Conception des Vehicules Mobiles Planetaires", Toulouse, Cépaduès, 1993

[Keppler-1986] Keppler, E., et al., "Neutral Gas Measurements of Comet Halley from Vega 1", Nature, 321, 1986, 273–274

[Kerr-1979] Kerr, R.A., "Planetary Science on the Brink Again", Science, 206, 1979, 1288–1289

[Kerr-1984] Kerr, R.A., "Probing the Long Tail of the Magnetosphere", Science, 226, 1984, 1298–1299

[Kerr-1985] Kerr, R.A., "New Plasma Physics Lab at Giacobini–Zinner", Science, 230, 1985, 51–52

[Kerr-1986] Kerr, R.A., "VEGA's 1 and 2 Visit Halley", Science, 231, 1986, 1366

[Kerr-1990] Kerr, R.A., "Will Magellan Find a Half-Sister of Earth's?", Science, 249, 1990, 742–744

[Kerr-1991] Kerr, R.A., "Magellan: No Venusian Plate Tectonics Seen", Science, 252, 1991, 213

[Kerr-1993] Kerr, R.A., "More Venus Science, or the Off Switch for Magellan?", Science, 259, 1993, 1696–1697

[Keszthelyi-1999] Keszthelyi, L., et al., "Revisiting the Hypothesis of a Mushy Global Magma Ocean on Io", Icarus, 141, 1999, 415

[Khurana-1997] Khurana, K.K., et al., "Absence of an Internal Magnetic Field at Callisto", Nature, 387, 1997, 262–264

[Khurana-1998] Khurana, K.K., et al., "Induced Magnetic Fields as Evidence for Subsurface Oceans in Europa and Callisto", Nature, 395, 1998, 777–780

[Kieffer-2000] Kieffer, S.W., et al. "Prometheus: Io's Wandering Plume", Science, 288, 2000, 1204-1208

[Killeen-1995] Killeen, T., Brace, L., "MUADEE: A Discovery-Class Mission for Exploration of the Upper Atmosphere of Mars", Acta Astronautica, 35, 1995, 377–386

[Kiseleva-2007] Kiseleva, T.P., Khrutskaya, E.V., "Pulkovo Astrometric Observations of Bodies in the Solar System from 1898 to 2005: Observational Database", Solar System Research, 41, 2007, 72–80

[Kissel-1986a] Kissel, J., et al., "Composition of Comet Halley Dust Particles from Giotto Observations", Nature, 321, 1986, 336–337

[Kissel-1986b] Kissel, J., et al., "Composition of Comet Halley Dust Particles from Vega Observations", Nature, 321, 1986, 280–282

[Kivelson-1993] Kivelson, M.G., et al., "Magnetic Field Signatures Near Galileo's Closest Approach to Gaspra", Science, 261, 1993, 331–334

[Kivelson-1996a] Kivelson, M.G., et al., "A Magnetic Signature at Io: Initial Report from the Galileo Magnetometer", Science, 273, 1996, 337–340

[Kivelson-1996b] Kivelson, M.G., et al., "Discovery of Ganymede's Magnetic Field by the Galileo Spacecraft", Nature, 384, 1996, 537–541

[Kivelson-1997] Kivelson, M.G., et al., "Europa's Magnetic Signature: Report from Galileo's Pass on 19 December 1996", Science, 276, 1997, 1239–1241

[Kivelson-2000] Kivelson, M.G., et al., "Galileo Magnetometer Measurements: A Stronger Case for a Subsurface Ocean at Europa", Science, 289, 2000, 1340–1343

[Klaes-1993] Klaes, L., "The Soviets and Venus – Part 3", Electronic Journal of the Astronomical Society of the Atlantic, April 1993

[Klimov-1986] Klimov, S., et al. "Extremely-Low-Frequency Plasma Waves in the Environment of Comet Halley", Nature, 321, 1986, 292–293

[Kolyuka-1991] Kolyuka, Yu, et al., "Phobos and Deimos Astrometric Observations from the Phobos Mission", Astronomy & Astrophysics, 244, 1991, 236–241

[Konopliv-1996] Konopliv, A.S., Sjogren, W.L., "Venus Gravity Handbook", Pasadena, JPL, 1996

[Korablev-2002] Korablev, O.I., "Solar Occultation Measurements of the Martian Atmosphere on the Phobos Spacecraft: Water Vapor Profile, Aerosol Parameters, and Other Results", Solar System Research, 36, 2002, 12–34

[Kotelnikov-1984] Kotelnikov, V.A., et al., "The Maxwell Montes Region, Surveyed by the Venera 15, Venera 16 Orbiters", Soviet Astronomy Letters, 10, 1984, 369–373

[Kovtunenko-1990] Kovtunenko, V.M., et al., "Unifitsirovanniy Avtomaticheskiy Kosmicheskiy Apparat Dlya Provedeniya Issledovaniy Dalniy Planet Solnechnoy Sistemy Meshplanetnogo Prostranstva i Solntsa (Proyekt 'Tsiolkovskii')", (Unified Space Probes to Observe the Distant Planets of the Solar System, the Interplanetary Space, and the Sun (Project 'Tsiolkovskii')), 1990 (?). (in Russian)

[Kovtunenko-1995] Kovtunenko, V.M., et al., "Opportunity to Create the System for Space Protection of the Earth Against Asteroids and Comets on the Base of Modern Technology", paper presented at the Planetary Defense Workshop, Lawrence Livermore National Laboratory, May 1995

[Kozicharow-1979] Kozicharow, E., "Timing, Budget Spur Solar Polar Mission", Aviation Week & Space Technology, 29 October 1979, 46–47

[Kraemer-2000a] Kraemer, R. S., "Beyond the Moon: A Golden Age of Planetary Exploration 1971–1978", Washington, Smithsonian Institution Press, 2000, 225

[Krasnopolsky-1986] Krasnopolsky, V.A., et al., "Spectroscopic Study of Comet Halley by the Vega 2 Three-Channel Spectrometer", Nature, 321, 1986, 269–271

[Krasnopolsky-1989] Krasnopolsky, V.A., et al., "Solar Occultation Spectroscopic Measurements of the Martian Atmosphere at 1.9 and 3.7 μm", Nature, 341, 1989, 603–604

[Kremnev-1986a] Kremnev, R.S., et al., "The Vega Balloons: A Tool for Studying Atmosphere Dynamics on Venus", Soviet Astronomy Letters, 12, 1986, 7–9

[Kremnev-1986b] Kremnev, R.S., et al., "VEGA Balloon System and Instrumentation", Science, 231, 1986, 1408–1411

[Kresak-1987] Kresak, L., "The 1808 Apparition and the Long-Term Physical Evolution of Periodic Comet Grigg–Skjellerup", Bulletin of the Astronomical Institute of Czechoslovakia, 38, 1987, 65–75

[Krimigis-1995] Krimigis, S.M., Veverka, J., "Foreword: Genesis of Discovery", Journal of Astronautical Sciences, 43, 1995, 345–347

[Krogh-2007] Krogh, K., "Iorio's 'High-Precision Measurement' of Frame-Dragging with the Mars Global Surveyor", Arxiv pre-print astro-ph/0701653

[Kronk-1984a] Kronk, G.W., "Comets: A Descriptive Catalog", Hillside, Henslow, 1984, 308–309

[Kronk-1984b] ibid., 254–255

[Kronk-1984c] ibid., 255–256

[Kronk-1984d] ibid., 248–249

[Kronk-1988a] Kronk, G.W., "Meteor Showers: A Descriptive Catalog", Hillside, Enslow, 1988, 189–194

[Kronk-1988b] ibid., 57–59

[Kronk-1999] Kronk, G.W., "Cometography: A Catalog of Comets" Volume 1: Ancient-1799, Cambridge University Press, 1999, 375

[Krüger-2005] Krüger, H., et al., "Dust Stream Measurements from Ulysses' Distant Jupiter Encounter", paper presented at the Dust in Planetary Systems conference, 26–28 September 2005, Kaua'i

[Krüger-2007] Krüger, H., et al., "Interstellar Dust in the Solar System", Arxiv astro-ph/0706.3310 preprint

[Krupp-2006] Krupp, E.C., "Lost in Space", Sky & Telescope, September 2006, 40–41

[Ksanfomality-1989] Ksanfomality, L.V., et al., "Spatial Variations in Thermal and Albedo Properties of the Surface of Phobos", Nature, 341, 1989, 588–591

[Kumar-1978] Kumar, S., "Science Strategy for Halley Flyby/Tempel-2 Rendezvous Mission", paper presented at the Workshop on Experimental Approaches to Comets, Houston, 11–13 September 1978

[Kurth-2002] Kurth, W.S., et al., "The Dusk Flank of Jupiter's Magnetosphere", Nature, 415, 2002, 991–994

[Kuznik-1985] Kuznik, F., "Visit to a Small Comet", Space World, July 1985, 23–26

[Lande-1995] Lande, A.L., "The Mars '94 Oxidant Experiment (MOx): Creation of Something from Nothing om 1 Year", Acta Astronautica, 35, 1995, 69–78

[Langevin-1983] Langevin, Y., "The New European Project for the Exploration of Asteroids: AGORA", paper presented at the XIV Lunar and Planetary Science Conference, Houston, 1983

[Lanzerotti-1996] Lanzerotti, L.J., et al., "Radio Frequency Signals in Jupiter's Atmosphere", Science, 272, 1996, 858–860

[Lanzerotti-1998] Lanzerotti, L.J., et al., "Spin Rate of Galileo Probe During Descent into the Atmosphere of Jupiter", Journal of Spacecraft and Rockets, 35, 1998, 100–102

[Laplace-1993] Laplace, H., Morelière, M., Gorse, C., "The Mars 96 Balloon Guiderope: an Autonomous System in Extreme Environment Conditions". In: "Missions, Technologies et Conception des Vehicules Mobiles Planetaires", Toulouse, Cépaduès, 1993

[Lardier-1992a] Lardier, C., "L'Astronautique Soviétique" (Soviet Astronautics), Paris, Armand Colin, 1992, 275 (in French)

[Lardier-1992b] ibid., 279–280

[Laros-1997] Laros, J.G., et al., "Gamma-Ray Burst Arrival Time Localizations: Simultaneous Observations by Mars Observer, Compton Gamma Ray Observatory, and Ulysses", The Astrophysical Journal Supplement Series, Vol. 110, May 1997, 157–161

[Lawler-2005] Lawler, A., "NASA Plans to Turn Off Several Satellites", Science, 307, 2005, 1541

[Lee-1996] Lee, W., "Mars Global Surveyor Project Mission Plan", JPL Document D-12088, November 1996

[Leertouwer-1990] Leertouwer, J.P., Eaton, D., "The Ulysses Launch Campaign", ESA Bulletin, 63, 1990, 56–59

[Lenorovitz-1985] Lenorovitz, J.M., "France Designing Spacecraft for Soviet Interplanetary Mission", Aviation Week & Space Technology, 7 October 1985, 50–51

[Lenorovitz-1986a] Lenorovitz, J.M., "Both Soviet Vega Spacecraft Relay New Data From Halley", Aviation Week & Space Technology, 17 March 1986, 18–20

[Lenorovitz-1986b] Lenorovitz, J.M., "Soviets Urge International Effort Leading to Manned Mars Mission", Aviation Week & Space Technology, 24 March 1986, 76–77

[Lenorovitz-1987a] Lenorovitz, J.M., "French Offer Balloon Platform for Use on Soviet Mars Mission", Aviation Week & Space Technology, 3August 1987, 63–65

[Lenorovitz-1987b] Lenorovitz, J.M., "Soviets Advance Definition Work on 1990s Unmanned Mars Mission", Aviation Week & Space Technology, 26 October 1987, 72–73

[Lenorovitz-1988] Lenorovitz, J.M., "Launch of Two Phobos Spacecraft Begins Ambitious Mission to Mars", Aviation Week & Space Technology, 18 July 1988, 16

[Lenorovitz-1989] Lenorovitz, J.M., "Soviets to Study Phobos Surface from Fixed-Site, Mobile Landers", Aviation Week & Space Technology, 29 August 1989, 48–49

[Lenorovitz-1993] Lenorovitz, J.M., "Asteroid Flyby Proposed Using LEAP Penetrators", Aviation Week & Space Technology, 28 June 1993, 27–28

[Lenorovitz-1994] Lenorovitz, J.M., "LEAP Lander Proposed", Aviation Week & Space Technology, 6 June 1994, 25–26

[Li-2002] Li, H., Robinson, M.S., Murchie, M., "Preliminary Remediation of Scattered Light in NEAR MSI Images", Icarus, 155, 2002, 244–252

[Linkin-1986] Linkin, V.M., et al, "Vertical Thermal Structure in the Venus Atmosphere from Provisional Vega 2 Temperature and Pressure Data" Soviet Astronomy Letters, 12, 1986, 40–42

[Logsdon-1989] Logsdon, J.M., "Missing Halley's Comet: The Politics of Big Science", Isis, 80, 1989, 254–280

[Lomberg-1996] Lomberg, J., "Visions of Mars", Sky & Telescope, December 1996, 30–34

[Lopes-Gaultier-2000] Lopes-Gaultier, R., et al., "A Close-Up Look at Io from Galileo's Near-Infrared Mapping Spectrometer", Science, 288, 2000, 1201–1204

[Lorenz-2006] Lorenz, R.D., "Spin of Planetary Probes in Atmospheric Flight", Journal of the British Interplanetary Society, 59, 2006, 273–282

[Lundin-1989] Lundin, R., et al., "First Measurements of the Ionospheric Plasma Escape from Mars", Nature, 341, 1989, 609–612

[Lundquist-2008] Lundquist, C.A., "Fred L. Whipple, Pioneer in the Space Program", Acta Astronautica, 62, 2008, 91–96

[Lusignan-1991] Lusignan, B., et al., (ed.), "The Stanford US-USSR Mars Exploration Initiative", Stanford University, 1991, Vol. 1, 721–725 and 734–743

[Luttmann-1992] Luttmann, H.W., "Russen Testen Sonnensegel" (The Russians to test solar sails), Flug Revue, October 1992, 86–87 (in German)

[Lyons-1999] Lyons, D.T., et al., "Mars Global Surveyor: Aerobraking Mission Overview", Journal of Spacecraft and Rockets, 36, 1999, 307–313

[Maeda-1993] Maeda, T., et al., "The Robotic Mars Rover". In: "Missions, Technologies et Conception des Vehicules Mobiles Planetaires", Toulouse, Cépaduès, 1993

[Maehl-1983] Maehl, R., "The TIROS-based Asteroid Mission", Spaceflight, December 1983, 430–435

[Maffei-1987a] For one of the best popular accounts of the history of Halley's comet see: Maffei, P., "La Cometa di Halley" (Halley's Comet), Milan, Mondadori, 1987, 149–315 (in Italian)

[Maffei-1987b] ibid., 362–363

[Malin-1991] Malin, M.C., et al., "Design and Development of the Mars Observer Camera", International Journal of Imaging Systems and Technology, 3, 1991, 76–91

[Malin-1998] Malin, M.C., et al., "Early Views of the Martian Surface from the Mars Orbiter Camera of Mars Global Surveyor", Science, 279, 1998, 1681–1685

[Malin-1999] Malin, M.C., "Visions of Mars", Sky & Telescope, April 1999, 42–49

[Malin-2000a] Malin, M.C., Edgett, K.S., "Sedimentary Rocks of Early Mars", Science, 290, 2000, 1927–1937

[Malin-2000b] Malin, M.C., Edgett, K.S., "Evidence for Recent Groundwater Seepage and Surface Runoff on Mars", Science, 288, 2000, 2330–2335

[Malin-2001a] Malin. M.C., Edgett, K.S., "Mars Global Surveyor Mars Orbiter Camera: Interplanetary Cruise through Primary Mission", Journal of Geophysical Research, 106, 2001, 23429–23570

[Malin-2001b] Malin, M.C., Caplinger, M.A., Davis, S.D., "Observational Evidence for an Active Surface Reservoir of Solid Carbon Dioxide on Mars", Science, 294, 2001, 2146–2148

[Malin-2003] Malin, M.C., Edgett, K.S., "Evidence for Persistent Flow and Aqueous Sedimentation on Early Mars", Science, 302, 2003, 1931–1934

[Malin-2006] Malin. M.C., et al., "Present-Day Impact Cratering Rate and Contemporary Gully Activity on Mars", Science, 314, 2006, 1573–1577
[Mama-1993] Mama, H.P., "An Indian Spacecraft to Mercury?", Spaceflight, June 1993, 211
[Maran-1985] Maran, S.P., "On the Trail of Comet G–Z", Sky & Telescope, September 1985, 198–203
[Marsden-1991] Marsden, R.G., Wenzel, K.-P., "First Scientific Results from the Ulysses Mission", ESA Bulletin, 67, 1991, 78–83
[Marsden-1992] Marsden, R.G., Wenzel, K.-P., "The Ulysses Jupiter Flyby – The Scientific Results", ESA Bulletin, 72, 1992, 52–59
[Marsden-1995] Marsden, R.G., "Ulysses Explores the South Pole of the Sun", ESA Bulletin, 82, 1995, 48–55
[Marsden-1996] Marsden, R.G., Smith, E.J., "Ulysses: Solar Sojourner", Sky & Telescope, March 1996, 24–30
[Marsden-1997] Marsden, R.G., Wenzel, K.-P., Smith, E.J., "The Heliosphere in Perspective – Key Results from the Ulysses Mission at Solar Minimum", ESA Bulletin, 92, 1997, 75–81
[Marsden-2000] Marsden, R.G., "Ulysses at Solar Maximum and Beyond", ESA Bulletin, 103, 2000, 41–47
[Marsden-2003] Marsden, R.G., Smith, E.J., "News from the Sun's Poles Courtesy of Ulysses", ESA Bulletin, 114, 2003, 61–67
[Martin-1995] Martin, T.Z., et al., "Observations of Shoemaker–Levy Impacts by the Galileo Photopolarimeter Radiometer", Science, 268, 1995, 1875–1879
[Mastal-1990] Mastal, E.F., Campbell, R.W., "RTGs – The Powering of Ulysses", ESA Bulletin, 63, 1990, 50–55
[Mazets-1986] Mazets, E.P., "Comet Halley Dust Environment from SP-2 Detector Measurements", Nature, 321, 1986, 276–278
[McBride-1997] McBride, N., et al., "The Inner Dust Coma of Comet 26P/Grigg–Skjellerup: Multiple Jets and Nucleus Fragments?", Monthly Notices of the Royal Astronomical Society, 289, 1997, 535–553
[McComas-2006] McComas, D.J., "Solar Probe: A Long Time Coming", Astronomy, December 2006, 47
[McCord-1998] McCord, T.B., et al., "Salts on Europa's Surface Detected by Galileo's Near Infrared Mapping Spectrometer", Science, 280, 1998, 1242–1245
[McCord-2001] McCord, T.B., Hansen, G.B., Hibbitts, C.A., "Hydrated Salt Minerals on Ganymede's Surface: Evidence of an Ocean Below", Science, 292, 2001, 1523–1525
[McCullogh-2007] McCullogh, M.E., "Can the Flyby Anomalies be Explained by a Modification of Inertia?" Arxiv pre-print astro-ph/0712.3022
[McCurdy-2005a] McCurdy, H.E., "Low-Cost Innovation in Spaceflight: the Near Earth Asteroid Rendezvous (NEAR) Shoemaker Mission", Washington, NASA, 2005, 6
[McCurdy-2005b] ibid., 18–19
[McCurdy-2005c] ibid., 14–15
[McCurdy-2005d] ibid., 35–37
[McCurdy-2005e] ibid., 47–49
[McDonnell-1986] McDonnell, J.A.M., et al., "Dust Density and Mass Distribution near Comet Halley from Giotto Observations", Nature, 321, 1986, 338–341
[McDonnell-1987] McDonnell, J.A.M., et al., "The Dust Distribution within the Inner Coma of Comet P/Halley 1982i: Encounter by Giotto's Impact Detectors", Astronomy and Astrophysics, 187, 1987, 719–741
[McDonnell-1993] McDonnell, J.A.M., "Dust Particle Impacts During the Giotto Encounter with Comet Grigg–Skjellerup", Nature, 362, 1993, 732–734

[McDonnell Douglas-1995] "Kilauea: A Terrestrial Analogue for Planetary Exploration", McDonnell Douglas brochure, Undated but probably 1995
[McEwen-1998] McEwen, A.S., et al., "High-Temperature Silicate Volcanism on Jupiter's Moon Io", Science, 281, 1998, 87–90
[McEwen-2000] McEwen, A.S., et al., "Galileo at Io: Results from High-Resolution Imaging", Science, 288, 2000, 1193–1198
[McEwen-2002] McEwen, A.S., "Active Volcanism on Io", Science, 297, 2002, 2220–2221
[McFadden-1993] McFadden, L.A., et al. "The enigmatic object 2201 Oljato: Is it an asteroid or an evolved comet?" Journal of Geophysical Research, 98, 1993, E2, p. 3031–3041
[McGarry-1997] McGarry, A., Angold, N., "Ulysses 7 Years On – Operational Challenges and Lessons Learned", ESA Bulletin, 92, 1997, 69–74
[McGarry-2004a] McGarry, A., Castro, F., Hodges, M., "Hydrazine Operations at Near-Freezing Temperatures During the Ulysses Extended Mission", paper presented at the 4th International Spacecraft Propulsion Conference, 2–9 June 2004 Chia Laguna
[McGarry-2004b] McGarry, A., Castro, F., Hodges, M.L., "Increasing Science with Diminishing Resources – Extending the Ulysses Mission to 2008", paper presented at the SpaceOps 2004 Conference, 17–21 May 2004, Montreal
[McInnes-2003] McInnes, C.R., "Solar Sailing: Mission Applications and Engineering Challenges", Philosophical Transactions of the Royal Society of London, 361, 2003, 2989–3008
[McKay-1996] McKay, D.S., et al., "Search for Past Life on Mars: Possible Relic Biogenic Activity in Martian Meteorite ALH84001", Science, 273, 1996, 924–930
[McKenna-Lawlor-2002] McKenna-Lawlor, S.M.P., "Overview of the Observations Made by the EPONA Instrument During the Giotto/GEM Mission", Space Chronicle, 55, 2002, 51–69
[McLaughlin-1984] McLaughlin, W.I., Randolph, J.E., "Starprobe: to Confront the Sun", Journal of the British Interplanetary Society, 37, 1984, 375–380
[McLaughlin-1985] McLaughlin, W., "Near Earth Asteroid Rendezvous", Spaceflight, December 1985, 440–441
[McLaughlin-1992] McLaughlin, W.I., "Ulysses Swings by Jupiter", Spaceflight, May 1992, 166–167
[Mecham-1989] Mecham, M., "Mars Observer Begins New Era Using Proven Spacecraft Design", Aviation Week & Space Technology, 9 October 1989, 79–82
[Meltzer-2007a] Meltzer, M., "Mission to Jupiter: A History of the Galileo Project", Washington, NASA, 2007, 9–36
[Meltzer-2007b] ibid., 37–59 and 65–66
[Meltzer-2007c] ibid., 61–62
[Meltzer-2007d] ibid., 118–148
[Meltzer-2007e] ibid., 66–68
[Meltzer-2007f] ibid., 71–84
[Meltzer-2007g] ibid., 94
[Meltzer-2007h] ibid., 96–103
[Meltzer-2007i] ibid., 151–152
[Meltzer-2007j] ibid., 171–179
[Meltzer-2007k] ibid., 180–181
[Meltzer-2007l] ibid., 195–197
[Meltzer-2007m] ibid., 202–209
[Meltzer-2007n] ibid., 209–221
[Mendillo-2006] Mendillo, M., et al., "Effects of Solar Flares on the Ionosphere of Mars", Science, 311, 2006, 1135–1138

[Michielsen-1968] Michielsen, H.F., "A Rendezvous with Halley's Comet in 1985–1986", Journal of Spacecraft, 5, 1968, 328–334
[Milazzo-2002] Milazzo, M.P., et al., "Eruption Temperatures at Tvashtar Catena, Io From Galileo I25 and I27", paper presented at the Lunar and Planetary Science Conference XXXIII, Houston, 2002
[Mishkin-2003a] Mishkin, A., "Sojourner: An Insider's View of the Mars Pathfinder Mission", New York, Berkeley Book, 2003, 13–37
[Mishkin-2003b] ibid., 38–51
[Mishkin-2003c] ibid., 65
[Mishkin-2003d] ibid., 57–58
[Mishkin-2003e] ibid., 66–81
[Mishkin-2003f] ibid., 95
[Mishkin-2003g] ibid., 134–144
[Mishkin-2003h] ibid., 97–123
[Mishkin-2003i] ibid., 248–249
[Mishkin-2003j] ibid., 282–301
[Mishkin-2003k] ibid., 301–303
[Mitchell-1998] Mitchell, R.T.., et al., "Project Galileo The Europa Mission", paper presented at the XLIX Congress of the International Astronautical Federation, Melbourne, 1998
[Morabito-2000] Morabito, D., et al., "The 1998 Mars Global Surveyor Solar Corona Experiment", JPL TDA Progress Report 42-142, 2000, 1–18
[Mordovskaya-2002a] Mordovskaya, V.G., Oraevsky, V.N., Styashkin, V.A., "The Peculiarities of the Interaction of Phobos with the Solar Wind are Evidence of the Phobos Magnetic Obstacle (from Phobos-2 Data)", Arxiv pre-print astro-ph/0212072
[Mordovskaya-2002b] Mordovskaya, V.G., Oraevsky, "In Situ Measurements of the Phobos Magnetic Field During the Phobos-2 Mission", Arxiv pre-print astro-ph/0212073
[Moreels-1986] Moreels, G., et al., "Near-Ultraviolet and Visible Spectrophotometry of Comet Halley from Vega 2", Nature, 321, 1986, 271–272
[Moshkin-1986] Moshkin, B.E., et al., "Vega 1, 2 Optical Spectrometry of Venus Atmospheric Aerosols at the 60–30 km Levels: Preliminary Results" Soviet Astronomy Letters, 12, 1986, 36–39
[MSSS-1996] "Mars 96", Malin Space Science Systems Internet site
[Mudgway-2001a] Mudgway, D.J., "Uplink-Downlink A History of the Deep Space Network 1957–1997", Washington, NASA, 2001, 216–219
[Mudgway-2001b] ibid., 280–281
[Mudgway-2001c] ibid., 324–326
[Mudgway-2001d] ibid., 329
[Muenger-1985] Muenger, E.A., "Searching the Horizon: A History of Ames Research Center 1940–1976", Washington, NASA, 1985, 250
[Münch-1986] Münch, R.E., Sagdeev, R.Z., Jordan, J.F., "Pathfinder: Accuracy Improvement of Comet Halley Trajectory for Giotto Navigation", Nature, 321, 1986, 318–320
[Murray-1989a] Murray, B., "Journey into Space", New York, W.W. Norton & C., 1989, 125–129
[Murray-1989b] ibid., 243–251
[Murray-1989c] ibid., 257–263
[Murray-1989d] ibid., 271–273
[Murray-1989e] ibid., 185–219
[Murray-1989f] ibid., 221–237
[Naeye-2007] Naeye, R., "Flowing Water on Today's Mars?", Sky & Telescope, March 2007, 17

[NASA-1966] "Space Flight Handbooks Vol. III Part 5: Trajectories to Jupiter, Ceres and Vesta", NASA, 1966
[NASA-1980] "To Explore Venus – Venus Orbiting Imaging Radar Mission", NASA Brochure 1060-145, July 1980
[NASA-1986] "Space Shuttle Mission STS-51L Press Kit", Washington, NASA, 1986
[NASA-1987] "A Preliminary Study of Mars Rover/Sample Return Missions", Washington, NASA, January 1987
[NASA-1993a] "Mars Observer Mars Orbit Insertion Press Kit", Washington, NASA, August 1993
[NASA-1993b] "Mars Observer Mission Failure Investigation Board Report", Washington, NASA, 31 December 1993
[NASA-1993c] "Discovery Program Workshop Summary Report", NASA TM-108233, 1993
[NASA-1994] Joint U.S./Russian Technical Working Groups, "Mars Together and Fire & Ice", NASA CR-19884, October 1994, 65–90
[NASA-1995] "Near-Earth Asteroid Returned Sample (NEARS) Final Technical Report", NASA CR-197297, 1995
[NASA-1997] "Mars Pathfinder Landing Press Kit", Washington, NASA, July 1997
[NASA-2003] "Galileo End of Mission Press Kit", Washington, NASA, September 2003
[NASA-2007] "Mars Global Surveyor (MGS) Spacecraft Loss of Contact", NASA Release, 13 April 2007
[Nasirov-1989] Nasirov, P.P., et al., "Unikal'nyi Eksperiment Pa Nevestoy Mekhanike" (A unique experiment in celestial mechanics), Zemliya i Vselennaya, 1989, 6, page unknown (in Russian)
[Naudet-1996] Naudet, C.J., Border, J.S., Woo, R., "Magellan Radio Scattering Measurements in the Solar Wind", paper presented at the Spring 1996 Meeting of the American Geophysical Union
[Nelson-1995] Nelson, R.M., et al., "Hermes Global Orbiter: A Discovery Mission in Gestation", Acta Astronautica, 35, 1995, 387–395
[Nelson-1997] Nelson, R.M., "Mercury: The Forgotten Planet", Scientific American, November 1997, 56–67
[Nelson-2001] Nelson, R.L., Whittenburg, K.E., Holdridge, M.E., "433 Eros Landing: Development of NEAR–Shoemaker's Controlled Descent Sequence", paper presented at the XV Annual AIAA/USU Conference on Small Satellites, Logan, 2001
[Neubauer-1986] Neubauer, F.M., et al., "First Results from the Giotto Magnetometer Experiment at Comet Halley", Nature, 321, 1986, 352–355
[Neugebauer-1983] Neugebauer, M., "Mariner Mark II and the Exploration of the Solar System", Science, 219, 1983, 443–449
[Neugebauer-2007] Neugebauer, M., et al., "Encounter of the Ulysses Spacecraft with the Ion Tail of Comet McNaught", Center for Solar-Terrestrial Research preprint, 2007
[Neukum-1996] Neukum, G., et al., "The Experiments HRSC and WAOSS on the Russian Mars 94/96 Missions", Acta Astronautica, 38, 1996, 713–720
[Niemann-1996] Niemann, H.B., et al., "The Galileo Probe Mass Spectrometer: Composition of Jupiter's Atmosphere", Science, 272, 1996, 846–849
[Nock-1987] Nock, K.T., "TAU – A Mission to a Thousand Astronomical Units", Paper AIAA-87-1049
[NRC-1998a] National Research Council, European Space Foundation, "U.S.-European Collaboration in Space Science", Washington, National Academy Press, 1998, 61–62
[NSSDC-2004] NASA NSSDC Internet site, Venera 16 proton flux data
[Oberg-1999] Oberg, J., "The Probe that Fell to Earth", New Scientist, 6 March 1999, 38

[Oberg-2000] Oberg, J., "The Strange Case of Fobos-2", Space.com website, 30 June 2000
[Oertel-1984] Oertel, D., et al., "Venera 15 and Venera 16 Infrared Spectrometry: First Results", Soviet Astronomy Letters, 10, 1984, 101–105
[Oglivie-1986] Oglivie, K.W., et al., "Ion Composition Results During the International Cometary Explorer Encounter with Giacobini–Zinner", Science, 232, 1986, 374–377
[Olson-1979] Olson, R.J.M., "Giotto's Portrait of Halley's Comet", Scientific American, 240, 1979, No. 5, 160–170
[O'Neil-1990] O'Neil, W.J., "Project Galileo", paper presented at the AIAA Space Programs and Technologies Conference, Huntsville, 25–28 September 1990
[O'Neil-1991] O'Neil, W.J., "Project Galileo Mission Status", paper presented at the XLII Congress of the International Astronautical Federation, Montreal, 1991
[O'Neil-1992] O'Neil, W.J., et al., "Galileo Completing VEEGA – A Mid-Term Report", paper presented at the XLIII Congress of the International Astronautical Federation, Washington, 1992
[O'Neil-1993] O'Neil, W.J., et al., "Performing the Galileo Jupiter Mission with the Low-Gain Antenna (LGA) and an Enroute Report", paper presented at the XLIV Congress of the International Astronautical Federation, Graz, 1993
[O'Neil-1994] O'Neil, W.J., et al., "Galileo Preparing for Jupiter Arrival", paper presented at the XLV Congress of the International Astronautical Federation , Jerusalem, 1994
[O'Neil-1995] O'Neil, W.J., et al., "Galileo on Jupiter Approach", paper presented at the XLVI Congress of the International Astronautical Federation, Oslo, 1995
[O'Neil-1996] O'Neil, W.J., et al., "Project Galileo at Jupiter", paper presented at the XLVII Congress of the International Astronautical Federation, Beijing, 1996
[O'Neil-1997] O'Neil, W.J., et al., "Project Galileo Completing its Primary Mission", paper presented at the XLVIII Congress of the International Astronautical Federation, Turin, 1997
[Orton-1996] Orton, et al., "Earth-Based Observations of the Galileo Probe Entry Site", Science, 272, 1996, 839–840
[Ostro-1985] Ostro, S.J., "Radar Observations of Asteroids and Comets", Publications of the Astronomical Society of the Pacific, 97, 1985, 877–884
[Ostro-1996] Ostro, S.J., et al., "Radar Observations of Asteroid 1620 Geographos", Icarus, 121, 1996, 46–66
[Otero-2000] Otero, S.A., Fieseler, P.D., Lloyd, C., "Delta Velorum is an Eclipsing Binary", Information Bulletin On Variable Stars No. 4999, 7 December 2000
[Page-1975] Page, D.E., "Exploratory Journey out of the Ecliptic Plane", Science, 190, 1975, 845–850
[Paige-2001] Paige, D.A., "Global Change on Mars?", Science, 294, 2001, 2107–2108
[Palluconi-1997] Palluconi, F.D., Albee, A.L., "Mars Global Surveyor: Ready for Launch in November 1996", Acta Astronautica, 40, 1997, 511–516
[Pappalardo-1998] Pappalardo, R.T., et al., "Geological Evidence for Solid-State Convection in Europa's Ice Shell", Nature, 391, 1998, 365–368
[Pardini-1990] Pardini, C., Anselmo, L., "Missione Piazzi: Importanza Scientifica e Fattibilità Tecnica" (The Piazzi mission: scientific importance and technical feasibility), CNUCE Internal report C90-36, 10 December 1990 (in Italian)
[Parker-1998] Parker, S., "Mars Global Surveyor: You Ain't Seen Nothin' Yet", Sky & Telescope, January 1998, 32–34
[Parker-1989] Parker, T.J. et al., "Transitional morphology in west Deuteronilus mensae, Mars: Implications for modification of the lowland/upland boundary", Icarus, 82, 1989, 111–145
[Parker-1993] Parker, T.J. et al., "Coastal geomorphology of the Martian northern plains", J. Geophys. Res., 98, 1993, 11061–11078

[Parker-2007a] Parker, T.J., et al., "HiRISE Captures the Viking and Mars Pathfinder Landing Sites", paper presented at the Lunar and Planetary Science Conference XXXVIII, Houston, 2007

[Parker-2007b] Parker, T., Manning, R., "Mars Litter Inventory: Using HiRISE to Find out Stuff", presentation dated 28 February 2007

[Perminov-1999] Perminov, V.G., "The Difficult Road to Mars: A Brief History of Mars Exploration in the Soviet Union", Washington, NASA, 1999, 76

[Perminov-2004] Perminov, V., "Perviye Otechestvyenniye Radiolokatsionniye Karti Veneri" (The first national radar maps of Venus), Novosti Kosmonavtiki, No. 9, 2004, page unknown (in Russian)

[Perminov-2005] Perminov, V., "Aerostaty v Nyeve Veneri: K 20-Letniyu Poleta AMS Vega" (Aerostats in the atmosphere of Venus: on the 20th Anniversary of the Flight of the Vega Probe), Novosti Kosmonavtiki, August 2005, 60–63 (in Russian)

[Perminov-2006] Perminov, V., "Vstretsa S Kometoy Galleya – K 20-Letniyu Poleta AMS Vega" (Encounter with Comet Halley: on the 20th Anniversary of the Flight of the Vega Probe), Novosti Kosmonavtiki, May 2006, 68–72 (in Russian)

[Petropoulos-1993] Petropoulos, B., Telonis, P., "Physical Parameters of the Atmosphere of Venus from Venera 15 and 16 Missions", Earth Moon and Planets, 63, 1993, 1–7

[Phillips-1991] Phillips, R.J., et al., "Impact Craters on Venus: Initial Analysis from Magellan", Science, 252, 1991, 288–297

[Powell-1959] Powell, B.W., "Solar Sail: Key to Interplanetary Voyaging?", Spaceflight, October 1959, 116–118

[Preston-1986] Preston, R.A., et al., "Determination of Venus Winds by Ground-Based Radio Tracking of the VEGA Balloons", Science, 231, 1986, 1414–1416

[Prialnik-1992] Prialnik, D., Bar-Nun, A., "Crystallization of Amorphous Ice as the Cause of Comet P/Halley's Outburst at 14 AU", Astronomy and Astrophysics, 258, 1992, L9–L12

[Ragent-1996] Ragent, B. et al., "Results of the Galileo Probe Nephelometer Experiment", Science, 272, 1996, 854–856

[Randolph-1978] Randolph, J.E., "Solar Probe Study". In: Neugebauer, M., Davies, R.W., "A Close-Up of the Sun", Pasadena, JPL, 1978, 521–534

[Rawal-1986] Rawal, J.J., "Possible Satellites of Mercury and Venus", Earth, Moon, and Planets, 36, 1986, 135–138

[Reinhard-1986a] Reinhard, R., "A Brief History of the Giotto Mission", ESA Bulletin, 46, 1986, 19–21

[Reinhard-1986b] Reinhard, R., "The Giotto Experiments", ESA Bulletin, 46, 1986, 41–51

[Rème-1986] Rème, H., et al., "Comet Halley-Solar Wind Interaction from Electron Measurements Aboard Giotto", Nature, 321, 1986, 349–352

[Rieder-1997] Rieder, R., et al., "The Chemical Composition of Martian Soil and Rocks Returned by the Mobile Alpha Proton X-Ray Spectrometer: Preliminary Results from the X-Ray Mode", Science, 278, 1997, 1771–1774

[Riedler-1986] Riedler, W., et al. "Magnetic Field Observations in Comet Halley's Coma", Nature, 321, 1986, 288–289

[Riedler-1989] Riedler, W., et l., "Magnetic Fields near Mars: First Results", Nature, 341, 1989, 604–607

[Robertson-1994] Robertson, D.F., "To Boldly Go...", Astronomy, December 1994, 34–41

[Robinson-2001] Robinson, M.S., et al., "The Nature of Ponded Deposits on Eros", Nature, 413, 2001, 396–400

[Rocard-1989] Rocard, F., et al., "French Participation in the Soviet Phobos Mission", Acta Astronautica, 22, 1990, 261–267

[Rogers-1996] Rogers, A., "Come in, Mars", Newsweek, 19 August 1996, 41–45
[Rokey-1993] Rokey, M.J., "Magellan Radar Special Flight Experiments", Journal of Spacecraft and Rockets, 30, 1993, 715–723
[Rosenbauer-1989] Rosenbauer, H., et al., "Ions of Martian Origin and Plasma Sheet in the Martian Magnetosphere: Initial Results of the TAUS Experiment", Nature, 341, 1989, 612–614
[Rosengren-1990] Rosengren, M., "Orbit Design and Control for Ulysses", ESA Bulletin, 63, 1990, 66–69
[Rossman-2002] Rossman, I.P. III, et al., "A Large Paleolake Basin at the Head of Ma'adim Vallis, Mars", Science, 296, 2002, 2209–2212
[Rover Team-1997] Rover Team, "Characterization of the Martian Surface Deposits by the Mars Pathfinder Rover, Sojourner", Science, 278, 1997, 1765–1767
[Russell-2000] Russell, C.T., Kivelson, M.G., "Detection of SO in Io's Exosphere", Science, 287, 2000, 1998–1999
[Russo-2000a] Russo, A., "The Definition of ESA's Scientific Programme for the 1980s". In: Krige, J., Russo, A., Sebesta, L. (eds.) , "A History of the European Space Agency 1958–1987", Vol. 2, Noordwijk, ESA, 2000, 138–179
[Russo-2000b] Russo, A., "Towards the Turn of the Century". Ibid., 189–195 and 210–217
[Russo-2000c] Russo, A., "The Scientific Programme between ESRO and ESA (1973–1977)". Ibid., 109
[Russo-2000d] Russo, A., "Towards the Turn of the Century". Ibid., 189
[Rust-2005] Rust, D.M., et al., "Comparison of Interplanetary Disturbances at the NEAR Spacecraft with Coronal Mass Ejections at the Sun", The Astrophysical Journal, 621, 2005, 524–536
[Rust-2006] Rust, T. III, "Galileo Probe Thermal Control", paper presented at the 4th International Planetary Probe Workshop, Pasadena, 2006
[Sagan-1993a] Sagan, C., et al., "A Search for Life on Earth from the Galileo Spacecraft", Nature, 365, 1993, 715–721
[Sagan-1993b] Sagan, C., "Return to the Wonder World: Mars Observer in Perspective", The Planetary Report, November/December 1993, 6–7
[Sagdeev-1986a] Sagdeev, R.Z., et al., "Television Observations of Comet Halley from Vega Spacecraft", Nature, 321, 1986, 262–266
[Sagdeev-1986b] Sagdeev, R.Z., et al., "Vega Spacecraft Encounters with Comet Halley", Nature, 321, 1986, 259–262
[Sagdeev-1986c] Sagdeev, R.Z., et al., "Overview of VEGA Venus Balloon In Situ Meteorological Measurements", Science, 231, 1986, 1411–1414
[Sagdeev-1986d] Sagdeev, R.Z., et al., "The VEGA Balloon Experiment", Science, 231, 1986, 1407–1408
[Sagdeev-1989] Sagdeev, R.Z., Zakharov, A.V., "Brief History of the Phobos Mission", Nature, 341, 1989, 581–585
[Sagdeev-1994a] Sagdeev, R.Z., "The Making of a Soviet Scientist", New York, John Wiley & Sons, 1994, 275–276
[Sagdeev-1994b] ibid., 280
[Sagdeev-1994c] ibid., 282–283
[Sagdeev-1994d] ibid., 283–284
[Sagdeev-1994e] ibid., 313–314
[Sagdeev-1994f] ibid., 315–316
[Saito-1986] Saito, T., et al., "Interaction Between Comet Halley and the Interplanetary Magnetic Field observed by Sakigake", Nature, 321, 1986, 303–307

[Santo-1995] Santo, A.G., Lee, S.C., Gold, R.E., "NEAR Spacecraft and Instrumentation", Journal of Astronautical Sciences, 43, 1995, 373–397

[Saunders-1951] Saunders, R. (i.e. Wiley, C.), "Clipper Ships of Space", Astrounding Science Fiction, May 1951, 136–143

[Saunders-1991] Saunders, R.S., Pettengill, G.H., "Magellan: Mission Summary", Science, 252, 1991, 247–249

[Saunders-1999] Saunders, R.S., "Venus". In: Kelly Beatty, J., Petersen, C.C., Chaikin, A. (eds.), "The New Solar System", Cambridge University Press, 4th edition, 1999, 97–110

[Savich-1986] Savich, N.A., et al., "Dual-Frequency Vega Radio Sounding of Comet Halley", Soviet Astronomy Letters, 12, 1986, 283–286

[Sawyer-2006a] Sawyer, K., "The Rock from Mars: a Detective Story on Two Planets", New York, Random House, 3–21

[Sawyer-2006b] ibid., 161

[Sawyer-2006c] ibid., 132–133

[Scarf-1986] Scarf, F.L., et al., "Plasma Wave Observations at Comet Giacobini–Zinner", Science, 232, 1986, 377–381

[Schaber-1986] Schaber, G.G., Kozak, R.C., "Venera 15/16 and Arecibo Radar Images of Venus: Complementary Data Sets", paper presented at the XVII Lunar and Planetary Science Conference, Houston,1986

[Schaefer-2007] Schaefer, D.H., Paddack, S.J. Rubincam, D.P., "Explorer XII: Spinning Faster than Expected", Science, 317, 2007, 898–899

[Schenk-2001] Schenk, P.M., et al., "Flooding of Ganymede's Bright Terrains by Low-Viscosity Water-Ice Lavas", Nature, 410, 2001, 57–60

[Schofield-1997] Schofield, J.T, et al., "The Mars Pathfinder Atmospheric Structure Investigation/Meteorology (ASI/MET) Experiment", Science, 278, 1997, 1752–1757

[Schubert-1996] Schubert, G., et al., "The Magnetic Field and Internal Structure of Ganymede", Nature, 384, 1996, 544–545

[Schulze-Makuch-2002] Schulze-Makuch, D., Irwin, L.N., Irwin, T., "Astrobiological relevance and feasibility of a sample collection mission to the atmosphere of Venus", In: "Proceedings of the First European Workshop on Exo-Astrobiology, 16–19 September 2002, Graz", 247–250

[Schwaiger-1971] Schwaiger, L.-E., et al., "Solar Electric Propulsion Asteroid Belt Mission", Journal of Spacecraft and Rockets, 8, 1971, 612–617

[Schwehm-1992] Schwehm, G.H., "The Giotto Estended Mission to Comet Grigg–Skjellerup: Summary of Preliminary Results", ESA Bulletin, 72, 1992, 61–65

[Scoon-1993] Scoon, G.E.N., "Mission and System Concepts from Mars Robotic Precursor Missions". In: "Missions, Technologies et Conception des Vehicules Mobiles Planetaires", Toulouse, Cépaduès, 1993

[Scott-1996a] Scott, W.B., "Clementine 2 to Fire Probes at Asteroids", Aviation Week & Space Technology, 27 May 1996, 46–47

[Scott-1996b] Scott, W.B., "MGS Completing Prelaunch Checks", Aviation Week & Space Technology, 16 September 1996, 49

[Sedbon-1989] Sedbon, G., "Rosetta – Key to the Solar System", Flight International, 2 September 1989, 28–29

[Seiff-1996] Seiff,A., et al., "Structure of the Atmosphere of Jupiter: Galileo Probe Measurements", Science, 272, 1996, 844–845

[Seiff-1997] Seiff, A., et al., "Thermal Structure of Jupiter's Upper Atmosphere Derived from the Galileo Probe", Science, 276, 1997, 102–104

[Sekanina-1985] Sekanina, Z., "Precession Model for the Nucleus of Periodic Comet Giacobini–Zinner", The Astronomical Journal, 90, 1985, 827–845

[Sekanina-1986] Sekanina, Z., Larson, S.M., "Dust Jets in Comet Halley Observed by Giotto and from the Ground", Nature, 321, 1986, 357–361

[Sekanina-1987] Sekanina, Z., "Nucleus of Comet Halley as a Torque-Free Rigid Rotator", Nature, 325, 1987, 326–328

[Selivanov-1989] Selivanov, A.S., et al., "Thermal Imaging of the Surface of Mars", Nature, 341 1989, 593–595

[Selivanov-1990] Selivanov, A.S., et al., "The TERMOSKAN Experiment: A Thermal Survey of the Surface of Mars from Phobos 2", Soviet Astronomy Letters, 16, 1990, 147–150

[Shutte-1989] Shutte, N.M., et al., "Observations of Electron and Ion Fluxes in the Vicinity of Mars with the HARP Spectrometer", Nature, 341, 1989, 614–616

[Siddiqi-2002a] Siddiqi, A.A., "Deep Space Chronicle: A Chronology of Deep Space and Planetary Probes 1958–2000", Washington, NASA, 2002, 131–132

[Siddiqi-2002b] ibid., 137–139

[Simpson-1986] Simpson, J.A., et al., "Dust Counter and Mass Analyser (DUCMA) Measurements of Comet Halley's Coma from Vega Spacecraft", Nature, 321, 1986, 278–280

[Simpson-1994] Simpson, R.A., Pettengill, G.H., Ford, P.G., "The Magellan Quasi-Specular Bistatic Radar Experiment", paper presented at the Lunar and Planetary Science Conference XXV, Houston, 1994

[Simpson-2000] Simpson, R.A., Tyler, G.L., "MGS Bistatic Radar Probing of the MPL/DS2 Target Area", paper presented at the 2000 Division for Planetary Science Meeting, Pasadena, 23–27 October 2000

[Simpson-2002] Simpson, R.A., "Highly Oblique Bistatic Radar Observations Using Mars Global Surveyor", paper presented at the 2002 General Assembly of the Union Radio-Scientifique Internationale

[Sjogren-1992a] Sjogren, W.L., "Venus Gravity: Status and New Data Acquisition", paper presented at the Lunar and Planetary Science Conference XXIII, Houston, 1992

[Sjogren-1992b] Sjogren, W.L., "Venus Gravity: Summary and Coming Events", paper presented at the International Colloquium on Venus, 1992

[Sjogren-1993] Sjogren, W.L., Konopliv, A.S., Borderies, N., "Venus Gravity: New Magellan Low Altitude Data", paper presented at the Lunar and Planetary Science Conference XXIV, Houston, 1993

[Sjogren-1994] Sjogren, W.L., Konopliv, A.S., "Venus Gravity Field Determination: Progress and Concern", paper presented at the Lunar and Planetary Science Conference XXV, Houston, 1994

[Sjogren-1997] Sjogren, W.L., "Venus: Gravity". In: Shirley, J.H., Fairbridge, R.W., "Encyclopedia of Planetary Sciences", Dordrecht, Kluwer, 1997, 904–905

[Slyuta-1988] Slyuta, E.N., Nikolaeva, O.V., "Distribution of Small Domes on Venus: Venera 15/16 Data", paper presented at the XIX Lunar and Planetary Science Conference, Houston, 1988

[Smith-1982] Smith, B.A., "JPL Attempting to Revive Venus Radar Imaging Plan", Aviation Week & Space Technology, 15 March 1982, 18–19

[Smith-1984] Smith, B.A., "New Radar Unit Cuts Venus Mapper Costs", Aviation Week & Space Technology, 16 April 1984, 141–145

[Smith-1986] Smith, E.J., et al., "International Cometary Explorer Encounter with Giacobini–Zinner: Magnetic Field Observations", Science, 232, 1986, 382–385

[Smith-1987a] Smith, B.A., et al., "Rejection of a Proposed 7.4-day Rotation Period of the Comet Halley Nucleus", Nature, 326, 1987, 573–574

[Smith-1987b] Smith, B.A., "Future Soviet Space Exploration to Focus on Mars, Asteroids", Aviation Week & Space Technology, 22 June 1987, 81–85
[Smith-1989] Smith, B.A., "Missions Mark Resurgence of U.S. Planetary Exploration", Aviation Week & Space Technology, 9 October 1989, 44–54
[Smith-1992] Smith, E.J., Wenzel, K.-P., Page, D.E., "Ulysses at Jupiter: An Overview of the Encounter", Science, 257, 1992, 1503–1507
[Smith-1994] Smith, B.A., "Mars Global Surveyor Faces Tight Timetable", Aviation Week & Space Technology, 8 August 1994, 63–64
[Smith-1997a] Smith, B.A., "MGS Settling into Mars Orbit", Aviation Week & Space Technology, 6 October 1997, 33
[Smith-1997b] Smith, P.H., et al., "Results from the Mars Pathfinder Camera", Science, 278, 1997, 1758–1764
[Smith-1998] Smith, D.E., et al., "Topography of the Northern Hemisphere of Mars from the Mars Orbiter Laser Altimeter", Science, 279, 1998, 1686–1692
[Smith-1999a] Smith, B.A., "Antenna Problem Stalls Mars Mapping Mission", Aviation Week & Space Technology, 26 April 1999, 85
[Smith-1999b] Smith, D.E., et al., "The Global Topography of Mars and Implications for Surface Evolution", Science, 284, 1999, 1495–1503
[Smith-1999c] Smith, D.E., et al., "The Gravity Field of Mars: Results from Mars Global Surveyor", Science, 286, 1999, 94–97
[Smith-2001] Smith, D.E., et al., "Seasonal Variations of Snow Depth on Mars", Science, 294, 2001, 2141–2144
[Snyder-1997] Snyder, C.W., "Phobos Mission". In: Shirley, J.H., Fairbridge, R.W., "Encyclopedia of Planetary Sciences", Dordrecht, Kluwer, 1997, 574–576
[Sobel-1993] Sobel, D., "The Last World", Discover, May 1993, 68–76
[Sobel'man-1990] Sobel'man, I.I., et al., "Images of the Sun Obtained with the TEREK X-Ray Telescope on the Spacecraft Phobos 1", Soviet Astronomy Letters, 16, 1990, 137–140
[Somogyi-1986] Somogyi, A.J., et al. "First Observations of Energetic Particles near Comet Halley", Nature, 321, 1986, 285–288
[Spaceflight-1977] "Solar Sailing", Spaceflight, April 1977, 124–125
[Spaceflight-1992a] "What Became of the Other Halley Explorers?" Spaceflight, June 1992, 212
[Spaceflight-1992b] "NASA Unveils Lean Budget for 1993", Spaceflight, March 1992, 93–95
[Spaceflight-1992c] "Indian Space Probe", Spaceflight, November 1992, 345
[Spaceflight-1992d] "Magellan Probe Suffers Major Failure", Spaceflight, February 1992, 38
[Spaceflight-1992e] "Magellan Resumes Venus Mapping Following Transmitter Failure", Spaceflight, March 1992, 78
[Spaceflight-1992f] "Crucial Mars Launch Delayed", Spaceflight, October 1992, 336
[Spaceflight-1992g] "Mars Observer Launched", Spaceflight, November 1992, 342
[Spaceflight-1992h] "Space Probe Diary", Spaceflight, December 1992, 393
[Spaun-2001] Spaun, N.A., et al., "Scalloped Depressions on Ganymede from Galileo (G28) Very High Resolution Imaging", paper presented at the XXXII Lunar and Planetary Science Conference, Houston, 2001
[Spencer-1992] Spencer, J.R., et al., "Volcanic Activity on Io at the Time of the Ulysses Encounter", Science, 257, 1992, 1507–1510
[Spencer-1995] Spencer, J.R., Mitton, J. (eds.), "The Great Comet Crash: The Impact of Comet Shoemaker–Levy 9 on Jupiter", Cambridge University Press, 1995, 75–76
[Spencer-1998] Spencer, D.A., et al., "Mars Pathfinder Atmospheric Entry Reconstruction", paper AAS 98-146

[Spencer-1999a] Spencer, J.R., et al., "Temperatures on Europa form Galileo Photopolarimeter-Radiometer: Nighttime Thermal Anomalies", Science, 284, 1999, 1514–1516
[Spencer-1999b] Spencer, D.A., et al., "Mars Pathfinder Entry, Descent, and Landing Reconstruction", Journal of Spacecraft and Rockets, 36, 1999, 357–366
[Spencer-2000a] Spencer, J.R., et al., "Discovery of Gaseous S2 in Io's Pele Plume", Science, 288, 2000, 1208–1210
[Spencer-2000b] Spencer, J.R., et al., "Io's Thermal Emission from the Galileo Photopolarimeter-Radiometer", Science, 288, 2000, 1198–1201
[Spencer-2001] Spencer, J., "Galileo's Closest Look at Io", Sky & Telescope, May 2001, 40–46
[Spudis-1994] Spudis, P. D., Plescia, J.B., Stewart, A.D., "Return to Mercury: The Discovery-Mercury Polar Flyby Mission", paper presented at the Lunar and Planetary Science Conference XXV, Houston, March 1994
[Sromovsky-1996] Sromovsky, L.A., et al., "Solar and Thermal Radiation in Jupiter's Atmosphere: Initial Results of the Galileo Probe Net Flux Radiometer", Science, 272, 1996, 851–854
[ST-1946] "Meteorites and Space Travel", Sky & Telescope, November 1946, 7. (Reprinted in: Page, T, Page, L.W., "Wanderers in the Sky", New York, Macmillan, 1965, 206–207, replacing "space vessel" with "space probe")
[ST-1995] "Metal 'Frost' on Venus?" Sky & Telescope, August 1995, 13
[ST-1998] "Cydonia Defaced", Sky & Telescope, July 1998, 20
[ST-1999] "A Shot in the Dark", Sky & Telescope, November 1999, 17
[ST-2000a] "Ganymede's Snows", Sky & Telescope, March 2000, 24
[ST-2000b] "Recent Volcanism on Mars", Sky & Telescope, October 2000, 34
[Staehle-1994] Staehle, R.L., et al., "Last but not Least – Trip to Pluto", Spaceflight, March 1994, 101–104, April 1994, 140–143
[Staehle-1999] Staehle, R.L., et al., "Ice & Fire: Missions to the Most Difficult Solar System Destinations... on a Budget", Acta Astronautica, 45, 1999, 423–439
[Steffes-1992] Steffes, P.G., et al., "Preliminary Results from the October 1991 Magellan Radio Occultation Experiment", paper presented at the 1992 24th Annual DPS Meeting
[Steffy-1983] Steffy, D.A., "The Mars Geoscience Climatology Orbiter", paper presented at the Lunar and Planetary Science Conference XIV, Houston, March 1983
[Stephenson-1994] Stephenson, R.R., Bernard, D.E., "JPL Mars Observer In-Flight Anomaly Investigation (With Emphasis on Attitude Control Aspects)", Draft dated 25 January 1994
[Stern-1993] Stern, S.A., et al., "A Low-Cost Mission to 2060 Chiron Based on the Pluto Fast Flyby", 1993
[Stern-1998] Stern, A., Mitton, J., "Pluto and Charon", New York, John Wiley & Sons, 1998,171–202
[Stern-2007] Stern, S.A., "The New Horizons Pluto Kuiper Belt Mission: An Overview with Historical Context", Arxiv pre-print astro-ph/0709.4417
[Stofan-1993] Stofan, E.R., "The New Face of Venus", Sky & Telescope, August 1993, 22–31
[Stooke-2000] Stooke, P.J., "The Pathfinder Landing Area in MGS/MOC Images", paper presented at the Lunar and Planetary Science Conference XXXI, Houston, 2000
[Stuhlinger-1970] Stuhlinger, E., "Planetary Exploration with Electrically Propelled Vehicles", paper presented at the Third Conference on Planetology and Space Mission Planning, New York, October 1970
[Stuhlinger-1986] Stuhlinger, E., et al., "Comet Nucleus Sample Return Missions with Electrically Propelled Spacecraft", Journal of the British Interplanetary Society, 39, 1986, 273–281

[Sukhanov-1985] Sukhanov, A.A., "Otchet o Nauchno-Issledovatelskoy Rabote 'Issledovaniye Vozmoshhostey Osutschestvleniya Nekatorikh Perspektivniykh Kosmicheskikh Proyektov'" (Relation on the Scientific Research Work 'Feasibility Study of Some Long Term Space Projects'), Moscow, IKI, 1985 (in Russian)
[Surkov-1986a] Surkov, Yu.A.., et al., "Vega 1 Mass Spectrometry of Venus Cloud Aerosols: Preliminary Results" Soviet Astronomy Letters, 12, 1986, 44–45
[Surkov-1986b] Surkov, Yu.A., et al., "Vega 1, 2 Humidity Profiles for the Venus Atmosphere", Soviet Astronomy Letters, 12, 1986, 31–33
[Surkov-1986c] Surkov, Yu.A., et al., "Vega 2 Lander Analysis of Rock Composition in Northern Aphrodite Terra", Soviet Astronomy Letters, 12, 1986, 28–31
[Surkov-1986d] Surkov, Yu.A., et al., "Uranium, Thorium, Potassium Abundances in Venus Rocks", Soviet Astronomy Letters, 12, 1986, 46–48
[Surkov-1989] Surkov, Yu.A., et al., "Determination of the Elemental Composition of Martian Rocks from Phobos 2", Nature, 341 1989, 595–598
[Surkov-1993] Surkov, Yu.A., "Discovery Venera Surface – Atmosphere Geochemistry Experiments Mission Concept", paper presented at the Lunar and Planetary Science Conference XXIV, Houston, 1993
[Surkov-1997a] Surkov, Yu. A., "Exploration of Terrestrial Planets from Spacecraft", Chichester, Wiley–Praxis, 1997, 406–408 and 371–373
[Surkov-1997b] ibid., 387–392
[Surkov-1997c] ibid., 212–220 and 378–381
[Surkov-1997d] ibid., 381–382
[Surkov-1997e] ibid., 383–386
[Surkov-1997f] ibid., 396–400 and 419–427
[Surkov-1997g] ibid. 433–436
[Tehilig-2001] Theilig, E.E., Bindschadler, D.L., Vandermey, N., "Project Galileo: From Ganymede Back to Io", paper presented at the LII Congress of the International Astronautical Federation, Toulouse, 2001
[Tehilig-2002] Theilig, E.E., et al., "Project Galileo: Farewell to the Major Moons of Jupiter", paper presented at the LIII Congress of the International Astronautical Federation, Houston, 2002
[Thomas-Keprta-2002] Thomas-Keprta, K.L., "Magnetofossils from Ancient Mars: A Robust Biosignature in the Martian Meteorite ALH84001", Applied and Environmental Microbiology, 68, 2002, 3663–3672
[Thomson-1982a] Thomson, A.A., "Off to the Asteroids", Spaceflight, January 1982, 7–9
[Thomson-1982b] Thomson, A.A., "Exploring Mars with Kepler", Spaceflight, 24, April 1982, 151–153
[Time-1977] "Sailing to Halley's Comet", Time, 14 March 1977, 22
[Tolson-1995] Tolson, R.H., Patterson, M.T., Lyons, D.T., "Magellan Windmill and Termination Experiments". In: "Mécanique Spatiale/Spaceflight Mechanics", Toulouse, Cépaduès, 1995
[Tolson-1999] Tolson, R.H., et al., "Utilization of Mars Global Surveyor Accelerometer Data for Atmospheric Modeling", AAS 99-386
[Treiman-1999] Treiman, A., "Microbes in a Martian Meteorite?", Sky & Telescope, April 1999, 52–58
[Trombka-2000] Trombka, J.I., et al., "The Elemental Composition of Asteroid 433 Eros: Results of the NEAR–Shoemaker X-ray Spectrometer", Science, 289, 2000, 2101–2105
[Tsander-1924] Tsander, F.A., "Report of the Engineer F.A. Tsander Concerning Interplanetary Voyages", 1924?. In: Tsander, F.A., "From a Scientific Heritage", Washington, NASA, 1969

[Tsou-1985] Tsou, P., Brownlee, D.E., Albee, A.L., "Comet Coma Sample Return Via Giotto II", Journal of the British Interplanetary Society, 38, 1985, 232–239
[Tsou-1985b] Tsou, P., Albee, A., "Comet Flyby Sample Return", Paper AIAA-85-0465
[Turtle-2001] Turtle, E.P., Pierazzo, E., "Thickness of a Europan Ice Shell from Impact Crater Simulations", Science, 294, 2001, 1326–1328
[Turtle-2004] Turtle, E.P., et al., "The Final Galileo SSI Observations of Io: Orbits G28-I33", Icarus, 169, 2004, 3–28
[Tyler-1991] Tyler, G.L., et al., "Magellan: Electrical and Physical Properties of Venus' Surface", Science, 252, 1991, 265–270
[Tytell-2000] Tytell, D., "Martian Mudflows", Sky & Telescope, September 2000, 56–57
[Tytell-2001a] Tytell, D., "Ancient Martian Lakes? Perhaps.", Sky & Telescope, March 2001, 20–21
[Tytell-2001b] Tytell, D., Kelly Beatty, J., "Other Ways to Make Martian Gullies", Sky & Telescope, July 2001, 26
[Tytell-2001c] Tytell, D., "A Greener, Drier Mars", Sky & Telescope, February 2001, 20–21
[Tytell-2001d] Tytell, D., "Dust Storm Clouds Out Mars", Sky & Telescope, November 2001, 22
[Tytell-2004] Tytell, D., "When Mars Had an Icy Equator", Sky & Telescope, July 2004, 26
[Uesugi-1986] Uesugi, K., "Collision of Large Dust Particles with Suisei Spacecraft", In: ESA Proceedings of the 20th ESLAB Symposium on the Exploration of Halley's Comet. Volume 2: Dust and Nucleus, 1986, 219–222
[Uesugi-1988] Uesugi, K., et al., "Follow-On Missions of Sakigake and Suisei", Acta Astronautica, 18, 1988, 241–246
[Uesugi-1995] Uesugi, K, Kawaguchi, J., Tsou, P., "SOCCER (Sample of Comet Coma Earth Return) Mission", Acta Astronautica, 35, 1995, 171–179
[Ulivi-2004] Ulivi, P., with Harland, D.M., "Lunar Exploration: Human Pioneers and Robotic Surveyors", Chichester, Springer–Praxis, 2004, 257–264
[Ulivi-2006] Ulivi, P., "ESRO and the deep space: European Planetary Exploration Planning before ESA", Journal of the British Interplanetary Society, 59, 2006, 204–223
[Ulivi-2008] Ulivi, P., "Europe's 'Arrows to the Sun': Two Gravity and Solar Probe proposals from ESRO and ESA", Journal of the British Interplanetary Society, 61, 2008, 98-112
[Vaisberg-1986] Vaisberg, O.L., "Dust Coma Structure of Comet Halley from SP-1 Detector Measurements", Nature, 321, 1986, 274–276
[Vekshin-1999] Vekshin, B., "Pisma Zhitateley" (reader's letters), Novosti Kosmonavtiki, No. 5, 1999, 53 (in Russian)
[Verigin-1999] Verigin, V., "9 Let Granata" (9 years of Granat), Novosti Kosmonavtki, No.2 1999, 38–40 (in Russian)
[Veverka-1997a] Veverka, J.F., Farquhar, R.W., "NEAR Views of Mathilde", Sky & Telescope, October 1997, 30–32
[Veverka-1997b] Veverka, J., et al., "NEAR's Flyby of 253 Mathilde: Images of a C Asteroid", Science, 278, 1997, 2109–2114
[Veverka-1999] Veverka, J., et al., "Imaging of Asteroid 433 Eros During NEAR's Flyby Reconnaissance", Science, 285, 1999, 562–564
[Veverka-2000] Veverka, J., et al., "NEAR at Eros: Imaging and Spectral Results", Science, 289, 2000, 2088–2097
[Veverka-2001a] Veverka, J., et al., "Imaging of Small-Scale Features on 433 Eros from NEAR: Evidence for a Complex Regolith", Science, 2001, 292, 484–488
[Veverka-2001b] Veverka, J., et al., "The Landing of NEAR–Shoemaker on Asteroid 433 Eros", Nature, 413, 2001, 390–393

[VnIITransmash-1999] VnIITransmash, "Specimens of Space Technology, Earth Based Demonstrators of Planetary Rovers, Running Mock-ups", Saint Petersburg, 1999
[VnIITransmash-2000] "Pages of history of VNIITransmash", Saint Petersburg, VnIITransmash, pages unknown (in Russian)
[Volare-1989] "Primi Accordi Italiani con la NASA dell'Est" (First Italian agreements with the Eastern NASA), Volare, June 1989, 14 (in Italian)
[von Rosenvinge-1986] von Rosenvinge, T.T., Brandt, J.C., Farquhar, R.W., "The International Cometary Explorer Mission to Comet Giacobini–Zinner", Science, 232, 1986, 353–356
[von Zahn-1996] von Zahn, U., Hunten, D.M., "The Helium Mass Fraction in Jupiter's Atmosphere", Science, 272, 1996, 849–851
[Vorontsov-1989] Vorontsov, V.A., et al., "Mars Exploration: Balloons and Penetrators", Acta Astronautica, 19, 1989, 843–845
[Waldrop-1981a] Waldrop, M.M., "Down the Wire with Halley", Science, 214, 1981, 35
[Waldrop-1981b] Waldrop, M.M., "Planetary Science *in Extremis*", Science, 214, 1981, 1322–1324
[Waldrop-1982] Waldrop, M.M., "Planetary Science: Up from the Ashes?", Science, 218, 1982, 665–666
[Waldrop-1989] Waldrop, M.M., "Phobos at Mars: A Dramatic View – And Then Failure", Science, 245, 1989, 1044–1045
[Weinberger—1984] "Defense Space Launch Strategy", Memorandum from Secretary of Defense to Secretaries of the Military Departments, *et al.*, 7 February 1984
[Weinstein-1993] Weinstein, S., et al., "Follow on Missions for the Pluto Spacecraft", paper presented at the IAA International Conference on on Low Cost Missions, 1993
[Weisbin-1993] Weisbin, C.R., Montemerlo, M., Whittaker, W., "Evolving Directions in NASA's Planetary Rover Requirements and Technology". In: "Missions, Technologies et Conception des Vehicules Mobiles Planetaires", Toulouse, Cépaduès, 1993
[Weissman-1995] Weissman, P.R., et al., "Galileo NIMS Direct Observations of the Shoemaker–Levy 9 Fireballs and Fall Back", paper presented at the Lunar and Planetary Science Conference XXVI, Houston, March 1995
[Weissman-1999] Weissman, P.R., "Cometary Reservoirs". In: Kelly Beatty, J., Petersen, C.C., Chaikin, A. (eds.), "The New Solar System", Cambridge University Press, 4th edition, 1999, 59–68
[Wenzel-1990a] Wenzel, K.-P., Eaton, D., "Ulysses – A Brief History", ESA Bulletin, 63, 1990, 10–12
[Wenzel-1990b] Wenzel, K.-P., et al., "The Scientific Mission of Ulysses", ESA Bulletin, 63, 1990, 21–27
[West-1986] West, R.M., et al., "Post Perihelion Imaging of Comet Halley at ESO", Nature, 321, 1986, 363–365
[Westwick-2007a] Westwick, P.J., "Into the Black: JPL and the American Space Program 1976–2004", New Haven, Yale University Press, 2007, 42–58
[Westwick-2007b] ibid., 108–110
[Westwick-2007c] ibid., 96–97
[Westwick-2007d] ibid., 70
[Westwick-2007e] ibid., 175–177
[Westwick-2007f] ibid., 175–185
[Westwick-2007g] ibid., 268
[Westwick-2007h] ibid., 160
[Westwick-2007i] ibid., 198–201

[Westwick-2007j] ibid., 195
[Westwick-2007k] ibid., 183–185
[Westwick-2007l] ibid., 258–260
[Westwick-2007m] ibid., 227
[Westwick-2007n] ibid., 48
[Westwick-2007o] ibid., 142–154 and 207–227
[Westwick-2007p] ibid., 218–219
[Westwick-2007q] ibid., 149
[Westwick-2007r] ibid., 142–154 and 263
[Whipple-1966] Whipple, F.L., interviewed by Caras, R.A. on 6 May 1966 in: Frewin, A., "Are We Alone? The Stanley Kubrick Extraterrestrial-Intelligence Interviews", Elliot & Thompson, 2005
[Whipple-1987] Whipple, F.L., "The Cometary Nucleus: Current Concepts", Astronomy & Astrophysics, 187, 1987, 852–858
[Wilkins-1986] Wilkins, D.E.B., Parkes, A., Nye, H., "The Giotto Encounter and Post-Encounter Operations", ESA Bulletin, 46, 1986, 66–70
[Willcockson-1999] Willcockson, W.H., "Mars Pathfinder Heathshield Design and Flight Experience", Journal of Spacecraft and Rockets, 36, 1999, 374–379
[Williams-2005] Williams, D., personal communication with the author, 27 September 2005
[Wilmoth-1999] Wilmoth, R.G., et al., "Rarefied Aerothermodynamic Predictions for Mars Global Surveyor", Journal of Spacecraft and Rockets, 36, 1999, 314–322
[Wilson-1985] Wilson, K.T., "The CRAF Mission", Spaceflight, December 1985, 452–453
[Wilson-1986a] Wilson, A., "Sampling the Snowballs", Flight International, 7 June 1986, 45–46
[Wilson-1986b] Wilson, A., "Comet Workshop", Flight International, 6 September 1986, 44–45
[Wilson-1986c] Wilson, A., "Missions to Mars", Flight International, 12 July 1986, 35–37
[Wilson-1987a] Wilson, A., "Solar System Log", London, Jane's Publishing, 1987, 112–113
[Wilson-1987b] ibid., 117–118 and 122–124
[Wilson-1987c] ibid., 118–122
[Wilson-1987d] ibid., 114–117
[Wilson-1987e] ibid., 106–107
[Wilson-1987f] Wilson, A., "Comets Loom Closer", Flight International, 8 August 1987, 33–35
[Wilson-1987g] Wilson, A., "Return to Mercury", Flight International, 19 September 1987, 46–49
[Woerner-1998] Woerner, D.F., "Revolutionary Systems and Technologies for Missions to the Outer Planets", paper presented at the Second IAA Symposium on Realistic Near-Term Advanced Scientific Space Missions, Aosta, 29 June–1 July 1998
[Wood-1981] Wood, L.J., "Navigation Accuracy Analysis for a Halley Intercept Mission", Journal of Guidance, 5, 1981, 300–306
[Yeomans-1997] Yeomans, D.K., et al., "Estimating the Mass of Asteroid 253 Mathilde from Tracking Data During the NEAR Flyby", Science, 278, 1997, 2106–2109
[Yeomans-1999] Yeomans, D.K., et al., "Estimating the Mass of Asteroid 433 Eros During the NEAR Spacecraft Flyby", Science, 285, 1999, 560–561
[Yeomans-2000] Yeomans, D.K., et al., "Radio Science Results During the NEAR–Shoemaker Spacecraft Rendezvous with Eros", Science, 289, 2000, 2085–2088
[Yoder-2003] Yoder, C.F., et al., "Fluid Core Size of Mars from Detection of the Solar Tide", Science, 300, 2003, 299–303

[Young-1990] Young, C. (ed.), "The Magellan Venus Explorer's Guide", Pasadena, JPL, 1990, 51–68

[Young-1996] Young, R.E., Smith, M.A., Sobeck, C.K., "Galileo Probe: In Situ Observations of Jupiter's Atmosphere", Science, 272, 1996, 837–838

[Zaitsev-1989] Zaitsev, Yu., "The Successes of Phobos-2", Spaceflight, November 1989, 374–377

[Zak-2004] "Planetary: Projects and Concepts", Anatoly Zak website

[Zhulanov-1986] Zhulanov, Yu.V., Mutkin, L.M., Nenarokov, D.F., "Aerosol Counts in the Venus Clouds: Preliminary Vega 1, 2 Density Profiles, H = 63–47 km", Soviet Astronomy Letters, 12, 1986, 49–52

[Zuber-1998] Zuber, M.T., "Observations of the North Polar Region of Mars from the Mars Orbiter Laser Altimeter", Science, 282, 1998, 2053–2060

[Zuber-2000a] Zuber, M., et al., "The Shape of 433 Eros from NEAR–Shoemaker Laser Rangefinder", Science, 289, 2000, 2097–2101

[Zuber-2000b] Zuber, M.T., et al., "Internal Structure and Early Thermal Evolution of Mars from Mars Global Surveyor Topography and Gravity", Science, 287, 2000, 1788–1793

Further reading

BOOKS

Godwin, R., (editor), "Deep Space: The NASA Mission Reports", Burlington, Apogee, 2005

Godwin, R., (editor), "Mars: The NASA Mission Reports", Burlington, Apogee, 2000

Godwin, R., (editor), "Mars: The NASA Mission Reports Volume 2", Burlington, Apogee, 2004

Kelly Beatty, J., Collins Petersen, C., Chaikin, A. (editors), "The New Solar System", 4th edition, Cambridge University Press, 1999

Shirley, J.H., Fairbridge, R.W., "Encyclopedia of Planetary Sciences", Dordrecht, Kluwer Academic Publishers, 1997

Surkov, Yu.A., "Exploration of Terrestrial Planets from Spacecraft", Chichester, Wiley–Praxis, 1994

MAGAZINES

Aerospace America

l'Astronomia (in Italian)

Aviation Week & Space Technology

ESA Bulletin

Espace Magazine (in French)

Flight International

Novosti Kosmonavtiki (in Russian)

Science

Scientific American

Sky & Telescope

Spaceflight

INTERNET SITES

Don P. Mitchell's "The Soviet Exploration of Venus" (www.mentallandscape.com/V_Venus.htm)

Encyclopedia Astronautica (www.astronautix.com)
Jonathan's Space Home Page (planet4589.org/space/space.html)
JPL (www.jpl.nasa.gov)
Malin Space Science Systems (www.msss.com)
NASA NSSDC (nssdc.gsfc.nasa.gov)
Novosti Kosmonavtiki (www.novosti-kosmonavtiki.ru)
NPO Imeni S.A. Lavochkina (www.laspace.ru)
Space Daily (www.spacedaily.com)
Spaceflight Now (www.spaceflightnow.com)
The Planetary Society (planetary.org)

Previous volumes in this series:

Part 1: The golden age 1957–1982

Index

Made in the USA
Lexington, KY
22 January 2012